COOPERATIVE EFFECTS IN OPTICS

Superradiance and Phase Transitions

MALVERN PHYSICS SERIES

General Series Editor: **Professor E R Pike** FRS

COOPERATIVE EFFECTS IN OPTICS

Superradiance and Phase Transitions

by

A V Andreev, V I Emel'yanov and Yu A Il'inskiĭ

International Laser Center, Moscow State University, Russia

Institute of Physics Publishing

Bristol and Philadelphia

British Library Cataloguing-in-Publication Data

A catalogue record for this book is available from the British Library.

ISBN 0-7503-0219-4

Library of Congress Cataloging-in-Publication Data are available

Published by IOP Publishing Ltd, a company wholly owned by the Institute of Physics, London
IOP Publishing Ltd, Techno House, Redcliffe Way, Bristol BS1 6NX, England
US Editorial Office: IOP Publishing Inc, The Public Ledger Building, Suite 1035, Independence Square, Philadelphia, PA 19106, USA

Printed in Great Britain by Galliard (Printers) Ltd., Great Yarmouth, Norfolk

Contents

Preface

This book is dedicated to systematic treatment of the modern theory of cooperative optical phenomena, i.e., such processes in which the behavior of the many-body systems of radiators or absorbers is essentially determined by their collective interactions with each other.

Special attention is paid to the theory of the collective spontaneous radiation or superradiance. The superradiant system of N excited atoms ($N \gg 1$) radiates its energy in a time interval much shorter then T_1 — the time of radiative decay of a single excited atom. Such behavior of the system of atoms is affected by their mutual influence on each other through common radiation field. This enhancement of the decay rate of the atomic ensemble occurs gradually in the process of decay and appears under certain condition from the isotropic spontaneous decay state. The book gives the detailed physical explanation of the mechanism of the collective spontaneous emission.

The cooperative nature of the atomic system, interacting through the radiation field can reveal itself in other conditions and lead to different effects. The book considers a whole number of models of novel nonequilibrium light-induced phase transitions in a typical quantum electronics system: ensemble of two-level atoms, interacting through the radiation field, and also through the electrostatic or lattice displacement fields.

We tried to expose the connection of two aspects of cooperative behavior of the ensembles of radiators. From the viewpoint of quantum electronics the cooperative system of radiators enables us to obtain such a regime of emission, which is fundamentally different from the spontaneous and stimulated radiation. On the other hand, from the viewpoint of the theory of phase transitions, the ensemble of radiators is the system with the critical behavior determined by a special type of interatomic interaction — the interaction through the radiation field. While the phase transitions in the other systems have been thoroughly studied, the investigation of the optical phase transitions has begun comparatively recently (a well studied example is the laser system). Superradiance, resonatorless bistability, spontaneous breaking down of the saturation symmetry and a number of other nonequilibrium phase transitions, considered in the book essentially enlarge the class of optical phase transitions. These transitions give rise to sudden changes of the optical characteristics of the systems, that can essentially influence the operation of quantum electronics devices .

In the work over this book the authors have used as a basis their book "Cooperative Phenomena in Optics" (Moscow: Nauka), published in Russian in 1988, which was significantly changed and extended.

We greatly appreciate numerous discussions with Profs. S.A.Akhmanov,

L.V.Keldysh and Yu.L.Klimontovich. Professor S.A.Akhmanov suddenly died this summer. He was an outstanding pioneer in nonlinear optics and laser physics. We would like to extend our acknowledges the the coauthors of our publications Drs.R.V.Arutyunyan, N.A.Enaki, M.V.Fedotov, A.S.Mikoyan, V.N.Seminogov,M.O. Skaiymkulov, O.Yu.Tikhomirov and V.I.Yukalov.

Chapter 1 and Appendices have been written by Yu.A.Ilinskii, Chapters 2-4 by A.V.Andreev and Chapters 5-6 by V.I.Emelyanov.

INTRODUCTION

The appearance of lasers and their further intensive development allowed to carry out the unique physical experiments, which changed the ideas on the nature of the resonant light-matter interactions, in particularly, of the collective interactions of atoms in the process of emission of radiation. In fact, any multiatomic ensemble prepared initially in a phase-consistent state will be in this state for a time T_2, which is a phase memory time or a characteristic time of the transversal relaxation. In the phase-consistent state, the induced dipole moment of an atomic ensemble is proportional to the number of atom. Consequently, the emission intensity is proportional to the square of the number of atoms. Because the time T_2 is very short for the optical transitions of atoms, it became possible to observe this phenomenon only when the powerful lasers appeared. Further importance of the processes of a collective interaction of atoms can occur in the spontaneous decay of the multiatomic systems.

In his pioneering 1954 paper [1], Dicke showed that a system of N two-level atoms with a population inversion could spontaneously decay to the ground state for a time inversely proportional to the number of atoms, $\tau_c \quad N^{-1}$. This effect is due to the induction of correlations between the dipole moments of the spatially separated atoms interacting via common electromagnetic field. As a result, the atoms in a volume of macroscopic size emit coherently. Since the total energy radiated by an ensemble of N atoms is $N\hbar\omega_0$, where ω_0 is the transition frequency, then the emission intensity is $I \quad N\hbar\omega_0/\tau_c \quad N^2$. This emission has been called by a collective spontaneous emission or Dicke superradiance (SR) . In the process of usual spontaneous decay, when the atoms decay independently on each other, the decay time T_1 does not depend on the number of atoms and the intensity of emission is proportional to the number of atoms, $I \quad N\hbar\omega_0/T_1 \quad N$. (See, for example, Ref. [2] as an review of the spontaneous emission theory).

The effective self-induction of correlations between the atomic dipole moments can be possible only in a case, when the evolution time τ_c of the process is shorter than the phase relaxation time T_2 , and also shorter than T_1 (usually $T_2 < T_1$). Therefore, from the point of view of the atomic subsystem dynamics the superradiance is a transient process occurring for a time shorter than T_1 and T_2 . It should be stressed, that the correlation between the radiators appears spontaneously during the emission process. It is this feature which differs principally SR from other coherent transient effects such as a free induction decay, self-induced transparency or photon echo [3,4], in which the individual radiators are also in correlated states, and the emission intensity is also proportional to N^2, but this phasing is due to the coherent pumping.

The nature of SR and conditions under which it can be observed are qualitatively different from that for spontaneous emission, and for induced emission. Let us discuss the specific features of SR on the example of a typical experiment carried out to observe it [5]. We assume, that N two-level atoms are confined into the active volume which has a shape of cylinder with the length L, cross-section area Z ($L \gg \sqrt{Z}$, $V = LZ$, $n = N/V$). In the initial moment all of the atoms are excited into the upper state by the short pumping pulse ($\tau_{pump} < \tau_c$) so, that the initial state is incoherent, i.e., there are absent any correlations between the wave functions of the different atoms. Then, the atoms evolve independently in radiative decay, the nature of which is determined by the relationships between the following parameters: T_1, T_2, τ_c, and, also, $\tau = L/c$, which is the active volume transit time, L is an active volume length, c is the velocity of the light. Let the density of the number of atoms is so small, that the following conditions holds:

$$T_1 = \left(\frac{4}{3}\frac{\omega_0^3 d^2}{\hbar c^3}\right)^{-1} < \tau_c \quad \left(\frac{2\pi n d^2 \omega_0}{\hbar \tau^{-1}}\right)^{-1},$$

where d is a dipole matrix element of the resonant atomic transition. In this case, each atom decays independently and the atomic system transits from the upper state into the lower one for a characteristic time T_1, the emitted intensity is isotropic in all directions.

We now assume that

$$\tau \ll \tau_c \ll T_2, T_1 .$$

These conditions determine the superradiance regime. The right-hand inequality means that the collective processes occur faster than the relaxation processes in the individual atoms. The left-hand inequality means that the photons leave the active volume for a time, which is shorter than the characteristic time of evolution of interatomic correlations, hence, the stimulated processes are unimportant during the emission. Under these conditions, the system of N atoms emits the superradiance pulse, the peak intensity of which is a few orders of magnitude higher than the intensity of a spontaneous decay (approximately ten orders of magnitude higher in the experiments of Skribanoeitz et al [5]). The bulk of the emitted energy is radiated into the small solid angles along the active volume axis. This directionality is due to the interference effect and is determined by the active volume shape. If $\tau \sim \tau_c$ then the part of the emitted photons is resonantly reabsorbed by the atomic subsystem and the emission takes the form of the train of pulses of decreasing height (oscillatory superradiance).

An important characteristics of the SR pulse is a delay time t_0, which is determined as a time of achieving of a peak pulse intensity. This time is approximately an order of magnitude longer than the pulse temporal width ($t_0 \sim \tau_c \ln N$) . This delay of SR is explained by the fact that the process of a radiative decay starts with the isotropic spontaneous emission and then, subsequently, the interatomic correlations grow in a result of the interaction of atoms via common radiation field. These correlations reach their maximum exactly at the time t_0. The population inversion difference is equal to zero at this moment.

The directionality of SR in the direction of the active volume axis is reminiscent the corresponding property of the amplified spontaneous emission. In the transient case this regime is determined, in contrast to SR, by the conditions

$$T_2 < \tau'_c = \left[\frac{2\pi n d^2 \omega_0}{\hbar T_2^{-1}} \right]^{-1} \ll \tau \, .$$

The left-hand inequality means, that the polarization is rapidly adjusted to the field, whereas the right-hand inequality means that the photons remain inside the active volume for a time τ . This time is long enough for photons to be involved in process of avalanche increase of a stimulated emission, which occurs over a time τ'_c . It should be noted that the process of amplified spontaneous emission is also frequently referred to as "superradiance" (or "superluminescence"). SR , as a collective spontaneous emission process, differs from it in the following principle way, that the processes of a stimulated emission does not play any essential role in the formation of SR pulse.

The difference between SR and a stimulated emission can be demonstrated on the example of single-pass mirrorless laser with a geometric configuration and a energy-level spacing the same as for SR (see Fig.I.1). However, the pump acts now continuously. At a large gain

$\mu_0 L = 2\pi n d^2 \omega_0 L/(\hbar T_2^{-1} c) \gg 1$ the system of the two-level atoms will be in a saturated state (the populations of the upper and lower levels is equal to $N/2$). Since the rate of pumping is proportional to N then the emission intensity of a such laser will also be proportional to N . Therefore, if we increase the density of the resonant atoms, then the intensity of amplified spontaneous emission will also increase proportionally to the pressure, while the SR intensity varies proportionally to the square of the pressure. This circumstance is used for the experimental identification of SR.

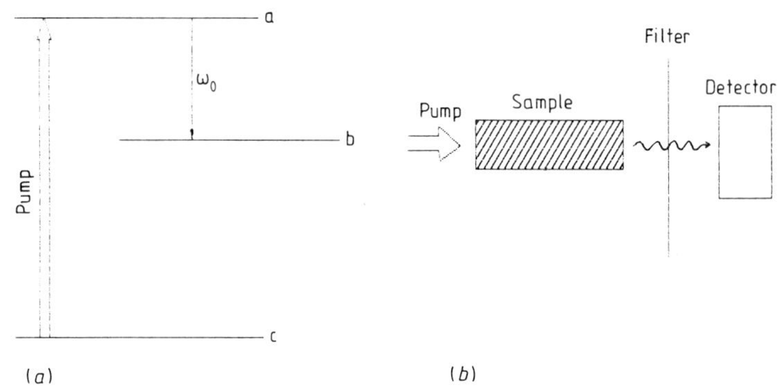

Fig.I.1. Energy-level diagram (a) and experimental set up (b) for typical experiment on SR.

The studies of SR effect are interesting both from the scientific and applied point of view. From the one hand SR is the example of the cooperative behavior of the system of N particles ($N \gg 1$), interacting though the electromagnetic field. The question about the nature of the superradiant correlated state in such many-body system and its formation, on the role of the geometry of the medium in the spatial coherence formation and also the connection of SR effect with the non-equilibrium phase transition are of the general physical interest.

From the applied point of view the SR effect is interesting as one of the methods of coherent radiation generation in mirrorless systems. The difficulties of the design of the reflecting mirrors in X-ray region do not permit to use for the generation of coherent radiation the usual stimulated radiation process. Because of that SR may turn out to be the possible mechanism of coherent radiation generation in this region (this is confirmed by the various estimates [6 - 8]). The SR can yield also the spectroscopic data, be used for the generation of ultra short pulses and so on.

After the pioneering Dicke work the theory of SR was developing only scarcely. The new stage of the intensive studies of SR effect has begun after the first experimental observation of SR in optical region. Up to now a rather large number of experiments have been performed on the SR observation. The experimental studies of SR effect has shown, that the existing theories in a number of important aspects do not yield the satisfactory agreement with experimental data. This has stimulated the further development of the theory of SR in extended systems. To the theoretical problems having been studied in the last time belong the taking into account of the retardation effects, transverse inhomogeneity, quantum fluctuations, polariton nature of SR effects in multiphoton processes.

It should be noted, that there exist several theoretical approaches to the SR effect description. This situation is the consequence of the complexity of the problem in its general form (the nonlinear and extended atomic system, the multimode electromagnetic field, the necessity of taking into account of the geometry of the system etc.). While taking into account of one aspect of the problem the different approaches can not to describe other important features. Thus we tried to use in this book the all basic approaches to the problem of SR description.

As it has been already said the phenomenon of SR is a striking example of the cooperative behavior of atomic system, interacting through the radiation field in strongly non-equilibrium conditions. Another equally striking effect observed in this system is the bistability of optical transmission. This effect consists in that under the action of the light wave with the input intensity I_{in} on the atomic system, the latter has two stationary states of the output intensity I_{out} at the same value of I_{in}. The optical bistability appears under the transmission of light through the material with non-linear absorption or dispersion. From the theoretical point of view it is the interesting example of the cooperative non-equilibrium behavior. From the practical viewpoint optical bistability suggest the possibility of realization of the memory elements, which can be used for the optical computers.

The theoretical consideration of bistability was for the first time carried out in Ref. [9]. The first observation of this effect was performed in Fabry-Perot resonator filled with sodium vapors [10]. Up to now the bistability was observed also in a number of solids (ruby, semiconductors) [11]. Experimental work stimulated the active theoretical and experimental investigations which can be divided into the two groups.

To the first belong in the main technological works dedicated to the devices developing and to the second - the works, developing the theoretical aspects of bistability and analogous phenomena in cooperative atomic systems [12]. In this book we consider a number of model of light-induced bistability in resonatorless systems.

Optical bistability in its manifestation (the jump like change of transmission, and hysteresis) is analogous to phase transition of the first kind. The dependence of the transmission on I_{in} is analogous to the van der Waals isotherm, describing liquid-vapor phase transition. The analogy of the laser generation threshold with the traditional phase transition goes very deeply. The basic notions of the phase transitions theory such as the order parameter, soft mode, the spontaneous symmetry breaking can be used also in description of the transition to laser generation state [13]. This is also the case with bistability (see Rev. [14]). It is interesting to study the SR effect from the point of view of the non-equilibrium phase transitions theory. This treatment is carried in this book.

The cooperative character of system of atoms, interacting through the radiation field can manifest itself also in a variety of other critical phenomena. In this book a whole number of models of novel light-induced instabilities and the corresponding non-equilibrium phase transitions are considered. These critical effects occur in the typical quantum electronics system - ensemble of two-level atoms, interacting through the radiation field. More complex systems - the three level systems and two-band semiconductors and other types of interatomic interactions (through the electrostatic and lattice displacement field) are also treated. Some of these models are used for the interpretation of experimentally observed light-induced critical phenomena.

This book consists of six chapters.

The chapter 1 served as an introduction. It discusses the model of two-level atom and its limits, the methods of description of two-level atoms interaction with the radiation field. In this chapter, in contrast all other, the Schrödinger representation is used. In the rest of chapters is used the Heisenberg representation since it is more close to the classical physics. In particular the quantum equations, describing the system of two-level atoms, interacting with the resonance field have in the operator form the same form as its semiclassical analogs.

In chapter 2 the derivation of operator equations of motion of the two-level atomic system, interacting with electromagnetic field is carried out. In accordance with said above we use three different approaches. Two types of atom-field representation are used: the first one exploits for the the field description the operators of vector potential and canonically conjugated generalized pulse, the second one uses the photon creation and annihilation operators, and the third approach uses the atomic representation in which the field variables are excluded with the help of solution of operator field equations. The integrals of motion of this system are obtained which play the essential role in collective processes description. Then, we derive the equations for the quantum mean values. The semiclassical system of equations is obtained. In this case, the decoupling procedure of the products of operators, entering the operator equations, is based on the assumption that the field is classical. The solution of the linearized system of semiclassical equations describing the system evolution from the initial state of isotropic spontaneous decay is presented. The systems of equations are obtained for the pair operator correlators such as the number of photons in a field mode, the field density operator, and the products of the collective atomic operators. These equations describe the effects of the back action of radiation field on the atom emitted it and consequently contain the terms, corresponding to the isotropic spontaneous decay. In derivation of these equations one encounters the problem of introduction of decoupling rules for the triple correlations. We used as the traditional decoupling with the neglect of the population difference fluctuations, as decoupling which take into account the symmetrical properties of Pauli spin matrices.

The chapter 3 is dedicated to the treatment of the quantum theory of SR effect. At first the more simple models are considered - the Markovian theories of SR in the point-source systems and in one-mode approximation. Further in the framework of the multimode theory the non-Markovian effects are taken into account, the conditions of the appearance of SR effect are determined, the SR regimes in weakly amplifying media, oscillatory SR regime and SR in weakly anisotropic media are studied. The effects of the frequency modulation in SR and the influence of the inhomogeneity on the intensity and shape of SR pulses are discussed. The investigation of the peculiarities of SR in the condition of Bragg diffraction is carried out. A number of SR models taking into account of the retardation effects and also the effects of transverse inhomogeneity of the medium are discussed. The peculiarities of SR in three level systems are studied. In the end of this chapter a survey of the experimental investigations of SR is given.

The chapter 4 is devoted to the semiclassical theory of SR. Here we introduce the dimensionless form of semiclassical equations for slowly varying amplitudes, and discuss the initial and boundary conditions. We analyze the linear stage of a pulse generation by ensembles of the two-level atoms, and show that the linear analysis enables us to determine the delay time, peak intensity, and influence of inhomogeneous broadening on the main pulse parameters. The process of coherent amplification is considered and the difference between the lethargic and exponential amplification is discussed. We study the soliton propagation in resonantly amplifying media of finite length and demonstrate the new soliton-like solutions for spatially inhomogeneous resonant media. On a basis of computer simulations of the generation dynamics, we present graphically the dependencies of the main parameters of generated pulses on the matter parameters. These graphics clearly demonstrate the difference between the superradiance, superluminescence, and lasing. The feasibility of optimization and control of generated pulse parameters is shown. The supersymmetric solutions of the general system of semiclassical equations are obtained and it is shown that these solutions correspond to the new regime of a radiative decay in extended systems of two-level atoms. The specific features of this new regime, called by a collective superradiance, are discussed.

The chapter 5 is devoted to the consideration of the non-equilibrium phase transitions in the cooperative systems of two-level atoms. The classification of the possible phase transitions in the cooperative systems of two-level atoms is carried out. A whole number of novel light-induced transitions in a typical quantum electronics system - cooperative system of two-level atoms interacting through the radiation field is considered including: the transition to the state of polariton generation in the open active medium, resonatorless bistability, ferromagnetic phase transition, orientational transition with threshold second harmonic generation and equilibrium ferroelectric phase transition.

In the chapter 6 the light-induced phase transitions are studied in the systems of atoms and semiconductor occurring due to the interaction through the permanent dipole-dipole and electrostatic fields or through the lattice displacement field. The special attention is paid to the light-induced transition with the spontaneous breaking of the atomic transitions saturation symmetry and threshold appearance of the static polarization or photovoltaic current in the centrosymmetric media. In non-centrosymmetric media this phase transition bears the character of bistability, which is considered in the atomic system with the dipole-dipole interaction and with interaction through the lattice displacement field. The orientational phase transition in the system of dipole centers with the threshold generation of optical second harmonic in initially centrosymmetric medium is studied. It is noted that the light-induced phase transition considered in Chaps.5 and 6 can essentially influence the performance of the quantum electronics devices. The consideration of novel electron-deformation thermal instability in interband transition under the action of light on the surfaces of semiconductors, leading to the non-equilibrium semiconductor-metal phase transition with the periodic superstructures formation is carried out.

On the basis of the theory developed a number of experimentally observed effects are interpreted. The chapter 6 is concluded with the brief comparison and survey of the resonant and non-resonant light-induced instabilities and phase transitions, considered in chapters 5 and 6 and in the literature. Two novel broad classes of laser-induced instabilities on the surface of solids - interferential instabilities of the surface relief and diffusion-deformational instabilities with the ordered structures formation are briefly discussed.

In the work over the book it was not our aim to write a review. In this there is no necessity from our point of view since, for example, there are a number of detailed reviews [15-20] and collections of papers [21-26]. We see our aim as the exposition of the physical essence of the theoretical models, developed for the description the cooperative effects in the multiatomic systems. In spite of the variety of the approaches adopted on the literature all they are reduced to a few models in accordance with the simplifying assumptions used.

CHAPTER 1

DICKE THEORY OF SUPERRADIANCE AND SOME GENERALIZATIONS

It is convenient to use the Heisenberg representation when we study the specific features of a collective spontaneous emission of a collection of inverted atoms or molecules or some related collective phenomena. This representation is used throughout the remainder of this book. But in this introductory chapter, both the Heisenberg and Schrödinger representations will be used in treating the Dicke superradiance.

In studying superradiance the Schrödinger representation has been used in the original paper by Dicke [1], as well as in more recent papers of Bonifacio [2] and some other authors. The main advantages of applying the Schrödinger representation are the clarity and simplicity of the physical treatment. In its simplest form, the theory is based on the calculation of probability of a spontaneous emission for a collection of the inverted atoms, which are in some specific correlated states. This probability differs from the sum of probabilities of a spontaneous emission for the individual atoms and it can be easily calculated by using perturbation theory. This seems to be the optimal way of making acquaintance with the superradiance theory. That is why we shall begin with such a treatment, all the more since the works [1,2] etc have not been widely discussed in textbooks or reviews.

Further, in this chapter we use the semiclassical treatment of superradiance, which is widely and extensively employed by specialists in the field of quantum radiophysics and laser physics.

1.1. Dicke model

The main specific features of a superradiance (a collective spontaneous emission) have been established by Dicke [1] on the basis of simplest model when all of the atoms are assumed to be located in a volume with dimensions much smaller than the wavelength of the emitted radiation. Some properties of this type of collective spontaneous emission can be explained from the classical point of view, because spontaneous emission is not the purely quantum effect, and it does not disappear in the classical limit.

1.1.1. *Classical treatment.* In classical electrodynamics we get the following formula for power of dipole emission at frequency ω:

$$I = \frac{4\omega^4}{3c^3} \, |\mathbf{d}_\omega|^2, \tag{1.1}$$

where $\mathbf{d}_\omega$ is the Fourier component of the classical dipole moment

$$\mathbf{d}(t) = \sum_\omega \mathbf{d}_\omega \exp(-\,\mathrm{i}\omega t).$$

We see that, according to (1.1), the radiation power is proportional to the dipole moment squared. Hence, if there are N identical radiators (atoms, molecules, oscillators) localized into the volume with the dimensions much smaller then the wavelength of the radiation with frequency ω, then the complete dipole moment is $\mathbf{D}(t) = N\mathbf{d}(t)$, and $\mathbf{D}_\omega = N\mathbf{d}_\omega$. So the emission power of collection of radiators is equal to $N^2 I_0$, where I_0 is the emission power for an individual radiator. This cooperative effect is explained by the fact that even in the absence of direct interaction of radiators, there is indirect interaction via the common radiation field. In this case, every radiator is exposed to its own field and the resulting field of all other radiators.

In the case of systems with linear dimensions much smaller than the radiation wavelength, the electric field, associated with the radiation and produced by all of the radiators, in any point inside the considered system is, determined by [3]

$$\mathbf{E} = \frac{2}{3c^3} \dddot{\mathbf{D}}. \tag{1.2}$$

This electric field is proportional to N, and the work per unit time, produced by it over all of the radiators, is

$$\mathbf{E}\,\dot{\mathbf{D}} = \frac{2}{3c^3}(\dot{\mathbf{D}}\;\dddot{\mathbf{D}}) = \frac{2}{3c^3}\frac{\mathrm{d}}{\mathrm{d}t}(\dot{\mathbf{D}}\;\ddot{\mathbf{D}}) - \frac{2}{3c^3}\ddot{\mathbf{D}}^2.$$

In averaging over time, the first term in the right-hand side of this equation disappears, so the averaged work per time is determined by

$$-\frac{2}{3c^3}\overline{\ddot{\mathbf{D}}^2} = -\frac{4\omega^4}{3c^3}|\mathbf{D}_\omega|^2.$$

We see that the last formula will coincide with the formula (1.1) for the emission power, if we substitute $\mathbf{d}_\omega$ by $\mathbf{D}_\omega$ in (1.1), where $\mathbf{D}_\omega$ is the Fourier component of the complete dipole moment:

$$\mathbf{D}(t) = \sum_\omega \mathbf{D}_\omega \exp(-\,\mathrm{i}\omega t), \qquad \mathbf{D}_{-\omega} = \mathbf{D}^*_\omega.$$

Thus, if there is a collection of N radiators, then the back action on each of them is enhanced N times in comparison with the case of an isolated radiator. At the same time, the net emission power, which is determined by the work of the radiation field over the whole collection of radiators (with the opposite sign), is enhanced N^2 times.

Let us consider a collection of N identical radiators. The ith oscillator displacement, $\mathbf{r}_i$, due to the electric field, obeys the following equation

$$m\ddot{\mathbf{r}}_i + m\omega^2\mathbf{r}_i = e\mathbf{E}_i, \qquad (1.3)$$

where ω is the frequency of an oscillator with mass m and charge e, and $\mathbf{E}_i$ is the electric field at the position of the ith oscillator, $\mathbf{R}_i$ ($i = 1, ..., N$).

Every oscillator is an electric dipole with moment $e\mathbf{r}_i$. Therefore, the electric field $\mathbf{E}_i$ at the position of the ith oscillator is produced by all other atoms and in the retardationless approximation it is determined by

$$\mathbf{E}_i = \sum_{i'\neq i} \frac{1}{|\mathbf{R}_i - \mathbf{R}_{i'}|^3}\left[\frac{3e(\mathbf{r}_{i'}(\mathbf{R}_i - \mathbf{R}_{i'}))(\mathbf{R}_i - \mathbf{R}_{i'})}{|\mathbf{R}_i - \mathbf{R}_{i'}|^2} - e\mathbf{r}_{i'}\right].$$

If this equation is substituted into (1.3), we get the set of linear equations for the displacements $\mathbf{r}_i$. As long as the field $\mathbf{E}_i$ is not taken into account, the equations (1.3) describe the vibrations at the frequency ω, and this frequency is $3N$-fold degenerate.

If $\mathbf{E}_i$ is accounted for, then in general the degeneracy is removed, and the normal frequencies of vibrations for the collection of coupled oscillators (the number of frequencies is equal to $3N$ or N for the case of polarized oscillators) will fill the band $\Delta\omega$ near the undisturbed frequency ω, where

$$\frac{\Delta\omega}{\omega} \propto \frac{e^2}{R^3 m\omega^2},$$

because $eE_i \propto e^2 r_i/R^3$. Here R is the mean distance between the oscillators. If the oscillators occupy the volume V_0, then $R \propto (N/V_0)^{-1/3}$. Thus, the direct dipole–dipole interaction of oscillators leads to splitting of the normal frequencies and, hence, to the broadening of the emission line.

We could improve our treatment if we took into account magnetic interactions between the moving charges. When the velocities of oscillators, $v \propto |\dot{\mathbf{r}}_i|$, are much smaller than the light velocity, c, it means that the interactions between the moving charges are accounted for to an accuracy of the order of $(v/c)^2$ [3]. The inclusion of magnetic interactions does not change the oscillator behaviour drastically, and the equation for the emission power will be supplemented by terms to the third power of the small value v/c. As we have mentioned earlier, in the case of a system with linear dimensions much smaller than the radiation wavelength, the third order contribution to the electric field due to the emission of radiation, is determined by $\mathbf{E}_r = (2/3c^3)\,\dddot{\mathbf{D}}$. In this case

$$\mathbf{D} = e \sum_i \mathbf{r}_i .$$

In accordance with the previous treatment, the ratio of values of the electric field $\mathbf{E}$ to the dipole–dipole field $\mathbf{E}_d$ is determined by

$$\frac{E_r}{E_d} \sim \frac{eNr_i\omega^3/c}{er_i/R^3} \sim \frac{N\omega^3R^3}{c^3} \sim \frac{N\omega^3}{c^3}\frac{V_0}{N} \sim \frac{V_0}{\lambda^3} .$$

We see that this ratio is below unity when $V_0 \ll \lambda^3$ (where λ is the radiation wavelength, $\lambda = 2\pi c/\omega$). Nevertheless, we shall often use in further consideration the point-system with the volume $V_0 \ll \lambda^3$, for which we are not taking into account the dipole–dipole interaction, i.e. the interaction via a longitudinal field, but taking into account the interaction via the transversal field only. That means we shall assume $\mathbf{E}_i = \mathbf{E}_r = (2/3c^3)\,\dddot{\mathbf{D}}$. This assumption underlies the Dicke model of superradiance.

As we have seen, for a real system with $V_0 \ll \lambda^3$ the condition $E_d/E_r \ll 1$ could not be met at any time. We shall discuss this problem later, but it should be noted now that the justification of the Dicke model efficiency lies in the fact that all the main features of the emission dynamics prescribed by this model are transferred to the case of extended media with linear dimensions much greater than the radiation wavelength.

Under this assumption we can replace the equation (1.3) by the following equation

$$m\ddot{\mathbf{r}}_i + m\omega^2\mathbf{r}_i = \frac{2e^2}{3c^3}\sum_k \dddot{\mathbf{r}}_k. \tag{1.4}$$

Summing the equations (1.4) over i, we obtain the following equation for the variable $\mathbf{r} = \sum_i \mathbf{r}_i/N$:

$$m\ddot{\mathbf{r}} + m\omega^2\mathbf{r} = \frac{2e^2}{3c^3} N \dddot{\mathbf{r}}. \tag{1.5}$$

Let us choose the set of arbitrary coefficients α_i, for which the condition $\sum_i \alpha_i = 0$ holds. Multiplying (1.4) by α_i and summing them over i we find the equation for $\mathbf{r}' = \sum_i \alpha_i\mathbf{r}_i$:

$$m\ddot{\mathbf{r}}' + m\omega^2\mathbf{r}' = 0. \tag{1.6}$$

The equation (1.5) has the following solution

$$\mathbf{r}(t) = \mathbf{r}_0(t)\exp(-i\omega t) + \text{c.c.},$$

where $\mathbf{r}_0(t) = \mathbf{r}_0(0)\exp(-Nt/2T)$ is a complex slowly varying amplitude of vibrations, and

$$T = \frac{3mc^3}{2e^2\omega^2}$$

is an individual oscillator decay time due to the interaction with the radiation field. It is supposed that $T^{-1} \ll \omega$. This solution describes the damping of oscillator vibrations with the characteristic damping time $\tau_c = T/N$. As a consequence of cooperative processes, the decay time is shortened N times, while the power of emission increases N^2 times in comparison with the emission power of an isolated oscillator whereas in the case of spontaneous uncorrelated decay of a collection of N atoms the intensity is proportional to the first power of N.

Using (1.5) and (1.6) the general solution of (1.4) can be written as

$$\mathbf{r}_k(t) = \mathbf{r}(t) + (\mathbf{r}_k(t) - \mathbf{r}(t)) = \mathbf{r}(t) + \mathbf{r}'_k(t) = (\mathbf{r}_0(t) + \mathbf{r}'_{k0})e^{-i\omega t} + \text{c.c.}$$

Here $\mathbf{r}'_k(t) = \mathbf{r}_k(t) - \mathbf{r}(t)$ satisfies (1.6), because

$$\mathbf{r}'_k = \sum_i \left(\delta_{ki} - \frac{1}{N}\right)\mathbf{r}_i,$$

and

$$\sum_i (\delta_{ki} - \frac{1}{N}) = 0.$$

As an example, for a pair of oscillators (N = 2) with complex amplitudes $\mathbf{r}_{10}$ and $\mathbf{r}_{20}$ at t = 0, the solution is

$$\mathbf{r}_1 = \frac{1}{2}(\mathbf{r}_{10} + \mathbf{r}_{20})e^{-t/T} e^{-i\omega t} + \frac{1}{2}(\mathbf{r}_{10} - \mathbf{r}_{20})\, e^{-i\omega t} + \text{c.c.}$$

for $\mathbf{r}_2(t)$.

With increasing time the solution tends to the undamped antiphase vibrations of oscillator pair

$$\mathbf{r}_1(t) = \frac{1}{2}(\mathbf{r}_{10} - \mathbf{r}_{20})e^{-i\omega t} + \text{c.c.}, \qquad \mathbf{r}_2(t) = -\mathbf{r}_1(t).$$

The undamped character of the residual vibrations is, of course, a result of our model, in which the emission of radiation is the only mechanism of energy dissipation. This solution describes the real process if the observation time is less than T_1, where T_1 is a net relaxation time for both the radiative and radiationless decay processes. Evidently, $T_1 \leq T$.

The different interactions of oscillators with the environment (thermostat) can cause not only the energy dissipation but also the jumps and drift of oscillator phases. The corresponding characteristic relaxation time is usually denoted by T_2 and it will be a relaxation time of the complete dipole moment of oscillators. Usually $T_2 < T_1$ and even $T_2 \ll T_1$. The above solution will be valid for $t \ll T_1, T_2$. If the conditions $\tau_c \ll T_1, T_2$ hold, then the light pulse with width τ_c should be emitted by the excited system of oscillators.

Another reason preventing the superradiative decay of a system of excited oscillators is an inhomogeneous line broadening, when there is a distribution of the oscillator transition frequencies ω inside some region with width $\Delta\omega$. In particular, the dipole–dipole interaction of oscillators is one of the reasons for such broadening. The superradiance can occur if the condition $\tau_c \ll T_2^* = 2\pi/\Delta\omega$ holds, where T_2^* is a characteristic inhomogeneous broadening relaxation time.

Thus, only the completely symmetric component of oscillator vibrations $\mathbf{r}(t)$ decays with the characteristic variation time τ_c. If all of the oscillators are in phase in the initial state and their initial amplitudes are equal, then the net energy of oscillators decreases with characteristic decay time τ_c. If the initial phases of oscillators are not correlated or if they change unpredictably due to relaxation processes than the decay time should not decrease and the power of emission, which is proportional to $|\mathbf{D}_\omega|^2$, should be determined by

$$\frac{4\omega^4}{3c^3}\sum_i |\mathbf{d}_\omega^{(i)}|^2,$$

because $\langle \mathbf{d}_\omega^{(i)}\mathbf{d}_\omega^{(k)}\rangle = 0$ for $k \neq i$. Here and below the brackets $<\ \ >$ mean the statistical and quantum mechanical averaging. The power of emission is proportional to N in this case, in contrast to N^2 dependence in the case of superradiance.

The emission power of a collection of oscillators is

$$I(t) = \frac{4\omega^4}{3c^3}|\mathbf{D}_\omega|^2 = \frac{4\omega^4}{3c^3}e^2N^2|\mathbf{r}_0(t)|^2,$$

and energy of the completely symmetrical vibrations is

$$E = 2m\omega^2 N|\mathbf{r}_0(t)|^2.$$

Hence, the equation $\mathrm{d}E/\mathrm{d}t = -I(t)$ takes the form

$$\frac{\mathrm{d}E}{\mathrm{d}t} = -\frac{NE}{T}$$

and its solution is

$$E\ \exp(-N(t-t_0)/T).$$

The exponential character of the decay profile follows from the fact that the square of the net dipole moment of oscillators $|\mathbf{D}_\omega|^2$ is proportional to the stored energy E:

$$|\mathbf{D}_\omega|^2 = \frac{e^2N}{2m\omega^2}E.$$

For some hypothetical systems of oscillators with some arbitrary relationship between $|\mathbf{D}_\omega|^2$ and E, the master equation should have the form

$$\frac{\mathrm{d}E}{\mathrm{d}t} = -\frac{4\omega^4}{3c^3}|\mathbf{D}_\omega|^2(E). \tag{1.7}$$

It is this form of the Dicke equation which will be derived in the next section.

1.1.2. *Quantum treatment.* A model of the two-level atom is widely used in the quantum treatment of a resonance light–matter interaction. The conditions under which this model holds will be discussed later, and we simply assume now that the atom has only two states — upper and lower levels. The wavefunctions of these states will be designed as $|+\rangle$ and $|-\rangle$ for upper and lower levels respectively. The energy level spacing difference, $E_2 - E_1$, is designated as $\hbar\omega_0$.

In the dipole approximation the interaction of this two-level atom with the electromagnetic field is determined by the value of the matrix elements of the atomic dipole moment operator **d**. We assume that only the off-diagonal matrix elements of this operator are different from zero

$$\langle + | \hat{\mathbf{d}} | - \rangle = \langle - | \hat{\mathbf{d}} | + \rangle^* = \mathbf{d}.$$

Let us consider now the two identical two-level atoms. If both of them are in the excited state (upper level), then their wavefunction is a product of the wavefunctions $|+\rangle^{(1)}$ and $|+\rangle^{(2)}$, where the superscript refers to the number of the atom. The ground state of this system, when both atoms are in the lower level, has the wavefunction $|-\rangle^{(1)}|-\rangle^{(2)}$.

If one atom is now in an excited state and the other is in the ground state, than there are two wavefunctions which correspond to the same energy $E_1 + E_2$. They are $|-\rangle^{(1)}|+\rangle^{(2)}$ and $|+\rangle^{(1)}|-\rangle^{(2)}$. These four states are all the possible linearly independent states of a pair of two-level atoms. The energy levels of these states are $2E_2$, $E_1 + E_2$ (double degenerate states), and $2E_1$.

The dipole moment operator of this pair is

$$\hat{\mathbf{D}} = \hat{\mathbf{d}}^{(1)} + \hat{\mathbf{d}}^{(2)}.$$

It is convenient to use the linear combinations of the degenerated wavefunctions. One of them is symmetric with respect to the atomic rearrangements:

$$|1\rangle_s = \frac{1}{\sqrt{2}}\left(|-\rangle^{(1)}|+\rangle^{(2)} + |+\rangle^{(1)}|-\rangle^{(2)} \right),$$

and the other is the antisymmetric in respect to the atomic rearrangements

$$|1\rangle_a = \frac{1}{\sqrt{2}}\left(|-\rangle^{(1)}|+\rangle^{(2)} - |+\rangle^{(1)}|-\rangle^{(2)} \right).$$

The factor $1/\sqrt{2}$ has been introduced to normalize the wavefunctions $|1>_{s,a}$ and the index 1 means that there is one excited atom in the system. It is not hard to make sure that the dipole moment operator $\hat{\mathbf{D}}$ has non-zero matrix elements only for transitions from the states $|2> = |+>^{(1)}|+>^{(2)}$ and $|0> = |->^{(1)}|->^{(2)}$ to the symmetric state $|1>$, because the states $|2>$ and $|0>$ are self-evidently the symmetric states in respect to the atomic rearrangements. These matrix elements are

$$<2|\hat{\mathbf{D}}|1>_s = {}_s<1|\hat{\mathbf{D}}|2>^* = {}_s<1|\hat{\mathbf{D}}|0> = <0|\hat{\mathbf{D}}|1>_s^* = \sqrt{2}\mathbf{d}.$$

According to the quantum mechanical rules and the correspondence principle, the equation for the spontaneous emission power follows from (1.1) if we substitute the dipole moment $\mathbf{d}_\omega$ by the matrix elements of the dipole moment operator. Therefore, the spontaneous emission power for the transition from the state $|2>$ to the state $|1>_s$ with the energy change $\hbar\omega_0$ is determined by

$$I\left(|2> \longrightarrow |1>_s\right) = \frac{4\omega_0^4}{3c^3}\,|{}_s<1|\hat{\mathbf{D}}|2>|^2 = \frac{4\omega_0^4}{3c^3}\,|\sqrt{2}\,\mathbf{d}|^2.$$

The probability of the photon emission with energy $\hbar\omega_0$ per unit time is determined by

$$w\left(|2> \longrightarrow |1>_s\right) = \frac{I}{\hbar\omega_0} = \frac{4\omega_0^4}{3c^3}\,\frac{2|\mathbf{d}|^2}{\hbar\omega_0} = 2w = 2T^{-1}.$$

Here

$$w_0 = \frac{1}{T} = \frac{4\omega_0^3|\mathbf{d}|^2}{3\hbar c^3}$$

is the probability of radiative decay of one atom per unit time, and T is a spontaneous emission decay time. Similarly

$$I(|1>_s \longrightarrow |0>) = \hbar\omega\, w(|1>_s \longrightarrow |0>) = 2I$$

is equal to twice the power of spontaneous emission for an isolated atom, and the probability of the radiative decay of the symmetric state, $|1>_s$, with one excited atom to the ground state, $|0>$, is equal to $2w_0$. It is not unusual that the state $|2>$ with the two excited atoms decays with probability twice that with respect to w_0. But it is a surprise that the symmetric state $|1>_s$ with the one excited atom decays with the probability which is also equal to $2w_0$. At the same time it is important that the antisymmetric state $|1>_a$ with one excited atom is not coupled with the states $|2>$ and $|0>$ by radiative transitions; the matrix elements of the dipole moment operator for these transitions are equal to zero. The probability of the state $|1>_a$ decay equals zero.

In the same way we can calculate the probabilities of the dipole transitions for a collection of N two-level atoms. It is essential that the state of complete inversion $|N>$ and ground state $|0>$, which are intrinsically symmetric with respect to the atomic rearrangements, are coupled radiatively only with the symmetric states $|N-1>$ and $|1>_s$, respectively. Here

$$|1>_s = N^{-1/2}(|->^{(1)}...|->^{(N-1)} |+>^{(N)} +...+ |+>^{(1)}|->^{(2)}...|->^{(N)}).$$

In turn, the state $|N-1>_s$ is coupled with the state $|N-2>_s$ only and so on. Therefore, if the system is initially in the state of complete inversion then only the symmetric states evolve by the radiative decay. The same holds true for induced transitions starting with the ground state $|0>$.

Further we will be interested in the transitions between the symmetric states only and bearing this in mind we will omit the index s. We will also omit an operator sign $\wedge$ in any case when we are sure that no misunderstanding will arise. As in the case $N = 2$ we can determine the matrix elements of the net dipole moment operator

$$\mathbf{D} = \sum_i \mathbf{d}_i.$$

These matrix elements differ from zero only for transitions with $\Delta N = \pm 1$:

$$<n-1|\mathbf{D}|n>^* = <n|\mathbf{D}|n-1> = \sqrt{n(N-n+1)}\mathbf{d}.$$

It means that the probability of the transition from the state $|n>$ to the state $|n-1>$ is

$$w(|n> \longrightarrow |n-1>) = w_0 n(N - n + 1) = T^{-1} n(N - \text{n} + 1). \quad (1.8)$$

Notice that the state $|N>$ with N excited atoms decays with the probability Nw_0. This is quite natural. The probability w is a maximum for $n \approx N/2$ when $w_{max} \approx N^2/4$. We see that maximal probability is proportional to N^2.

Referring to (1.8) for transition probability we can easily write the equation of motion for the mean value of a number of excited atoms:

$$\frac{\mathrm{d}n}{\mathrm{d}t} = -\frac{n(N - n + 1)}{T}.$$

It is convenient to rewrite it in the form of the equation for population difference. We will widely use later the operator R_3 of population difference; its mean value is designated as $\langle R_3 \rangle$. It is evident that

$$\langle R_3 \rangle = n - (N - n) = 2n - N.$$

Hence,

$$\frac{\mathrm{d}}{\mathrm{d}t} \langle R_3 \rangle = -\frac{2}{T}\left(\frac{N}{2} + \frac{\langle R_3 \rangle}{2}\right)\left(\frac{N}{2} - \frac{\langle R_3 \rangle}{2} + 1\right),$$

or

$$\frac{\mathrm{d}}{\mathrm{d}t} \langle R \rangle = -\frac{1}{T}\left[\frac{N^2}{2} - \frac{\langle R_3 \rangle^2}{2} + N + \langle R_3 \rangle\right]. \tag{1.9}$$

This is the Dicke equation for population difference. If we consider the radiative decay of a collection of the two-level atoms which are completely inverted initially, then we should solve the equation (1.9) with the following initial condition

$$\langle R_3 \rangle \Big|_{t=0} = N.$$

Equation (1.9) can be solved analytically. The right–hand side of this equation can be rewritten as

$$\frac{1}{2}\,[(N + 1) - (\langle R_3 \rangle - 1)^2].$$

Using the substitution $\langle R_3 \rangle - 1 = -(N + 1)\cos\vartheta$ we obtain the following equation for ϑ:

$$\dot{\vartheta} = -\frac{N + 1}{2T}\sin\vartheta. \tag{1.10}$$

This is an equation for the overdamped physical pendulum with the equilibrium point $\vartheta = 0$ and the unstable equilibrium point $\vartheta = \pi$.

The solution of (1.10) is

$$\cos\vartheta = \tanh\left[\frac{N + 1}{2T}(t - t_0)\right],$$

i.e.

$$\langle R_3 \rangle = 1 - (N + 1)\tanh\left[\frac{N + 1}{2T}(t - t_0)\right]$$

and according to the initial condition, $\langle R_3(0) \rangle = N$, we get

$$t_0 = \frac{T}{N + 1} \ln N. \tag{1.11}$$

The emitted radiation power as a function of the time can be determined from (1.9)

$$I(t) = -\frac{\hbar\omega_0}{2} <\dot{R}_3> = I_0\left(\frac{N + 1}{2}\right)^2 \sin^2\vartheta$$

$$= I_0\left(\frac{N + 1}{2}\right)^2 \operatorname{sech}^2\left[\frac{N + 1}{2T}(t - t_0)\right]. \tag{1.12}$$

Equation (1.12) shows that the maximal power of the superradiance is achieved at $t = t_0$. This peak value is proportional to N^2 ($N \gg 1$ and we can omit the unity everywhere in (1.12)). The time t_0 is a pulse delay time, which is

$$t_0 = \tau_c \ln N,$$

and τ_c is a pulse width,

$$\tau_c = \frac{T}{N + 1}.$$

Since $N \gg 1$, then $t_0 = \tau_c \ln N \gg \tau_c$. In real physical systems $\ln N \sim 20$, therefore it is the time t_0 which determines the duration of the process of system de-excitation.

The characteristic power profile $I(t)$ is shown in Fig. 1.1.

Following this preliminary treatment of the Dicke model let us enumerate its basic assumptions.

1. There are N identical two-level elementary systems (atoms) with $E_2 - E_1 = \hbar\omega_0$.

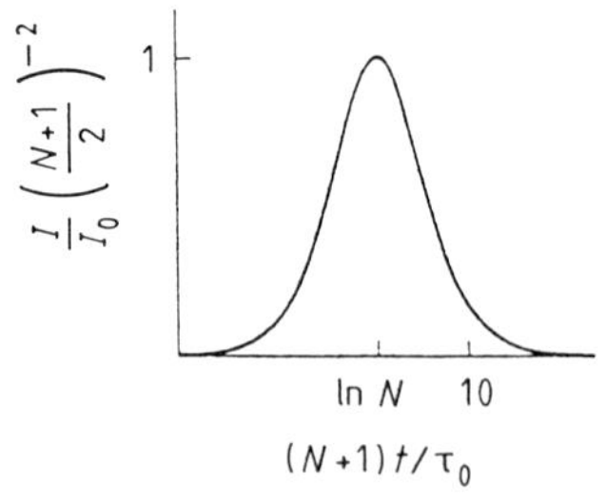

Fig. 1.1. Characteristic SR intensity profile given by (2.12) for $N = 10^3$ ($\ln N = 6.904$).

2. At the initial moment all of the atoms are in the upper excited state (with the energy E_2).

3. The atoms are confined to a volume with dimensions much smaller than the radiation wavelength, $\lambda = 2\pi c/\omega_0$.

4. Atoms interact only via the common radiation field; all other mechanisms of the interaction can be ignored.

The third assumption is used in calculations of the decay probabilities in the dipole approximation. The fourth assumption means that the relaxation times T_1, T_2 and $T_2^* = 2\pi/\Delta\omega$ must be greater than the characteristic decay time t_0 (see (1.11)).

As has been discussed in section 1.1.1 the third and fourth assumptions are inconsistent with one another. In fact, the value of $\hbar\Delta\omega$ is approximately equal to the energy of the dipole–dipole interaction, $|d|^2/R^3$, where R is the mean distance between the atoms, which is $(N/V_0)^{-1/3}$ and V_0 is the volume filled by the atoms. Therefore

$$T_2^* \sim \frac{\hbar R^3}{|\mathbf{d}|^2} \sim \frac{\hbar V_0}{N|\mathbf{d}|^2}.$$

At the same time, according to (1.11) we get

$$t_0 = T\frac{\ln N}{N+1} \sim \frac{\hbar c^3 \ln N}{\omega_0^3|\mathbf{d}|^2\, N}.$$

Hence, the condition $t_0 \ll T_2^*$ means that

$$V_0 \gg \frac{c^3}{\omega_0^3}\ln N = \left(\frac{\lambda}{2\pi}\right)^3 \ln N.$$

Thus the condition $t_0 \ll T_2^*$ does not hold if the active volume is much smaller λ^3. In this case the dipole–dipole interaction of atoms leads to the destruction of the SR state in time which is much smaller than t_0, while t_0 is the time of SR formation from the initial state. Therefore the point source Dicke model, where the atoms interact only via the transverse field, is useful only as a simplest model of a superradiance. Its advantages are the simplicity, and the possibility of carrying out an almost complete investigation. It is important that the one-mode model, which will be discussed in section 1.3 and which is free from the intrinsic contradictions belonging to the point-source model, leads to approximately the same results and equations as the Dicke model.

1.2. Dicke state and coherent atomic states

The systems which are usually considered in quantum electronics are collections of a macroscopic number of atoms (molecules) interacting with the electromagnetic field and thermostat, the latter, among other things, being responsible for creation of population inversion. These are very complex systems, especially if we use quantum theory to describe the electromagnetic field and thermostat (pumping mechanism) just as in the treatment of atomic transitions. The resonance light-matter interactions are at the basis of an action of the quantum electronics devices. In the resonance case the interaction of an electromagnetic field with matter is not described properly by the perturbation theory, which is one of the basic quantum mechanical methods. At the same time, the two level models or, in general, the models with a small number of levels can be used with great success. The conditions under which these models can be applied are evident. They are applicable only when some definite pair of atomic levels interacts effectively with the radiation. This occurs when the energy-level difference $\hbar\omega_0$ is close to $\hbar\omega$, where ω is a radiation frequency.

It is well known and we shall discuss it later in more detail, that quasiclassical systems exhibit equidistant or quasi-equidistant spectra, when the energy-level difference of the two adjacent levels does not depend on the quantum numbers of these levels or slowly depends on them. Such systems cannot be treated by models with a small number of levels. Harmonic or weakly unharmonic oscillators, which are widely used for a description of the molecular vibrations, are also out of the sphere of application of these models. Only atoms or molecules with a pair of levels (at least one of them should be populated substantially), for which the energy-level difference is far essentially from the energy-level difference for any other transitions, can be described with the help of the two-level atom model.

If we consider an isolated atom, then its wavefunction the in energy representation is a function of the discrete index s. This is a vector with the infinite number of components ($s = 1,2,\ldots$). In the two-level approximation the dimensionality of this vector is equal to two ($s = 1,2$). The wavefunction becomes a function of a variable which takes only the two values. All of the physical quantity operators referring to an atom can be treated as the reduced operators acting in the two-dimensional space of states $\sum_{s=1}^{2} a_s |\psi_s\rangle$.

Let A be an operator acting in the Hilbert space of an atom; then it can be represented as the two-column matrix $a_{ss'} = \langle\psi_s|A|\psi_{s'}\rangle$. The operator A corresponding to a physical quantity is a Hermitian operator and, hence the matrix $a_{ss'}$ should be the Hermitian matrix, i.e. $a_{ss'} = a^*_{s's}$. As a result, there appears a formal analogy between an arbitrary two-level system and a spin system with the spin value of 1/2. A particle with spin 1/2 can have the two values of a spin projection on an arbitrary axis: ±1/2. The physical quantity operator acting on the spin variables can be represented by a two-column matrix. Therefore any operator A of the two-level system can be associated with some operator of the spin system.

As a result of this analogy it is useful to treat the two-level atomic system with the help of the Pauli spin matrices [4]

$$\sigma_1 = \begin{pmatrix} 0 & 1 \\ 1 & 0 \end{pmatrix}, \qquad \sigma_2 = \begin{pmatrix} 0 & -i \\ i & 0 \end{pmatrix} \qquad \sigma_3 = \begin{pmatrix} 0 & 1 \\ -1 & 0 \end{pmatrix} \tag{2.1}$$

The Pauli matrices are closely related to the quantum mechanical theory of angular momentum and the operators $J_i = \sigma_i/2$ obey the commutation rules for the angular momentum operators:

$$[J_j, J_k] = i\varepsilon_{jkl} J_{l'} \tag{2.2}$$

where ε_{jkl} is a completely antisymmetric tensor with the components equal to +1, if jkl are the even rearrangements of the figures 123, and equal to − 1, if jkl are the odd rearrangements of these figures.

The vectors $\begin{pmatrix} 1 \\ 0 \end{pmatrix}$ and $\begin{pmatrix} 0 \\ 1 \end{pmatrix}$ are the eigenvectors of the operator J_3 with the eigenvalues $m = \pm 1/2$. In quantum mechanics of the $S = 1/2$ spin these operators correspond to the states with the ±1/2 projection of a spin moment on the axis with index 3. For a two-level system these vectors correspond to the upper and lower atomic states for which the angular momentum quantum number can be undefined in general, or not exist at all. Nevertheless, it is convenient to designate the upper and lower states of a two-level atomic system as $|\pm 1/2\rangle$ or even $|\pm\rangle$.

Bearing in mind an analogy with the spin systems, the concept of a quasispin or an energetical spin of a two-level system is often introduced. The last term follows from the fact that the energy of an isolated two-level atom is represented by $H = \hbar\omega_0 J_3$. Exercising freedom in a the choice of the energy scale origin we have assumed that the energy of the upper atomic level is equal to $\hbar\omega_0/2$ and the energy of the lower level is $-\hbar\omega_0/2$, where $\hbar\omega_0$ is the energy-level difference.

The Pauli matrices, σ_1, σ_2, σ_3, and the identity matrix

$$I = \begin{pmatrix} 1 & 0 \\ 0 & 1 \end{pmatrix}$$

are linear independent. Hence an arbitrary Hermitian matrix of a two-level system, $a_{ss'}$, can be represented as a linear combination of the matrices σ_1, σ_2, σ_3 and I. In particular, we will widely use the expression for the dipole moment operator of a two-level system in terms of the Pauli matrices.

1.2.1. *Collection of two-level atoms.* We turn now to a treatment of a collection of the two-level atoms. Let us assume that all of the atoms are identical but distinguishable. The last property can be due to the localization of atoms and besides it can be assumed that their wavefunctions do not overlap. In this case the rearrangement symmetry of the complete wavefunction, incorporating the centre of mass positions of the each atom, can be neglected and we need take into account only the internal atomic degrees of freedom. The vector of the state or the wavefunction of a collection of the N two-level atoms will be the function $\psi(m_1,...,m_N)$, depending on the variables, which account the two values $m_l = \pm 1/2$. The scalar product of two such vectors is

$$<\phi|\psi> = \sum_{m_l} \phi^*(m_1,...,m_N)\, \psi(m_1,...,m_N),$$

where $\{m_l\} = (m_1,...,m_N)$.

Consider the states in which there are n atoms in the upper state and $N - n$ atoms in the lower state. The wavefunction of one of these states is

$$\psi(m_1,...,m_N) = \psi_+(m_1)...\psi_+(m_n)\psi_-(m_{n+1})...\psi_-(m_N), \qquad (2.3)$$

where $\psi_+(m) = \delta_{1/2,m}$ and $\psi_-(m) = \delta_{-1/2,m}$. The other states with n atoms in the upper state can be produced from the wavefunction (2.3) by the rearrangement of the atoms, i.e. by the rearrangement of the variables $m_1,...,m_N$ in the wavefunction. The whole number of such different states is

$$C_N^n = \frac{N!}{n!(N-n)!},$$

and the net number of the states is, evidently,

$$\sum_n C_N^n = 2^N, \quad n = 0,...,N.$$

Further developing the analogy with the spin systems, let us introduce the summary energetical spin of a collection of the two- level atoms

$$\mathbf{J} = \sum_{l} \mathbf{J}^{(l)}, \ l = 1,...,N,$$

where $\mathbf{J}_i^{(l)} = \sigma_i^{(l)}/2$ are the above introduced operators for the lth atom. It is seen that the states (2.3) are the eigenvectors of J_3 operator, corresponding to the eigenvalues

$$(n - (N - n))/2 = (2n - N)/2.$$

The operators of the summary energetical spin components, J_i, obey the usual commutation rules for the components of the angular momentum (see (2.2)). We can introduce the operator $\mathbf{J}^2 = J_1^2 + J_2^2 + J_3^2$, which commutes with the operator J_3 along with the other operators J_1 and J_2. Therefore we can introduce the $|jm\alpha\rangle$ states of a collection of the two-level atoms, for which

$$\mathbf{J}^2|jm\alpha\rangle = j(j + 1)|jm\alpha\rangle, \ J_3|jm\alpha\rangle = m|jm\alpha\rangle, \ - j \leq m \leq j. \quad (2.4)$$

The states $|jm\alpha\rangle$ are named by the Dicke states. They are the eigenvectors of the operators $\mathbf{J}^2$ and J_3. In these states the summary energetical spin and its projection on an axis have the definite values. The quantum numbers j and m do not describe these states unambiguously; there is an additional third quantum number α, the physical sense of which will be discussed later.

For a set of non-interacting two-level atoms, the Dicke states $|jm\alpha\rangle$ are the states with a definite value of energy E, because the Hamiltonian of the free atomic system is

$$H = \sum_{l} H^{(l)} = \hbar\omega_0 \sum_{l} J_3^{(l)} = \hbar\omega_0 J_3 = \hbar\omega_0 R_3/2.$$

Here we have introduced the population difference operator $R_3 = 2J_3$. Its value in the state (2.3) is equal to n $-$ $(N - n)$, that is why we can define it as an operator of the population difference for the upper and lower states.

Fig. 1.2 shows the equidistant energy-level spacing scheme for a collection of the two-level atoms in the case of the states with the integer and half-integer quantum number j, when $E = \hbar\omega_0 m$, and $|m| \leq j$.

The quantum numbers j and m are integer or half-integer in dependence on the parity of the number of atoms N because $m = (2n - N)/2$ in the states (2.3).

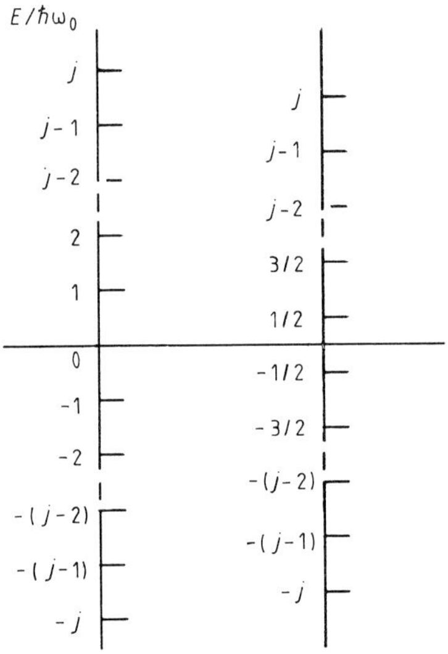

Fig. 1.2. Equidistant energy-level spacing of an ensemble of two-level atoms, when j is an integer (on the left) or a half-integer (on the right)

When all of the atoms are in the upper excited state ($n = N$ in (2.3)), then $m = N/2$. This is the maximal value of the m and, hence, of the j for a given number N. The ground state of a collection of N atoms has the following values of the quantum numbers: $j = N/2$ and $m = -N/2$. In the ground state all of the atoms are in the lower state and $n = 0$. These two states are completely symmetric with respect to the atomic rearrangements. The other symmetric states can be produced from the state (2.3) with the help of the symmetrization procedure. All of these symmetric states correspond to $j = N/2$ and $m = (2n - N)/2$.

Under the action of the operators J_1, J_2, and J_3 act on the states $|j{=}N/2,m\rangle$, they are transformed into the symmetric states with the same value of $j = N/2$. It follows from the form of operator: $J_i = \sum_l J_i^{(l)}$. We see that J_i commutes with any operator of the atomic rearrangements, hence, the symmetry of the wavefunction $J_i|jm\rangle$ is the same as the symmetry of the wavefunction $|jm\rangle$.

Thus, the symmetric states comprise a set of the states of the N atomic system with the maximal value of the quantum number j, which is

$N/2$. They can be rewritten in the following normalized form

$$|N/2,m\rangle = \frac{1}{\sqrt{N!(N/2 + m)!(N/2 - m)!}} \Sigma P_N |+\rangle^{(1)} \ldots$$

$$\ldots|+\rangle^{(N/2+m)} |-\rangle^{(N/2+m+1)} \ldots |-\rangle^{(N)}, \qquad (2.5)$$

where the summation in (2.5) is performed over $N!$ rearrangements of the operators P_N. Only the states $|N/2,m,\alpha\rangle$ are not degenerated in respect to the third quantum number α, i.e. they are completely and unambiguously determined by the quantum number m, which is a population difference, and in this case the index α can be omitted.

For the other values of j $N/2$ there are n_j vectors for the given quantum numbers j and m. It can be easily found that

$$n_j = C_N^{N/2+j} - C_N^{N/2+j+1}, \; j < N/2,$$

and the net number of vectors is

$$\sum_j n_j(2j + 1) = 2^N, \; j \leq N/2.$$

Really, the states (2.3) are not degenerated for $n = N$ and $n = 0$ and correspond to $j = N/2$, so we get $n_{N/2} = 1 = C_N^N$. The number of states (2.3) for $n = N - 1$ (or $n = 1$) is $C_N^N = C_N^1 = N$, but a symmetric linear combination of them belongs to the set with $j = N/2$, so it remains the following number of vectors with $|m| = N/2 - 1$:

$$n_{N/2-1} = C_N^{N-1} - C_N^N.$$

It is evidently that they belong to $j = N/2 - 1$. In the same way for $n = N/2 - 2$ (or $n = 2$) and $|m| = N/2 - 2$ the number of the vectors (2.3) is equal to C_N^{N-2}, but C_N^{N-1} their linear combinations correspond to the quantum numbers $j = N/2$ and $j = N/2 - 1$, so, it remains

$$n_{N/2-2} = C_N^N - C_N^{N-1},$$

etc.

Therefore we should use the index α to differ n_j states $|jm\alpha\rangle$. The complete classification of the Dicke states is given in the Appendix 1, because we shall not use it directly in the future treatments. Notice only that in addition to the value of the energetical spin the quantum number j defines simultaneously the transpositional symmetry of states $|jm\alpha\rangle$. This transpositional symmetry does not depend on the value of the quantum number m, and, as we have mentioned, there is a complete transpositional symmetry for $j = N/2$. The operators J_1, J_2, J_3 couple the states with the same values of the quantum numbers j and α, and, as it is known from the quantum mechanical theory of an angular momentum, the operators $R^{\pm} = J^{\pm} = J_1 \pm J_2$ have only the following non-zero matrix elements [4]:

$$\langle j,m,\alpha|J^{+}|j,m-1,\alpha\rangle = \langle j,m-1,\alpha|J^{-}|j,m,\alpha\rangle = \sqrt{(j+m)(j-m+1)}.$$

Otherwise speaking

$$J^{+}|j,m-1,\alpha\rangle = \sqrt{(j+m)(j-m+1)}\;|j,m,\alpha\rangle,$$

$$J^{-}|j,m,\alpha\rangle = \sqrt{(j+m)(j-m+1)}\;|j,m-1,\alpha\rangle, \tag{2.6}$$

i.e., the action of the operators $J^{\pm}$ lead to increase (decrease) of the quantum number m by unity and these operators are the quantum number m increasing (lowering) operators.

As we have mentioned above, the energy of an atomic collection, $E = m\hbar\omega$, depends on the quantum number m. Under statistical treatment of the atomic systems, we can neglect the energy fluctuations, when N is a big number, and assume that the states with $m = E/\hbar\omega$ corresponds to the energy E. As we have mentioned earlier, the degeneracy multiplicity of this state is

$$C_N^{N/2+m} = C_N^{N/2-m} = \frac{N!}{(N/2+m)!\,(N/2-m)!}.$$

It is believed that the states with the given m are uniformly distributed. Then the dominant statistical weight will have the states $|jm\alpha\rangle$ with the minimal j under given m, i.e. $j = |m|$. This is true for the cases, when almost all of the atoms are excited or they are in the ground state, i.e. when $N - 2|m| \ll N$. This follows from the fact, that the states $j = |m|$ have the statistical weight

$$n_{|m|} = C_N^{N/2-|m|} - C^{N/2-|m|-1},$$

while the states with the other j, but the same m, have the statistical weight

$$\sum_j n_j = C^{N/2-|m|-1}, \; j<|m|.$$

The ratio of these statistical weights becomes significant when $(N - 2|m|)/N \ll 1$. The last remark enables to assume that for the large number of atoms N the system is in the states $||m|m\alpha\rangle$ in which the energy or m has a definite value and $|m|$ is close to $N/2$.

1.2.2. *Collection of two-level atoms interacting with electromagnetic field.* To treat the quantum electronic problems, in particularly to study the dynamics of a superradiance, it is necessary to incorporate into the theory the effects of the interaction of the two-level atoms with the electromagnetic field.

The operator of the electromagnetic field has in Schrödinger representation the following form [5]:

$$\mathbf{E}(\mathbf{r}) = i \sum_{\mathbf{k}\nu} \sqrt{\frac{2\pi\hbar\omega_{\mathbf{k}\nu}}{W}} \left(\mathbf{e}_{\mathbf{k}\nu} e^{i\mathbf{kr}} a_{\mathbf{k}\nu} - \mathbf{e}^*_{\mathbf{k}\nu} e^{-i\mathbf{kr}} a^+_{\mathbf{k}\nu}\right), \tag{2.7}$$

where W is a quantization volume, $a_{\mathbf{k}\nu}$ $(a^+_{\mathbf{k}\nu})$ is an operator of creation (annihilation) of the photons with the wavevector $\mathbf{k}$ and polarization index ν $(\nu = 1,2)$; $\mathbf{e}_{\mathbf{k}\nu}$ is a unit polarization vector and $\omega_{\mathbf{k}\nu} = c|\mathbf{k}|$ is a photon frequency.

The Hamiltonian of the free electromagnetic field is

$$H_F = \sum_{\mathbf{k}\nu} \hbar\omega_{\mathbf{k}\nu} a^+_{\mathbf{k}\nu} a_{\mathbf{k}\nu}, \tag{2.8}$$

and the Hamiltonian of the atom-field interaction in the dipole approximation is

$$H_{AF} = -\sum_l \mathbf{d}_l \mathbf{E}(\mathbf{r}_l) = \sum_{\mathbf{k}\nu} \left(g_{\mathbf{k}\nu} a_{\mathbf{k}\nu} R^+_{\mathbf{k}} + g^*_{\mathbf{k}\nu} a^+_{\mathbf{k}\nu} R^-_{\mathbf{k}}\right). \tag{2.9}$$

Here the index l enumerates the atoms, and the radius-vector $\mathbf{r}_l$ defines the position of lth atom. In (2.9) we have used the following relation between the dipole moment operator $\mathbf{d}_l$ and the Pauli spin matrices $\sigma_\pm$:

$$\mathbf{d}_l = \sigma^{(l)}_+ \mathbf{d} + \sigma^{(l)}_- \mathbf{d}^*, \tag{2.10}$$

where $\sigma^{(l)}_\pm = (\sigma^{(l)}_1 \pm i\sigma^{(l)}_2)/2$ are the increasing and lowering operators for lth atom. They have the following matrix forms

$$\sigma_+ = \begin{pmatrix} 0 & 1 \\ 0 & 0 \end{pmatrix}, \qquad \sigma_- = \begin{pmatrix} 0 & 0 \\ 1 & 0 \end{pmatrix}. \tag{2.11}$$

Notice, that $\sigma_+ |-1/2\rangle = |1/2\rangle$, $\sigma_+ |1/2\rangle = 0$ and $\sigma_- |1/2\rangle = |-1/2\rangle$, $\sigma_- |-1/2\rangle = 0$.

The equation (2.10) shows that the mean values of the dipole moment in the upper and lower states are equal zero. And vector **d** is a matrix element of the operator $\mathbf{d}_l$ for a transition between the lower and upper states:

$$\mathbf{d} = \langle 1/2 | \mathbf{d}_l | -1/2 \rangle.$$

It is supposed that atoms are identical and **d** does not depend on index l.

The coefficients $g_{\mathbf{k}\nu}$ in (2.9) are the interaction constants

$$g_{\mathbf{k}\nu} = -i \sqrt{\frac{2\pi\hbar\omega_{\mathbf{k}\nu}}{W}}\, \mathbf{d}\mathbf{e}_{\mathbf{k}\nu}, \tag{2.12}$$

which defines the degree of interaction of the atomic system with the $\mathbf{k}\nu$ mode of the electromagnetic field. And, finally, we have introduced in (2.9) the modified collective atomic operators

$$R^{\pm}_{\mathbf{k}} = \sum_l \sigma^{(l)}_{\pm}\, e^{\pm i\mathbf{k}\mathbf{r}_l}. \tag{2.13}$$

The Hamiltonian (2.9) corresponds to the "rotating waves approximation", when only the secular terms, aR^+ and a^+R^-, are included into the Hamiltonian. These terms lead to the energy conservation in the case ω_0 $\omega_{\mathbf{k}\nu}$. And we have omitted in the Hamiltonian (2.9) the non-secular terms aR^- and a^+R^+, which in the interaction representation oscillate at the frequency $2\omega_0$. The term "rotating waves" was introduced in the theory of a spin resonance, where the only one from the two circularly polarized waves effectively interacts with the spin system. The second wave makes a rapidly varying contribution to the process of interaction.

The general form of the Hamiltonian describing the interaction of a system of the two-level atoms with the radiation is

$$H = H_A + H_F + H_{AF}, \tag{2.14}$$

where

$$H_A = \frac{1}{2} \sum_l \hbar\omega_0 \sigma_3^{(l)} = \frac{1}{2} \hbar\omega_0 R_3 = \hbar\omega_0 J_3$$

is the Hamiltonian of the free atomic system.

1.2.3. *Probabilities of spontaneous and induced transitions.* As we have mentioned, the main specific features of the collective atom-field interactions have been investigated by Dicke [1] with the help of the simplest model based upon the assumption that atoms are confined into the volume with dimensions much smaller than the wavelength of emitted radiation, i.e. when the active volume, V_0, satisfies the condition $V \ll \lambda^3 = (2\pi c/\omega_0)^3$, where ω_0 is the atomic transition frequency. In this case the collective atomic operators $R_{\mathbf{k}}^{\pm}$ are transformed into the operators $R^{\pm} = J^{\pm} = J_1 \pm iJ_2$, that were introduced in section 1.2.1, and, in its turn, the Hamiltonian depends on the operators J_1 J_2, J_3 or $J^{\pm}$, J_3, which satisfy the commutation rules (2.2). Using the Hamiltonian of the atom-field interaction,

$$H_{AF} = \sum_{\mathbf{k}\nu} H_{\mathbf{k}\nu} = \sum_{\mathbf{k}\nu} (g_{\mathbf{k}\nu} a_{\mathbf{k}\nu} J^{+} + g^{*}_{\mathbf{k}\nu} a^{+}_{\mathbf{k}\nu} J^{-}), \tag{2.15}$$

we can define the probability of the atomic system transition from the initial state $|jm\alpha\rangle$ into the final state $|j'm'\alpha'\rangle$ with the emission or absorption of photons.

The operator $H_{\mathbf{k}\nu}$, as it follows from (2.15), has the non-zero matrix elements only for transitions between the states $|jm\alpha\ n_{\mathbf{k}\nu}\rangle$ and $|j'm'\alpha'\ n'_{\mathbf{k}\nu}\rangle$, for which

$$j = j',\ m = m' \pm 1,\ \alpha = \alpha',\ n_{\mathbf{k}\nu} = n'_{\mathbf{k}\nu} \mp 1.$$

As a result, the non-stationary perturbation theory generates in the lowest order the following equation for per time probability of a transition with increasing by unity of the quantum number $n_{\mathbf{k}\nu}$:

$$w_{\mathbf{k}\nu} = \frac{2\pi}{\hbar} \langle j,m-1,\alpha,n_{\mathbf{k}\nu}+1 | H_{\mathbf{k}\nu} | j,m,\alpha,n_{\mathbf{k}\nu}\rangle \delta(\hbar\omega_0 - \hbar\omega_{\mathbf{k}\nu}).$$

The summary per time probability of a transition from the state $|j,m,\alpha\rangle$ to the state $|j,m-1,\alpha\rangle$ with a photon emission is

$$w^{-} = \sum_{\mathbf{k}\nu} w_{\mathbf{k}\nu} = \frac{W}{(2\pi)^3} \int d^3k \sum_{\nu} w_{\mathbf{k}\nu}$$

$$= \frac{W}{(2\pi)^2 \hbar} \int d^3k \sum_{\nu} |g_{\mathbf{k}\nu}|^2$$

$$\times (n_{\mathbf{k}\nu} + 1)(j + m)(j - m + 1)\delta(\hbar\omega_0 - \hbar\omega_{\mathbf{k}\nu})$$

$$= \frac{1}{T}(j + m)(j - m + 1)(\bar{n} + 1). \qquad (2.16)$$

Analogically, the summary per time probability of a transition from the state $|j,m,\alpha\rangle$ to the state $|j,m + 1,\alpha\rangle$ with a photon absorption is

$$w^+ = \frac{1}{T}(j + m + 1)(j - m)\,\bar{n} \qquad (2.17)$$

where T is an radiative lifetime of the excited atomic state:

$$\frac{1}{T} = \frac{4\omega_0^3}{3\hbar c^3} |\mathbf{d}|^2$$

and $\bar{n}$ is a mean photon number in the field modes $\mathbf{k}\nu$ for which

$$\omega_{\mathbf{k}\nu} \approx \omega_0.$$

In calculating of the probabilities w^+ and w^- we have substituted the summation by the integration over the field modes (the field modes density in the $\mathbf{k}$-space is equal to $W/(2\pi)^3$). And we have used (2.6) for the matrix elements of the collective atomic operators $J^{\pm}$. As it is well known [5], the matrix elements of the photon creation (annihilation) operator $a^+_{\mathbf{k}\nu}$ ($a_{\mathbf{k}\nu}$) are

$$\langle n_{\mathbf{k}\nu}+1|\,a^+_{\mathbf{k}\nu}\,|n_{\mathbf{k}\nu}\rangle=\sqrt{n_{\mathbf{k}\nu}+1}\,,$$

$$\langle n_{\mathbf{k}\nu}-1|\,a_{\mathbf{k}\nu}\,|n_{\mathbf{k}\nu}\rangle=\sqrt{n_{\mathbf{k}\nu}}\,. \qquad (2.18)$$

The transition probability w^- consists of the two components. The first one is the probability of the spontaneous emission

$$w^-_{sp} = \frac{1}{T}(j + m)(j - m + 1) \qquad (2.19)$$

and the second one is the probability of the induced transition

$$w^-_i = \frac{1}{T}(j + m)(j - m + 1)\,\bar{n}. \qquad (2.20)$$

The power of the induced emission by the atomic collection, which is in the state $|jm\alpha\rangle$, is

$$I_i = \hbar\omega_0(w_i^- - w^+) = \frac{\hbar\omega_0}{T}[(j + m)(j - m + 1)\,\bar{n}$$
$$- (j + m + 1)(j - m)\,\bar{n}\,] = 2I_0 m\,\bar{n}. \qquad (2.21)$$

An insight into the implication of this equation can be obtained, if we take into account that the population difference is equal to $2m$, and $I_0 = \hbar\omega_0/T$ is a power of emission by an isolated atom. It is essential, that the power of the induced emission depends on the population difference $2m$ and the mean occupation number of photon states $\bar{n}$, but it does not depend on the cooperative quantum number j (and on the quantum number α also).

In contrast to this, the power of the spontaneous emission, which is

$$I_{sp} = \hbar\omega_0 \bar{w}_{sp} = I_0(j + m)(j - m + 1), \qquad (2.22)$$

depends significantly on quantum number j and for the symmetrical states, when $j = j_{max} = N/2 \gg 1$, we get $I_{sp} \propto I_0(N^2/4 - m^2)$. It is seen that I_{sp} N^2 when $m \ll N$. Notice that for the completely inverted system, when $j = m = N/2$, (2.22) becomes $I_{sp} = NI_0$.

It will be shown later, that the induced processes do not play any essential role in resonatorless systems with small dimensions. The emitted photons leave the active volume for the time $\tau = L/c$, where L is a length of active volume. As it will be shown later, for $\tau \ll T_2$, where T_2 is the homogeneous broadening relaxation time, we can neglect the induced processes. In another way we can say that the active volume is so small in comparison with the quantization volume W that the mean photon number $\bar{n}$ is very small even when all of the excited atoms have emitted the photons.

Equation (2.22) for the intensity of a spontaneous radiation enables to discuss the kinetics of collective decay of a collection of the excited atoms.

If all of the atoms were initially in the excited state then the atomic system was in the symmetric state $j = j_{max} = N/2$ and $m = N/2$, i.e. in the state $|N/2,N/2\rangle$. As we have mentioned, in process of a radiative decay the quantum number j does not change, while the population difference $2m$ changes from the value $2(N/2) = N$ at the moment $t = 0$ up to the value $2(-N/2) = -N$ at the moment $t = \infty$, when all of the atoms are in the ground state.

In the process of emission the atomic system passes through the state $m = 0$, when the emitted intensity is a maximum $I_{max} = I_0 N^2/4$, i.e. the peak intensity is proportional to N^2. The state $m = 0$ is designated as SR state. It can be excited from the ground state of atomic system $|N/2,-N/2>$ by the irradiation of an intense resonance pulse which can equalize the population of the atomic levels for the time much smaller than the dephasing time T_2.

However, we will be interested here by another process. It is a radiative decay of the completely excited initially collection of atoms which comes spontaneously from the state $|N/2,N/2>$ into the SR state. The equation (2.22) for spontaneous emission power can be written in the form [1]

$$I_{sp} = I_0 <jm\alpha|J^+J^-|jm\alpha>.$$

Taking into account that the operator J^+J^- is a diagonal operator with respect to the quantum numbers j, m, α, we can easily generalize the last formula for the case of an arbitrary initial atomic state:

$$I_{sp} = I_0 <J^+J^->. \tag{2.23}$$

Using the commutation rules (2.2) we can get

$$[J^+,J^-] = 2J_3.$$

With the help of the following identical equation

$$\mathbf{J}^2 = J_1^2 + J_2^2 + J_3^2 = \frac{1}{2}(J^+J^- + J^-J^+) + J_3^2$$

we can represent the operator J^+J^- in the form

$$J^+J^- = \mathbf{J}^2 - J_3^2 + J_3. \tag{2.24}$$

If we use the energy conservation law, we get the following equation

$$\hbar\omega_0 \frac{d}{dt} <J_3> = -\frac{\hbar\omega_0}{T}(<\mathbf{J}^2> - <J_3^2> + <J_3>). \tag{2.25}$$

The left-hand side of this equation is the time derivative of the mean energy of the atomic system, and the right-hand side of (2.25) determines the power of a collective spontaneous emission. The operator $2J_3$ coincides with the population difference operator R_3, and $<\mathbf{J}^2>$ is an integral of motion because the operator $\mathbf{J}^2$ commutes with the Hamiltonian H. For the initial state $|N/2,N/2>$ the mean value of $<\mathbf{J}^2>$ is

$$\langle \mathbf{J}^2 \rangle = \frac{N}{2}\ \frac{N}{2} + 1\ .$$

If we neglect the fluctuations of a population difference, i.e.,if we assume $\langle R_3^2 \rangle = \langle R_3 \rangle^2$, than (2.25) becomes the equation of motion for the mean value of the population difference $\langle R_3 \rangle = 2\langle J_3 \rangle$:

$$\frac{d}{dt}\langle R_3 \rangle = -\ \frac{1}{T}\ 2\langle \mathbf{J}^2 \rangle - \frac{1}{2}\langle R_3 \rangle^2 + \langle R_3 \rangle \Big] . \tag{2.26}$$

For the initial state $|N/2,N/2\rangle$ corresponding to the complete inversion of an atomic system this equation transforms into the Dicke equation (1.9). For the initial state $|jj\alpha\rangle$ we must simply change N by $2j$ in the equation (1.9).

The conservation of $\mathbf{J}^2$ is a specific feature of the Dicke model, which simplifies essentially the treatment of a system dynamics. Unfortunately, this law loses its validity when we take into account the dipole–dipole interaction of atoms or when we consider the radiative decay of an atomic collection with the dimensions comparable with the wavelength of radiation or much greater of it.

1.2.4. *Quasiclassical and semiclassical approaches.* The Dicke equation (1.9) is an equation of the type (1.7), where the square of dipole moment, $|\mathbf{D}_\omega|^2$, in terms of energy E can be found by the quasiclassical method.

For an quasiclassical system frequencies of the the adjacent transitions differ only slightly. Therefore for such systems the energy-level spacing is always equidistant. The transition frequencies are equal to the classical frequency of motion $\tilde{\omega}$ [4], so the energy-level spacing is $\hbar\tilde{\omega}$ for any two adjacent levels.

There is a general correspondence (the correspondence principle) between the quantum matrix elements and the Fourier-components of classical variables. According to this principle, in quasiclassical limit the matrix elements f_{mn}, where indexes m and n enumerate the energy levels, transform into the Fourier-components of the classical variables at the frequency $\omega = (n - m)\tilde{\omega}$. In this case the matrix elements decrease rapidly with increasing $|n - m|$ and at the same time they remain the slowly varying functions of the numbers n and m . Particularly the matrix elements of the dipole moment operator, $\mathbf{d}_{mn} = \langle m|\hat{\mathbf{d}}|n\rangle$, are transformed into the Fourier components, $\mathbf{d}_\omega$, of the classical dipole moment $\mathbf{d}(t)$.

We can illustrate this fact if we take into account that $\mathbf{d}(t)$

coincides with the mean value of a dipole moment operator in the quasiclassical approximation, when the state of an atomic system is represented by the wave-packet of the states with n near to $\bar{n}$ and when the width of the wave-packet Δn satisfies the condition $1 \ll \Delta n \ll \bar{n}$. For such wave-packets we can assume that

$$\omega_{mn} = \frac{1}{\hbar} (E_{\bar{m}} - E_{\bar{n}}) = (m - n)\, \tilde{\omega} \, .$$

In the quasiclassical region the quantum equation for the power of a spontaneous emission

$$I = \frac{4\omega_{mn}^4}{3c^3} \, |d_{mn}|^2 \tag{2.27}$$

is transformed into the classical equation (1.1) for the power of the dipole emission. In the case of collections of the two-level atoms with $N \gg 1$ the dipole moment is determined according to (2.10) by

$$\mathbf{D} = \Sigma\, \mathbf{d}_l = \mathbf{d}\, J^+ + \mathbf{d}^* J^- = \mathbf{D}^+ + \mathbf{D}^- \, .$$

The operator $\mathbf{D}^+$ has the following nonzero matrix elements (see (2.6))

$$\langle j,m,\alpha | \mathbf{D}^+ | j,m - 1,\alpha \rangle = \mathbf{d} \sqrt{(j + m)(j - m + 1)} \, .$$

The operator $\mathbf{D}^-$ is the operator Hermitian-conjugated to $\mathbf{D}^+$. The state $|jm\alpha\rangle$ has the energy $E = \hbar\omega_0 m$. Thus the operator $\mathbf{D}$ has the nonzero matrix elements only for transitions between the adjacent energetical levels and it has all properties of an operator of a quasiclassical physical variable.

Let us consider the following wave packet

$$|\psi\rangle = \sum_m a_m |jm\alpha\rangle \, , \tag{2.28}$$

where m lays inside an interval Δm near the mean value $\bar{m}$ and the width Δm of which satisfies the condition $1 \ll \Delta m \ll j$. The mean value of the operator $\mathbf{D}^+$ for this packet is given by

$$\langle \mathbf{D}^+ \rangle = \mathbf{d} \sum_m a_m^* a_{m-1} \sqrt{(j + m)(j - m + 1)}$$

$$\mathbf{d}\sqrt{(j+\bar{m})(j-\bar{m}+1)}\ , \tag{2.29}$$

as for the quasiclassical packet we get $a_{m-1} \sim a_m$, hence,

$$\sum_m a^*_m a_{m-1} \sim 1.$$

As it will be shown later the coherent atomic states are an example of such normalized packets, for which the coefficients a_m are slightly changed with m increasing by unity and the wave-packet width Δm is much smaller then j . This is true for the atomic systems with $N \gg 1$ and, hence, with $j \gg 1$. As it follows from (2.29) the mean value of $\mathbf{D}^+$ is great enough and equals approximately $\mathbf{d}N/2$ for $j = N/2$ and $\bar{m} \ll j$.

The mean value of the operator $\mathbf{D}$ is given by

$$<\mathbf{D}> = <\mathbf{D}^+>\exp(i\omega_0 t) + <\mathbf{D}^->\exp(-\, i\omega_0 t)\ ,$$

where we have included the time depended exponents of the wave functions. It is seen that $<\mathbf{D}^->$ acts as the Fourier-component $\mathbf{D}_\omega$ of the classical dipole moment of a system for the frequency ω_0 . According to (2.29) we get

$$|\mathbf{D}_{\omega_0}|^2 = |\mathbf{d}|^2(j+\bar{m})(j-\bar{m}+1)$$

$$= \left(\frac{|\mathbf{d}|}{\hbar\omega_0}\right)^2 (E+E_0)(E_0-E+\hbar\omega_0)\ , \tag{2.30}$$

because the energy of the system is $E = m\hbar\omega_0$, and $j\hbar\omega_0$ determines the energy of the atomic system under complete excitation.

If we substitute the Eq. (2.30) into the equation (1.7), than we get the Dicke equation (1.9) in which the mean value $<R_3>$ is replaced by $2E/\hbar\omega_0$.

Of course, the derivation of the Dicke equation discussed above can not be justified near the state of the complete inversion $m = j$, where $\Delta m = 0$ and the condition $\Delta m \gg 1$ does not hold. Therefore the last term, $\hbar\omega_0$, in the right-hand side of the Eq. (2.30) exceeds the accuracy of the derivation. This additional term is very essential for the Dicke equation because it describes the spontaneous decay of a completely inverted atomic system. In this case the emission power is equal to the product of the emission power for an individual atom I_0 and the number of atoms N. The emission power for $m = j$ is given by (2.22) and equals $I = 2jI_0 = NI_0$ for the symmetrical state $j = N/2$.

According to quantum treatment of the effect the emission power is given by (2.23)

$$I = I_0 \langle I^+ I^- \rangle = I_0 (\langle \mathbf{J}^2 \rangle - \langle J_3^2 \rangle + \langle J_3 \rangle) .$$

If the atomic system is in the state (2.28), we get

$$I = \frac{4\omega_0^4}{3c^3} |\mathbf{d}|^2 [j(j+1) - \sum_m (m^2 - m)|a_m|^2]$$

$$= \frac{4\omega_0^4}{3c^3} |\mathbf{d}|^2 [j(j+1) - \overline{m^2} + \bar{m}] .$$

This equation will coincide with the classical equation

$$I = \frac{4\omega_0^4}{3c^3} |\mathbf{D}_{\omega_0}^-|^2 ,$$

$$\mathbf{D}_{\omega_0} = \langle \mathbf{D}^- \rangle = \mathbf{d} \sqrt{(j + \bar{m})(j - \bar{m} + 1)} ,$$

if we neglect the difference between $\overline{m^2}$ and $\bar{m}^2$. It is really the case for the packets (2.28) when $\Delta m \ll j$, where $\Delta m^2 = \overline{m^2} - \bar{m}^2$ is a dispersion of the quantum number m, i.e., the dispersion of J_3 in the state (2.28). It is this approximation which has been used under derivation of the Dicke equation (2.26). We assumed in Sec. 1.2.3 that we could neglect the fluctuations of the population difference and replace $\langle R_3^2 \rangle$ by $\langle R_3 \rangle^2$. Notice, that $R_3 = 2J_3$.

In contrast to this the semiclassical method enables to determine not only the energetical characteristics of a SR pulse but also the amplitude and even spectral ones. The semiclassical method is widely used in the laser physics. It is assumed in this method that the electromagnetic field is a classical c-number field while the atoms are treated from the quantum mechanical point of view. The atomic equation of motion is the Schrödinger equation with the perturbation due to the c-number electromagnetic field which is treated as a given field. The field is determined in the self-consistent way by the Maxwell equations with the electrical currents which are calculated as the mean currents or dipole moments in the atomic system.

The essential simplification consists in the fact that the vector of the atomic system state or its density matrix are represented as a product of the vectors of the individual atom states or their density matrices because the perturbation operator is a sum of the operators referring to the different atoms.

For the case of a system with the dimensions much smaller than the wavelength of radiation the Hamiltonian of the atom-field interaction is given in the dipole approximation by

$$H' = \sum_l H'_l = -\sum_l (\mathbf{d}_l\mathbf{E}) = -(\mathbf{D}\mathbf{E}(t)),$$

where the electric field $\mathbf{E}$ equals $\dfrac{2}{3c^3}\dddot{\mathbf{D}}$ according to (1.2), and $\mathbf{d}_l$ and $\mathbf{D}$ are the dipole moment operators for lth atom and the whole system respectively.

The density matrix ρ_l of the individual atom satisfies the equation of motion

$$i\hbar \frac{\partial \rho_l}{\partial t} = [H', \rho_l]$$

in the interaction representation. For the case of the two-level atom this equation can be replaced by the two equations for the difference of the diagonal matrix elements $p = \rho_{1/2,1/2} - \rho_{-1/2,-1/2}$ and for the non-diagonal matrix element $\mu = \rho_{-1/2,1/2}$. Using $H'_l = -(\mathbf{d}_l\mathbf{E})$ and the definition (2.10) of the operator $\mathbf{d}_l$, we get

$$\frac{dp}{dt} = -\frac{2N\,|\mu|^2}{T}, \quad \frac{d\mu}{dt} = \frac{N\mu p}{2T}. \tag{2.31}$$

The field $\mathbf{E}(t)$ has been excluded from these equations, and we have accounted that

$$\mathbf{E}(t) = \frac{2N}{3c^3}\frac{d^3}{dt^3}[\mathbf{d}\mu \exp(i\omega_0 t) + \mathbf{d}^*\mu^* \exp(-i\omega_0 t)]$$

$$\mathbf{E}_{\omega_0} \exp(-i\omega_0 t) + \mathbf{E}^*_{\omega_0} \exp(i\omega_0 t) ,$$

where $\mathbf{E}_{\omega_0} = 2iN\omega_0^3 \mu^* \mathbf{d}^* / 3c^3$. The time T in the right-hand side of the Eqs. (2.31) is the spontaneous lifetime of an individual atom

$$T = \frac{3c^3\hbar}{4\omega_0^3 |\mathbf{d}|^2} .$$

In obtaining the above equations we have used the rotating wave approximation, i.e. we have ignored the rapidly varying terms in Hamiltonian H'_l .

Multiplying the second equation (2.31) by μ^* and adding it to the complexly conjugated equation we get

$$\frac{d|\mu|^2}{dt} = \frac{N|\mu|^2 p}{T} .$$

Comparing this equation with the first equation (2.31) we see that there is the following integral of motion of Eqs. (2.31):

$$p^2 + 4|\mu|^2 = \text{const} .$$

The atoms are in the upper state at $t = 0$, so $p(0) = 1$ and $\mu(0) = 0$. Hence $p^2 + 4|\mu|^2 = 1$ and if we apply this equation to the first equation (2.31) we get

$$\frac{dp}{dt} = -\frac{N(1 - p^2)}{2T} .$$

Introducing the new variable ϑ , according to $p = -\cos\vartheta$, we get the following equation

$$\dot{\vartheta} = -\frac{N}{2T} \sin\vartheta ,$$

which coincides practically with the Eq. (1.10). The difference between $N + 1$ and N is essentially negligible because we assume that $N \gg 1$. The main difference between the above equation and equation (1.10) consists in the form of the initial condition. Here we assume that $\vartheta(0) = \pi$. It is a position of the unstable equilibrium for the equation under consideration. To get the solution in the form (1.11), (1.12) we need to assume that the initial angle ϑ_0 differs slightly from π, i.e. $\vartheta \neq \pi$ at $t = 0$. This uncertainty in the value of the initial angle is the main disadvantage of a semiclassical method. The value of the initial angle ϑ_0 determines the delay time of a superradiance pulse.

In the previous section we obtained the Eq. (1.10) of the same type, but the initial value of the Bloch angle ϑ was determined there from the initial conditions

$$\vartheta_0 = \arccos(-(N - 1)/N + 1) \neq \pi ,$$

because for $t = 0$ we have $<R_3> = N$.

The Eqs. (2.31) and their integral of motion $p^2 + 4|\mu|^2 = 1$ enable to define $|\mu|^2$ and μ as a result of solution of equation for p. The mean dipole moment of the atomic system,

$$<\mathbf{D}> = N<\mathbf{d}_1> = N(\mathbf{d}^* \mu^* \exp(-\,i\omega_0 t) + \text{c.c.}) ,$$

is known if we have determined $\mu(t)$. Substituting $<\mathbf{D}>$ in the traditional formulas for the dipole emission, we can determine the emitted field and its angular and spectral characteristics. As it is seen from the equation for $<\mathbf{D}>$, the spectrum of the emitted pulse coincides with the spectrum of the function $\mu(t)$.

In conclusion, let us discuss the properties of the wave functions of the atomic collections. As we have mentioned, if the initial state is the completely inverted state, for which the vector of the state is a product of the individual atomic state vectors, then this factorization is conserved during the whole decay. In any moment

$$|\psi> = \prod_l |\psi>^{(l)} .$$

It means that $|\psi>$ will be always the completely symmetrical vector and it can be expressed in terms of the Dicke vectors $|N/2,m>$.

On the other hand, all of the vectors $|\psi>^{(l)}$ are identical for the point source (all of the atoms are in the same field $\mathbf{E}(t)$). The arbitrary state $|\psi>^{(l)}$ of an individual two-level atom is determined by the two parameters if we are not interested in the phase factor. We can choose the angles ϑ and ϕ as such parameters

$$|\psi>^{(l)} = \begin{pmatrix} e^{-i\phi} \sin(\vartheta/2) \\ \cos(\vartheta/2) \end{pmatrix}^{(l)} = |\vartheta,\phi>^{(l)} . \tag{2.32}$$

In this case the vector of the state of a collection of the two-level atoms $|\psi>$ is given by $|\vartheta,\phi> = \prod_l |\vartheta,\phi>^{(l)}$.

The states $|\vartheta,\phi>$ for a collection of N atoms are called by the coherent atomic states and widely used in different works on superradiance. Thus, in the semiclassical treatment the atomic system is always in the coherent atomic state and its evolution can be described by the time evolution of the angles ϑ and ϕ . The physical sense of the angles ϑ and ϕ , which were introduced here formally, will be given in Appendix 2.

It is easy to assure that the coherent atomic state $|\vartheta,\phi>$ for the big number N of atoms is just a quasiclassical packet of the type (2.28). It is shown in Appendix 2 that the state $|\vartheta,\phi>$ is expressed in terms of the Dicke states $|N/2,m>$ in the following form

$$|\vartheta,\phi> = \sum_m a_m |N/2,m> ,$$

$$a_m = \sqrt{C_N^{N/2+m}} \sin^{N/2+m}(\vartheta/2) \cos^{N/2-m}(\vartheta/2) e^{-i(N/2+m)\phi} . \tag{2.33}$$

The probability distribution for the different m is

$$P_m = |a_m|^2 = C_N^{M/2+m} p^{N/2+m} q^{N/2-m} ,$$

where $p = \sin^2(\vartheta/2)$, $q = 1 - p = \cos^2(\vartheta/2)$. This is a binomial distribution, which for the big number N , according to the Laplace theorem of the probability theory, is approximated by the normal distribution

$$P_m \quad \sqrt{2\pi Npq} \; \exp\left[-\frac{1}{2}\left(\frac{N/2 + m - Np}{\sqrt{Npq}}\right)^2\right]$$

with the mean value $\bar{m} = Np - N/2 = N(p - 1/2)$ and the dispersion $\Delta m^2 = \overline{m^2} - \bar{m}^2 = Npq$. The relative dispersion

$$\Delta m^2/N^2 = pq/N \ll 1$$

for $N \gg 1$. Therefore for the coherent atomic state the deviation Δm is small and this state is an example of a quasiclassical packet of the type (2.28).

So far we considered the point Dicke model. Its restrictions and contradictions have been discussed. In the following paragraph we begin to study the next simple, but important model, that is the one-mode model, which leads to the same equations as the Dicke model.

1.3. One-mode model of superradiance

The simplest model of a superradiative system of a large extent (when the active volume dimensions are greater than wavelength) is the one-mode model.

The sphere of its applications will be discussed later and the main results derived from it will be compared with the results of the multimode theories. Here we discuss the limited case of a pure superradiance in the one-mode model and compare it with the point-source Dicke model.

The one-mode model of SR is very close to the simplest one-mode laser models. It is considered a collection of atoms, which can be excited by an external source, and an one-mode optical resonator, the field in which interacts with the atomic collection. The main difference of this model from the laser models consists in the fact that resonator q-factor is assumed to be so low that the emitted photons leave the resonator very fastly and have no possibility to be absorbed by atoms or to induce the stimulated transitions. More precisely speaking, it is assumed that the photon lifetime in resonator τ is much smaller than the characteristic atomic times, such as the radiative lifetime T of the excited atomic state, and relaxation times T_1 and T_2 (T_1 is a relaxation time of the atomic population difference).

In contrast, in the case of the laser system it is assumed that $\tau \gg T$, T_1 , T_2 . This occurs when the reflectively coefficients of the resonator mirrors have the high values and when the resonator length is big enough. If these conditions hold we can adiabatically exclude the atomic variables because they are simply expressed in terms of the field variables. As a result we get the closed equations for the field variables or the field density matrix. Such procedure of solution has been used, for example, by Scully and Lamb [6]. In this way we get for the field density matrix in the coherent field representation (the

Glauber representation) the equation of the Fokker Plank type (see [7]).

If the conditions $\tau \ll T$, T_1, T_2 hold we can adiabatically exclude the field variables and get the equations for the atomic density matrix. Let us discuss this procedure in the frames of the semiclassical approximation. The atoms are described by the following equation for the density matrix

$$i\hbar \frac{d\rho}{dt} = [H',\rho] \ .$$

We have omitted the index l , which enumerates the atoms, and ignored the relaxation processes, i.e. we have assumed $T_1 = T_2 = \infty$. The simplest models will be considered only.

The classical equation for electromagnetic field has the form

$$\text{rot rot } \mathbf{E} + \frac{1}{c^2}\frac{\partial^2 \mathbf{E}}{\partial t^2} = -\frac{4\pi}{c^2}\frac{\partial^2 \mathbf{P}}{\partial t^2} ,$$

where **E** is an electric field, and **P** is a polarization, i.e. the net dipole moment of the active medium unit volume.

Let us assume that there is essential only the one mode of the electromagnetic field

$$\mathbf{E}(\mathbf{r},t) = \mathbf{e}(\mathbf{r})\, e^{-i\omega_0 t} a(t) + \mathbf{e}^*(\mathbf{r})\, e^{i\omega_0 t} a^*(t) , \qquad (3.1)$$

where the function $\mathbf{e}(\mathbf{r})$ describes the distribution of the field in this mode and this function is normalized by the condition

$$\int \mathbf{e}^*(\mathbf{r})\, \mathbf{e}(\mathbf{r})\, d\mathbf{r} = 1 \ .$$

The function $a(t)$ in Eq. (3.1) is a slowly varying complex amplitude of the mode.

Substituting (3.1) into the equation for $\mathbf{E}(\mathbf{r},t)$, and next, multiplying it by $\mathbf{e}^*(\mathbf{r})$ and integrating over the spatial variables we get the following equation for the amplitude $a(t)$:

$$\dot{a} = 2\pi i \omega_0 b ,$$

where $b(t)$ is the polarization amplitude. We have ignored the second derivative of amplitude $\ddot{a}$ in the last equation. In analogy with (3.1) the polarization $\mathbf{P}(\mathbf{r},t)$ can be written in the form

$$\mathbf{P}(\mathbf{r},t) = \mathbf{P}_0(\mathbf{r},t)\exp(-\,i\omega_0 t) + \mathbf{P}_0^*(\mathbf{r},t)\exp(i\omega_0 t)\ ,$$

where $\mathbf{P}_0(\mathbf{r},t)$ is the slowly varying function of time, and the function $b(t)$ is

$$b(t) = \int \mathbf{e}^*(\mathbf{r})\,\mathbf{P}_0(\mathbf{r},t)\,d\mathbf{r}\ .$$

The radiative loss for field mode at the active volume boundaries can be incorporated phenomenologically by introducing the relaxation term with the characteristic mode decay time τ . In this case the equation for the field amplitude $a(t)$ becomes

$$\dot{a} + a/\tau = 2\pi i\omega_0 b\ . \tag{3.2}$$

The time τ can be estimated as L/c , where L is the active volume length.

In the semiclassical approach the amplitude $\mathbf{P}_0(\mathbf{r},t)$ is obtained in result of quantum mechanical averaging of the atomic dipole moments:

$$\mathbf{P}_0(\mathbf{r},t) = \mathbf{d}^*\sum_i \delta(\mathbf{r}-\mathbf{r}_i)\,\langle\sigma_-^{(i)}\rangle = \mathbf{d}^*\sum_i \delta(\mathbf{r}-\mathbf{r}_i)\,\mu_i^*\ .$$

The nondiagonal element $\rho^{(i)}_{-1/2,1/2}$ of the ith atom density matrix is designated here by μ_i . We shall assume that $\mathbf{e}(\mathbf{r}_i) = \mathbf{e}_0\, e^{i\phi_i}$, i.e. it is assumed that each atom feels the field with the same amplitude. In this case we can write $\mu_i = \mu\, e^{i\phi_i}$, and the diagonal elements of the atomic density matrix are not depend on the atomic index i .

The polarization amplitude is now

$$b = \int \mathbf{e}^*(\mathbf{r})\,\mathbf{P}_0(\mathbf{r},t)\,d\mathbf{r} = N\,(\mathbf{d}^*\mathbf{e}_0^*)\,\mu^*\ ,$$

and the equations for the functions μ , p , b are given by

$$i\hbar\,\frac{d\mu}{dt} = -\,(\mathbf{d}^*\mathbf{e}_0^*)\,a^*p\ ,$$

$$i\hbar\,\frac{dp}{dt} = -\,2(\mathbf{d}\mathbf{e}_0)\,\mu a + 2(\mathbf{d}^*\mathbf{e}_0^*)\,\mu^* a^*\ , \tag{3.3}$$

$$\frac{da}{dt} + \frac{a}{\tau} = 2\pi i\omega_0 N(\mathbf{d}^*\mathbf{e}_0^*)\,\mu^* .$$

The field amplitude can be expressed in terms of μ^* by

$$a = 2\pi i\omega_0 N(\mathbf{d}^*\mathbf{e}_0^*) \int_0^t \mu^*(t') \exp[(t' - t)/\tau]\, dt' . \tag{3.4}$$

If (3.4) is applied to the first two equations (3.3) we get the closed set of equations for the density matrix of an arbitrary atom. As a result of (3.4) the obtained equations will be non-Markovian equations, because the evolution of the density matrix is determined not only by its value in the given moment of time but also its value in the previous stages. In other words, the time derivative of the density matrix is not determined by its value in the same moment of time. If we assume, in addition to the previous assumptions, that μ , p , and a vary slowly for a period of time τ , then we can let $da/dt = 0$ or, what is the same, we can let $\mu^*(t')$ $\mu^*(t)$ and after that integrate over t' in the Eq. (3.4). If we substitute the relationship

$$a = 2\pi i\omega_0 N\mu^*(\mathbf{d}^*\mathbf{e}_0^*)$$

into the first two equations (3.3) we get the same Markovian equations as (2.31), which were obtained for point-source model. We should only replace T^{-1} by

$$\tilde{T}^{-1} = 4\pi\tau\omega_0\hbar^{-1}|\mathbf{d}\mathbf{e}_0|^2 .$$

The equations, obtained by this way, with the help of substitution $p = -\cos\vartheta$ can be transformed to the following form

$$\dot{\vartheta} = -\frac{N}{2\tilde{T}} \sin\vartheta . \tag{3.5}$$

Thus in the limiting case of the very small resonator relaxation time (when $\tau \ll \tilde{T}/N = \tilde{\tau}_c$) the equations for the one-mode model of superradiance coincide with the equations for the point-source model. The analysis of the radiative decay dynamics, made above for the point-source model, can be applied for the one-mode superradiative decay.

1.3.1. *Bonifacio one-mode theory.* The quantum theory of the one-mode model of a superradiance was developed by Bonifacio, Schwendimann, Haake [2]. The simplicity of this model permitted them to promote essentially in the study of the dynamic and statistical properties of the field-atomic interactions.

Starting from the master equation for the density matrix of an atomic system interacting with the resonance one-mode field Bonifacio et al obtained the closed equation for the density matrix of the atomic subsystem in the SR limit, when the field variables excluded adiabatically.

For the density matrix of the atom-field system comprising N atoms they used the equation

$$i\hbar \frac{d\rho}{dt} = [H_{AF},\rho] + \frac{i\hbar}{\tau} ([a\rho,a^+] + [a,\rho a^+]) , \qquad (3.6)$$

where ρ is a density matrix in the interaction representation, H_{AF} is the Hamiltonian of the atom-field interaction, τ is a resonator decay time, a (a^+) is a photon annihilation (creation) operator for the accounted field mode.

In the one-mode case the interaction Hamiltonian H_{AF} has the form

$$H_{AF} = gaR^+_{\mathbf{k}} + g^* a^+ R^-_{\mathbf{k}} ,$$

where g is the coupling constant defined by the Eq. (2.12), and

$$R^{\pm}_{\mathbf{k}} = \sum_l \sigma^{(l)}_{\pm} \exp(\pm i\mathbf{kr}_l) .$$

It is assumed that the field frequency is equal exactly to the atomic transition frequency ω_0 .

Applying the unitary transformation to the atomic states

$$\begin{aligned} |+1/2\rangle'^{(l)} &= e^{i\mathbf{kr}_l/2} \, |+1/2\rangle^{(l)} , \\ |-1/2\rangle'^{(l)} &= e^{-i\mathbf{kr}_l/2} \, |-1/2\rangle^{(l)} , \end{aligned} \qquad (3.7)$$

we can transform the operators $R^{\pm}_{\mathbf{k}}$ into the operators $R^{\pm} = J^{\pm} = J_1 \pm J_2$. Then the Hamiltonian of the atom-field interaction takes the form

$$H_{AF} = gaJ^+ + g^* a^+ J^- . \qquad (3.8)$$

This transformation demonstrates that the inclusion of the retardation effects, i.e. the factors $\exp(\pm i\mathbf{k}\mathbf{r}_l)$, in the case of the one-mode field does not complicate the appropriate equations and we can easily remove these factors from the operator equations. After that we get the equations coinciding completely with the equations for the point-source model.

The Dicke states and the coherent atomic states, constructed from the new monoatomic states $|\pm 1/2\rangle'^{(l)}$ can be used in the one-mode models as effectively as the states constructed from the vectors $|\pm 1/2\rangle^{(l)}$ are used in the point-source model. In semiclassical description we simply introduce the factors $\exp(\pm i\phi_l)$. When the atomic subsystem interacts with the one-mode field, there will be conserved $\mathbf{J}^2$ and quantum numbers j and α , if the Dicke states $|jm\alpha\rangle$ will be constructed from the monoatomic basic vectors $|\pm 1/2\rangle'$. It is the main simplification of the one-mode model. The conservation of j and α means the conservation of the atomic states symmetry in respect to the atomic rearrangements. It follows from the fact that the Hamiltonian H_{AF} is expressed in terms of the completely symmetrical operators $J^{\pm}$. This symmetry breaks in the multimode case because the operators $R^{\pm}_{\mathbf{k}}$ does not conserve the transpositional symmetry when $\mathbf{k}$ 0 . In the case of the one-mode field the application of the unitary transformation enables to transform the equations to the form of the equations for the case $\mathbf{k} = 0$.

If we exclude adiabatically the field variables than we get the equation for the atomic density matrix ρ_A [2]:

$$\frac{d\rho_A}{dt} = \frac{|g|^2\tau}{\hbar^2}([R^-,\rho_A R^+] + [R^-\rho_A,R^+]) . \tag{3.9}$$

This equation was derived for the Markovian approximation and it holds for $t \gg \tau$. In addition to this condition it was assumed that the following conditions were met: T_1, $T_2 \gg \tau_c \gg \tau$, where $\tau_c = \hbar^2/(|g|^2\tau N)$ is a characteristic width of the superradiance pulse. The first condition assures that we can ignore the processes of the spontaneous incoherent decay and the atomic relaxation, and the second condition is a condition of the applicability of the adiabatic exclusion and Markovian approximation.

If we have determined the solution of the Eq. (3.9), we can define the mean values of any atomic operators. Additionally, in this approximation the field operator moments are expressed in terms of the atomic operator moments and can be defined with the help of ρ_A [2]:

$$\langle a^{+l} a^{m} \rangle = (ig\tau/\hbar)^{l} \, (-ig^{*}\tau/\hbar)^{m} \, \langle R^{+l} R^{-m} \rangle \, . \tag{3.10}$$

Multiplying the Eq. (3.9) by R_3 and performing the trace operation we get the equation for the population difference mean value $\langle R_3 \rangle$:

$$\langle \dot{R}_3 \rangle = -\frac{4|g|^2\tau}{\hbar^2} \langle R^+ R^- \rangle$$

$$= -\frac{4|g|^2\tau}{\hbar^2} \left(j(j+1) - \frac{1}{4} \langle R_3^2 \rangle + \frac{1}{2} \langle R_3 \rangle \right) .$$

In obtaining this equation we have used the commutation rules ,

$$[R^{\pm}, R_3] = \pm \, 2R^{\pm} \, ,$$

and expressed the operator R^+R^- in terms of the operator R_3 with the help of the relationship (2.24). The obtained equation is an immediate consequence of the Eq. (3.9). If we neglect the fluctuations of the population difference and assume $\langle R_3^2 \rangle = \langle R_3 \rangle^2$, then we get the equation for $\langle R_3 \rangle$ coinciding with the equations (1.9) or (2.26) for the point-source model, in which the time T^{-1} should be replaced by $2|g|^2\tau/\hbar^2$.

Thus, if we neglect the fluctuations of the population difference $\langle \Delta R_3^2 \rangle = \langle R_3^2 \rangle - \langle R_3 \rangle^2$, then we get the well known solution for $\langle R_3 \rangle$ and because the emission power, which is proportional to $\langle a^+ a \rangle$, can be expressed in terms of $\langle R^+ R^- \rangle$, which is

$$\langle R^+ R^- \rangle = j(j+1) - \frac{1}{4} \langle R_3^2 \rangle + \frac{1}{2} \langle R_3 \rangle \, ,$$

then, neglecting the fluctuations of R_3 , we get the well-known equation for the emission intensity, which will coincide with the Eq. (1.12) if we replace T^{-1} by $2|g|^2/\hbar^2$ in it.

At the same time the Bonifacio equation (3.9), we started with, takes into account the quantum fluctuations of the population difference. In the Dicke representation the equation for the diagonal matrix elements is given by [2]:

$$\frac{d}{dt}<jm\alpha|\rho_A|jm\alpha> =$$

$$= \frac{2|g|^2\tau}{\hbar^2}((j+m+1)(j-m)<j,m+1,\alpha|\rho_A|j,m+1,\alpha>$$

$$-(j+m)(j-m+1)<jm\alpha|\rho_A|jm\alpha>) . \qquad (3.11)$$

In deriving of (3.11) we have used the equations (2.6) for the matrix elements of the operators $R^{\pm} = J^{\pm}$.

The implication of the Eq. (3.11) is clear enough: $<jm\alpha|\rho_A|jm\alpha>$ is a probability for atom to be in the state $|jm\alpha>$ at the moment t. Therefore the Eq. (3.11) is the balance equation with the probability of the spontaneous transitions from the state $|j,m+1,\alpha>$ to the state $|jm\alpha>$ equaled to

$$\frac{2|g|^2\tau}{\hbar^2}(j+m+1)(j-m) .$$

The similar master equations could be written immediately for the point-source model as a result of the equations (2.19) for the spontaneous transition probabilities calculated with the help of the perturbation theory.

It is important to notice that in the equations (3.11) there are only the downwards transitions while the upwards transition $(m \longrightarrow m+1)$ are prohibited because there is no the photon back action on atoms in the SR limit.

We can also write the equations for the nondiagonal elements of the density matrix. But these equations have only the trivial solutions if all nondiagonal elements of the density matrix were equal to zero in the initial state at $t = 0$. This situation takes place in the case of the noncoherent excitation of atoms. In any case, even for the coherent excitation of the atomic system the equations (3.11) for the diagonal elements of the density matrix do not include the nondiagonal elements, and therefore the equations for the diagonal and nondiagonal elements of the density matrix can be solved separately.

The set of equations (3.11) is a hierarchy of the coupled equations. The number of them is equal to $2j + 1$, the order of magnitude of this value is approximately N . These solutions can be solved numerically for the moderate values of N . The results of the numerical calculations for the case $N = 10^2$ are presented in [2].

The calculations show that if we take the coherent atomic state as the initial state then the dynamics of the radiative decay of the system is very close to that described by the quasiclassical equations or the equations (2.26) with the approximation $<R_3^2> = <R_3>^2$. The initial binomial distribution over m is slightly changed, because the coherent atomic state with the angle ϑ depending on time does not satisfy the Bonifacio equation (3.9). However the dispersion of R_3 remains small and the difference between the calculated emission intensity and the predicted one by the quasiclassical equation does not exceed 0.5% (this value coincides with the accuracy of the numerical calculations). The relative dispersion of R_3 is of the order of N^{-1} during the main body of the SR pulse.

In contrast to the above case the dispersion of R_3 becomes essential if the initial state is the state $|jj\alpha>$, when all of the atoms are in the excited state for a given j. At the intensity maximum the relative dispersion of R_3,

$$\sigma^2(R_3) = \frac{<R_3^2> - <R_3>^2}{(2j)^2},$$

is equal to 0.24, hence the intensity at a maximum is in the order of 77% of its value in the quasiclassical limit, when we can neglect the fluctuation of the population difference.

In the case. when the atomic subsystem contains a few atoms or a few tens of atoms, the more exact calculations can be made. Such exact solution for the collective spontaneous emission from an ensemble of atoms placed into a resonant damped cavity and excited to a symmetrical Dike state $|j = N/2, m = -N/2 + 1>$ with only one atom in the excited state, was derived by Seke in [11]. The comparisons of the numerical results with the calculations [2] enabled to determine the values of parameters for which both theories gave the same results.

1.3.2. *Fokker-Plank equation.* It is convenient to consider the equation (3.9) for the atomic density matrix not in the Dicke representation, but in the coherent atomic representation [8]. As it is shown in the Appendix 2, the atomic density matrix has the following diagonal representation

$$\rho_A = \int |\vartheta,\phi><\vartheta,\phi| \; P(\vartheta,\phi) \sin\vartheta \, d\vartheta \, d\phi ,$$

where $P(\vartheta,\phi)$ is a c-number function, describing the distribution of the quasiprobabilities of the different states $|\vartheta,\phi>$ on a surface of the sphere.

The function $P(\vartheta,\phi,t)$ is normalized by the condition

$$\mathrm{Sp}\ \{\rho_A(t)\} = \int^{\pi} \int^{2\pi} P(\vartheta,\phi) \sin\vartheta \, d\vartheta \, d\phi = 1 ,$$

using this equation we can find the mean value of any atomic operator of the type $R^{+l} R_3^n R^{-m}$. Really

$$\langle R^{+l} R_3^n R^{-m} \rangle = \left(\frac{\partial}{\partial\alpha}\right)^l \left(\frac{\partial}{\partial\beta}\right)^n \left(\frac{\partial}{\partial\gamma}\right)^m \langle e^{\alpha R^+} e^{\beta R_3} e^{\gamma R^-} \rangle \Big|_{\alpha=\beta=\gamma=0} .$$

For the coherent state the mean value of

$$\langle \vartheta,\phi | e^{\alpha R^+} e^{\beta R_3} e^{\gamma R^-} | \vartheta,\phi \rangle = \chi(\alpha,\beta,\gamma;\vartheta,\phi)$$

can be calculated [8]:

$$\chi(\alpha,\beta,\gamma;\vartheta,\phi) = [e^{\beta}\sin(\vartheta/2) + e^{-\beta}(\alpha e^{i\phi}\sin(\vartheta/2)$$

$$+ \cos(\vartheta/2))(\gamma e^{-i\phi}\sin(\vartheta/2) + \cos\,(\vartheta/2))]^{2j} .$$

Then

$$\langle R^{+l} R_3^n R^{-m} \rangle = \int \sin\vartheta \, d\vartheta \, d\phi \, P(\vartheta,\phi,t)$$

$$\times \left(\frac{\partial}{\partial\alpha}\right)^l \left(\frac{\partial}{\partial\beta}\right)^n \left(\frac{\partial}{\partial\gamma}\right)^m \chi(\alpha,\beta,\gamma;\vartheta,\phi) \Big|_{\alpha=\beta=\gamma=0} .$$

In the case of the large values of j this equation is essentially simplified for $l = m \ll j$ and $n \ll j$:

$$\langle R^{+l} R_3^n R^{-l} \rangle = \int \sin\vartheta \, d\vartheta \, d\phi \, P(\vartheta,\phi,t)(j \sin\vartheta)^{2l}(-2j\cos\vartheta)^n . \quad (3.12)$$

The mean value for the case n m is equal zero, if $P(\vartheta,\phi,t)$ does not depend on ϕ . This holds for the incoherent atomic excitation, when $P(\vartheta,\phi,0)$ does not depend on ϕ .

The equation for the function

$$Q(\vartheta,t) = \sin\vartheta \int_0^{2\pi} P(\vartheta,\phi,t) \, d\phi$$

is [8]

$$\frac{\partial}{\partial t} Q(\vartheta,t) = \frac{2|g|^2\tau}{\hbar^2}\left\{-\frac{\partial}{\partial\vartheta}\left[-\left(j \sin\vartheta + \frac{\sin\vartheta}{2(1+\cos\vartheta)}\right) Q(\vartheta,t)\right]\right.$$

$$\left. + \frac{\partial^2}{\partial\vartheta^2}\left[\frac{1-\cos\vartheta}{2} Q(\vartheta,t)\right]\right\} . \qquad (3.13)$$

This equation is equivalent to the Eq. (3.9) for ρ_A if at $t = 0$ the function $P(\vartheta,\phi,0)$ does not depend on ϕ . If we have solved this equation, for example numerically, then applying the Eq. (3.12) and (3.10) we can find the moments of the atomic and field operators. In particular, the emission power is

$$I = \frac{2\hbar\omega_0}{\tau} \langle a^* a\rangle = 2\hbar^{-1}|g|^2\tau \langle R^+R^-\rangle = 2\hbar^{-1}\omega_0|g|^2\tau j^2 \int Q(\vartheta,t)\sin^2\vartheta \, d\vartheta .$$

The Eq. (3.13) has a form of the classical Fokker-Plank equation for the Markovian stochastic process. It is well known, that the first term in the right-hand side of this equation is determined by the mean rate of the ϑ angle variation for a given ϑ :

$$\lim_{\Delta t \to 0} \frac{\langle\Delta\vartheta\rangle}{\Delta t} = -\frac{2|g|^2\tau}{\hbar^2}\left(j\cos\vartheta + \frac{\sin\vartheta}{2(1+\cos\vartheta)}\right) .$$

For large N and for angles ϑ π we can ignore the second term in the right-hand side of the Eq. (3.13), and we get the equation

$$\frac{d\vartheta}{dt} = -\frac{2|g|^2\tau}{\hbar^2} j \sin\vartheta . \qquad (3.14)$$

This equation coincides with the equation (3.5) which was obtained in semiclassical approximation for the one-mode model. Really, if $j = N/2$ then

$$\tilde{T}^{-1} = 4\pi\tau\omega_0|\mathbf{de}_0|^2/\hbar^2 = 2\tau|g|^2/\hbar^2 ,$$

where we have used the normalization condition

$$|\mathbf{de}_0|^2 = \frac{|g|^2}{2\pi\hbar\omega} ,$$

and we see that the Eqs. (3.5) and (3.14) coincide exactly.

The Eq. (3.14) shows that if we take the coherent atomic state with $\vartheta \quad \pi$ as the initial state then the state of the atomic system will be the coherent atomic state during the whole process of system evolution if $N \gg 1$. The remainder members in the Fokker-Plank equation (the additional term, $-(1 + \cos \vartheta)\sin \vartheta / 2\tilde{T}$, in the equation for the rate of the ϑ angle drift, and the term with the diffusion coefficient $(1 - \cos \vartheta)/2\tilde{T}$) are essential only in the vicinity of the angle $\vartheta = \pi$, where the main term, proportional to j ($j \gg 1$), disappears. This angle $\vartheta = \pi$ corresponds to the state of the complete inversion of the atomic system.

The analysis of the Eq. (3.13) performed in [8] shows that the mean values of the atomic operators $\langle R^{+l} R_3^n R^{-l} \rangle$ are factorized as $\langle R^{+l} R^{-l} \rangle \langle R_3 \rangle^n$ in any region of the angles ϑ excepting the vicinity of the complete inversion region, where $\vartheta = \pi$. It was shown in [8,9] that the closer the initial state to the state of the complete inversion $\vartheta = \pi$, the less the validity of the semiclassical approximation. When $\vartheta(t = 0)$ tends to π , then the peak intensity I_{max} and the delay time t_0 can differ from these values calculated with the quasiclassical equations (1.11) and (1.12) by 20% .

Gronchi and Lugiato [10] improved the Eq. (3.13) by introducing the additional term into the diffusion coefficient of the Fokker-Plank equation:

$$\frac{1}{\tilde{T}} \left[\frac{1 - \cos \vartheta}{2} + \frac{\tau}{\tilde{T}} j^2 \sin^2 \vartheta \right] .$$

This additional term is essential during the whole process of the SR pulse formation.

1.4. Superradiance of collection of three-level atoms.

The lower level of a superradiant transition need not to be the atomic ground level. In such a situation, the radiative decay of the lower SR level can generate the SR pulse in the adjacent transition. The two SR pulse with the frequencies corresponding to the frequencies of the adjacent atomic transitions will correlate. In the general case, the decay time and the width of the second SR pulse will depend on these parameters for the first SR pulse.

The simplest model, which enables to discuss the specific features of the cascade superradiative decay, is a collection of the three-level atoms confined into the volume with the dimensions much smaller than the wavelengths of any transition. Let us begin with the quasiclassical treatment of the three-level superradiance. The atomic levels are enumerated by the index $s = 1,2,3$ and it is assumed the energy-level spacing is such that $E_3 > E_2 > E_1$.

The Hamiltonian of the atom-field interaction is

$$H'_{ss'} = -\,(\mathbf{d}_{ss'}\mathbf{E}_{s's}(t)) \; .$$

Here $\mathbf{d}_{ss'}$ are the matrix elements of the dipole moment operator (it is assumed that $\mathbf{d}_{ss} = 0$), and $\mathbf{E}_{s's}$ are the slowly varying Fourier-components of the electric field:

$$\mathbf{E}(t) = \sum_{\omega_{s's}} \mathbf{E}_{s's}(t)\exp(\mathrm{i}\omega_{s's}t) \; , \qquad \hbar\omega_{s's} = E_{s'} - E_s \; .$$

According to (1.2), we get

$$\mathbf{E}_{s's} \quad -\frac{2N\omega^3_{s's}\,\mathrm{i}}{3c^3}\,\mathbf{d}_{s's}\,a_s a^*_{s'} \; , \tag{4.1}$$

where a_s is a wave function of an atom in the interaction representation: $|\psi\rangle = \sum_s a_s |s\rangle$, and $|s\rangle$ are the unperturbed wave functions of an atom ($s = 1,2,3$).

All of the atoms decay synchronously and if the initial state of an atom is the pure state it will remain the pure state during the whole process of a system dynamics, because it is assumed that the atom interacts only with the c-number radiation field, and any other interactions can be ignored. Among other things, the relaxation processes are also ignored, what can be true if the characteristic times of radiative process are smaller than the relaxation time.

If (4.1) is applied to the Schrödinger equation for an atom,

$$i\hbar \frac{da_s}{dt} = \sum_{s'} H'_{ss'} a_{s'} ,$$

the resulting equation for a_s is

$$\frac{da_s}{dt} = - \sum_{s'} \frac{|a_{s'}|^2 a_s N}{2\tau_{s's}} , \tag{4.2}$$

where $\tau^{-1}_{s's} = 4\omega^3_{ss'} |\mathbf{d}_{s's}|^2 / 3\hbar c^2$ is the spontaneous decay rate for the transition $s — s'$. Notice that, by definition, $\tau^{-1}_{ss'} = - \tau^{-1}_{s's} < 0$ at $s > s'$.

Multiplying (4.2) by a^*_s we get the equation for the level population $n_s = N|a_s|^2$:

$$\frac{dn_s}{dt} = - \sum_{s'} \frac{n_{s'} n_s}{\tau_{s's}} . \tag{4.3}$$

It is not difficult also to get the Dicke-type equations for the population differences. In the case of an atomic system with the dimensions much smaller than wavelength of radiation, the rearrangement symmetry is conserved, because the Hamiltonian is the completely symmetric operator commuting with any rearrangement operator. If the initial state is the state of the complete inversion, when all of the atoms are in the upper level with energy E_3, then this state is completely symmetric state and this symmetry will be conserved. Thus we can restrict ourselves to the case when the atomic state is the symmetric state in any moment of time.

Such symmetric states can be defined by the occupation numbers: $|n_1 n_2 n_3\rangle$ is the state with n_1 atoms in the level $|1\rangle$, n_2 atoms in the level $|2\rangle$, and n_3 atoms in the level $|3\rangle$. Any symmetric operator can be expressed in terms of the operators c_s and c^+_s, which are the operators of the annihilation and creation of the atom in the state $|s\rangle$. The operators c_s and c^+_s obey the commutation rules

$$[c_s, c^+_{s'}] = \delta_{ss'} \ .$$

In particular, the operator of the interaction of the atoms with the field (2.9) and (2.15) can be written in the form

$$H_{AF} = \sum_{\mathbf{k}\nu, s'>s} (g^{s's}_{\mathbf{k}\nu} a_{\mathbf{k}\nu} c^+_{s'} c_s + g^{s's*}_{\mathbf{k}\nu} a^+_{\mathbf{k}\nu} c^+_s c_{s'}) \ , \qquad (4.4)$$

where

$$g^{s's}_{\mathbf{k}\nu} = - i \sqrt{\frac{2\pi\omega_{\mathbf{k}\nu}}{W}} (\mathbf{d}_{s's} \mathbf{e}_{\mathbf{k}\nu}) \ .$$

Using (4.4) we can easily find the matrix elements of the interaction operator H_{AF} and calculate, in a manner like the Eqs. (2.16) and (2.19), the probability of the spontaneous transition from the state $|n_1 n_2 n_3>$ to the state $|n'_1 n'_2 n'_3>$. We get the following equation for the probability of the transition from the second level to the first one, when $n'_1 = n_1 + 1$, $n'_2 = n_2 - 1$:

$$w_{12} = n_2(n_1 + 1)/\tau_{12} \ ,$$

and, in the same way, for the transition $n'_2 = n_2 + 1$, $n'_3 = n_3 - 1$:

$$w_{23} = n_3(n_2 + 1)/\tau_{23} \ ,$$

and for the transition $n'_1 = n_1 + 1$, $n'_3 = n_3 - 1$:

$$w_{13} = n_3(n_1 + 1)/\tau_{13} \ .$$

Finally, we get the following master equations for the population of the atomic levels

$$\frac{dn_3}{dt} = - w_{23} - w_{13} = - \frac{n_3(n_2 + 1)}{\tau_{23}} - \frac{n_3(n_1 + 1)}{\tau_{13}} \ ,$$

$$\frac{dn_2}{dt} = w_{23} - w_{12} = \frac{n_3(n_2 + 1)}{\tau_{23}} - \frac{n_2(n_1 + 1)}{\tau_{12}} \ , \qquad (4.5)$$

$$\frac{dn_1}{dt} = w_{12} + w_{13} = \frac{n_2(n_1 + 1)}{\tau_{12}} + \frac{n_3(n_1 + 1)}{\tau_{13}} .$$

It follows from these equations, that $\frac{d}{dt}(n_1 + n_2 + n_3) = 0$, therefore $n_1 + n_2 + n_3 = N$. Hence, it is natural to consider only the two equations out of the three equations (4.5), for example, the first two equations with n_1 in them replaced by $n_1 = N - n_2 - n_3$.

If in the initial state $n_3 = 0$ and $n_2 = N$, then the equations (4.5) are reduced to the one equation

$$\frac{dn_2}{dt} = - \frac{n_2(N - n_2 + 1)}{\tau_{12}} .$$

Substituting $n_2 = N/2 + m$, we get

$$\frac{dm}{dt} = \frac{(N/2 + m)(N/2 - m + 1)}{\tau_{12}} .$$

We see that this equation exactly coincides with the Dicke equation (2.26), because $\langle R_3 \rangle = 2\bar{m}$. Thus, the equations (4.5) are the generalization of the Dicke equation (2.26) for the case of the three-level atoms. In contrast to the semiclassical equations (4.3), these equations incorporate the effect of the spontaneous emission from the completely inverted state.

The power of emission at the frequency $\omega_{ss'}$ is determined by

$$I_{ss'} = \hbar\omega_{ss'} w_{s's} \quad (s > s')$$

and it can be found from the solution of the equations (4.5).

We can generalize the one-mode model discussed in 1.3 if we assume that there is only the one mode of the field for each frequency $\omega_{ss'}$. Such one-mode model for the three level atoms can be transformed to the point-source model, described by equations (4.5), if the synchronism condition holds

$$\mathbf{k}_{32} + \mathbf{k}_{21} = \mathbf{k}_{31} ,$$

where $\mathbf{k}_{ss'}$ is the wave vector of the wave with the frequency $\omega_{ss'}$. We have assumed that all three transitions are the allowed transitions.

To transform this model to the point-source model we should make the unitary transformation of the basis state vectors $|s\rangle^{(l)}$ of the lth atom and construct the symmetric multiatomic states on the new vectors. If the transition $3 \longrightarrow 1$ is forbidden transition, while the transitions $3 \longrightarrow 2$ and $2 \longrightarrow 1$ are allowed ones, then the synchronism condition is not required. The Eqs. (4.5), along with the Dicke equation (2.26), do not take into account the fluctuations of the populations n_1, n_2, n_3. If the transition $3 \longrightarrow 1$ is a forbidden transition, then the equations (4.5) become

$$\frac{dn_3}{dt} = -\frac{n_3(N - n_3 - n_1 + 1)}{\tau_{23}},$$
$$\frac{dn_1}{dt} = \frac{(n_1 + 1)(N - n_3 - n_1)}{\tau_{12}}. \qquad (4.6)$$

In (4.6) we have replaced n_2 by $n_2 = N - n_3 - n_1$. We can rewrite the Eqs. (4.6) in the following form

$$\frac{d}{dt}\ln n_3 = -\frac{n_2 + 1}{\tau_{23}}, \qquad \frac{d}{dt}\ln(n_1 + 1) = \frac{n_2}{\tau_{12}}.$$

It is seen that there is the following first integral of motion

$$\tau_{23}\ln n_3 + \tau_{12}\ln(n_1 + 1) + t = \text{const},$$

hence

$$e^{t/\tau_0}\, n_3(n_1 + 1)^{\beta} = \text{const},$$

where $\tau_0 = \tau_{23}$, $\beta = \tau_{12}/\tau_{23}$. The initial conditions $n_3(0) = N$, $n_1(0) = 0$ enable to define the constant, which is $\text{const} = N$. Using the first integral, the Eqs. (4.6) can be transformed to the fist order nonlinear equation, the solution of which can be reduced to the quadrature.

Fig. 1.3 shows results of the numerical calculation for the emission power profiles for the successive SR transitions. It is seen that in the case $\tau_{23} = \tau_{12}$ and $\beta = 1$ the two SR pulses at the frequencies ω_{32} and ω_{31} are emitted subsequently. Initially the SR emission occurs in the upper transition, and then in the lower one. The pulses are not overlapped almost, and the process of SR emission coincides essentially with that for the two-level atoms. The first SR pulse prepares the population inversion for the second transition. The emission process has the same behavior in the case $\beta < 1$.

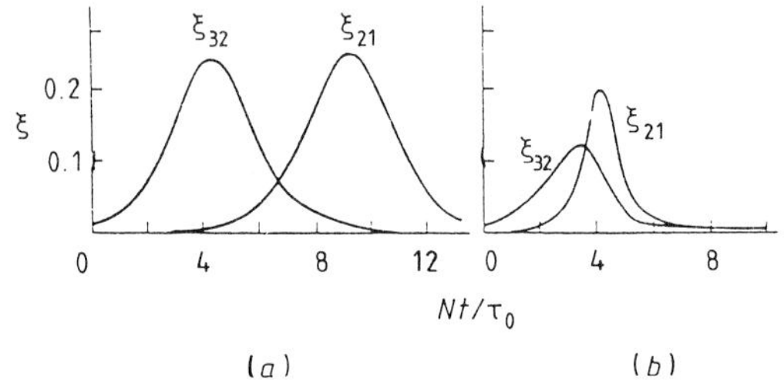

Fig.1.3. Normalized intensity profiles $\xi_{21} = I_{21}\tau_0/(\hbar\omega_{21}N^2)$ and $\xi_3 = I_{32}\tau_0/(\hbar\omega_{32}N^2)$. The normalized time is $t\ N/\tau_0$. The initial populations of levels are $n_3(0) = N = 10^2$, $n_1(0) = n_2(0) = 0$. The parameter β is $\beta = 1$ (a), $\beta = 10$ (b).

At $\beta \gg 1$ (see Fig. 1.3b) the essential decrease of the intensity peak value takes place for the first SR pulse in comparison with its value $\xi_{max} = 0.25$ for the individual transition. This is because the intermediate level depopulates when SR in the lower transition is developed. The pulse becomes non-symmetric, its tail is distorted hardly. There appears the steep fall stage, which is followed by the very slow decay stage.

CHAPTER 2

OPERATOR AND MOMENTUM EQUATIONS OF RESONANT LIGHT-MATTER INTERACTION

In chapter 1 the model of a two–level atom and the mathematical aspects of a treatment of the two–level atoms interaction with resonant electromagnetic field have been presented. Here, this treatment will be continued further. The dynamics of the collective processes is determined primarily by the nonlinear matter–field interactions and, as a result, it is described by the nonlinear equations. Unfortunately the general solution of these equations is not known now, and any analyses of them meats with the serious difficulties. Therefore,in treating of some specific aspects of complex dynamics of radiative decay in multiatomic systems it is more convenient to use some approximate methods which are adequate to considered problem but not overcomplicated by accounting of the unessential features. For example, it is not necessary, at least during the first acquaintance with the subject, to complicate the study of the angular radiation spectra by the inclusion of the spatial inhomogeneity effects. The use of the different approaches is especially useful in the textbook because it enables to unify the parameters of the different sets of equation, to demonstrate the generality of the different approaches, and to help in studying of a special literature.

Here, we introduce the spatial–temporal dynamics equations in the atom–field representation (see Sec. 2.1), the field modes dynamics equations in the atom–field representation (see Sec. 2.3), and the dynamics equations of atomic subsystem in the atomic representation (see Sec. 2.4), i.e., in the frames of the boson operators exclusion method. In Sec. 2.2, and Sec. 2.3, the main integrals of motion are obtained, which are valuable for the physical interpretation of the processes under consideration.

As we have mentioned earlier the operator equations are very complex for analyses because of their nonlinearity. Therefore, to alleviate the atom–field interaction dynamics treatment it is useful to try to simplify the appropriate equations of motion. One of such simplifications consists in transformation of operator equations into the equations for mean values. In this case we should not pay any attention to the order of cofactors in the products of the unknown functions, and we can use some additional information about the mean values of the variables at the different stages of a system dynamics. Unfortunately, the nonlinearity of operator equations again creates some complications. Namely,the equations for mean values of the product of n operators will always depend on the mean value of the product of $n + 1$ operators. Hence we must always use the procedure of correlations decoupling. Here we get the equations for the mean values which include the correlators of the first order (see Sec. 2.5) and the four types of equations for the second order correlators (see Secs. 2.6 – 2.9).

The theories based on the first order correlators lead to the equations coinciding in appearance with the appropriate operator equations. Thus, in this case we simply substitute the quantum operators by the classical variables. It means that the fluctuations of the quantum variables are not taken into account. As a result, in this case we lose the effects of the photon back action on the atom emitted it, and hence, we lose in our equations the terms which describe the sources of a spontaneous emission.

The theories based on the second order correlators includes the spontaneous emission effects. Considering that the triple correlations can be represented as the products of the lower order correlations in a variety of ways, it is clear that in such theories there is some latitude in the choice of the type of decoupling. Hence we can put forward certain additional demands to the correlator properties. As an example, in Sec.2.9 we show that the decoupling which takes into account symmetrical properties of Pauli matrix leads to the equations incorporating the fluctuations of the operator σ_3 .

We will obtain the equation in the form suitable for a treatment of the collective spontaneous emission effect. But it should be noted that these equations are applicable not only for a description of the radiative decay of the highly–excited atomic ensembles. They can be applied to the treatment of any other process of interaction of electromagnetic field with a system of resonant atoms. The main distinctions in the description of the different effects consist in the form of the boundary and initial conditions.

2.1. Operator wave equation

Consider an ensemble of radiators (i.e. atoms, molecules, nuclei) in which each radiator interacts radiatively with any other and, probably, with the external resonance electromagnetic field. For the sake of simplicity, we will further speak about the atomic transitions but it does not restrict the generality of a consideration because we use the Pauli spin matrices $(\sigma_{i+}, \sigma_{i-}, \sigma_{i3})$ as the atomic operators. So we will speak about the radiative transitions in molecules and nuclei as needed. The Pauli spin matrices obey the commutation rules discussed in Chapter 1 and defined by

$$[\sigma_{i+}, \sigma_{i-}] = \delta_{ij}\, \sigma_{i3}, \qquad [\sigma_{i\pm}, \sigma_{j3}] = \mp 2\, \delta_{ij}\, \sigma_{i\pm}\,, \tag{1.1}$$

where $\sigma_{i\pm} = (\sigma_{i1} + \sigma_{i2})/2$. The Hamiltonian of atomic subsystem is

$$H_a = \frac{1}{2} \sum_{i=1}^{N} \hbar\omega_{0i}\sigma_{i3}, \tag{1.2}$$

where ω_{0i} is the i–th atom radiative transition frequency, and N is a number of atoms in ensemble.

To describe the electromagnetic field the operator of the vector potential $\mathbf{A}(\mathbf{r},t)$ and canonically conjugated generalized pulse [1]

$$\mathbf{B}(\mathbf{r},t) = (4\pi c^2)^{-1} \partial \mathbf{A}(\mathbf{r},t)/\partial t \tag{1.3}$$

are usually used. The operators $\mathbf{A}(\mathbf{r},t)$ and $\mathbf{B}(\mathbf{r},t)$ obey the following commutation rules

$$[A_\alpha(\mathbf{r},t), A_\beta(\mathbf{r}',t)] = [B_\alpha(\mathbf{r},t), B_\beta(\mathbf{r}',t)] = 0\,, \tag{1.4a}$$

$$[A_\alpha(\mathbf{r},t), B_\beta(\mathbf{r}',t)] = i\hbar\delta_{\alpha\beta}\; \delta(\mathbf{r} - \mathbf{r}'), \quad \alpha,\beta = 1,2,3. \tag{1.4b}$$

The Hamiltonian of the free electromagnetic field (i.e. in the absence of the atomic subsystem) is

$$H_f = \int \{2\pi c^2 \mathbf{B}^2(\mathbf{r},t) + (8\pi)^{-1}[\mathrm{rot}\ \mathbf{A}(\mathbf{r},t)]^2\}\, d\mathbf{r}\ . \tag{1.5}$$

The interaction of the atomic subsystem with the e.m. field is described by the following Hamiltonian

$$H_{int} = -\frac{1}{c} \int \mathbf{j}(\mathbf{r},t)\ \mathbf{A}(\mathbf{r},t)\ d\mathbf{r}\ , \tag{1.6}$$

where $\mathbf{j}(\mathbf{r},t)$ is an operator of atomic transition current density.

There is the following relationship between the current density operator $\mathbf{j}(\mathbf{r},t)$ and the Pauli spin matrices $\sigma_{i\pm}$:

$$\mathbf{j}(\mathbf{r},t) = \sum_{i=1}^{N} \{<+|\mathbf{j}_0^+(\mathbf{r}-\mathbf{r}_i)|->\sigma_{i+}(t) + <-|\mathbf{j}_0^-(\mathbf{r}-\mathbf{r}_i)|+>\sigma_{i-}(t)\}, \qquad (1.7)$$

where $\mathbf{j}_0^+(\mathbf{r},t)$ is the positive frequency part of the current density operator for an individual atom. $|+>$ and $|->$ are the atom wave functions in the excited and ground states, respectively. We assume that atom is point radiator, i.e. the radius of its electron shells is much smaller than the wavelength λ. In this case the transition matrix elements can be represented in the form

$$<+|\mathbf{j}_0^+(\mathbf{r}-\mathbf{r}_i)|-> = \mathbf{m}_i\delta(\mathbf{r}-\mathbf{r}_i) \qquad (1.8)$$

where $\mathbf{m}_i$ is the matrix element of the i–th atom transition current.

The Hamiltonian

$$H = H_a + H_f + H_{int} \qquad (1.9)$$

together with the commutation rules (1.1) and (1.4) generate the following set of coupled nonlinear operator equations

$$\frac{\partial \mathbf{A}(\mathbf{r},t)}{\partial t} = 4\pi c^2\mathbf{B}(\mathbf{r},t), \qquad (1.10a)$$

$$\frac{\partial \mathbf{B}(\mathbf{r},t)}{\partial t} = -(4\pi)^{-1}\mathrm{rot\ rot\ }\mathbf{A}(\mathbf{r},t) + c^{-1}\mathbf{j}(\mathbf{r},t), \qquad (1.10b)$$

$$\frac{\partial \sigma_{i+}}{\partial t} = i\omega_{0i}\sigma_{i+} + \frac{i}{\hbar c}\mathbf{m}_i^*\mathbf{A}(\mathbf{r}_i,t)\,\sigma_{i3}, \qquad (1.10c)$$

$$\frac{\partial \sigma_{i3}}{\partial t} = \frac{2i}{\hbar c}(\mathbf{m}_i\sigma_{i+} - \mathbf{m}_i^*\sigma_{i-})\mathbf{A}(\mathbf{r},t). \qquad (1.10d)$$

The Heisenberg equations (1.10) describe the dynamics of the atoms and field operators development in process of matter–field interaction. These equations provide us a basis for the subsequent treatment of a superradiance and the related phenomena.

2.2. Integrals of motion

The general solution of equations (1.10) has not been found in analytical form so far. Because of this, try to find the integrals of motion of these equations and determine the implication of derived integrals. This will help us in analyses of equations and in physical interpretation of the processes under consideration.

2.2.1. *Monoatomic integral of motion.* Multiply equation(1.10c) by σ_{i-} from the left, and the Hermitian–conjugated equation by σ_{i+} from the right. Next, adding the derived equations we get

$$\frac{d}{dt}(\sigma_{i+}\sigma_{i-}) = \frac{i}{\hbar c}(m_i^*\sigma_{i3}\sigma_{i-} - m_i\sigma_{i+}\sigma_{i3})A(r_i,t) \ . \tag{2.1}$$

In (2.1) we have taken into account that the operator $A(r_i,t)$ commutes with the all atomic operators referring to the same moment t, because the field and atomic variables belong to the different degrees of a system freedom. The equation for the operator $\sigma_{i-}\sigma_{i+}$ can be derived by the same way. It is

$$\frac{d}{dt}(\sigma_{i-}\sigma_{i+}) = \frac{i}{\hbar c}(m_i^*\sigma_{i-}\sigma_{i3} - m_i\sigma_{i3}\sigma_{i+})A(r_i,t) \ . \tag{2.2}$$

Adding Eqs. (2.1) and (2.2) and taking into account Eq. (1.10d) we get

$$\frac{d}{dt}(\sigma_{i+}\sigma_{i-} + \sigma_{i-}\sigma_{i+}) = \frac{1}{2}\left[\sigma_{i3}\frac{d\sigma_{i3}}{dt} + \frac{d\sigma_{i3}}{dt}\sigma_{i3}\right] . \tag{2.3}$$

Hence

$$\frac{d}{dt}\left[\sigma_{i+}\sigma_{i-} + \sigma_{i-}\sigma_{i+} + \frac{1}{2}\sigma_{i3}^2\right] = 0 \tag{2.4}$$

and we have the following integral of motion

$$\sigma_i^2 = 2(\sigma_{i+}\sigma_{i-} + \sigma_{i-}\sigma_{i+}) + \sigma_{i3}^2 = \text{const.} \tag{2.5}$$

This equation can be transformed to the following form

$$\sigma_i^2 = \sigma_{i1}^2 + \sigma_{i2}^2 + \sigma_{i3}^2 = 3I, \tag{2.6}$$

where I is a identity operator.

Thus the integral of motion (2.5) which has been derived from the equations of motion (1.10), reflects simply the following properties of the Pauli matrices [2]

$$\sigma_{i1}^2 = \sigma_{i2}^2 = \sigma_{i3}^2 = I.$$

Equation (2.4) can be rewritten in the form of Heisenberg equation

$$\frac{d\sigma_i^2}{dt} = \frac{1}{i\hbar}[\sigma_i^2, H] = 0\ , \tag{2.7}$$

where H is determined by Eq. (1.9).

2.2.2. *Collective integral of motion.* Let introduce the collective atomic operator R^2 in analogy with the Eq. (2.5)

$$R^2 = 2(R^+R^- + R^-R^+) + R_3^2, \tag{2.8}$$

where collective operators $R^{\pm}$ and R_3 are determined by

$$R^{\pm} = \sum_{i=1}^{N} \sigma_{i\pm}, \quad R_3 = \sum_{i=1}^{N} \sigma_{i3}, \tag{2.9}$$

To determine whether operator R^2 is integral of motion or not, let calculate the commutator $[R^2, H]$. As operator R^2 depends on the atomic variables only, then

$$[R^2, H] = [R^2, H_a + H_{int}].$$

In calculating of this operator we can use the following rules

$$[R^+R^-, H] = R^+[R^-, H] + [R^+, H]R^-,$$

$$[R_3^2, H] = R_3[R_3, H] + [R_3, H]R_3. \tag{2.10}$$

If (1.7) and (1.8) are applied to (1.6), the interaction Hamiltonian H_{int} takes the form

$$H_{int} = \frac{1}{c}\sum_i (\mathbf{m}_i\sigma_{i+} + \mathbf{m}_i^*\sigma_{i-})\mathbf{A}(\mathbf{r}_i,t). \tag{2.11}$$

Using (1.2), (2.9) and (2.11), we get ($i = 1,\ldots,N$)

$$[R_3, H_a] = 0,$$

$$[R_3, H_{int}] = -\frac{2}{c}\sum_i (\mathbf{m}_i\sigma_{i+} - \mathbf{m}_i^*\sigma_{i-})\mathbf{A}(\mathbf{r}_i,t),$$

$$[R^{\pm}, H_a] = \mp\sum_i \hbar\omega_{0i}\sigma_{i\pm}, \tag{2.12}$$

$$[R^+, H] = \frac{1}{c}\sum_i \mathbf{m}_i^*\sigma_{i3}\mathbf{A}(\mathbf{r}_i,t),$$

$$[R^-, H_{int}] = \frac{1}{c}\sum_i \mathbf{m}_i\sigma_{i3}\mathbf{A}(\mathbf{r}_i,t).$$

And so, the desired commutator $[R^2, H]$ has the following form

$$[R^2, H] = -2\sum_{i,j}\hbar(\omega_{0i} - \omega_{0j})(\mathbf{m}_i\mathbf{A}(\mathbf{r}_i,t) - \mathbf{m}_j\mathbf{A}(\mathbf{r}_j,t)$$

$$-\frac{2}{c}\sum_{ij}(\sigma_{i+}\sigma_{j3} + \sigma_{j3}\sigma_{i+})(\mathbf{m}_i\mathbf{A}(\mathbf{r}_i,t) - \mathbf{m}_j\mathbf{A}(\mathbf{r}_j,t)) \tag{2.13}$$

$$+\frac{2}{c}\sum_{i,j}(\sigma_{i-}\sigma_{j3} + \sigma_{j3}\sigma_{i-})(\mathbf{m}_i^*\mathbf{A}(\mathbf{r}_i,t) - \mathbf{m}_j^*\mathbf{A}(\mathbf{r}_j,t)).$$

Hence, operator R^2 commutes with the Hamiltonian H , if the following two conditions hold

$$\omega_{0i} = \omega_{0j}, \tag{2.14}$$

$$\mathbf{m}_i \mathrm{A}(\mathbf{r}_i,t) = \mathbf{m}_j \mathrm{A}(\mathbf{r}_j,t) \tag{2.15}$$

The first condition means that the atomic system should be free from the inhomogeneous broadening, i.e., characteristic inhomogeneous broadening time $\Delta\omega_{ij}^{-1}$ should be much longer than the characteristic time of the field– matter energy exchange.

The condition (2.15) means that

$$\mathbf{m}_i = \mathbf{m}_j \, , \tag{2.15a}$$

$$\mathrm{A}(\mathbf{r}_i,t) = \mathrm{A}(\mathbf{r}_j,t). \tag{2.15b}$$

The first condition can be satisfied if there is a preferred quantization axis in a considered system. For example, in the strong magnetic field, when the splitting of the superfine sublevels of an atomic optical transition exceeds essentially the width of any sublevel, we can consider the atomic transitions between the sublevels with a definite angular momentum projection on the quantization axis. In this case we have $\mathbf{m}_i =$ $\mathbf{m}_0$.

It can easily be shown that relationship (2.15b) is satisfied in one important particular case. Namely, when the linear dimensions of a system are much smaller than the radiation wavelength. In fact, expanding $\mathrm{A}(\mathbf{r}_i,t)$ in the plane waves

$$\mathrm{A}(\mathbf{r}_i,t) = \sum_{\mathbf{k}} \mathrm{A}(\mathbf{k},t)\, \mathrm{e}^{i\mathbf{k}\mathbf{r}_i} \, ,$$

in the case $|\mathbf{k}|r_i \quad (\omega_0/c)r_i \le 2\pi d/\lambda \ll 1$ we get

$$\mathbf{A}(\mathbf{r}_i,t) \simeq \sum_{\mathbf{k}} \mathbf{A}(\mathbf{k},t) \simeq \mathbf{A}(\mathbf{r}_j,t).$$

Thus, for the systems with linear dimensions much smaller than the radiation wavelength, the collective atomic operator,

$$R^2 = R_1^2 + R_2^2 + R_3^2 , \tag{2.16}$$

is the integral of motion, if $\omega_{0i} = \omega_{0j}$ and $\mathbf{m}_i = \mathbf{m}_j$. The operator R^2 determines the length of the Bloch vector

$$\mathbf{R} = \mathbf{e}_1 R_1 + \mathbf{e}_2 R_2 + \mathbf{e}_3 R_3 \tag{2.17}$$

where $(\mathbf{e}_1, \mathbf{e}_2, \mathbf{e}_3)$ are a triad of the orthogonal unit vectors.

2.3. Plane waves representation

2.3.1. *General equations.* The vector potential $\mathbf{A}(\mathbf{r},t)$, appearing in Eqs. (1.10), can be expended into series in eigenmodes of a some quantization volume

$$\mathbf{A}(\mathbf{r},t) = \sum_n c_n[\mathbf{u}_n(\mathbf{r})a_n(t) + \mathbf{u}_n^*(\mathbf{r})a_n^+(t)], \tag{3.1}$$

where $\mathbf{u}_n(\mathbf{r})$ are the orthonormal eigenmodes of the quantization volume W,

$$\int_W \mathbf{u}_{n\lambda}^*(\mathbf{r})\, \mathbf{u}_{n'\lambda'}(\mathbf{r})\, d\mathbf{r} = \delta_{nn'}\delta_{\lambda\lambda'} , \tag{3.2}$$

and c_n are the normalization constants. In Heisenberg representation the photon creation a_n^+ and annihilation a_n operators for mode "n" obey the following commutation rules [3]:

$$[a_n , a_{n'}] = [a_n^+ , a_{n'}^+] = 0, \quad [a_n , a_{n'}^+] = \delta_{nn'} . \tag{3.3}$$

The quantization volume W is usually chosen in the form of the cube with the length L_W along the each axis. In this case

$$\mathbf{u}_n(\mathbf{r}) = \mathbf{e}_\lambda \exp(i\mathbf{k}_n\mathbf{r}), \quad c_\mathbf{k} = (2\pi\hbar c^2/W\omega_\mathbf{k})^{1/2}, \tag{3.4}$$

where

$$\mathbf{k}_n = (2\pi/L_W)(n_1\mathbf{e}_1 + n_2\mathbf{e}_2 + n_3\mathbf{e}_3), \tag{3.5}$$

n_1, n_2, n_3 are the integers. Another choice of the quantization volume will be used in Sec. 4.4.5.

If the linear dimensions of the generated pulse is much shorter than the active volume length, $\tau_p c \ll L$ then it is convenient to choose the active volume as a quantization volume. Actually, in this case there is a wide variety of the harmonics of the type (3.5) inside the spatial spectrum of a generated pulse:

$$2\pi/L \ll 2\pi/\tau_p c.$$

However, not restricting the generality of consideration we assume that the quantization volume W exceeds essentially the active medium volume V.

If (3.1) and (3.4) are applied to (1.9) and the rules (3.2) and (3.3) are used, we get for the Hamiltonian,

$$H = \frac{1}{2}\sum_i \hbar\omega_{0i}\sigma_{i3} + \sum_\mathbf{k} \hbar\omega_\mathbf{k}\, a_\mathbf{k}^+ a_\mathbf{k}$$

$$- \sum_\mathbf{k}\sum_i (g_\mathbf{k}\sigma_{i+} + g_\mathbf{k}^*\sigma_{i-})(a_\mathbf{k}\exp(i\mathbf{k}\mathbf{r}_i) + a_\mathbf{k}^+\exp(-i\mathbf{k}\mathbf{r}_i), \tag{3.6}$$

where

$$g_\mathbf{k} = (2\pi\hbar/W\omega_\mathbf{k})^{1/2}(\mathbf{e}_\lambda\mathbf{m}).$$

Summing over $\mathbf{k}$ includes summing over (n_1, n_2, n_3) (see Eq. (3.5)) and summing over the polarizations. In the free field Hamiltonian

$$H_f = \sum_\mathbf{k} \hbar\omega_\mathbf{k}\, a_\mathbf{k}^+ a_\mathbf{k} \tag{3.7}$$

we have omitted the term, $\hbar\omega_{\mathbf{k}}/2$, corresponding to the energy of the vacuum fluctuations. This is done by the appropriate choice of the energy scale origin.

Using the commutation rules (1.1) and (3.3) and Hamiltonian (3.6) we can obtain the following operator equations of motion

$$\frac{da_{\mathbf{k}}}{dt} = -i\omega_{\mathbf{k}}a_{\mathbf{k}} + \frac{1}{\hbar}\sum_{i}(g_{\mathbf{k}}\sigma_{i+} + g^{*}_{\mathbf{k}}\sigma_{i-})\exp(-i\mathbf{k}\mathbf{r}_{i}), \tag{3.8a}$$

$$\frac{d\sigma_{i+}}{dt} = i\omega_{0i}\sigma_{i+} + \frac{i}{\hbar}\sum_{\mathbf{k}} g^{*}_{\mathbf{k}}(a_{\mathbf{k}}\exp(i\mathbf{k}\mathbf{r}_{i}) + a^{+}_{\mathbf{k}}\exp(-i\mathbf{k}\mathbf{r}_{i}))\sigma_{i3}, \tag{3.8b}$$

$$\frac{d\sigma_{i3}}{dt} = \frac{2i}{\hbar}\sum_{\mathbf{k}}(g_{\mathbf{k}}\sigma_{i+} - g^{*}_{\mathbf{k}}\sigma_{i-})(a_{\mathbf{k}}\exp(i\mathbf{k}\mathbf{r}_{i}) + a^{*}_{\mathbf{k}}\exp(-i\mathbf{k}\mathbf{r}_{i})). \tag{3.8c}$$

2.3.2. *Rotating-wave approximation.* The interaction representation operators $\rho_{i\pm}$, $\mathrm{a}_{\mathbf{k}}$ and $\mathrm{a}^{+}_{\mathbf{k}}$ are connected with the Heisenberg representation operators $\sigma_{i\pm}$, $a_{\mathbf{k}}$ and $a^{+}_{\mathbf{k}}$ by the following relations:

$$\sigma_{i\pm}(t) = \rho_{i\pm}(t)\exp(\pm i\omega_{0i}t),$$

$$a_{\mathbf{k}}(t) = \mathrm{a}_{\mathbf{k}}(t)\exp(-i\omega_{\mathbf{k}}t), \tag{3.9}$$

$$a_{\mathbf{k}}(t) = \mathrm{a}_{\mathbf{k}}(t)\exp(i\omega_{\mathbf{k}}t).$$

In the rotating–wave approximation (RWA) we neglect in the terms oscillating at the summary frequency $\omega_{0} + \omega_{\mathbf{k}}$:

$$H_{\mathrm{int}} = -\sum_{\mathbf{k}}\sum_{i}[g_{\mathbf{k}}\rho_{i+}\exp(i\omega_{0i}t) + g^{*}_{\mathbf{k}}\rho_{i-}\exp(-i\omega_{0i}t)]$$

$$\times\{\mathrm{a}_{\mathbf{k}}\exp[i(\mathbf{k}\mathbf{r}_{i}-\omega_{\mathbf{k}}t)] + \mathrm{a}^{+}_{\mathbf{k}}\exp[-i(\mathbf{k}\mathbf{r}_{i}-\omega_{\mathbf{k}}t)]$$

$$-\sum_{\mathbf{k}}\sum_{i}\{g_{\mathbf{k}}\rho_{i+}\mathrm{a}_{\mathbf{k}}\exp[i\mathbf{k}\mathbf{r}_{i} + i(\omega_{0i} - \omega_{\mathbf{k}})t]$$

$$+ g_{\mathbf{k}}^{*} a_{\mathbf{k}}^{+} \rho_{i-} \exp[-i\mathbf{k}\mathbf{r}_i - i(\omega_{0i} - \omega_{\mathbf{k}})t]\}. \qquad 3.10)$$

The relative value of the omitted terms is evaluated, for example, in Ref.[4]. In RWA the Hamiltonian (3.6) takes form

$$H_{\text{RWA}} = \hbar\omega_0 R_3/2 + \sum_{\mathbf{k}} \hbar\omega_{\mathbf{k}} a_{\mathbf{k}}^{+} a_{\mathbf{k}} - \sum_{\mathbf{k}} (g_{\mathbf{k}} R_{\mathbf{k}}^{+} a_{\mathbf{k}} + g_{\mathbf{k}}^{+} a_{\mathbf{k}}^{+} R_{\mathbf{k}}^{-}) \quad (3.11)$$

where

$$R_{\mathbf{k}}^{\pm} = \sum_{i=1}^{N} \sigma_{i\pm} \exp(\pm i\mathbf{k}\mathbf{r}_i). \qquad (3.12)$$

In (3.11) we have ignored in inhomogeneous broadening ($\omega_{0i} = \omega_0$), and we have used normal operator ordering. In normal operator ordering the annihilation operators $a_{\mathbf{k}}$ are always on the right side of the operator products whereas the creation operators $a_{\mathbf{k}}^{+}$ are always on the left side of products. This operator ordering is not essential in the atom–field representation, but it is very important in the boson operator exclusion method, when we average the operator equations of motion over the initial quantum states.

The operator equations of motion in RWA have the following form

$$\frac{da_{\mathbf{k}}}{dt} = -i\omega_{\mathbf{k}} a_{\mathbf{k}} + \frac{i}{\hbar} g_{\mathbf{k}}^{*} R_{\mathbf{k}}^{-}, \qquad (3.13a)$$

$$\frac{dR_{\mathbf{k}}^{+}}{dt} = i\omega_0 R_{\mathbf{k}}^{+} + \frac{i}{\hbar} \sum_{\mathbf{k}'} g_{\mathbf{k}'}^{*} a_{\mathbf{k}'}^{+} R_{3,\mathbf{k}-\mathbf{k}'}, \qquad (3.13b)$$

$$\frac{dR_{3,\Delta\mathbf{k}}}{dt} = \frac{2i}{\hbar} \sum_{\mathbf{k}'} (g_{\mathbf{k}'} R_{\mathbf{k}'+\Delta\mathbf{k}}^{+} a_{\mathbf{k}'} - g_{\mathbf{k}'}^{*} a_{\mathbf{k}'}^{+} R_{\mathbf{k}'-\Delta\mathbf{k}}^{-}), \qquad (3.13c)$$

where

$$R_{3,\Delta\mathbf{k}} = \sum_{i=1}^{N} \sigma_{i3} \exp(i\Delta\mathbf{k}\mathbf{r}_i). \qquad (3.14)$$

2.3.3. *Integrals of motion in rotating wave approximation.* Let us multiply the equation (3.13b) by $R_{\mathbf{k}}^{-}$ from the right, and add it to the Hermitian conjugated equation multiplied by $R_{\mathbf{k}}^{+}$ from the left. In consequence, we get

$$\frac{d}{dt}(R^+_{\mathbf{k}}R^-_{\mathbf{k}}) = \frac{i}{\hbar}\sum_{\mathbf{k}'}(g^*_{\mathbf{k}'}a^+_{\mathbf{k}'}R_{3,\mathbf{k}-\mathbf{k}'}R^-_{\mathbf{k}} - g_{\mathbf{k}'}R^+_{\mathbf{k}}R_{3,\mathbf{k}'-\mathbf{k}}\,a_{\mathbf{k}'}). \quad (3.15)$$

The equation for the operator product $R^-_{\mathbf{k}}R^+_{\mathbf{k}}$ can be obtained analogically:

$$\frac{d}{dt}(R^-_{\mathbf{k}}R^+_{\mathbf{k}}) = \frac{i}{\hbar}\sum_{\mathbf{k}'}(g^*_{\mathbf{k}'}a^+_{\mathbf{k}'}R^-_{\mathbf{k}}R_{3,\mathbf{k}-\mathbf{k}'} - g_{\mathbf{k}'}R_{3,\mathbf{k}'-\mathbf{k}}R^+_{\mathbf{k}}a_{\mathbf{k}'}). \quad (3.16)$$

Adding Eqs. (3.15) and (3.16) and summing them over $\mathbf{k}$ we get the following relationship

$$\frac{d}{dt}\sum_{\mathbf{k}}(R^+_{\mathbf{k}}R^-_{\mathbf{k}} + R^-_{\mathbf{k}}R^+_{\mathbf{k}}) = -\frac{1}{2}\sum_{\Delta\mathbf{k}}\frac{d}{dt}(R_{3,\Delta\mathbf{k}}R_{3,-\Delta\mathbf{k}}). \quad (3.17)$$

Summing over $\Delta\mathbf{k}$ in the right hand of Eq. (3.17) appeared in result of the following transformation

$$\sum_{\mathbf{k}}\sum_{\mathbf{k}'} \Rightarrow \sum_{\Delta\mathbf{k}}\sum_{\mathbf{k}''}$$

where $\Delta\mathbf{k} = \pm(\mathbf{k} - \mathbf{k}')$. It should be noted that $\sigma_{i\pm}$ are the rapidly varying functions of $\mathbf{r}_i$ with the oscillation period λ, whereas σ_{i3} are the slowly varying functions of $\mathbf{r}_i$. Hence the functions $R^{\pm}_{\mathbf{k}}$ and $R_{3\mathbf{k}}$ have their maxima in the different regions of $|\mathbf{k}|$. To stress this we put

$$\Delta\mathbf{k} = \mathbf{k} - \mathbf{k}_0.$$

When we are concerned with the investigations of superradiance of the needle shaped active volumes, it is naturally to assume that the wave vector $\mathbf{k}_0$ is parallel to a sample axis and that its value is determined by $|\mathbf{k}_0| = \omega/c$.Finally, we get the following integral of motion which is exact in the RWA approximation and is valid for the active volumes of the arbitrary dimensions:

$$\sum_{\mathbf{k}}\{2(R^+_{\mathbf{k}}R^-_{\mathbf{k}} + R^-_{\mathbf{k}}R^+_{\mathbf{k}}) + R_{3,\mathbf{k}-\mathbf{k}_0}R_{3,-\mathbf{k}+\mathbf{k}_0}\} = \sum_{\mathbf{k}}R^2_{\mathbf{k}} = \text{const}, (3.18)$$

where

$$R_{\mathbf{k}}^2 = \sum_i \sum_j \{2(\sigma_{i+}\sigma_{j-} + \sigma_{j-}\sigma_{i+}) \exp(i\mathbf{k}\mathbf{r}_{ij})$$

$$+ \sigma_{i3}\sigma_{j3}\exp[i(\mathbf{k}-\mathbf{k}_0)\mathbf{r}_{ij}]\}, \quad i,j = 1,...,N. \tag{3.19}$$

The second integral of motion is inherently related to the energy conservation law and it follows from the Eqs. (3.13):

$$\frac{d}{dt}(a_{\mathbf{k}}^{+}a_{\mathbf{k}}) = \frac{i}{\hbar}(g_{\mathbf{k}}^{*}a_{\mathbf{k}}^{+}R_{\mathbf{k}}^{-} - g_{\mathbf{k}}R_{\mathbf{k}}^{+}a_{\mathbf{k}}). \tag{3.20}$$

Summing (3.20) over $\mathbf{k}$ we get

$$\frac{d}{dt}(\sum_{\mathbf{k}} a_{\mathbf{k}}^{+}a_{\mathbf{k}}) = \frac{i}{\hbar}\sum_{\mathbf{k}}(g_{\mathbf{k}}^{*}a_{\mathbf{k}}^{+}R_{\mathbf{k}}^{-} - g_{\mathbf{k}}R_{\mathbf{k}}^{+}a_{\mathbf{k}}) = -\frac{1}{2}\frac{dR_3}{dt}, \tag{3.21}$$

or

$$\frac{d}{dt}\left[\frac{R_3}{2} + \sum_{\mathbf{k}} a_{\mathbf{k}}^{+} a_{\mathbf{k}}\right] = 0. \tag{3.22}$$

Hence

$$\frac{R_3}{2} + \sum_{\mathbf{k}} a_{\mathbf{k}}^{+}a_{\mathbf{k}} = \text{const.} \tag{3.23}$$

Thus, in the RWA approximation we have the two integrals of motion (see Eq. (3.18) and Eq. (3.23)). The first one reflects the conservation of the modified Bloch vector length (3.19) and it is valid only under conditions (2.14) and (2.15a). It should be emphasized that we have assumed the validity of these conditions in deriving of Eqs. (3.8) and (3.13). The second integral of motion shows that the net number of the emitted photons and excited atoms is a constant. In fact, we can rewrite it in the form

$$n + N_2 = \text{const}, \tag{3.24}$$

where $n = \sum_{\mathbf{k}} n_{\mathbf{k}}$,and $n_{\mathbf{k}} = a_{\mathbf{k}}^{+}a_{\mathbf{k}}$ is an operator of the number of photons in the field mode with the wave vector $\mathbf{k}$, and $N_2 = (N + R_3)/2$ is an operator of the number of excitations, the mean value of the latter operator is equal to the number of the atoms in the upper state.

2.4. The boson operator exclusion method.

Up to now we have derived the equations both for the atomic and field subsystems. However, the dynamics of the field subsystem in the process of the radiative decay is in strong correlation with the dynamics of the atomic subsystem. Therefore, it is often convenient to exclude the variables referring to the field subsystem and to study only the behavior of the atomic subsystem. The radiators interact via the retarded electromagnetic field. However, the nature of this interaction will manifest itself only in the form of the coefficients in the interaction matrix.

Let us consider again Eqs. (1.10). We will take into account only the transversal part of the electromagnetic field which obeys to the condition of the Coulomb gauge:

$$\text{div}\ \mathbf{A}_{\perp} = 0. \tag{4.1}$$

In this case, we get from (1.10a,b) the following equation

$$\left[\Delta - \frac{1}{c^2} \frac{\partial^2}{\partial t^2} \right] \mathbf{A}_{\perp}(\mathbf{r},t) = - \frac{4\pi}{c} \mathbf{j}_{\perp}(\mathbf{r},t) \ , \tag{4.2}$$

where Δ is an Laplace operator. The solution of the Eq. (4.2) can be written in the form [5]

$$\mathbf{A}(\mathbf{r},t) = \mathbf{A}_0(\mathbf{r},t) + \frac{1}{c} \int \frac{1}{|\mathbf{r} - \mathbf{r}'|} \mathbf{j}\left(\mathbf{r}', t - \frac{|\mathbf{r} - \mathbf{r}'|}{c} \right) d\mathbf{r}' \tag{4.3}$$

where $\mathbf{A}_0(\mathbf{r},t)$ is a vacuum field. The vector–potential operator of the vacuum field can be expended in the plane waves, similarly to Eqs. (3.1) – (3.4):

$$\mathbf{A}_0(\mathbf{r},t) = \sum_{\mathbf{k}} (2\pi\hbar c^2/W\omega_{\mathbf{k}})^{1/2} \mathbf{e}_{\mathbf{k}} [b_{\mathbf{k}}(t)\, e^{i\mathbf{kr}} + b^+_{\mathbf{k}}(t)\, e^{-i\mathbf{kr}}], \tag{4.4}$$

where the creation, $b^+_{\mathbf{k}}$, and annihilation, $b_{\mathbf{k}}$,operators under averaging over the vacuum field state $|0>$ obey the following rules

$$<0|b^+_{\mathbf{k}}b_{\mathbf{k}'}|0> = 0, \qquad <0|b_{\mathbf{k}}b^+_{\mathbf{k}'}|0> = \delta_{\mathbf{k},\mathbf{k}'} \,. \tag{4.5}$$

The second term in Eq. (4.3) represents the component of the electromagnetic field associated with the radiation sources. Substituting the Eqs. (1.7), (1.8) in (4.3), we get for this field component [6]

$$A_{s\lambda}(\mathbf{r}_i,t) = \frac{1}{c}\sum_j r_{ij}^{-1}\,[(\mathbf{e}_\lambda\mathbf{m}_j)\sigma_{j+}(t - r_{ij}/c)$$

$$+ (\mathbf{e}_\lambda\mathbf{m}^*_j)\sigma_{j-}(t - r_{ij}/c)], \tag{4.6}$$

where $\mathbf{e}_\lambda$ ($\lambda = 1,2$) are the polarization vectors. These vectors comprise with the wave vector $\mathbf{k}$ an orthogonal vector triad.

If (4.3) and (4.6) are applied to (1.10) the following operator equations of motion are obtained -

$$\frac{d\sigma_{i+}}{dt} = \frac{i}{\hbar c}\sum_\lambda m^*_{i\lambda}A_{o\lambda}\sigma_{i3} + i\omega_{oi}\sigma_{i+} \tag{4.7a}$$

$$+ \frac{i}{\hbar c^2}\sum_\lambda m^*_{i\lambda}\sum_j \frac{1}{r_{ij}}\left[m_{j\lambda}\sigma_{j+}\left(t - \frac{r_{ij}}{c}\right) + m^*_{j\lambda}\sigma_{j-}\left(t - \frac{r_{ij}}{c}\right)\right]\sigma_{i3}\,,$$

$$\frac{d\sigma_{i3}}{dt} = \frac{2i}{\hbar c}\sum_\lambda (m_{i\lambda}\sigma_{i+} - m^*_{i\lambda}\sigma_{i-})A_{0\lambda}(\mathbf{r}_i,t) + \frac{2i}{\hbar c^2}\sum_\lambda (m_{i\lambda}\sigma_{i+} - m^*_{i\lambda}\sigma_{i-})$$

$$\times \sum_j \frac{1}{r_{ij}}\left[m_{j\lambda}\sigma_{j+}\left(t - \frac{r_{ij}}{c}\right) + m^*_{j\lambda}\sigma_{j-}\left(t - \frac{r_{ij}}{c}\right)\right], \tag{4.7b}$$

where $m_{i\lambda} = (\mathbf{e}_\lambda\mathbf{m}_i)$. These equations describe the processes of the interaction of an ensemble of the two–level atoms with the transversal electromagnetic field. In RWA these equations take the form ($j = 1,\ldots,N$)

$$\frac{d\sigma_{i+}}{dt} = \frac{i}{\hbar c}\sum_\lambda m^*_{i\lambda}A^*_{0\lambda}(\mathbf{r}_i,t)\sigma_{i3} + i\omega_{0i}\sigma_{i+}$$

$$+\frac{i}{\hbar c^2}\sum_{\lambda}\sum_{j}\frac{1}{r_{ij}}m^{*}_{i\lambda}m_{j\lambda}\sigma_{j+}\left(t-\frac{r_{ij}}{c}\right)\sigma_{i3}(t), \tag{4.8a}$$

$$\frac{d\sigma_{i3}}{dt}=\frac{2i}{\hbar c}\sum_{\lambda}\left[m_{i\lambda}\sigma_{i+}A^{-}_{0\lambda}(\mathbf{r}_i,t)-m^{*}_{i\lambda}A^{+}_{0\lambda}(\mathbf{r}_i,t)\sigma_{i-}\right]$$

$$+\frac{2i}{\hbar c^2}\sum_{\lambda}\sum_{j}\frac{1}{r_{ij}}\left[m_{i\lambda}m^{*}_{j\lambda}\sigma_{i+}(t)\sigma_{j-}\left(t-\frac{r_{ij}}{c}\right)\right.$$

$$\left.-m^{*}_{i\lambda}m_{j\lambda}\sigma_{i-}(t)\sigma_{j+}\left(t-\frac{r_{ij}}{c}\right)\right]. \tag{4.8b}$$

We see that the dynamics of the atomic subsystem is determined by the interatomic interactions and its interaction with the vacuum field.

2.5. Semiclassical System of Equations

2.5.1. *Three dimensional model.* Consider again Eqs. (1.10). Let us average these equations over the initial density matrix corresponding to the vacuum state of the field and the given initial state of the atomic subsystem. The first two equations, (1.10a) and (1.10b), can be combined in one equation taking the form

$$\text{rot rot }\mathbf{A}(\mathbf{r},t)+\frac{1}{c^2}\frac{\partial^2\mathbf{A}(\mathbf{r},t)}{\partial t^2}=\frac{4\pi}{c}\mathbf{j}(\mathbf{r},t), \tag{5.1}$$

where

$$\mathbf{A}(\mathbf{r},t)=\langle\hat{\mathbf{A}}(\mathbf{r},t)\rangle,\quad \mathbf{j}(\mathbf{r},t)=\langle\hat{\mathbf{j}}(\mathbf{r},t)\rangle.$$

Further we will consider only the transversal e.m. fields satisfying Eq. (4.1). In this case Eq. (5.1) can be rewritten in the following form

$$\left[\Delta-\frac{1}{c^2}\frac{\partial^2}{\partial t^2}\right]A_{\lambda}=-\frac{4\pi}{c}j_{\lambda}, \tag{5.2}$$

where

$$\mathbf{A}=\sum_{\lambda}\mathbf{e}_{\lambda}A_{\lambda},\quad j_{\lambda}=(\mathbf{e}_{\lambda}\mathbf{j}),\quad \lambda=1,2,$$

and the unit polarization vectors $\mathbf{e}_\lambda$ form with the wave vector $\mathbf{k}$ the right vector triad. To incorporate the inhomogeneous broadening effects we will use a normalized inhomogeneous line broadening function $G(\omega_0)$, which obeys the following normalization rule

$$\int_0^\infty G(\omega_0)d\omega_0 = 1, \tag{5.3}$$

or the detuning distribution function

$$f(\Delta) = G(<\omega_0> + \Delta) \tag{5.4}$$

with the normalization rule

$$\int_{-\infty}^\infty f(\Delta)d\Delta = 1. \tag{5.5}$$

The procedure of incorporation of the inhomogeneous broadening effects described by the function $f(\Delta)$ can be done by the following manner. Let us introduce the new operators $\rho_{i\pm}$:

$$\sigma_{i\pm} = \rho_{i\pm}\, e^{\pm i\omega_0 t}. \tag{5.6}$$

In this case the current density can be rewritten in the following form

$$\mathrm{j}(\mathbf{r},t) = \mathrm{j}^+(\mathbf{r},t)\, e^{i\omega_0 t} + \mathrm{j}^-(\mathbf{r},t)\, e^{-i\omega_0 t}$$

$$= \sum_i \left[\mathrm{m}_i <\rho_{i+}(t)> e^{i\omega_0 t} + \mathrm{m}_i^* <\rho_{i-}(t)> e^{-i\omega_0 t} \right] \delta(\mathbf{r} - \mathbf{r}_i)$$

$$= \int_{-\infty}^\infty \mathrm{j}^+(\Delta,\mathbf{r},t) f(\Delta)d\Delta\; e^{i\omega_0 t} + \text{c.c.} \tag{5.7}$$

where

$$\mathrm{j}^+(\Delta,\mathbf{r},t) = \sum_{i'} \mathrm{m}_{i'} <\rho_{i'+}>\, \delta(\mathbf{r} - \mathbf{r}_{i'}) \tag{5.8}$$

Only the atoms with the frequencies $\omega_{0i'} = \omega_0 + \Delta$ are taken into account in summing over i' in Eq. (5.8). The equation of motion for $\mathrm{j}^+(\Delta,\mathbf{r},t)$ can be obtained from Eq. (1.10c) and it has the form

$$\frac{\partial}{\partial t}\, \mathbf{j}^{+}(\Delta,\mathbf{r},t) = i\Delta \mathbf{j}^{+}(\Delta,\mathbf{r},t)$$

$$+ \frac{i}{\hbar c}\, \mathbf{m}(\mathbf{m}^{*}\mathbf{A}(\mathbf{r},t))\ \rho(\Delta,\mathbf{r},t)\ \exp(-i\omega_{0}t), \tag{5.9}$$

where

$$\rho(\Delta,\mathbf{r},t) = \sum_{i'} <\sigma_{i'3}>\delta(\mathbf{r} - \mathbf{r}_{i'}), \quad \omega_{0i'} = \omega_{0} + \Delta, \tag{5.10}$$

and we have assumed for a simplicity, that $\mathbf{m}_{i} = \mathbf{m}$. Analogically, from Eq. (1.10d) we get the following equation of motion for $\rho(\Delta,\mathbf{r},t)$:

$$\frac{\partial}{\partial t}\, \rho(\Delta,\mathbf{r},t) = \frac{2i}{\hbar c}\, [\mathbf{j}^{+}(\Delta,\mathbf{r},t)\exp(i\omega_{0}t)$$

$$-\mathbf{j}^{-}(\Delta,\mathbf{r},t)\exp(-i\omega_{0}t)]\mathbf{A}(\mathbf{r},t). \tag{5.11}$$

Under derivation of Eqs. (5.9), (5.11) we have used the following type of decoupling of correlators

$$<A\ \sigma_{i3}> = <A><\sigma_{i3}>, \quad <A\ \sigma_{i\pm}> = <A><\sigma_{i\pm}>. \tag{5.12}$$

This type of decoupling assumes the weakness of the correlations between the field and atomic subsystems. In this case the field and atomic variables manifest themselves as the classical variables, and their fluctuations are supposed to be negligible. As a result of these assumptions the semiclassical equations of motion do not include the members describing the spontaneous radiation process.

Up to now we have averaged the equations over the density matrix of a system including the resonant atoms and field only. If we take into account the interaction of this system with a thermostat then we get the additional relaxation terms in Eqs. (5.5), (5.10). The resulting equations of motion are

$$\left[\Delta - \frac{1}{c^{2}}\frac{\partial^{2}}{\partial t^{2}}\right]A_{\lambda} = -\frac{4\pi}{c}\int [j_{\lambda}^{+}(\Delta)e^{i\omega_{0}t} + j_{\lambda}^{-}(\Delta)e^{-i\omega_{0}t}]f(\Delta)d\Delta \tag{5.13a}$$

$$\frac{\partial}{\partial t} j_{\lambda}^{+}(\Delta) + \left(-i\Delta + \frac{1}{2T_{2}}\right)j_{\lambda}^{+}(\Delta) = \frac{i}{\hbar c}\, m_{\lambda} \sum_{\lambda'} m_{\lambda'}^{*}A_{\lambda'}e^{-i\omega_{0}t}\rho(\Delta), \tag{5.13b}$$

$$\frac{\partial}{\partial t}\rho(\Delta) + \frac{\rho(\Delta)-\rho_e(\Delta)}{T_1} = \frac{2i}{\hbar c}\sum_\lambda [j^+_\lambda(\Delta)e^{i\omega_0 t} - j^-_\lambda(\Delta)e^{-i\omega_0 t}]A_\lambda , \quad (5.13c)$$

where T_2 is a homogeneous transversal relaxation time, T_1 is a longitudinal relaxation time, $\rho_e(\Delta)$ is an equilibrium value of $\rho(\Delta)$.

The equations (5.13) must be supplemented by the initial and boundary conditions. The boundary conditions, corresponding to the continuity of the vector potential $A(\mathbf{r},t)$ and its normal derivative at the active volume boundary, are

$$A(\mathbf{r},t)\big|_{\mathbf{r}\in Z} = A^{(e)}(\mathbf{r},t)\big|_{\mathbf{r}\in Z}, \quad (5.14a)$$

$$\mathbf{n}(\mathbf{r})\frac{\partial}{\partial \mathbf{r}}A(\mathbf{r},t)\big|_{\mathbf{r}\in Z} = \mathbf{n}(\mathbf{r})\frac{\partial}{\partial \mathbf{r}}A^{(e)}(\mathbf{r},t)\big|_{\mathbf{r}\in Z}, \quad (5.14b)$$

where $Z(\mathbf{r})$ is an active volume surface, and $\mathbf{n}(\mathbf{r})$ is the unit vector of the outer normal to the boundary at the point with the radius –vector $\mathbf{r}$, $A^{(e)}(\mathbf{r},t)$ is a solution of the equation

$$\left[\Delta - \frac{1}{c^2}\frac{\partial^2}{\partial t^2}\right]A^{(e)}(\mathbf{r},t) = 0 \quad (5.15)$$

describing the field outside the volume V bounded by the surface $Z(\mathbf{r})$. The field $A^{(e)}(\mathbf{r},t)$ must satisfy the radiation condition at the infinity ($|\mathbf{r}| \longrightarrow \infty$).

The initial conditions for SR should correspond to the absence of the field and polarization at the initial moment and some excited state of the atomic subsystem. However, for these initial conditions the equations (5.13) have only the trivial solution. This is a result of decoupling (5.12), which does not take into account the spontaneous decay of a system of excited atoms. Thus, we should introduce the phenomenological sources of a spontaneous emission or use the nonzero initial conditions for the field and polarization. For example, the initial conditions can be taken in the following form

$$A(\mathbf{r},0) = 0, \quad j^+(\mathbf{r},t) = j_0, \quad \rho(\mathbf{r},0) = \rho_0. \quad (5.16)$$

Here, j_0 is determined by the initial value of Bloch vector (see Chapter 1), and $\rho_0(\mathbf{r})$ is the initial distribution of a population inversion density. The choice of j_0 is simply an artificial way of instigating the semiclassical calculations, and reasonable variations in its value do not strongly affect the results.

2.5.2. *One-dimensional model.* The active volume of a superradiant medium has usually the form of an extended cylinder with the Fresnel number of unity, approximately. Because of this, the field $A(\mathbf{r},t)$ can be represented by

$$A(\mathbf{r},t) = A^{+}(x,t)\exp[i(\omega_0 t - \mathbf{k}_0\mathbf{r})]$$
$$+ A^{-}(x,t)\exp[-i(\omega_0 t - \mathbf{k}_0\mathbf{r})], \qquad (5.17)$$

where the direction of the wave vector $\mathbf{k}_0$ coincides with the cylinder axis (axis x), and its value is determined by the following relationship $|\mathbf{k}_0| = \omega_0/c = \kappa$. The slowly varying amplitudes, $A^{\pm}(x,t)$, satisfy the following conditions

$$|\partial A^{\pm}/\partial x| \ll \kappa|A^{\pm}|, \quad |\partial A^{\pm}/\partial t| \ll \omega_0|A^{\pm}|. \qquad (5.18)$$

If we neglect, in accordance with (5.18), the second order derivatives of $A^{\pm}(x,t)$, and average the equations (5.13) over volume V_1, with the dimensions L_1 much longer than the wavelength of radiation and much shorter than the characteristic amplification length, then we get from (5.13) the following equations of motion in RWA

$$\left(\frac{\partial}{\partial t} + c\frac{\partial}{\partial x}\right)A_{\lambda}^{+} = -i\frac{2\pi}{\kappa}m_{\lambda}\int \mathcal{R}(\Delta)f(\Delta)d\Delta\ , \qquad (5.19a)$$

$$\frac{\partial \mathcal{R}^{+}(\Delta)}{\partial t} + \left(-i\Delta + \frac{1}{2T_2}\right)\mathcal{R}(\Delta) = \frac{i}{\hbar c}\sum_{\lambda} m_{\lambda}^{*}A_{\lambda}^{+}\rho(\Delta)\ , \qquad (5.19b)$$

$$\frac{\partial\rho(\Delta)}{\partial t} + \frac{\rho(\Delta)-\rho_e(\Delta)}{T_1} = \frac{2i}{\hbar c}\sum_{\lambda}[m_{\lambda}\mathcal{R}^{+}(\Delta)A_{\lambda}^{-} - m_{\lambda}^{*}A_{\lambda}^{+}\mathcal{R}^{-}(\Delta)], \qquad (5.19c)$$

where

$$\mathcal{R}^{\pm}(\Delta) = V_1^{-1}\sum_i \langle\rho_{i\pm}\rangle\exp(\pm i\mathbf{k}_0\mathbf{r}_i)\ , \qquad \rho(\Delta) = V_1^{-1}\sum_i \langle\sigma_{i3}\rangle.$$

A summation over i is performed over the atoms which are in the averaging volume, i.e., $\mathbf{r}_i \in V_1(x)$, and the frequencies of which satisfy the relationship $\omega_{0i} = \omega_0 + \Delta$.

Eq. (5.17) corresponds to the one–wave approximation. In fact, we have taken into account only the wave travelling in the positive direction of the x axis and ignore the oncoming wave. In the case of SR both waves are equivalent and it is more correct to write the vector potential $A(\mathbf{r},t)$ in the following form

$$A(\mathbf{r},t) = \sum_l A_l(x,t)\exp[i(\omega_0 t - \mathbf{k}_l\mathbf{r})] + \text{c.c.}, \tag{5.20}$$

where

$$\mathbf{k}_l = (-1)^l\,\mathbf{k}_0 , \quad l = 1,2. \tag{5.21}$$

In the two–wave approximation the equations of motion for the resonant light–matter interaction become:

$$\left[\frac{\partial}{\partial t} + (-1)^l\, c\,\frac{\partial}{\partial x}\right] A_{l\lambda} = -i\,\frac{2\pi}{\kappa}\, m_\lambda \int \mathcal{R}^{+}_l(\Delta) f(\Delta)d\Delta\ , \tag{5.22a}$$

$$\frac{\partial}{\partial t}\,\mathcal{R}^{+}_l(\Delta) + \left[-i\Delta + \frac{1}{2T_2}\right]\mathcal{R}^{+}_l(\Delta) = \frac{i}{\hbar c}\sum_\lambda m^*_\lambda A_{l\lambda}\rho(\Delta)\ , \tag{5.22b}$$

$$\frac{\partial}{\partial t}\,\rho(\Delta) + \frac{\rho(\Delta)-\rho_e(\Delta)}{T_1} = \frac{2i}{\hbar c}\sum_l\sum_\lambda\,[m_\lambda \mathcal{R}^{+}_l(\Delta)A^*_{l\lambda} - m^*_\lambda A_{l\lambda}\,\bar{\mathcal{R}}_l(\Delta)]. \tag{5.22c}$$

The boundary conditions for the left, $x = 0$, and right, $x = L$, boundaries are

$$A_{2\lambda}(0,t) = A_{1\lambda}(L,t) = 0. \tag{5.23}$$

The initial conditions coincide with (1.16):

$$A_{l\lambda}(x,0) = 0, \quad \mathcal{R}^{+}_l(x,0) = \mathcal{R}^{+}_0, \quad \rho(x,0) = \rho_0. \tag{5.24}$$

2.5.3. *Solution of linearized semiclassical equations.* Let consider first the initial stage of a radiative decay, when

$$\rho_0(\Delta) - \rho(\Delta) \ll \rho_0(\Delta). \tag{5.25}$$

If we use this approximation and substitute the solution of Eq. (5.22b) into Eq. (5.22a) then the equation describing the dynamics of the right wave takes the form

$$\left(\frac{\partial}{\partial t} + c\frac{\partial}{\partial x}\right)A(x,t) = \int d\Delta\, f(\Delta)\,\frac{2\pi}{\hbar\omega}\,|m|^2\rho_0(\Delta)$$

$$\times \int_0^t A(x,t')\exp\left[\left(\mathrm{i}\Delta - \frac{1}{2T_2}\right)(t - t')\right]dt'$$

$$- \mathrm{i}\,\frac{2\pi}{\kappa}\,|m|^2 \int d\Delta\, f(\Delta)\,\mathcal{R}_0^{+}(\Delta)\exp\left[\left(\mathrm{i}\Delta - \frac{1}{2T_2}\right)t\right], \qquad (5.26)$$

where

$$A(x,t) = \sum_\lambda m_\lambda^* A_\lambda(x,t)\,, \qquad |m|^2 = \sum_\lambda |m_\lambda|^2.$$

Let us apply the Laplace transformation to the variable $A(x,t)$:

$$a(x,p) = \int_0^\infty \mathrm{e}^{-pt}A(x,t)\,dt. \qquad (5.27)$$

The inverse transformation is

$$A(x,t) = \frac{1}{2\pi \mathrm{i}}\int_{\delta-\mathrm{i}\infty}^{\delta+\mathrm{i}\infty} \mathrm{e}^{pt}a(x,p)dp\,. \qquad (5.28)$$

The Laplace transform of $A(x,t)$ satisfies the equation

$$\frac{da}{dx} + \frac{1}{c}\left(p - \frac{2\pi}{\hbar\omega}\,|m|^2\rho_0 G(p)\right)a = -\mathrm{i}\,\frac{2\pi}{\omega}\,|m|^2 G(p)\,\mathcal{R}_0^{+}\,, \qquad (5.29)$$

where the following designations have been used

$$G(p) = \int_{-\infty}^{\infty} f(\Delta)\left(p + \frac{1}{2T_2} - \mathrm{i}\Delta\right)^{-1} d\Delta.$$

Integrating (5.29) and applying the inverse Laplace transformation (5.28), we get

$$A(x,t) = \mathrm{i}\,\frac{2\pi}{\kappa}\,|m|^2\mathcal{R}_0^{+}\,\frac{1}{2\pi\mathrm{i}}\int_{\delta-\mathrm{i}\infty}^{\delta+\mathrm{i}\infty} dp\; G(p)\left(p - \frac{2\pi}{\hbar\omega}\,|m|^2\rho_0 G(p)\right)^{-1}$$

$$\times \left\{\exp\left[p\left(t - \frac{x}{c}\right) + \frac{2\pi}{\hbar\omega}\,|m|^2\rho_0 G(p)\,\frac{x}{c}\right] - \exp(pt)\right\}. \qquad (5.30)$$

$$\times \left\{ \exp\left[p\left(t - \frac{x}{c} \right) + \frac{2\pi}{\hbar\omega} |m|^2 \rho_0 G(p) \frac{x}{c} \right] - \exp(pt) \right\} . \qquad (5.30)$$

If we restrict ourselves, for a simplicity, to the case of the homogeneously broaden systems, when $f(\Delta)$ is determined by $f(\Delta) = \delta(\Delta)$, then the expression (5.30) takes the more compact form

$$A(x,t) = i \frac{2\pi}{\kappa} |m|^2 \mathcal{R}_0 \frac{1}{2\pi i} \int_{\delta - i\infty}^{\delta + i\infty} dp \left\{ \exp\left[p\left(t - \frac{x}{c} \right) + \frac{\alpha}{p - p_0} \frac{x}{c} \right] - \exp(pt) \right\}$$

$$\times \left[(p - p_1)(p - p_2) \right]^{-1} , \qquad (5.31)$$

where

$$p_{1,2} = \frac{1}{2} p_0 \pm \left(\frac{1}{4} p_0^2 + \alpha \right)^{1/2} , \quad p_0 = -\frac{1}{2T_2} , \quad \alpha = \frac{2\pi}{\hbar\omega} |m|^2 \rho_0 .$$

Performing the inverse Laplace transformation in (5.31), finally, we get for $A(x,t)$ the following equation

$$A(x,t) = i \frac{2\pi}{\kappa(p_2 - p_1)} |m|^2 \mathcal{R}_0 \int_0^{t - x/c} d\tau \, I_0\left[2\sqrt{\alpha \frac{x}{c} \tau} \right]$$

$$\times \left\{ p_2 \exp\left[p_1\left(t - \frac{x}{c} \right) + p_2 \tau \right] - p_1 \exp\left[p_2\left(t - \frac{x}{c} \right) + p_1 \tau \right] \right\} . \qquad (5.32)$$

2.6. Dynamics of Number of Photons

2.6.1. *General case.* The main disadvantage of a semiclassical theory of the electromagnetic waves interaction with the resonant media consists in application of the decoupling procedure (5.12). This type of decoupling, as we discussed earlier, results in the essential weakness of the correlations between the individual atoms. If we apply this decoupling we lose the processes of a back action of the photon on the atom which emitted it. It means that we lose the processes of a spontaneous radiative decay. We can illustrate this by the following manner. As we can see from Eq. (4.8b), the process of the photon back action is described by the operator $\sigma_{i+}\sigma_{i-}$ in the right hand side of this equation. If we average this operator product over the pure quantum state $|+\rangle_i$, then, in accordance with (5.12), we get

$$\langle + | \sigma_+ \sigma_- | + \rangle = \langle + | \sigma_+ | + \rangle \langle + | \sigma_- | + \rangle = 0 .$$

At the same time, according to the properties of the Pauli matrices, we obtain

$$<+|\sigma_+\sigma_-|+> = 1/2 <+|(I + \sigma_3)|+> = 1.$$

In order to incorporate in our treatment the processes of the spontaneous decay and, hence, not to lose the photon back action effects, we try to obtain the equations of motion for the second order correlators of the atomic and field operators. The most important example of the second order correlators is given by the mean value of the operator of the number of photons in the field mode **k**:

$$n_{\mathbf{k}} = <a^+_{\mathbf{k}} a_{\mathbf{k}}>.$$

Using Eqs. (3.13) we can get the following equations [7]:

$$\frac{dn_{\mathbf{k}}}{dt} = F_{\mathbf{k}}, \tag{6.1a}$$

$$\frac{dF_{\mathbf{k}}}{dt} = i(\omega_{\mathbf{k}} - \omega_0)G_{\mathbf{k}} + \frac{2|g_{\mathbf{k}}|^2}{\hbar^2} <R^+_{\mathbf{k}} R^-_{\mathbf{k}}>$$

$$+ \hbar^{-2} \sum_{\mathbf{k}'} (g^*_{\mathbf{k}} g_{\mathbf{k}'} <a^+_{\mathbf{k}} R_{3,\mathbf{k}'-\mathbf{k}} a_{\mathbf{k}'}> + g_{\mathbf{k}} g^*_{\mathbf{k}'} <a^+_{\mathbf{k}'} R_{3,\mathbf{k}-\mathbf{k}'} a_{\mathbf{k}}>), \tag{6.1b}$$

$$\frac{dG_{\mathbf{k}}}{dt} = i(\omega_{\mathbf{k}} - \omega_0)F_{\mathbf{k}}$$

$$+ \hbar^{-2} \sum_{\mathbf{k}'} (g^*_{\mathbf{k}} g_{\mathbf{k}'} <a^+_{\mathbf{k}} R_{3,\mathbf{k}'-\mathbf{k}} a_{\mathbf{k}'}> + g_{\mathbf{k}} g^*_{\mathbf{k}'} <a^+_{\mathbf{k}'} R_{3,\mathbf{k}-\mathbf{k}'} a_{\mathbf{k}}>), \tag{6.1c}$$

$$\frac{d}{dt}<R^+_{\mathbf{k}} R^-_{\mathbf{k}}> = \frac{i}{\hbar} \sum_{\mathbf{k}'} (g^*_{\mathbf{k}'} <a^+_{\mathbf{k}'} R_{3,\mathbf{k}-\mathbf{k}'} R^-_{\mathbf{k}}> - g_{\mathbf{k}'} <R^+_{\mathbf{k}} R_{3,\mathbf{k}'-\mathbf{k}} a_{\mathbf{k}'}>), \tag{6.1d}$$

$$\frac{dR_3}{dt} = -2 \sum_{\mathbf{k}} F_{\mathbf{k}}, \tag{6.1e}$$

where

$$F_{\mathbf{k}} = (i/\hbar)(g^*_{\mathbf{k}} <a^+_{\mathbf{k}} R^-_{\mathbf{k}}> - g_{\mathbf{k}} <R^+_{\mathbf{k}} a_{\mathbf{k}}>),$$

$$G_{\mathbf{k}} = (i/\hbar)(g^*_{\mathbf{k}} <a^+_{\mathbf{k}} R^-_{\mathbf{k}}> + g_{\mathbf{k}} <R^+_{\mathbf{k}} a_{\mathbf{k}}>),$$

$$R_3 = \sum_i <\sigma_{i3}> , \qquad i = 1,...,N.$$

The obtained equations are not self-consistent because of the following reasons. First of all, the right–hand sides of equations (6.1) contain the correlators of the third order. And, secondly, in addition to the operators of the numbers of photons $a^+_{\mathbf{k}}a_{\mathbf{k}}$ in the mode $\mathbf{k}$ and the population difference operator R_3 , there are the new operators $a^+_{\mathbf{k}}a_{\mathbf{k}'}$ and $R_{3,\mathbf{k}-\mathbf{k}'}$ in the right–hand sides of these equations.

Now we discuss the approximation that enable one to get the equations in the consistent form. We assume, that the mean values $<\sigma_{i3}>$ do not practically depend on the number i . In this case the Fourier transform of a population inversion $R_{3\mathbf{k}}$ can be represented in the following form

$$R_{3,\mathbf{k}'-\mathbf{k}} = \sum_i <\sigma_{i3}> \exp[i(\mathbf{k}' - \mathbf{k})\mathbf{r}_i] \simeq <\sigma_3> \sum_i \exp[i(\mathbf{k}' - \mathbf{k})\mathbf{r}_i]$$

$$= N <\sigma_3>V^{-1}\int_V \exp[i(\mathbf{k}' - \mathbf{k})\mathbf{r}]d\mathbf{r} = R_3 P(\mathbf{k}' - \mathbf{k}). \qquad (6.2)$$

We have introduced the following function

$$P(\mathbf{k}' - \mathbf{k}) = V^{-1}\int_V \exp[i(\mathbf{k}' - \mathbf{k})\mathbf{r}]d\mathbf{r}.$$

It is a Fourier–transform of the active volume V . In fact,

$$P(\mathbf{k}'-\mathbf{k}) = V^{-1}\int_V \exp[i(\mathbf{k}'-\mathbf{k})\mathbf{r}]d\mathbf{r} = V^{-1}\int_W P(\mathbf{r})\exp[i(\mathbf{k}'-\mathbf{k})\mathbf{r}]d\mathbf{r}, \qquad (6.3)$$

where W is an quantization volume, and

$$P(\mathbf{r}) = \begin{cases} 1, & \text{if } \mathbf{r} \in V \\ 0, & \text{if } \mathbf{r} \notin V. \end{cases}$$

The function $P(\mathbf{k}' - \mathbf{k})$ for a cylindrical active volume has the form

$$P(\Delta\mathbf{k}) = 2\,\frac{J_1(\Delta k_\perp R)}{\Delta k_\perp R}\,\frac{\sin(\Delta k_x L/2)}{\Delta k_x L/2}, \qquad (6.4)$$

where $\Delta\mathbf{k} = \mathbf{k}' - \mathbf{k}$, the axis x coincides with the cylinder axis, and $\Delta\mathbf{k}_\perp$ is a projection of vector $\Delta\mathbf{k}$ on a plane perpendicular to the cylinder axis. If $L,R \gg \lambda$ then the function $P(\Delta\mathbf{k})$ is a highly directional function with the peak value of unity and with the characteristic maximum width of $\Delta k_{\parallel} \simeq 1/L$, $\Delta k_{\perp} \simeq 1/R$. Taking these features into account the right–hand side of (6.1b) can be transformed to the following form

$$\sum_{\mathbf{k}'} g^*_{\mathbf{k}} g_{\mathbf{k}'} \langle a^+_{\mathbf{k}} R_{3,\mathbf{k}'-\mathbf{k}} a_{\mathbf{k}'} \rangle \simeq \sum_{\mathbf{k}'} g^*_{\mathbf{k}} g_{\mathbf{k}'} \langle a^+_{\mathbf{k}} R_3 a_{\mathbf{k}'} \rangle P(\mathbf{k}' - \mathbf{k})$$

$$\simeq |g_{\mathbf{k}}|^2 \langle a^+_{\mathbf{k}} R_3 a_{\mathbf{k}} \rangle \sum_{\mathbf{k}'} P(\mathbf{k}' - \mathbf{k}) = |g_{\mathbf{k}}|^2 \langle a^+_{\mathbf{k}} R_3 a_{\mathbf{k}} \rangle W/V. \qquad (6.5)$$

The right–hand side of Eq. (6.1e) is transformed in an analogous fashion.

Let us now turn to the problem of triple correlators decoupling. In decoupling of right–hand side of equations (6.1b,c) we will use again the assumption on the weakness of the correlations between the atomic and field variables. In this case we can use the following form of decoupling

$$\langle a^+_{\mathbf{k}} R_3 a_{\mathbf{k}} \rangle = \langle a^+_{\mathbf{k}} a_{\mathbf{k}} \rangle \langle R_3 \rangle. \qquad (6.6)$$

It is more complex to carry out the procedure of decoupling for the right–hand side of Eq. (6.1d). Here we must decouple the products of operators relating to the atomic subsystem. If we apply the decoupling defined by

$$\langle a^+_{\mathbf{k}'} R_{3,\mathbf{k}-\mathbf{k}'} R^-_{\mathbf{k}} \rangle = \langle R_{3,\mathbf{k}-\mathbf{k}'} \rangle \langle a^+_{\mathbf{k}'} R^-_{\mathbf{k}} \rangle, \qquad (6.7)$$

then we ignore the following properties of Pauli spin matrices [2]:

$$\sigma_{i3}\sigma_{i-} = -\sigma_{i-}\,, \qquad \sigma_{i+}\sigma_{i3} = -\sigma_{i+}. \qquad (6.8)$$

Because of this, we exclude firstly in the triple correlators of the right–hand side (6.1d) the terms in which both atomic operators refer to the same atom:

$$\langle a^+_{\mathbf{k}'} R_{3,\mathbf{k}-\mathbf{k}'} R^-_{\mathbf{k}} \rangle = \sum_i \sum_j \langle a^+_{\mathbf{k}'} \sigma_{i3} \sigma_{j-} \rangle \exp[i(\mathbf{k} - \mathbf{k}')\mathbf{r}_i - i\mathbf{k}\mathbf{r}_j]$$

$$= \sum_i {\sum_j}' \langle a^+_{\mathbf{k}'} \sigma_{i3} \sigma_{j-} \rangle \exp[i(\mathbf{k} - \mathbf{k}')\mathbf{r}_i - i\mathbf{k}\mathbf{r}_j] - \sum_i \langle a^+_{\mathbf{k}'} \sigma_{i-} \rangle \exp(-i\mathbf{k}'\mathbf{r}_i).$$

Here, the primed summation index means that summation is performed over $i \neq j$. The terms with $i \neq j$ are decoupled by the following manner

$$<a^+_{\mathbf{k}'}\sigma_{i3}\sigma_{j-}>|_{i\,j} = <\sigma_{i3}><a^+_{\mathbf{k}'}\sigma_{j-}>. \tag{6.9}$$

The second–order correlators $<R^+_{\mathbf{k}}R^-_{\mathbf{k}}>$ should also be divided into the two parts with $i \neq j$ and $i = j$:

$$<R^+_{\mathbf{k}}R^-_{\mathbf{k}}> = \sum_i \sum_j{}' <\sigma_{i+}\sigma_{j-}>\exp[i(\mathbf{k}(\mathbf{r}_i - \mathbf{r}_j)]$$

$$+ \sum_i <\sigma_{i+}\sigma_{i-}> = s_{\mathbf{k}} + s_0 , \tag{6.10}$$

where $s_0 = (N + R_3)/2$.

The resultant equations of motion, based on the assumptions discussed above, are:

$$\frac{dn_{\mathbf{k}}}{dt} = F_{\mathbf{k}} , \tag{6.11a}$$

$$\frac{dF_{\mathbf{k}}}{dt} + \frac{F_{\mathbf{k}}}{2T_2} = i(\omega_{\mathbf{k}} - \omega_0)G_{\mathbf{k}} + \frac{1}{T^2_{\mathbf{k}}}(S_{\mathbf{k}} + nR_3 + S_0), \tag{6.11b}$$

$$\frac{dG_{\mathbf{k}}}{dt} + \frac{G_{\mathbf{k}}}{2T_2} = i(\omega_{\mathbf{k}} - \omega_0)F_{\mathbf{k}} , \tag{6.11c}$$

$$\frac{dS_{\mathbf{k}}}{dt} + \frac{S_{\mathbf{k}}}{2T_2} = F_{\mathbf{k}}R_3 , \tag{6.11d}$$

$$\frac{dR_3}{dt} = -2 \sum_{\mathbf{k}} F_{\mathbf{k}} , \tag{6.11e}$$

where

$$T^2_{\mathbf{k}} = \hbar^2 V/2|g_{\mathbf{k}}|^2 W = \hbar\omega_{\mathbf{k}} V/4\pi|m|^2 ,$$

$$S_{\mathbf{k}} = s_{\mathbf{k}} \Big/ \sum_{\mathbf{k}'} P(\mathbf{k} - \mathbf{k}') = (V/W)s_{\mathbf{k}} , \quad S_0 = (V/W)s_0 . \tag{6.12}$$

The set of the equations (6.11a–e) is self-consistent. The initial conditions corresponding to the superradiant decay are

$$n_{\mathbf{k}}(0) = 0\ ,\quad F_{\mathbf{k}}(0) = 0\ ,\quad S_{\mathbf{k}}(0) = 0\ ,\quad R_3(0) = R_{30}\ . \qquad (6.13)$$

2.6.2. *One-mode model of superradiance.* An extremely important role for developing of the theory of SR has played the one–mode model. Its importance is explained by the fact, that in the frames of this theory an approximate analytical solution of the nonlinear equations describing the dynamics of SR was obtained. This model is based on the following two main assumptions:

$$\omega_{\mathbf{k}} = \omega_0\ , \qquad (6.14)$$

$$g_{\mathbf{k}} = g_{\mathbf{k}_0}\delta_{\mathbf{k},\mathbf{k}_0}\ . \qquad (6.15)$$

In this case the Hamiltonian (3.11) takes in RWA the following sufficiently simple form [8,9]

$$H_0 = \hbar\omega_0(R_3/2 + a^+_{\mathbf{k}_0}a_{\mathbf{k}_0}) - (g_{\mathbf{k}_0}R^+_{\mathbf{k}_0}a_{\mathbf{k}_0} + g^*_{\mathbf{k}_0}a^+_{\mathbf{k}_0}R^-_{\mathbf{k}_0})\ . \qquad (6.16)$$

In the one–mode approximation the quantization volume W loses any physical meaning. Therefore it is natural to substitute the quantization volume W in the expressions for the coefficients $g_{\mathbf{k}}$ by the active volume V. It follows from the fact that the volume V is the only parameter with the dimension of a volume distinguishing from the physical point of view. Thus the coefficient $g_{\mathbf{k}_0}$ in Eq. (6.16) becomes now

$$g_{\mathbf{k}_0} = (2\pi\hbar/V\omega_{\mathbf{k}_0})^{1/2}(\mathbf{e}_\lambda\mathbf{m})\ . \qquad (6.17)$$

Using the Hamiltonian (6.16) and commutation rules (1.1), and (3.3) and then averaging the obtained equations over the initial state, we can get the following equations of dynamics for the one–mode superradiance

$$\frac{dn}{dt} + \frac{n}{\tau} = F\ ,$$

$$\frac{dF}{dt} + \frac{1}{2}\left[\frac{1}{\tau} + \frac{1}{T_2}\right]F = \frac{1}{T_2}(nR_3 + S + S_0)\ ,$$

$$\frac{dS}{dt} + \frac{S}{T_2} = FR_3 , \tag{6.18}$$

$$\frac{dR_3}{dt} + \frac{N + R_3}{T_1} = -2F .$$

The initial conditions are

$$n(0) = 0 , \quad F(0) = 0 , \quad S(0) = 0 , \quad R_3(0) = R_{30} . \tag{6.19}$$

In (6.18) we have introduced the new relaxation terms with the characteristic relaxation time $\tau = L/c$, where L is an active volume length. This terms take into account the finiteness of the time of the field–matter interaction. Thus an active medium, which is spatially distributed in reality, is described as the point–source system. And the process of the radiative transfer is substituted by the process of effective field relaxation.

By analogy with the point–source model (for which $L \ll \lambda$) we can introduce the following collective atomic operator

$$R_{\mathbf{k}}^2 = 2(R_{\mathbf{k}}^+ R_{\mathbf{k}}^- + R_{\mathbf{k}}^- R_{\mathbf{k}}^+) + R_3^2 . \tag{6.20}$$

This operator defines the Bloch vector length and it commutes with the Hamiltonian (6.16) as it can be easily checked.

2.7. Equations for Spatio-Temporal Dynamics of Superradiance

Let us consider again the equations (1.10) and introduce a new set of the field variables defined by

$$\mathbf{A}(\mathbf{r},t) = \mathbf{A}^+(\mathbf{r},t)e^{i\omega t} + \mathbf{A}^-(\mathbf{r},t)e^{-i\omega t} , \tag{7.1}$$

$$\mathbf{B}(\mathbf{r},t) = (i\omega/4\pi c^2)[\mathbf{A}^+(\mathbf{r},t)e^{i\omega t} - \mathbf{A}^-(\mathbf{r},t)e^{-i\omega t}]. \tag{7.2}$$

The transformation (7.1)–(7.2) is a canonical transformation from the vector potential $\mathbf{A}(\mathbf{r},t)$ and generalized pulse $\mathbf{B}(\mathbf{r},t)$ operators to the new canonically conjugated variables $\mathbf{A}^+(\mathbf{r},t)$ and $\mathbf{B}^-(\mathbf{r},t) = -(i\omega/2\pi c^2)\mathbf{A}^-(\mathbf{r},t)$. The generating function of this transformation is

$$P = \int Q d\mathbf{r} = -\frac{i\omega}{2\pi c^2}\int\Big[\frac{1}{4}\mathbf{A}^2 + \frac{1}{2}\mathbf{A}^+\mathbf{A}^+ \exp (2i\omega t)$$

$$- A^+A^- \exp(i\omega t)\Big] dr \,. \tag{7.3}$$

Really it can be easily checked that

$$\frac{\partial Q}{\partial \mathbf{A}} = \mathbf{B} \,, \qquad \frac{\partial Q}{\partial \mathbf{A}^+} = -\mathbf{B}^- \,.$$

Hence, we must add to the Hamiltonian (1.9) the following term

$$\frac{\partial P}{\partial t} = -(\omega^2/2\pi c^2) \int A^+A^- dr \,. \tag{7.4}$$

As a result the Hamiltonian (1.9) in RWA takes the form

$$H = (8\pi)^{-1}\int [-(\omega^2/c^2)(A^+A^- + A^-A^+) + \text{rot}A^+\text{rot}A^- + \text{rot}A^-\text{rot}A^+]dr$$

$$+ (\hbar\omega_0/2)\sum_{i=1}^{N} \sigma_{i3} - \frac{1}{c}\sum_{i=1}^{N} [m\rho_{i+}A^-(r_i,t) + m^*A^+(r_i,t)\rho_{i-}]. \tag{7.5}$$

Introducing the density of the field Hamiltonian by

$$H_f = \int \mathcal{H}_f \mathrm{d}\mathbf{r} \,, \tag{7.6}$$

we can get the following equation of motion for $\mathcal{H}_f$:

$$\frac{\partial \mathcal{H}_f}{\partial t} = -(i\omega/4\pi)[A^+\text{rot rot}A^- - (\text{rot rot}A^+)A^-]$$

$$-(i\omega/c)\sum_{i=1}^{N} [m\rho_{i+}A^-(r_i) - m^*A^+(r_i)\rho_{i-}]\delta(r - r_i) \,. \tag{7.7}$$

As we have mentioned earlier, the active volume of superradiant media has usually the shape of an extended cylinder. Therefore $A^{\pm}(\mathbf{r},t)$ can be represented by

$$A^{\pm}(\mathbf{r},t) = \sum_{l} A_l^{\pm}(\mathbf{r},t)\exp(\mp i\mathbf{k}_l\mathbf{r}) \,, \quad l = 1,2 \,, \tag{7.8}$$

where $\mathbf{k}_l$ are determined by (2.1), and the amplitudes $A^{\pm}$ satisfy the conditions (5.18). If (7.8) is applied to (7.7) and equations (1.10c,d) are used, we get the following set of the coupled nonlinear equations for spatio–temporal dynamics of SR [10]:

$$\frac{\partial n_l}{\partial t} + (-1)^l c \frac{\partial n_l}{\partial x} = F_l \,, \tag{7.9a}$$

$$\frac{\partial F_l}{\partial t} + (-1)^l c \frac{1}{2} \frac{\partial F_l}{\partial x} + \frac{F_l}{2T_2} = \frac{1}{T_0^2} (n_l R_3 + S_l + S_0) \,, \tag{7.9b}$$

$$\frac{\partial S_l}{dt} + \frac{S_l}{T_2} = F_l R_3 \,, \tag{7.9c}$$

$$\frac{\partial R_3}{dt} + \frac{R_3 - R_{3e}}{T_1} = -2 \sum_l F_l \,, \tag{7.9d}$$

where

$$n_l = <\mathcal{H}_l> V/\hbar\omega N \,,$$

$$S_l = \sum_i {\sum_j}' <\rho_{i+}\rho_{j-}> \exp[i\mathbf{k}_l(\mathbf{r}_i - \mathbf{r}_j)]/N_1^2 \,,$$

$$S_0 = \sum_i (1 + <\sigma_{i3}>)/2N_1^2 \,, \qquad R_3 = \sum_i <\sigma_{i3}>/N_1 \,,$$

$$\frac{1}{T_0^2} = \frac{4\pi |\mathbf{m}|^2}{\hbar\omega} \frac{N}{V} \,, \qquad N_1 = \frac{N}{V} V_1 \,, \qquad i,j \in V_1(\mathbf{r}) \,. \tag{7.10}$$

The primed summation index means that $i \neq j$. In the case of a spontaneous radiative decay of an atomic system the initial and boundary conditions are chosen in the form

$$n_l(x,0) = 0 \,,\; F_l(x,0) = 0 \,,\; S_l(x,0) = 0 \,,\; R_3(x,0) = R_{30}(x) \tag{7.11}$$

$$n_2(0,t) = n_1(L,t) = 0 \,, \quad F_2(0,t) = F_1(L,t) = 0 \,. \tag{7.12}$$

In deriving of (7.9a–d) we have used the following assumptions

$$\frac{\partial A^+_{l\lambda}}{\partial x} \bar{\mathcal{R}}_l \simeq A^+_{l\lambda} \frac{\partial \bar{\mathcal{R}}_l}{\partial x} \simeq \frac{1}{2} \frac{\partial}{\partial x} (A^+_{l\lambda} \bar{\mathcal{R}}_l) \,. \tag{7.13}$$

Let us define the conditions of the validity of these approximate equalities. The variables $\bar{\mathcal{R}}_l$ in (7.13) are defined by

$$\bar{\mathcal{R}}_l(x,t) = V_1^{-1} \sum_i \sigma_{i-}(t)\exp(-i\mathbf{k}_l\mathbf{r}_i) \ , \quad \mathbf{r}_i \in V_1(\mathbf{r}) \ , \qquad (7.14)$$

and the amplitudes $A^{-}_{l\lambda}(x,t)$, according (7.1) and (7.8), are

$$A^{-}_{l\lambda}(x,t) = V_1^{-1} \int_{V_1} A_\lambda(\mathbf{r},t)\exp[i(\omega_0 t - \mathbf{k}_l\mathbf{r})]d\mathbf{r} \ . \qquad (7.15)$$

Integrating equation (3.13a) we get

$$a_\mathbf{k}(t) = (i/\hbar)g^*_\mathbf{k} \int_0^t \sum_i \sigma_{i-}(t')\exp[-i\mathbf{k}\mathbf{r}_i - i\omega_\mathbf{k}(t - t')]dt' \ . \qquad (7.16)$$

In (3.16) we have kept a component of the field associating with the radiation sources and omitted the free vacuum part of the field.

Substituting (7.16) into (3.1)–(3.4) we get

$$A_\lambda(\mathbf{r},t) = (i/\hbar) \sum_\mathbf{k} (2\pi\hbar c^2/W\omega_\mathbf{k})^{1/2} g^*_\mathbf{k}\exp(i\mathbf{k}\mathbf{r})$$

$$\times \int_0^t \sum_i \sigma_{i-}(t')\exp[-i\mathbf{k}\mathbf{r}_i - i\omega_\mathbf{k}(t - t')]dt'$$

$$= im^*_\lambda c(2\pi)^{-2} \int d\mathbf{k} \ \omega_\mathbf{k}^{-1} \exp(i\mathbf{k}\mathbf{r}) \int_0^t \sum_i \sigma_{i-}(t')\exp[-i\mathbf{k}\mathbf{r}_i - i\omega_\mathbf{k}(t - t')]dt'.$$

Hence, the amplitude $A^{-}_{l\lambda}(x,t)$ is determined by

$$A^{-}_{l\lambda}(x,t) = im^*_\lambda c(2\pi)^{-2} \int d\mathbf{k} \ \omega_\mathbf{k}^{-1}(V_1^{-1} \int_{V_1} d\mathbf{r} \ \exp[i(\mathbf{k} - \mathbf{k}_l)\mathbf{r}_i])$$

$$\times \int_0^t \sum_i \sigma_{i-}(t')\exp[-i\mathbf{k}\mathbf{r}_i - i\omega_\mathbf{k}(t - t')]dt' \ . \qquad (7.17)$$

If we can neglect the retardation time at a distance L_1 , then

$$\rho_{i-}(t - L_1/c) \quad \rho_{i-}(t) \ , \qquad (7.18)$$

and (7.17) can be transformed to the form

$$A^-_{l\lambda}(x,t) = i(2\pi c m^*_\lambda/\omega_0)V_1^{-1} \sum_i \rho_{i-} \exp(-i\mathbf{k}_l\mathbf{r}_i)$$

$$= i(2\pi c m^*_\lambda/\omega_0)\bar{\mathcal{R}}_l(x,t)\ , \quad \mathbf{r}_i \in V_1(\mathbf{r})\ . \tag{7.19}$$

Thus, the equalities (7.13) hold if the polarization amplitude does not vary considerably throughout the time $\tau_1 = L_1/c$, where, in accordance with its definition, the length L_1 should be much longer than the wavelength of radiation and much shorter than the characteristic length of the field amplitude variation.

2.8. Dynamics of Field Spatial Correlations

The main approximation we have made in obtaining Eqs. (4.9a–d) is the application of Eqs. (4.13). These equations have a very wide region of the applicability. Nevertheless this region is limited, and it is useful to develop a theory which is free from this approximation. The variables $n_l(x,t)$ determine the energy flux densities of the counterpropagating waves. The energy flux density is the one–point spatial correlation function. In fact, $n(x,t) = \langle a^+(x,t)a(x,t)\rangle$. The more precise description of the spatial dynamics of the generated fields can be obtained if we will use the two–point spatial correlation functions, like the following one

$$n(x,x',t) = \langle a^+(x,t)a(x',t)\rangle\ . \tag{8.1}$$

In this case we can get the self–consistent equations without application of the equalities (7.13).

If (7.1), (7.2), and (7.8) are applied to (5.10a–b) and (5.10c–d) are used we get the following hierarchy of the coupled nonlinear equations for the two–point spatial correlation functions

$$\frac{\partial}{\partial t} n(x,x',t) + c\left(\frac{\partial}{\partial x} + \frac{\partial}{\partial x'}\right) n(x,x',t) = \frac{1}{2}\left[F(x,x',t) + F^*(x',x,t)\right]\ ,$$

$$\frac{\partial}{\partial t} F(x,x',t) + c\,\frac{\partial}{\partial x'} F(x,x',t) + \frac{1}{2T_2} F(x,x',t)$$

$$= \frac{1}{T_0^2} [n(x,x',t)\, R_3(x) + S(x,x',t) + S_0(x,t)\, \delta_1(x - x')] , \tag{8.2}$$

$$\frac{\partial}{\partial t} S(x,x',t) + \frac{S(x,x',t)}{T_2} = \frac{1}{2} [F(x,x',t)\, R_3(x',t)$$

$$+ F^*(x,x',t)\, R_3(x,t)] ,$$

$$\frac{\partial R_3(x,t)}{\partial t} + \frac{R_3(x,t) - R_{3e}(x,t)}{T_1} = [F(x,x,t) + F^*(x,x,t)].$$

It is convenient to normalize the field amplitudes in (8.1) in such a way that, if $x = x'$, then the variable $n(x,x',t)$ will coincide with $n(x,t)$ defined by (7.10). To do this we should use the following relationship

$$A^+(x,t) = \sqrt{\frac{2\pi\hbar c^2}{\omega} \frac{N}{V}}\; a^+(x,t) . \tag{8.3}$$

The variables, used in (8.2), are expressed as follows in terms of the spin Pauli matrices:

$$F(x,x',t) = -\,i \frac{2\pi}{\kappa} m <R^+(x,t)a(x',t)> \left[\frac{2\pi\hbar c^2}{\omega} \frac{V}{N} \right]^{-1/2} ,$$

$$R^+(x,t) = \frac{1}{N_1} \sum_{i \in V_1(x)} <\rho_{i+}> e^{i\kappa x_i} ,$$

$$S(x,x',t) = \frac{1}{N_1^2} \sum_{i \in V_1(x)}{}' \sum_{j \in V_1(x')} <\rho_{i+}\rho_{j-}> e^{i\kappa(x_i - x_j)} , \tag{8.4}$$

$$R_3(x,t) = \frac{1}{N_1} \sum_{i \in V_1(x)} <\sigma_{i3}> ,$$

$$S_0(x,t) = \frac{1}{N_1^2} \sum_{i \in V_1(x)} < \frac{1 + \sigma_{i3}}{2} > .$$

We have introduced the following auxiliary function

$$\delta_1(x) = \frac{L_1}{2\pi}\int_{-\pi/L_1}^{\pi/L_1} e^{ikx}\,dk = \frac{\sin(\pi x/L_1)}{\pi x/L_1}, \tag{8.5}$$

where L_1 is a length of the physically small averaging volume V_1. We see that $\delta_1(x)$ tends to δ – function when $L_1 \to 0$:

$$\delta_1(x) \xrightarrow[L_1 \to 0]{} L\ \delta_1(x) .$$

2.9. Quantum Fluctuations

2.9.1. *Decoupling of correlations.* As we have mentioned in Sec.2.6 the decoupling procedure (6.6) overcomes the main disadvantages of the semiclassical theories and allows to take into account the effect of a radiative spontaneous decay. Nevertheless, such type of decoupling is not free from any limitations. For example, the decoupling procedure (6.6) does not take into account the fluctuations of the population difference. In fact, according to Eq. (6.6), we get

$$\langle\sigma_{i3}^2\rangle = \langle\sigma_{i3}\rangle\langle\sigma_{i3}\rangle = \langle\sigma_{i3}\rangle^2 . \tag{9.1}$$

It becomes clear if we represent σ_{i3}^2 in the form

$$\sigma_{i3}^2 = (2\sigma_{i+}\sigma_{i-} - 1)\sigma_{i3} ,$$

and, applying the decoupling procedure (6.6), we get

$$\langle\sigma_{i3}^2\rangle = 2\langle\sigma_{i+}\sigma_{i-}\sigma_{i3}\rangle - \langle\sigma_{i3}\rangle = 2\langle\sigma_{i+}\sigma_{i-}\rangle\langle\sigma_{i3}\rangle - \langle\sigma_{i3}\rangle$$

$$= (1 + \langle\sigma_{i3}\rangle)\langle\sigma_{i3}\rangle - \langle\sigma_{i3}\rangle = \langle\sigma_{i3}\rangle^2 . \tag{9.2}$$

Here we shall present an example of decoupling, which incorporates the effects of the population difference fluctuations, and demonstrate the consequences of such incorporation. We shall use in our treatment the boson operator exclusion method.

If quantum averaging is applied to (4.8b) and the equations (4.4), (4.5) are used, we get for $\langle\sigma_{i3}\rangle$ the following equation of motion

$$\frac{d\langle\sigma_{i3}\rangle}{dt} = \frac{2i}{\hbar c^2}\sum_j\sum_\lambda \frac{1}{r_{ij}}\,[m_{i\lambda}m^*_{j\lambda}\langle\sigma_{i+}(t)\sigma_{j-}(t - \tau_{ij})\rangle$$

$$- m^{*}_{i\lambda} m_{j\lambda} \langle \sigma_{j+}(t - \tau_{ij}) \sigma_{i-}(t) \rangle] \ . \tag{9.3}$$

The equations for the correlators in the right–hand side of (9.3) follow from Eq. (4.8a):

$$\begin{aligned} \frac{\mathrm{d}}{\mathrm{d}t} \langle \sigma_{i+}(t) \sigma_{j-}(t - \tau_{ij}) \rangle &= \mathrm{i}(\omega_{0i} - \omega_{0j}) \langle \sigma_{i+}(t) \sigma_{j-}(t - \tau_{ij}) \rangle \\ &+ \frac{\mathrm{i}}{\hbar c^2} \sum_{l} \sum_{\lambda} \frac{1}{r_{il}} m^{*}_{i\lambda} m_{l\lambda} \langle \sigma_{l+}(t - \tau_{il}) \sigma_{i3}(t) \sigma_{j-}(t - \tau_{ij}) \rangle \\ &- \frac{\mathrm{i}}{\hbar c^2} \sum_{l} \sum_{\lambda} \frac{1}{r_{jl}} m_{j\lambda} m_{l\lambda} \langle \sigma_{i+}(t) \sigma_{j3}(t - \tau_{ij}) \sigma_{l-}(t - \tau_{ij} - \tau_{jl}) \rangle . \end{aligned} \tag{9.4}$$

We could easily truncate this hierarchy of the coupled equations if we had a small parameter, usually the interaction Hamiltonian. Unfortunately, in the case of a superradiant decay there are no any grounds for such assumption, especially, in the region where the pulse intensity reaches its maximum, and where $\langle \sigma^2_{i3} \rangle$ and $\langle \sigma_{i3} \rangle^2$ differ considerably from one another. Some intrinsic figures of decoupling merit are not known now. Nevertheless, it is clear that the symmetry properties of the Pauli operators should not be broken under any decoupling procedures.

Notice that the operators σ_{i+} and σ_{i-} are the Fermi operators, when they refer to the same atom:

$$\sigma_{i+} \sigma_{i-} + \sigma_{i-} \sigma_{i+} = 1 \ , \tag{9.5}$$

and they are Bose operators when they relate to the different atoms. In fact, when $i \neq j$ they satisfy the following relationships

$$\sigma_{i+} \sigma_{j-} - \sigma_{j-} \sigma_{i+} \Big|_{i \neq j} = 0 \ . \tag{9.6}$$

We can see that in (9.4) the operators are in the normal ordering form. We shall decouple correlators without breaking in ordering of operators. According to the properties of the Pauli matrices we get

$$\sigma_{j3} = \sigma_{j+} \sigma_{j-} - \sigma_{j-} \sigma_{j+} \ . \tag{9.7}$$

If (9.7) is applied to the last correlator in the right–hand side of (9.4), this correlator can be rewritten in the form

$$<\sigma_{i+}(t)\sigma_{j3}(t')\sigma_{l-}(t'')> \; = \; <\sigma_{i+}(t)\sigma_{j+}(t')\sigma_{j-}(t')\sigma_{l-}(t'')>$$

$$- <\sigma_{i+}(t)\sigma_{j-}(t')\sigma_{j+}(t')\sigma_{l-}(t'')> . \tag{9.8}$$

Bearing in mind the commutation rules (9.5), (9.6) we can decouple the correlators in the right–hand side of (9.8) in analogy with the Wick theorem for the Bose and Fermi operators

$$<\sigma_{i+}(t)\sigma_{j+}(t')\sigma_{j-}(t')\sigma_{l-}(t'') \; = \; <\sigma_{j+}(t')\sigma_{j-}(t')><\sigma_{i+}(t)\sigma_{l-}(t'')>$$

$$- <\sigma_{i+}(t)\sigma_{j-}(t')><\sigma_{j+}(t')\sigma_{l-}(t'')> . \tag{9.9a}$$

$$<\sigma_{i+}(t)\sigma_{j-}(t')\sigma_{j+}(t')\sigma_{l-}(t'')> \; = \; <\sigma_{j-}(t')\sigma_{j+}(t')><\sigma_{i+}(t)\sigma_{l-}(t'')>$$

$$+ <\sigma_{i+}(t)\sigma_{j-}(t')><\sigma_{j+}(t')\sigma_{l-}(t'')> . \tag{9.9b}$$

As a consequence, we get [11]

$$<\sigma_{i+}(t)\sigma_{j3}(t')\sigma_{l-}(t'')> \; = \; <\sigma_{j3}(t')><\sigma_{i+}(t)\sigma_{l-}(t'')>$$

$$- 2<\sigma_{i+}(t)\sigma_{j-}(t')><\sigma_{j+}(t')\sigma_{l-}(t'')> . \tag{9.10}$$

We can see that decoupling (9.10) differs from traditional decoupling by the presence of the second term in the right–hand side of (9.10).

Let us now consider the difference between the new and old decoupling procedures. First of all, let us define whether the symmetrical properties of the Pauli operators conserve under decoupling of the third–order correlators. According to the Pauli matrix properties, we get for $i \neq j$

$$\sigma_{i+}(t)\sigma_{j3}(t - \tau_{ij})\sigma_{l-}(t - \tau_{ij} - \tau_{jl})\Big|_{i=j}$$

$$= \sigma_{i+}(t)\sigma_{i3}(t)\sigma_{l-}(t - \tau_{il}) = - \sigma_{i+}(t)\sigma_{l-}(t - \tau_{il}) . \tag{9.11}$$

On the other hand it follows from (9.10), that

$$<\sigma_{i+}(t)\sigma_{j3}(t - \tau_{ij})\sigma_{l-}(\mathrm{t} - \tau_{ij} - \tau_{jl})> \; = \; <\sigma_{i3}(t)><\sigma_{i+}(t)\sigma_{l-}(t - \tau_{il})>$$

$$-2<\sigma_{i+}(t)\sigma_{i-}(t)><\sigma_{i+}(t)\sigma_{l-}(t - \tau_{il})> \; = -<\sigma_{i+}(t)\sigma_{l-}(t - \tau_{il})> . \tag{9.12a}$$

Similarly, for $j = l$ it follows from (9.10), that

$$<\sigma_{i+}(t)\sigma_{j3}(t - \tau_{ij})\sigma_{l-}(t - \tau_{ij} - \tau_{jl})>\Big|_{j=l}$$

$$= -<\sigma_{i+}(t)\sigma_{j-}(t - \tau_{ij})> . \qquad (9.12b)$$

Comparing (9.11) and (9.12a,b) we can see that the decoupling procedure (9.10) takes into account the basic properties of Pauli matrices.

In the case $i = j = l$, the triple products of the Pauli matrices are transformed to the following form

$$\sigma_{i+}\sigma_{i3}\sigma_{i-} = -\sigma_{i+}\sigma_{i-} = -(1 + \sigma_{i3})/2 .$$

On the other hand, assuming in (9.12a) and (9.12b) $i = j = l$ we obtain

$$<\sigma_{i+}(t)\sigma_{j3}(t - \tau_{ij})\sigma_{l-}(t - \tau_{ij} - \tau_{jl})>\Big|_{i=j=l}$$

$$= -(1 + <\sigma_{i3}>)/2 . \qquad (9.13)$$

Thus, in contrast to decoupling (6.6) the decoupling procedure (9.10) incorporates the symmetrical properties of the Pauli matrices under transformations of the third–order correlators into the first– and second–order ones. It is especially important for the superradiative decay, because the fluctuations of a population difference are the most important in the time domain near the maximum of the pulse intensity. And it is this region where the most strong phase correlations between the individual atoms are established.

Let us now derive the self–consistent equations for the population difference and second–order correlators. Before applying the decoupling procedure (9.10) we discuss the main approximation which will be used here. We shall assume that $\mathbf{m}_i = \mathbf{m}$ and then we shall average the equations (9.3), (9.4) over the all possible directions of vector $\mathbf{m}$. After such averaging the quantity $\sum_\lambda |m_\lambda|^2$ becomes

$$(4\pi)^{-1}\int d\Omega_{\mathrm{m}} \sum_\lambda |m_\lambda|^2 = (4\pi)^{-1}\int d\Omega_{\mathrm{m}} |m|^2 \sin^2\vartheta$$

$$= (1/2)|m|^2\int_0^\pi \sin^3\vartheta \, d\vartheta = (2/3)|m|^2 . \qquad (9.14)$$

Thus, if (9.10) and (9.14) are applied to (9.3) and (9.4), we obtain

$$\frac{d}{dt}<\sigma_{i3}> + \Gamma_0(1 + <\sigma_{i3}>)$$

$$= i\Gamma_0 \sum_{j(\neq i)} [a_{ij}(t) S_{ij}(t, t - \tau_{ij}) - a^*_{ij}(t) S^*_{ij}(t, t - \tau_{ij})] \, , \tag{9.15}$$

$$\frac{d}{dt} S_{ij}(t, t - \tau_{ij}) + \Gamma_2 S_{ij}(t, t - \tau_{ij})$$

$$= \frac{i\Gamma_0}{2} \sum_{l(\neq i)} a^*_{il}(t) \langle \sigma_{i3}(t) \rangle S^*_{jl}(t - \tau_{ij}, t - \tau_{il})$$

$$- \frac{i\Gamma_0}{2} \sum_{l(\neq j)} a_{jl}(t - \tau_{ij}) \langle \sigma_{j3}(t - \tau_{ij}) \rangle S_{il}(t, t - \tau_{ij} - \tau_{jl})$$

$$+ i\Gamma_0 \{ \sum_{l(\neq j)} a_{jl}(t - \tau_{ij}) S_{jl}(t - \tau_{ij}, t - \tau_{jl})$$

$$- \sum_{l(\neq i)} a^*_{il}(t) S^*_{il}(t, t - \tau_{il}) \} \, S_{ij}(t, t - \tau_{ij}) \, . \tag{9.16}$$

We have used the following designations:

$$\Gamma_0 = 4\omega_0 |m|^2 / 3\hbar c^3 \tag{9.17}$$

is a spontaneous decay rate,

$$a_{ij}(t) = \exp[i(\omega_{0i} - \omega_{0j})t] \exp(i\omega_{0j} r_{ij}/c)(kr_{ij})^{-1} \, , \tag{9.18}$$

$$S_{ij}(t, t - \tau_{ij}) = \langle \rho_{i+}(t) \rho_{j-}(t - \tau_{ij}) \rangle \, , \tag{9.19}$$

and $\Gamma_2 = \Gamma_0 + 1/T_2$. It follows from (9.19), that the correlators S_{ij} obey the following rules

$$S^*_{ij}(t, t - \tau_{ij}) = S_{ji}(t - \tau_{ij}, t) \, . \tag{9.20}$$

Notice that the correlators S_{ij} include the atomic operators in the interaction representation.

2.9.2. *Dynamic equation for population difference in retardationless approximation.* The retardation time τ_{ij} appearing in the equations (9.15) and (9.16) is the time taken by a photon to cross the distance between the *i*–th and *j*–th atoms. All the intervals τ_{ij} satisfy the condition

$$\tau_{ij} \leq \tau = L/c\ , \tag{9.21}$$

where L is an active volume length. It is assumed that the characteristic time of $\langle\rho_{i\pm}(t)\rangle$ variation is essentially longer than τ. In this case we can ignore the retardation in the arguments of the variables in Eqs. (9.15), (9.16).

In the absence of the inhomogeneous broadening effects, i.e., when $\omega_{0i} = \omega_0$ for $i = 1, 2,..., N$, the Eq.(9.15), after the summation over index "i", becomes:

$$\frac{dR_3}{dt} + \Gamma_0(N + R_3) = -2\Gamma_0 \sum_i {\sum_j}' c_{ij} S_{ij}\ , \tag{9.22}$$

where

$$R_3 = \sum_i \langle\sigma_{i3}\rangle\ , \qquad c_{ij} = \frac{\sin(kr_{ij})}{kr_{ij}}\ . \tag{9.23}$$

If we differentiate (9.22) with respect to time and apply the equation (9.16), we get

$$\frac{d^2R_3}{dt^2} + (\Gamma_0 + \Gamma_2)\frac{dR_3}{dt} + \Gamma_0\Gamma_2(N + R_3)$$

$$= -2\Gamma_0^2 \sum_i {\sum_j}'' {\sum_l}' c_{ij}c_{il}\langle\sigma_{i3}\rangle S_{jl} - \Gamma_0^2 \sum_i {\sum_j}' c_{ij}^2\langle\sigma_{i3}\rangle(1 + \langle\sigma_{j3}\rangle)$$

$$+ 4\Gamma_0^2 \sum_i {\sum_j}' \sum_l c_{ij}c_{il}S_{il}S_{ij}\ , \tag{9.24}$$

where the notation $\sum_i {\sum_j}'' {\sum_l}'$ means that we should suppose $i \neq j \neq l$ in summation over these indexes, and $\sum_i {\sum_j}' \sum_l$ means that $i \neq j$ and $j \neq l$.

In obtaining (9.24) we have neglected the imaginary part ia''_{ij} of the interaction matrix. It corresponds to neglecting of the cooperative frequency shift effects (these effects will be discussed in Sec. 3.1.5). It is to be noted that the variables $\langle\sigma_{i3}\rangle$ in the right–hand side of (9.24) are the slowly varying functions of coordinates $\mathbf{r}_i$ in contrast to the coefficients c_{ij} which oscillate at the spatial frequency κ =

$2\pi/\lambda$. If we factor out the slowly varying terms $<\sigma_{i3}>$ in the right–hand side of (9.24), we get

$$\sum_i{}' \sum_j c_{ij}^2 <\sigma_{i3}>(1 + <\sigma_{j3}>) = \frac{R_3}{N} \frac{N + R_3}{N} \sum_i{}' \sum_j c_{ij}^2 \,, \tag{9.25}$$

$$\sum_i{}' \sum_j \sum_l c_{ij} c_{il} S_{il} S_{ij} = - \frac{1}{2\Gamma_0} \sum_i{}' \sum_j \left[\frac{d<\sigma_{i3}>}{dt} + \Gamma_0 (1 + <\sigma_{i3}>) \right] c_{ij} S_{ij}$$

$$= \frac{1}{2\Gamma_0 N} \left[\frac{dR_3}{dt} + \Gamma_0 (N + R_3) \right] \frac{1}{2\Gamma_0} \left[\frac{dR_3}{dt} + \Gamma_0 (N + R_3) \right]$$

$$= \frac{1}{4\Gamma^2 N} \left[\frac{dR_3}{dt} + \Gamma_0 (N + R_3) \right]^2 , \tag{9.26}$$

$$\sum_i{}'' c_{ij} c_{il} <\sigma_{i3}> = R_3 \gamma c_{jl} \,, \quad i \neq j \neq l \,. \tag{9.27}$$

Here γ, as we shall show later, has the following limiting values [12]:

$$\gamma = \begin{cases} \pi c / L\omega_0 & \text{if } \Phi \ll 1 \\ \pi c / Z\omega_0 & \text{if } \Phi \gg 1 \end{cases} , \tag{9.28}$$

where $\Phi = Z/\lambda L$ is a Fresnel number, Z is an active volume cross–section area, L is an active volume length, and λ is a wavelength.

Substituting (9.25) – (9.28) into (9.24) we get the following self–consistent equation describing the population difference dynamics:

$$\frac{d^2 R_3}{dt^2} + (\Gamma_0 + \Gamma_2) \frac{dR_3}{dt} + \Gamma_0 \Gamma_2 (N + R_3)$$

$$= \gamma \Gamma_0 R_3 \left[\frac{dR_3}{dt} + \Gamma_0 (N + R_3) \right] - \Gamma_0^2 \frac{R_3}{N} \frac{N + R_3}{N} \sum_i{}' \sum_j c_{ij}^2$$

$$+ \frac{1}{N} \left[\frac{dR_3}{dt} + \Gamma_0 (N + R_3) \right]^2 . \tag{9.29}$$

If we take into account that the following conditions hold for a superradiative decay:

$$\Gamma_0 N \ll \Gamma_2 N \ll |dR_3/dt| ,$$

then the Eq.(9.29) can be simplified:

$$\frac{d^2R_3}{dt^2} + \Gamma_2 \frac{dR_3}{dt} - \gamma\Gamma_0 R_3 \frac{dR_3}{dt} - \frac{1}{N}\frac{dR_3}{dt}\frac{dR_3}{dt}$$

$$= \Gamma_0^2 R_3 (N + R_3)\left(\gamma - \frac{1}{N^2} \sum_i {}' \sum_j c_{ij}^2 \right) . \tag{9.30}$$

Thus, in the frames of the retardationless approximation we get the self–consistent second–order differential equation for the population difference dynamics.

CHAPTER 3

QUANTUM THEORY OF SUPERRADIANCE

As we have discussed above, an effective self–induction of correlations between the dipole moments of the different atoms forms the basis for the superradiance effect. It is this self-organization, i.e., the spontaneous transition from an initially uncorrelated state to a macroscopically correlated one, which differs SR from the other transient coherent effects, such as the self-induced transparency, photon echo, optical nutation effect, etc. In the latter cases, the appearance of the macroscopic induced dipole moment is imposed, first of all, by the strong external resonant electromagnetic wave, and, only secondly, due to the Bloch integral of motion for an individual atom. In fact, if all of the atoms are initially in the similar states, then the strong resonant wave can induce a macroscopic polarization in the atomic system in result of the processes of phasing of the individual atoms with the external field.

There is no any external phasing in the process of SR. Therefore, the induction of the macroscopic polarization is impossible here without the mutual influence of the atoms. This mutual influence is realized via the transversal electromagnetic field. As an individual atom emits the spherical wave, therefore, the effect should be sensitive to the mean interatomic distance, i.e., there exist the threshold values for the density of the excited atoms.

The treatment of the nature of the induction of macroscopic correlations in an system of radiators, the study of the influence of an active volume shape on the angular spectra of emission, the role of the non-linearity and distributed nature of the atomic system – all these topics demonstrate the complexity of the problem in general formulation. That is why the various approaches have been developed in the previous chapters. Approaches which focus on only one aspect of the phenomenon may not describe other important aspects. Here, we will accordingly discuss or briefly summarize all the basic approaches to a description of SR, and we will compare the theoretical results with the experimental data.

3.1. Markovian theory of superradiance.

3.1.1. *Field intensity in terms of collective atomic operators.* The conservation of the Bloch vector length is a main distinctive feature of superradiance and other related cooperative phenomena. Let us try to use this fact to describe the profile of the emitted pulse. But we should first establish the relationship between the field intensity and the collective atomic operators $R_{\mathbf{k}}^{\pm}$. This relationship follows, for example, from the equation (2.3.8a). Using this equation we can show that the rate of the variation of the number of photons in the field mode $\mathbf{k}$:

$$n_{\mathbf{k}} = \langle a^{+}_{\mathbf{k}} a_{\mathbf{k}} \rangle \,, \tag{1.1}$$

is determined by

$$\frac{dn_{\mathbf{k}}}{dt} = \frac{|g_{\mathbf{k}}|}{2} \int^{t} dt' \{ \langle R^{+}_{\mathbf{k}}(t) R^{-}_{\mathbf{k}}(t') \rangle \exp[-\mathrm{i}(\omega_{\mathbf{k}} - \mathrm{i}\varepsilon_{\mathbf{k}})(t - t')]$$

$$+ \langle R^{+}_{\mathbf{k}}(t') R^{-}_{\mathbf{k}}(t) \exp[\mathrm{i}(\omega_{\mathbf{k}} + \mathrm{i}\varepsilon_{\mathbf{k}})(t - t')] \} \,. \tag{1.2}$$

where $\varepsilon^{-1}_{\mathbf{k}} = \tau_{\mathbf{k}}$, and $\tau_{\mathbf{k}}$ is the time taken by a photon to cross the active volume along the direction of the field mode $\mathbf{k}$. This equation can be simplified essentially if we use the following transformation

$$R^{\pm}_{\mathbf{k}} = \mathcal{R}^{\pm}_{\mathbf{k}} \exp(\pm \mathrm{i}\omega_0 t) \,, \tag{1.3}$$

and then assume that the characteristic variation time of $\mathcal{R}^{\pm}_{\mathbf{k}}(t)$ is much longer than $\varepsilon^{-1}_{\mathbf{k}}$. The characteristic time of $\mathcal{R}^{\pm}_{\mathbf{k}}(t)$ variation coincides with the emitted pulse width τ_p . Whereas $\varepsilon^{-1}_{\mathbf{k}} = \tau_{\mathbf{k}}$. Hence the above mentioned condition can be written in the form

$$\tau_p \gg \tau_{\mathbf{k}} = L_{\mathbf{k}}/c \,, \tag{1.4}$$

where $L_{\mathbf{k}}$ is an active volume length along the direction of the field mode $\mathbf{k}$. With these assumptions, the equation (1.2) can be rewritten in the form

$$\frac{dn_{\mathbf{k}}}{dt} = \frac{2\pi |g_{\mathbf{k}}|^2}{\hbar^2} \langle R^{+}_{\mathbf{k}} R^{-}_{\mathbf{k}} \rangle \, \delta(\omega_{\mathbf{k}} - \omega_0) \,. \tag{1.5}$$

Finally, we get the following equation for the integral intensity of emission, i.e., for the intensity averaged over the all possible directions of the wave vector $\mathbf{k}$,

$$I = \sum_{\mathbf{k}} I_{\mathbf{k}} = \frac{\omega_0^2}{2\pi c^3} \int d\Omega_{\mathbf{k}_0} \sum_{\lambda} |m_{\lambda}|^2 \langle R^{+}_{\mathbf{k}_0} R^{-}_{\mathbf{k}_0} \rangle, \tag{1.6}$$

where $I_{\mathbf{k}} = \hbar\omega_{\mathbf{k}} dn_{\mathbf{k}}/dt$, $|\mathbf{k}_0| = \omega_0/c$. Further insight into the implication of (1.6) can be obtained by considering the emission of radiation by the system containing a one atom. In this case

$$<R^+_{\mathbf{k}_0} R^-_{\mathbf{k}_0}> = <\sigma_+\sigma_-> = (1 + <\sigma_3>)/2 = N_2 . \tag{1.7}$$

The polarization factor in (1.6) can be transformed to the following form

$$\sum_\lambda |m_\lambda|^2 = |m|^2 \sin^2\vartheta_{\mathbf{m}} , \tag{1.8}$$

where $\vartheta_{\mathbf{m}}$ is an angle between the vectors $\mathbf{m}$ and $\mathbf{k}$. If (1.7) and (1.8) are applied to (1.6), we get

$$I = \hbar\omega_0 \frac{dn}{dt} = \frac{4\omega_0^2 |m|^2}{3c^3} N_2 = \frac{\hbar\omega}{T_1} N_2 , \tag{1.9}$$

where T_1 is a spontaneous decay time. Bearing in mind that in the one–atom case

$$\frac{d}{dt}\frac{R_3}{2} = -\frac{N_2}{T_1} , \tag{1.10}$$

we get the following equation

$$\frac{d}{dt}\left[n + \frac{R_3}{2} \right] = 0 , \tag{1.11}$$

from which the energy conservation law is followed

$$n + R_3/2 = 1/2 , \tag{1.12a}$$

or

$$n = \frac{1 - R_3}{2} . \tag{1.12b}$$

Let us compare the Eqs.(1.6), and (1.2). In Eq.(1.2) the operators relate to the different moments of time, whereas in (1.6) the arguments of both operators coincide. Because of this, it is customary to consider the equation (1.6) as the equation for the emitted intensity in the Markovian approximation.

3.1.2. *SR of point-source systems.* Thus, we have obtained the following equation for the emission intensity in the Markovian approximation:

$$I = \frac{\hbar\omega_0}{T_1}\frac{3}{8\pi}\int d\Omega_{\mathbf{k}_0} \langle R^+_{\mathbf{k}_0} R^-_{\mathbf{k}_0} \rangle \sin^2\vartheta_{\mathrm{m}} \ . \tag{1.13}$$

Let us consider now the superradiance of systems with the dimensions much smaller than the wavelength of radiation: $a \ll \lambda$. In this case

$$R^{\pm}_{\mathbf{k}} = \sum_{i=1}^{N} \sigma_{i\pm} e^{\pm i\mathbf{k}\mathbf{r}_i} \sum_{i=1}^{N} \sigma_{i\pm} = R^{\pm}_0 \ . \tag{1.14}$$

And, as we have shown in Sec.2.2 , the length of the Bloch vector,

$$\mathbf{R} = \mathbf{e}_1 R_1 + \mathbf{e}_2 R_2 + \mathbf{e}_3 R_3 \ ,$$

is an integral of motion. It means that operator

$$R^2_0 = 2(R^+_0 R^-_0 + R^-_0 R^+_0) + R^2_3 \tag{1.15}$$

commutes with the Hamiltonian (2.1.9). Substituting (1.14) into (1.13), we get

$$I = \Gamma_0 \hbar\omega_0 \langle R^+_0 R^-_0 \rangle \ . \tag{1.16}$$

If we average both sides of the Eq. (1.15) over the initial quantum state, which corresponds to the complete inversion in the atomic subsystem and to a vacuum state of the field, we get for the square of the Bloch vector length:

$$2\langle R^+_0 R^-_0 + R^-_0 R^+_0 \rangle + \langle R^2_3 \rangle = 4N + N^2 \ . \tag{1.17}$$

Applying the equation

$$R^-_{\mathbf{k}} R^+_{\mathbf{k}} = R^+_{\mathbf{k}} R^-_{\mathbf{k}} - R_3 \ , \tag{1.18}$$

and substituting (1.17) into (1.16), we obtain

$$I = \Gamma_0 \hbar\omega_0 [(N^2 - \langle R^2_3 \rangle)/4 + (N + \langle R_3 \rangle)/2] \ . \tag{1.19}$$

The right–hand side of this equation depends on the mean value of the population difference operator, whereas the left–hand side depends on the mean value of the photon number operator. Using the relationship between these operators in RWA (see Eq.(2.3.22)), we can rewrite equation (1.19) in the following form

$$\frac{dR_3}{dt} = -\Gamma_0[(N^2 - \langle R_3^2 \rangle)/2 + (N + R_3)] . \qquad (1.20)$$

If the fluctuations of the population difference are negligible, we can assume

$$\langle R_3^2 \rangle = \langle R_3 \rangle^2 = R_3^2 . \qquad (1.21)$$

In this case we get the self–consistent equation of SR dynamics [1]:

$$\frac{dR_3}{dt} = -\Gamma_0[(N^2 - R_3^2)/2 + (N + R_3)] . \qquad (1.22)$$

The solution of this equation, with the accuracy $1/N$, is

$$R_3 = -N \tanh[(t - t_0)/2\tau_c] , \qquad (1.23)$$

where [1,2]

$$\tau_c = T_1/N , \qquad (1.24)$$

$$t_0 = \tau_c \ln N = (T_1/N) \ln N . \qquad (1.25)$$

The intensity of the emitted radiation (with the accuracy $\propto 1/N$) is determined, according to (1.20), by the following equation

$$I = \frac{1}{4} N^2 \frac{\hbar\omega}{T_1} \operatorname{sech}^2 \frac{t - t_0}{2\tau_c} . \qquad (1.26)$$

Thus, an ensemble of N atoms, emits in the arbitrary direction $\mathbf{k}$ the pulse of radiation with the peak intensity proportional to N^2 (superradiance). The peak pulse intensity is reached at moment t_0 (delay time), and the pulse width is equal to τ_c . As $N \gg 1$, hence $\ln N \gg 1$ (for example $\ln N \simeq 20$ for $N = 10^8 \div 10^9$), consequently $t_0 \gg \tau_c$. So, it is the delay time t_0 which determines the duration of the radiative decay. It will be recalled that in deriving (1.26) we used the conservation law of the Bloch vector length (1.17). Hence, in accordance with Sec.2.2, the superradiance occurs, only, when the condition $t_0 \ll T_2$ holds.

3.1.3. *SR of extended atomic systems.* As we have shown in Sec.2.3 in the case of the extended atomic systems the Bloch integral of motion is substituted by

$$\sum_{\mathbf{k}} \{2(R^{+}_{\mathbf{k}}R^{-}_{\mathbf{k}} + R^{-}_{\mathbf{k}}R^{+}_{\mathbf{k}}) + R_{3,\mathbf{k}-\mathbf{k}_0}R_{3,-\mathbf{k}+\mathbf{k}_0}\} = \text{const.} \tag{1.27}$$

It is seen that this new integral of motion includes the summation over the all modes of the radiation field. To simplify the analysis of the emission dynamics we can assume that the different field modes **k** do not interact with each other. Let us define the conditions when this assumption holds. At the initial stage of the emission, when $N - R_3 \ll N$, we get

$$R_{3,\mathbf{k}-\mathbf{k}'} = NP(\mathbf{k} - \mathbf{k}') \quad N\delta_{\mathbf{k},\mathbf{k}'}. \tag{1.28}$$

Thus, in the case of the uniformly excited active medium the different field modes do not interact with each other at the initial stage of the process of a radiative decay. If (1.28) is applied to (2.3.13) then the right–hand side of equation (2.2.13b) will depend only on the variables with the index **k** coinciding with that in the left–hand side. Therefore, we can assume that at the initial stage the each term of the sum in (1.27) is conserved:

$$R^2_{\mathbf{k}} = 2(R^{+}_{\mathbf{k}}R^{-}_{\mathbf{k}} + R^{-}_{\mathbf{k}}R^{+}_{\mathbf{k}}) + R_{3,\mathbf{k}-\mathbf{k}_0}R_{3,-\mathbf{k}+\mathbf{k}_0} . \tag{1.29}$$

Averaging (1.29) over the initial state we get

$$4\langle R^{+}_{\mathbf{k}}R^{-}_{\mathbf{k}}\rangle + \langle R_{3,\mathbf{k}-\mathbf{k}_0}R_{3,-\mathbf{k}+\mathbf{k}_0}\rangle - 2R_3$$

$$= \langle R_{3,\mathbf{k}-\mathbf{k}_0}R_{3,-\mathbf{k}+\mathbf{k}_0}\rangle\Big|_{t=0} + 2N . \tag{1.30}$$

We assume again that initially all of the atoms are in the excited state and the field is in the vacuum state. The first term in the right–hand side of (1.30) takes after averaging the form

$$\langle R_{3,\mathbf{k}-\mathbf{k}_0}R_{3,-\mathbf{k}+\mathbf{k}_0}\rangle\Big|_{t=0} = \sum_i \sum_j \langle 0|\sigma_{i3}\sigma_{j3}|0\rangle \exp[i(\mathbf{k} - \mathbf{k}_0)\mathbf{r}_{ij}]$$

$$= \sum_i \sum_j \exp[i(\mathbf{k} - \mathbf{k}_0)\mathbf{r}_{ij}] = \Big| \sum_i \exp[i(\mathbf{k} - \mathbf{k}_0)\mathbf{r}_i] \Big|^2 . \tag{1.31}$$

Let us suppose that the active volume has the form of an extended cylinder with the length L along the x axis and with the radius R. In this case, the equation (1.31) takes the form

$$\langle R_{3,\mathbf{k}-\mathbf{k}_0} R_{3,-\mathbf{k}+\mathbf{k}_0} \rangle \Big|_{t=0}$$

$$= \left[2N \frac{\sin(1/2\ kL\ (1 - \cos\vartheta))}{1/2\ L\ (1 - \cos\vartheta)} \right]^2 \left[\frac{J_1(k\ R\ \sin\vartheta)}{k\ R\ \sin\vartheta} \right]^2 . \quad (1.32)$$

We determine the intensity $I_{\mathbf{k}}$ by

$$I = (1/4\pi) \int d\Omega_{\mathbf{k}}\ I_{\mathbf{k}}\ . \quad (1.33)$$

It is clear from (1.33) that $I_{\mathbf{k}}$ is the intensity of emission per unit solid angle. As we have shown in Sec. 3.1.2 $\langle\sigma_{i3}(t_0)\rangle \simeq 0$ at the moment t_0 at which the emitted radiation intensity is a maximum. Hence, from (1.30)–(1.32) we get

$$I_{\mathbf{k}}(t_0) = \frac{3\hbar\omega}{2} N^2 \Gamma_0 \sin^2\vartheta_{\mathbf{m}} \left[2 \frac{\sin(1/2\ kL(1 - \cos\vartheta))}{1/2\ kL(1 - \cos\vartheta)} \right]^2$$

$$\times \left[\frac{J_1(kR\ \sin\vartheta)}{kR\ \sin\vartheta} \right]^2 . \quad (1.34)$$

It will be recalled that $\vartheta_{\mathbf{m}}$ is an angle between the vectors $\mathbf{m}$ and $\mathbf{k}$, and angle ϑ is an angle between the wave vector $\mathbf{k}$ and the x axis coinciding with the cylinder axis. We have mentioned, that vector $\mathbf{k}_0$ should be parallel to the cylinder axis. We can assume $\vartheta_{\mathbf{m}} = \pi/2$, it simply shortens the formula but does not impose any restrictions. In this case Eq. (1.34) takes the form

$$I_{\mathbf{k}} = N^2 3\hbar\omega/2T_1 \left[\frac{2\ \sin(1/2\ kL(1 - \cos\vartheta))}{1/2\ kL(1 - \cos\vartheta)} \right]^2$$

$$\times \left[\frac{J_1(kR\ \sin\vartheta)}{kR\ \sin\vartheta} \right]^2 . \quad (1.35)$$

The intensity $I_{\mathbf{k}}$ as a function of ϑ has a maximum at $\vartheta = 0$ or $\vartheta = \pi$. The peak value of $I_{\mathbf{k}}(\vartheta)$ is defined by

$$I_{\mathbf{k}}(t_0)\Big|_{max} = \frac{3N^2\hbar\omega}{2T_1} . \tag{1.36}$$

The angle halfwidth of the first factor in (1.35) is

$$\Delta\vartheta_1 = (2\pi/kL)^{1/2} , \tag{1.37}$$

and the second factor has the half–width

$$\Delta\vartheta_2 = 2/kR . \tag{1.38}$$

The final half–width of the right–hand side of (1.35) depends on the value of the Fresnel number $\Phi = Z/\lambda L$. It is determined by (1.37) for $\Phi < 1$ and (1.38) for $\Phi > 1$. Intensity is proportional to the squared number of the atoms only within the region $0 < \vartheta < \Delta\vartheta_{1(2)}$. At $\vartheta > \vartheta_{1(2)}$

$$I_{\mathbf{k}}(t_0) \approx 6N^2\Gamma_0(2/k^2LR)^2 \leq N\Gamma_0 . \tag{1.39}$$

Thus, we have shown that, as it follows from the integral of motion (1.27), the peak intensity of the emitted radiation is proportional to the squared number of the atoms. The detailed discussion of the nature of the induction of the correlations in an extended system of radiators will given later.

3.1.4. *Dynamics of macroscopical correlation in process of superradiance.* The equations for the interatomic correlation functions were derived in Sec. 2.9.1, and with the help of these equations we derived in Sec. 2.9.2 the equation for the population difference dynamics in a process of the superradiative decay. Let us consider again the Eqs. (2.9.15), (2.9.16) and try to derive the equations describing the dynamics of evolution of the macroscopic correlation. In retardationless approximation the rate of variation of the population difference is determined by the Eq. (2.9.22):

$$\frac{dR_3}{dt} + \Gamma_0(N + R_3) = -2\Gamma_0 \sum_i {\sum_j}' c_{ij} S_{ij} . \tag{1.40}$$

The interatomic interaction matrix c_{ij} can be transformed to the following form

$$c_{ij} = \frac{\sin(kr_{ij})}{kr_{ij}} = \frac{1}{4\pi}\int d\Omega_{\mathbf{k}}\ \exp(i\mathbf{k}\mathbf{r}_{ij}) \ . \tag{1.41}$$

If transformation (1.41) is applied to (1.40), the resultant equation of motion for a population difference is

$$\frac{dR_3}{dt} + \Gamma_0(N + R_3) = -(\Gamma_0/2\pi)\int d\Omega_{\mathbf{k}}\ S(\mathbf{k}) \ , \tag{1.42}$$

where

$$S(\mathbf{k}) = \sum_i {\sum_j}' <\sigma_{i+}\sigma_{j-}> \exp(i\mathbf{k}\mathbf{r}_{ij}) \ . \tag{1.43}$$

If the Eq. (1.43) is applied to (2.9.16) the following equation of motion is obtained for $S(\mathbf{k})$:

$$\frac{dS(\mathbf{k})}{dt} + \Gamma_2 S(\mathbf{k}) = \frac{\Gamma_0(N + R_3)R_3}{2N^2}\sum_i {\sum_j}' c_{ij}\ \exp(i\mathbf{k}\mathbf{r}_{ij})$$

$$+ \frac{\Gamma_0}{2}\sum_i {\sum_j}'' {\sum_l}' c_{il}S_{lj}<\sigma_{i3}> \exp(i\mathbf{k}\mathbf{r}_{ij})$$

$$+ \frac{\Gamma_0}{2}\sum_i {\sum_j}'' {\sum_l}' c_{jl}S_{il}<\sigma_{j3}> \exp(i\mathbf{k}\mathbf{r}_{ij}) \ . \tag{1.44}$$

In deriving Eq.(1.44) we have ignored the imaginary part ia_{ij} of the matrix of the interatomic interactions, which is defined by (2.9.18). Notice, that we have already used this approximation under derivation of the Eq.(2.9.24). We have also used the semiclassical procedure of decoupling of the correlators

$$<\sigma_{i+}\sigma_{j3}\sigma_{l-}> = <\sigma_{i+}\sigma_{l-}><\sigma_{j3}> \ . \tag{1.45}$$

If transformation (1.41) is applied to (1.44), the last term in the right–hand side of (1.44) takes the form

$$\Sigma_1 = \sum_i \sum_j \sum_l c_{jl}S_{il}<\sigma_{j3}> \exp(i\mathbf{k}\mathbf{r}_{ij})$$

$$= \frac{1}{4\pi} \int d\Omega_{\mathbf{k}} \sum_i \sum_j \sum_l S_{il} \langle\sigma_{j3}\rangle \exp(i\mathbf{k}\mathbf{r}_{ij} + i\mathbf{k}'\mathbf{r}_{jl}) \ . \qquad (1.46)$$

We assume that $\langle\sigma_{j3}\rangle$ is practically independent on $\mathbf{r}_j$. Notice, that this assumption is very well consistent with the retardationless approximation. In this case the Eq. (1.46) can be transformed to the following form

$$\Sigma_1 = R_3 \sum_i \sum_l S_{il} \frac{1}{4\pi} \int d\Omega_{\mathbf{k}'} P(\mathbf{k}' - \mathbf{k}) \exp(i\mathbf{k}\mathbf{r}_i - i\mathbf{k}'\mathbf{r}_l) \ , \qquad (1.47)$$

where

$$P(\mathbf{k}' - \mathbf{k}) = N^{-1} \sum_j \exp[i(\mathbf{k}' - \mathbf{k})\mathbf{r}_j] = V^{-1} \int_V d\mathbf{r} \exp[i(\mathbf{k}' - \mathbf{k})\mathbf{r}] \ .$$

For active volume in the form of an extended cylinder the function $P(\mathbf{k}' - \mathbf{k})$ was already determined (see Eq. (2.6.4)):

$$P(\mathbf{k}' - \mathbf{k}) = \frac{\sin(s_{||}L/2)}{s_{||}L/2} \frac{J_1(s_\perp R)}{s_\perp R/2} \qquad (1.48)$$

where

$$\mathbf{s} = \mathbf{k}' - \mathbf{k} \ , \quad s_{||} = k(\cos\vartheta' - \cos\vartheta) \ ,$$

$$s_\perp = k[\sin^2\vartheta' + \sin^2\vartheta - 2\sin\vartheta'\cos(\phi' - \phi)]^{1/2} \ .$$

Here (ϑ',ϕ') are the polar and azimuthal angles of the wave vector $\mathbf{k}'$ in the Cartesian frame with the z axis coinciding with the cylinder axis, and (ϑ,ϕ) are the polar and azimuthal angles of the wave vector $\mathbf{k}$ in the same frame. Taking into account that for $\kappa L \gg 1$ and $\kappa R \gg 1$ the function $P(\mathbf{k}' - \mathbf{k})$ is a highly directional function, the Eq. (1.47) can be transformed to the following form

$$\Sigma_1 = S(\mathbf{k})R_3\gamma(\mathbf{k}) \ , \qquad (1.49)$$

where

$$\gamma(\mathbf{k}) = \frac{1}{4\pi} \int d\Omega_{\mathbf{k}'} P(\mathbf{k}' - \mathbf{k})$$

$$= \int_0^{\pi} d\vartheta' \sin\vartheta' \, \frac{\sin\,[kL(\cos\,\vartheta' - \cos\,\vartheta)/2]}{kL(\cos\,\vartheta' - \cos\,\vartheta)/2}$$

$$\times \left[\frac{\sin\,\vartheta' J_1(kR\,\sin\,\vartheta') J_0(kR\,\sin\,\vartheta)}{kR(\sin^2\vartheta' - \sin^2\vartheta)} \right.$$

$$\left. - \frac{\sin\,\vartheta\; J_0(kR\,\sin\,\vartheta') J_1(kR\,\sin\,\vartheta)}{kR(\sin^2\vartheta' - \sin^2\vartheta)} \right] . \qquad (1.50)$$

Similarly to (1.46)–(1.50) we can transform the second and first terms in the right–hand side of Eq. (1.44). After these transformations the Eq. (1.44) finally is

$$\frac{dS(\mathbf{k})}{dt} + \Gamma_2 S(\mathbf{k}) = \Gamma_0 \gamma(\mathbf{k}) S(\mathbf{k}) R_3 + \frac{1}{2}\Gamma_0 \gamma_0(\mathbf{k}) R_3 (N + R_3) \;, (1.51)$$

where

$$\gamma_0(\mathbf{k}) = \frac{1}{4\pi} \int d\Omega_{\mathbf{k}'} \left[\frac{\sin(s_{||}L/2)}{s_{||}L/2} \right]^2 \left[\frac{J_1(s_{\perp}R)}{s_{\perp}R/2} \right]^2 . \qquad (1.52)$$

It is clear from the Eq. (1.51) that $\gamma(\mathbf{k})$ determines the rate of evolution of the correlators $S(\mathbf{k})$. As we can see from (1.52) this rate depends essentially on the direction of the wave vector $\mathbf{k}$. For cylindrical active media, in result of the axial symmetry, the rate of $S(\mathbf{k})$ evolution depends only on the angle ϑ_0 between the wave vector $\mathbf{k}$ and the x–axis.

The peak value of the second term in the right–hand side of (1.51) is N times smaller than the peak value of the first one. The second term determines the intensity of the spontaneous emission sources as a function of the direction of the wave vector $\mathbf{k}$. It is clear from (1.52) that the angular dependence of $\gamma_0(\mathbf{k})$ coincides practically with the angular dependence of the emission intensity defined in the previous section (see Eq. (1.35)) with the difference that averaging over the all possible directions of the wave vector $\mathbf{k}_0$ is used in (1.52). This averaging is quite natural and it is for taking into account the all possible outgoing directions of the spontaneous quanta. As in Sec. 4.1.3 we supposed that wave vector $\mathbf{k}_0$ was parallel to the cylinder axis.

Further insight into the implication of (1.51) can be obtained by considering the linear stage of radiative decay. At the linear stage, when $N - R_3 \ll N$, the Eq. (1.51) has the following solution

$$S(\mathbf{k}) = \frac{\Gamma_0 N^2 \gamma_0(\mathbf{k})}{2(\Gamma_0 N\gamma(\mathbf{k}) - \Gamma_2)} \{ \exp[(\Gamma_0 N\gamma(\mathbf{k}) - \Gamma_2)t] - 1 \}$$

and, if $\Gamma_0 N\gamma(\mathbf{k}) >> \Gamma_2$, then

$$S(\mathbf{k}) \simeq \frac{N\gamma_0}{2\gamma} \exp[\Gamma_0 N\gamma(\mathbf{k})t] . \tag{1.53}$$

Let us define the explicit form of the functions $\gamma(\mathbf{k})$ and $\gamma_0(\mathbf{k})$. Two functions in integrand of Eq. (1.50) reach maxima at $\vartheta' = \vartheta$. In the case $\Phi < 1$ the angular width of the first function is less than the width of the second one. Hence, to evaluate the integral we can integrate only the first function of this product and then multiply the result by the maximal value of the second function. In this way we get

$$\gamma(\mathbf{k}) = \frac{\lambda}{2L} [J_0^2(kR \sin \vartheta) + J_1^2(kR \sin \vartheta)] . \tag{1.54}$$

Hence, $\gamma(\mathbf{k}) = \dfrac{\lambda}{2L}$ at $\vartheta < \dfrac{1}{\kappa R}$ and it sharply decreases in value at $\vartheta > \dfrac{1}{\kappa R}$, approaching to

$$\gamma(\mathbf{k})\Big|_{\vartheta>(kR)^{-1}} \simeq \frac{\lambda}{2L}\frac{\lambda}{R} .$$

If we transform in an analogous way the integrand of Eq. (1.52) the resulting equation for $\gamma_0(\mathbf{k})$ is

$$\gamma_0(\mathbf{k}) = \frac{\lambda}{4L} [J_0^2(kR \sin \vartheta) + J_1^2(kR \sin \vartheta)]^2 . \tag{1.55}$$

Thus, the self-consistent equations of dynamics for the correlators $S(\mathbf{k})$ and population difference R_3 are

$$\frac{dS(\mathbf{k})}{dt} + \Gamma_2 S(\mathbf{k}) = \Gamma_0\gamma(\mathbf{k})S(\mathbf{k})R_3 + \frac{\Gamma_0}{2}\gamma_0(\mathbf{k})R_3(N + R_3) , \tag{1.56a}$$

$$\frac{dR_3}{dt} + \Gamma_0(N + R) = -2\Gamma_0 \frac{1}{4\pi}\int d\Omega_{\mathbf{k}}\, S(\mathbf{k}) , \tag{1.56b}$$

where

$$\gamma(\mathbf{k}) = \frac{1}{4\pi} \int d\Omega_{\mathbf{k}'} P(\mathbf{k} - \mathbf{k}') \ , \tag{1.56c}$$

$$\gamma_0(\mathbf{k}) = \frac{1}{4\pi} \int d\Omega_{\mathbf{k}'} P^2(\mathbf{k} - \mathbf{k}') \ .$$

The parameters $\gamma(\mathbf{k})$ and $\gamma_0(\mathbf{k})$ can be approximated by the relationships (1.54) and (1.55), respectively.

Taking into account that $\gamma(\mathbf{k})$ and $\gamma_0(\mathbf{k})$ are the unimodal highly directional functions in the case of needle–shaped active volumes ($\Phi < 1$), the Eqs. (1.56) can be simplified further by the following transformation

$$\frac{dR_3}{dt} + \Gamma_0(N + R_3) = -2\Gamma_0 g\ S \ , \tag{1.57}$$

where

$$g = \frac{1}{4\pi} \int_0^{2\pi} d\phi \int_0^{\vartheta_c} d\vartheta \ \sin\vartheta = \vartheta_c^2/4 = (2kR)^{-2} \ . \tag{1.58}$$

In this case we get the following equation of motion for population difference, R_3:

$$\left[\frac{d}{dt} + \Gamma_2\right]\left[\frac{d}{dt} + \Gamma_0\right](N + R_3) - \Gamma_0\gamma R_3 \frac{dR_3}{dt}$$

$$= \Gamma_0^2\gamma(N + R_3)R_3(1 - g\frac{\gamma_0}{\gamma}) \ . \tag{1.59}$$

Comparing Eq. (1.59) with Eq. (2.9.30), we can see that there is an additional term in the left–hand side of (2.9.30):

$$\frac{1}{N}\frac{dR_3}{dt}\frac{dR_3}{dt} \ .$$

This term causes the decrease of the emitted intensity in the region of its maximum. The appearance of this additional term is due to decoupling (2.9.10), which have been used in deriving (2.9.30), whereas in this section we used decoupling (1.45). Fig. 3.1 gives a comparison of the superradiance pulse profiles for these two types of decoupling.

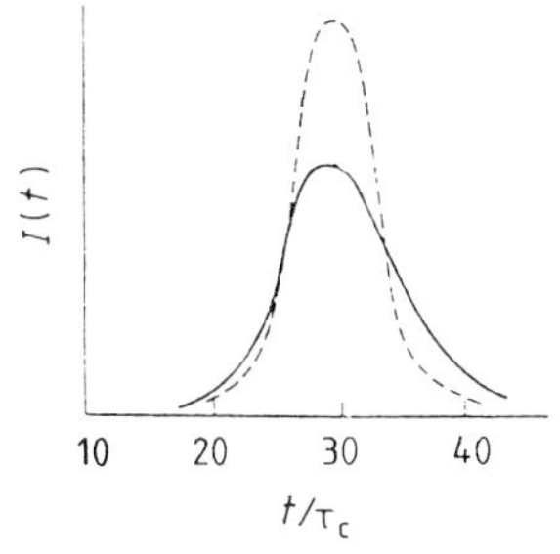

Fig.3.1. The SR intensity profiles given by Eq.(2.9.30) (solid curve), and by Eq.(1.59) (dashed curve).

3.1.5. *Cooperative frequency shift.* Only the real part of the interatomic coupling matrix, ia_{ij}, have been taken into consideration up to now. Here we discuss the effects which are due to the imaginary part of this matrix. Let us consider again the equations (2.4.8) and average them over the initial state with the application of the following decoupling

$$<\sigma_{i+}\sigma_{j3}> = <\sigma_{i+}><\sigma_{j3}> , \quad <\sigma_{i+}\sigma_{j-}> = <\sigma_{i+}><\sigma_{j-}> . \tag{1.60}$$

In this case, the (2.4.8b) takes, after summation over all atoms, the form

$$\frac{dR_3}{dt} + \Gamma_0(N + R_3) = -\frac{\Gamma_0}{2\pi}\int d\Omega_{\mathbf{k}}\ \mathcal{R}^{+}_{\mathbf{k}}\mathcal{R}^{-}_{\mathbf{k}} , \tag{1.61}$$

where

$$\mathcal{R}^{+}_{\mathbf{k}} = \sum_i <\rho_{i+}>\exp(i\mathbf{k}\mathbf{r}_i) , \quad \mathcal{R}^{-}_{\mathbf{k}} = (\mathcal{R}^{+}_{\mathbf{k}})^* .$$

In the same way Eq. (2.4.8a) can be transformed to the form

$$\frac{d\mathscr{R}_{\mathbf{k}}^{+}}{dt} + \frac{\Gamma_2 \mathscr{R}_{\mathbf{k}}^{+}}{2} = \frac{i\Gamma_0}{2} \sum_j <\rho_{j+}> \exp(i\mathbf{kr}_j) \sum_i <\sigma_{i3}> \frac{\exp(i\mathbf{kr}_{ij} - ikr_{ij})}{kr_{ij}} .(1.62)$$

If we factor the slowly varying function of index i on the right–hand side of (1.62) out of the summation sign and assume that the remaining sum,

$$\sum_i \frac{\exp(i\mathbf{kr}_{ij} - ikr_{ij})}{kr_{ij}} ,$$

does not almost depend on the index j, we get the following equation of motion for $\mathscr{R}_{\mathbf{k}}^{+}$:

$$\frac{d\mathscr{R}_{\mathbf{k}}^{+}}{dt} + \frac{\Gamma_2}{2} \mathscr{R}_{\mathbf{k}}^{+} = \frac{\Gamma_0}{2} \gamma(\mathbf{k}) R_3 \mathscr{R}_{\mathbf{k}}^{+} , \qquad (1.63)$$

where

$$\gamma(\mathbf{k}) = iV^{-1} \int d\mathbf{r} \frac{\exp(i\mathbf{kr} - ikr)}{kr}$$

$$= \frac{\lambda}{2L} [J_0^2(k_\perp R) + J_1^2(k_\perp R)] - \frac{i\lambda}{2L} [J_0(k_\perp R)\, N_0(k_\perp R) + J_1(k_\perp R)\, N_1(k_\perp R)] .(1.64)$$

The last equation have been derived under the assumption that the shape of the active volume (in a form of cylinder with the radius R and length L)satisfies the condition $\Phi < 1$. In addition to this we have assumed that $\kappa_\perp = \kappa \cos\vartheta$ has not been equal to zero precisely. Notice that the real part of $\gamma(\mathbf{k})$ defined by (1.64) coincides with $\gamma(\mathbf{k})$ defined by (1.54).

The equation (1.63) has the following solution for the initial stage of the superradiance dynamics:

$$R_{\mathbf{k}}^{+} = \mathscr{R}_{\mathbf{k}}^{+} \exp(i\omega_0 t) = \sum <\sigma_{i+}> \exp(i\mathbf{kr}_i)$$

$$= R_{\mathbf{k}}^{+}(0) \exp\{i[\omega_0 + \frac{1}{2}\Gamma_0 \gamma''(\mathbf{k})N]t + \frac{1}{2}[\Gamma_0 \gamma(\mathbf{k})N - \Gamma_2]t\} .$$

And at the final stage, when $R_3 = -N + U$, $U \ll N$, the solution of (1.63) is

$$R^+_{\mathbf{k}} = R^+_{\mathbf{k}}(0)\, \exp\{i[\omega_0 - \frac{1}{2}\Gamma_0\gamma''(\mathbf{k})N]t - \frac{1}{2}[\Gamma_0\gamma(\mathbf{k})N + \Gamma_2]t\} .$$

Comparing this two equations we see that the frequency of radiation changes from the value $\omega_0 + \Gamma_0\gamma''(\mathbf{k})N/2$ in the front of the SR pulse to $\omega_0 - \Gamma_0\gamma''(\mathbf{k})N/2$ in its tail [3,4]. The frequency shift $\Delta\omega = \Gamma_0\gamma''(\mathbf{k})N$ is proportional to the number of atoms, N.

3.2. Non-Markovian theory of superradiance

The boson operator exclusion method, which has been used in the previous paragraph, possesses the clearness and simplicity, and sharply displays the mechanism of the correlation induction. These features manifest themselves in the form of equations (2.9.15) and (2.9.16), and, in the most prominent form, show up in Eqs. (1.56). In fact, it is clear from the above mentioned equations that the angular spectrum of the emitted radiation depends on the active volume shape, and it is determined by the interatomic interaction matrix $\|a_{ij}\|$. In their turn, the coefficients a_{ij} , reflecting the manner of interaction of radiators, have a structure of the spherical wave emitted by the point source.

Thus, the manner in which the macroscopic polarization is induced is determined by the action on any given radiators of the summary field emitted by all other radiators of this ensemble. It should be emphasized that the final compact form of equations (1.56) is a result of serious approximations made in process of their deriving. One of these approximations is a retardationless approximation. It is this approximation which allows us to transform the equations of superradiance dynamics into the form of the Markov equations, where the electromagnetic field is excluded adiabatically.

Here, we use the atomic–field representation to analyse some distinctive features of superradiance dynamics and we shall compare the results presented here with the results presented above.

3.2.1. *Conditions for superradiance occurrence.* The equations describing the evolution of the number of photons $n_{\mathbf{k}}$ in the field mode **k** have been derived in Sec.2.6. In general case, the number of equations in the set of Eq. (2.6.11) tends to infinity, because the number of the field modes **k** is practically equal to infinity. With the help of some simplifications we can restrict the number of modes involved. Nevertheless, for the case of the N modes the number of equations is equal to $4N + 1$. Hence, for $N > 1$ it is difficult to analyze the equations analytically even at the linear stage of decay.

If the coefficients of equations (2.6.11) did not depend on **k** we could essentially simplify these equations by summing them over all possible **k** . The most essential dependence on **k** is in the coefficient $i(\omega_{\mathbf{k}} - \omega_0)$, because the frequency $\omega_{\mathbf{k}}$ in coefficient $1/T_{\mathbf{k}}^2$ in resonance approximation can be substituted by ω_0 with the high accuracy. The terms proportional to $i(\omega_{\mathbf{k}} - \omega_0)$ describe the transfer of radiation. Really, in the frames of the one–dimensional approximation we get from to (2.3.1), (2.3.4):

$$A^{+}(\mathbf{r},t) = \sum_{\mathbf{k}} \sqrt{\frac{2\pi\hbar c^2}{W\omega_{\mathbf{k}}}}\, \mathbf{e}_{\lambda} a_{\mathbf{k}}^{+}(t) \exp(-i\mathbf{kr})$$

$$= \sum_{\mathbf{k}} c_{\mathbf{k}} \mathbf{e}_{\lambda} a_{\mathbf{k}}^{+}(t) \exp[i\omega_{\mathbf{k}}(t - x/c) ,$$

i.e., the field amplitude depends both on the time t and the retarded time $t - x/c$. On the other hand, if we assume, that $\omega_{\mathbf{k}} = \omega_0$, then the field amplitude no longer depends on the retarded time. Nevertheless, we shall assume for the sake of simplicity that $\omega_{\mathbf{k}} = \omega_0$ and the process of the radiation transfer will be incorporated by including the phenomenological relaxation term $n_{\mathbf{k}}/\tau_{\mathbf{k}}$ (where $\tau_{\mathbf{k}}$ is the time taken by a photon to cross the active volume along the direction of the field mode **k**) into the equations of motion. Under these approximations, the equations (2.6.11) are significantly simplified and take the form [5,6]:

$$\frac{dn_{\mathbf{k}}}{dt} + \frac{n_{\mathbf{k}}}{\tau_{\mathbf{k}}} = F_{\mathbf{k}} , \tag{2.1a}$$

$$\frac{dF_{\mathbf{k}}}{dt} + \frac{1}{2}\left[\frac{1}{\tau_{\mathbf{k}}} + \frac{1}{T_2}\right] F_{\mathbf{k}} = \frac{1}{T_{\mathbf{k}}^2}(n_{\mathbf{k}}R_3 + S_{\mathbf{k}} + S_0) \text{ ,(2.1b)}$$

$$\frac{dS_{\mathbf{k}}}{dt} + \frac{S_{\mathbf{k}}}{T_2} = F_{\mathbf{k}}R_3 \text{ ,} \qquad (2.1c)$$

$$\frac{dR_3}{dt} = -2\sum_{\mathbf{k}} F_{\mathbf{k}} \text{ .} \qquad (2.1d)$$

Let us consider the right–hand side of Eq. (2.1b). The variable $F_{\mathbf{k}}$ determines the rate of the energy exchange between the active medium and the field mode **k** (in depend on its sign the energy is transferred from the active medium to the field or vice versa). This energy exchange is realized via the processes of the induced emission, $n_{\mathbf{k}}R_3$, collective spontaneous emission, $S_{\mathbf{k}}$, and isotropic spontaneous decay described by the term

$$S_0 = \frac{1}{2}(N + R_3) = \sum_{i=1}^{N} \langle\sigma_{i+}\,\sigma_{i-}\rangle \text{ .}$$

It is seen that S_0 is equal to the number of atoms in the excited state. The term $S_{\mathbf{k}}$ is a correlation function of the collective atomic operators associated with the field mode **k**. It is this correlation which can lead, under certain conditions, to a cooperative decay.

The first two equations in (2.1) show that the temporal evolution of $n_{\mathbf{k}}(t)$ is determined by the right–hand side of Eq. (2.1b), i.e., by the net effect of the induced, collective, and spontaneous processes. Let us compare the effect upon the emission intensity of the induced and collective processes. From the first and third equations of (2.1) we get

$$n_{\mathbf{k}}R_3 + S_{\mathbf{k}} = \int_0^t F_{\mathbf{k}}(t')\,\{R_3(t)\exp[-(t-t')/\tau]$$

$$+ R_3(t')\exp[-(t-t')/T_2]\}dt' \text{ .} \qquad (2.2)$$

It is seen that for atomic systems with $\tau = L/c \ll T_2$ the emission intensity is primarily determined by the collective processes $(S_{\mathbf{k}})$, since in this case the first term in the curly brackets is much less than the second one, when $t > \tau$. While in systems with $\tau \gg T_2$ the processes of the induced emission are predominant.

At the initial linear stage of a system dynamics, when the condition $N - R_3 \ll N$ holds, the solution of the linearized equations (2.1) has the form $n_{\mathbf{k}} = n_{0\mathbf{k}} \exp(\lambda_{\mathbf{k}} t)$. The solutions of appropriate characteristic equation are determined by

$$\lambda_{1\mathbf{k}} = -\frac{1}{2}\left[\frac{1}{\tau_{\mathbf{k}}} + \frac{1}{T_2}\right],$$

$$\lambda_{2,3\mathbf{k}} = \lambda_{1\mathbf{k}} \pm \left[\frac{1}{4}\left[\frac{1}{\tau_{\mathbf{k}}} - \frac{1}{T_2}\right]^2 + \frac{2N}{T_{\mathbf{k}}^2}\right]^{1/2} \qquad (2.3)$$

Thus, the exponentially increasing solutions are appeared only, when the following condition holds [5,6]:

$$\mu L_{\mathbf{k}} = \frac{2N\tau_{\mathbf{k}} T_2}{T_{\mathbf{k}}^2} = \frac{3\lambda^2}{4\pi} \frac{N}{V} \frac{T_2}{T_1} L_{\mathbf{k}} > 1 . \qquad (2.4)$$

Hence, for the case of the needle–shaped active volume the near–axial modes will be predominant, because for these modes the quantity $L_{\mathbf{k}}$ reaches its maximum equal to the active volume length: $L_{\mathbf{k}} = L$.

Let the number of the near–axial modes is equal to ν , for the designation of these modes we adopt the term "end–fire modes", which was introduced in Ref. 3. We shall designate the net number of photons in the end–fire modes by $n_1 = \sum_{\mathbf{k}_0} n_{\mathbf{k}_0}$, where $\mathbf{k}_0$ is a wave vector of an end–fire mode. Let us designate as n_2 the net number of photons in the remainder ("off–axial") modes. If we perform a separate summation over the end–fire and off–axial modes, we get

$$\frac{dn_1}{dt} + \frac{n_1}{\tau_1} = F_1 ,$$

$$\frac{dF_1}{dt} + \frac{1}{2}\left[\frac{1}{\tau} + \frac{1}{T_2}\right] F_1 = \frac{1}{T_0^2}(n_1 R_3 + S + \nu S_0) ,$$

$$\frac{dS}{dt} + \frac{S}{T_2} = F_1 R_3 ,$$

$$\frac{dn_2}{dt} + \frac{n_2}{\tau_2} = F_2 , \qquad (2.5)$$

$$\frac{dF_2}{dt} + \frac{1}{2}\left[\frac{1}{\tau_2} + \frac{1}{T_2}\right] F_2 = \frac{1}{T^2} S_0 ,$$

$$\frac{dR_3}{dt} = -2\,(F_1 + F_2) ,$$

where $\tau_1 = L/c$, $\tau_2 = L_2/c$, L_2 is an effective length (thickness) of the active volume for the off-axial modes. Notice, that the inequality $L_2 < L$ holds for a needle shaped active volume. F_1 and F_2 are the sums of $F_{\mathbf{k}}$ over the end-fire and off-axial modes, respectively. $1/T^2$ is a mean value of $1/T^2_{\mathbf{k}}$ for the off-axial modes.

In equation for F_2 we have omitted the terms describing the induced and collective processes, because they are negligibly small in comparison with S_0 . In fact, for systems with $\tau \ll T_2$ the induced term is much smaller than the collective term, as it follows from (2.2). Let us compare the sums $\sum_{\mathbf{k}'} S_0$ and $\sum_{\mathbf{k}'} S_{\mathbf{k}'}$, where $\mathbf{k}'$ is a wave vector of an off-axial mode. The first of these sums is equal to the product of the number of the off-axial modes (for a needle-shaped active volume this number is approximately equal to the total number of modes, see (2.30)) and the number of the excited atoms, N_2 :

$$\sum_{\mathbf{k}'} S_0 \simeq (Vk^2/2\pi L_2)N .$$

The value of the second sum can be evaluated by

$$\sum_{\mathbf{k}'} S_{\mathbf{k}'} \ll \sum_{i,j} \frac{\sin(kr_{ij})}{kr_{ij}} \left[\frac{N}{V}\right]^2 \frac{4\pi V}{k^2} .$$

Thus, in a wide range of variation in the density of atoms N/V , and in the wavelength of radiation λ , we get

$$\sum_{\mathbf{k}'} S_{\mathbf{k}'} \ll \sum_{\mathbf{k}'} S_0 \; .$$

As the emission into the off–axial modes is due to the spontaneous decay process, then $dn_2/dt \quad n_2/T_1$. Hence, for the superradiative decay the inequality $dn_2/dt \ll n_2/T_2$ holds, because in this case $\tau << T_2 << T_1$. Then it is not difficult to check that the expression for F_2 is

$$F_2 = S_0/T_1' \; ,$$

where $1/T_1'$ is the rate of the isotropic spontaneous–emission into a solid angle of 4π (after the effect of the end–fire modes has been subtracted). If (2.6) is applied to (2.5), we get the following self–consistent equations of motion for the number of photons in the end–fire modes $n = n_1$:

$$\frac{dn}{dt} + \frac{n}{\tau} = F \; ,$$

$$\frac{dF}{dt} + \frac{1}{2}\left(\frac{1}{\tau} + \frac{1}{T_2}\right) F = \frac{1}{T_0^2}\,(nR_3 + S + \nu S_0) \; ,$$

$$\frac{dS}{dt} + \frac{S}{T_2} = FR_3 \; , \qquad (2.7)$$

$$\frac{dR_3}{dt} + \frac{N + R_3}{T_1'} = -2\,F \; .$$

To determine the number of the end–fire modes, we sum the variable $S_{\mathbf{k}}$, which appears in (2.1) and which determines the angular spectrum of emission, over all possible directions of the wave vector $\mathbf{k}$:

$$\sum_{\mathbf{k}} S_{\mathbf{k}} = \sum_{\mathbf{k}} \sum_{i} {\sum_{j}}' <\sigma_{i+}\sigma_{j-}> \exp[i\mathbf{k}(\mathbf{r}_i - \mathbf{r}_j)] = \sum_{i} {\sum_{j}}' c_{ij} <\sigma_{i+}\sigma_{j-}> \; , \qquad (2.8)$$

where the interatomic interaction matrix c_{ij} is determined by (2.9.23).

Let us determine the number of modes which make a significant contribution to the emission process. We adopt the procedure of Ref. 7

and introduce the eigenfunctions $\psi_q(\mathbf{r}_i)$ and eigenvalues q of the matrix c_{ij} :

$$\sum_j c_{ij}\psi_q(\mathbf{r}_j) = q\ \psi_q(\mathbf{r}_i)\ ,\quad j = 1,...,N. \tag{2.9}$$

The eigenfunctions $\psi_q(\mathbf{r}_i)$ satisfy the completeness and orthonormality conditions:

$$\sum_i \psi_q(\mathbf{r}_i)\ \psi_{q'}(\mathbf{r}_i) = \delta_{qq'}\ ,\quad \sum_q \psi_q(\mathbf{r}_i)\ \psi_q(\mathbf{r}_j) = \delta_{ij}\ . \tag{2.10}$$

From (2.9) and (2.10) we get

$$c_{ij} = \sum_q q\ \psi_q(\mathbf{r}_i)\ \psi_q(\mathbf{r}_j)\ . \tag{2.11}$$

If (2.11) is applied to (2.8), we obtain

$$\sum_{\mathbf{k}} S_{\mathbf{k}} = \sum_q q \langle R^+(q)\ R^-(q)\rangle\ ,\quad R^{\pm}(q) = \sum_i \sigma_{i\pm}\ \psi_q(\mathbf{r}_i)\ .$$

The eigenfunctions ψ_q and eigenvalues q can be calculated for a cylindrical active volume in the limits of the small and large Fresnel numbers. For these cases, all the largest values of q_0 are ν–fold degenerated and are given by

$$\begin{aligned} q_0 &= \pi N/kL\ , \quad \nu = 1/\Phi \gg 1 \quad \text{for } \Phi \ll 1\ , \\ q_0 &= \pi N/k^2 Z\ , \quad \nu = 2\Phi \gg 1 \quad \text{for } \Phi \gg 1\ . \end{aligned} \tag{2.12}$$

The other eigenvalues satisfy the condition $q \ll q_0$. The eigenfunctions are

$$\psi_q(\mathbf{r}_i) = (1/N)^{1/2}(\cos \mathbf{k}_0\mathbf{r}_i + \sin \mathbf{k}_0\mathbf{r}_i)\ , \tag{2.13}$$

where $|\mathbf{k}_0| = \omega/c$. It follows from the comparison of the equations (2.13), and (2.12) with the formulas of the previous chapter, that the equations (2.12) and (2.13) correspond to the paraxial approximation, which have been discussed in Sec. 4.1.3.

It is clear from equations (2.1), (2.8), and (2.11), that the quantity $2\tau q/T_0^2$ determines the rate of a superradiative decay into the

q –th mode of the field. As we have demonstrated above, for cylindrical active volumes there are the ν modes with the eigenvalues $q = q_0$, whereas the eigenvalues associated with all other modes are essentially smaller. Thus, for the active volume with the form of a cylinder the number of the end–fire modes is determined by relationships (2.12).

From the mathematical point of view, the introduction of the functions $\psi_q(\mathbf{r}_i)$ corresponds to a transformation from the plane waves of a quantization volume to the eigenfunctions of the volume of the active medium.

3.2.2. *Dynamics of superradiance in multimode case.* The equations of motion (2.7) depends on the following main temporal parameters: the active volume transit time τ, the homogeneous relaxation time T_2, the spontaneous decay time T_1, defined by

$$\frac{1}{T_1} = \frac{4\ \omega |m|^2}{3\hbar c^3},$$

and the coherence time τ_c:

$$\frac{1}{\tau_c} = \frac{3\lambda^2}{4\pi} \frac{N}{V} \frac{L}{T_1}. \tag{2.14}$$

In the case when we can neglect the transversal and longitudinal relaxation processes, i.e., if we can assume that $T_2^{-1} = T_1'^{-1} = 0$ in the last two equations in (2.7), we obtain the following integral of motion

$$\sum_{\mathbf{k}_\nu} \langle R_{\mathbf{k}}^{+} R_{\mathbf{k}}^{-} \rangle + \frac{1}{4} R_3^2 = N(\frac{1}{4} N + \nu). \tag{2.15}$$

In the case $\tau \ll T_2$ we can assume that $dF/dt \ll F/\tau$, then we get the linear relationship between the variables F and S. Applying this relationship and using (2.15) we can easily get the self–consistent differential equation for R_3. To define the temporal profile of intensity we can express S in terms of a solution of differential equation for R_3 and then substitute it into the equation (1.6). The resultant equation for the integral intensity of the emission into the end–fire modes is [8 –14] :

$$I = (N\hbar\omega/4\tau_p)\,\mathrm{sech}^2((t - t_0)/2\tau_p)\ , \tag{2.16}$$

where

$$t_0 = \tau_p \ln(N/\nu)\ , \tag{2.17a}$$

$$\tau_p^{-1} = 2\tau(N + \nu)/T_0^2\ , \tag{2.17b}$$

and ν is determined by (2.12). The interpretation of (2.17a), (2.17b) will be given in Sec. 3.2.5.

It is clear from the equations (2.7) that in general case the order of the self–consistent differential equation for the variable R_3 is higher than the second order of the equations (1.56) and (2.9.30). Let us define the conditions under which the set of equations (2.7) can be transformed into a second order differential equation for R_3. We can approximate the time derivative of n by the following relationship

$$\frac{dn}{dt} \simeq \frac{n}{\tau_p}\ ,$$

where τ_p is the temporal width of the emitted pulse. As it follows from (2.16), at $\tau \ll \tau_p$ the first two equations in (2.7) can be transformed into the following relationships

$$n/\tau = F\ , \tag{2.18a}$$

$$F = (2\tau/T_0^2)(nR + S + \nu S_0)\ . \tag{2.18b}$$

On the other hand, at $\tau \ll T_2$, we can ignore the first term in the right–hand side of Eq. (2.18b) according to (2.2), because in this case the collective emission processes are predominant. As a result we easily obtain the following equation for the population difference R_3:

$$\left[\frac{d}{dt} + \Gamma_0'\right]\left[\frac{d}{dt} + \Gamma_2\right](N + R_3) - \frac{1}{N\tau} R_3 \frac{dR_3}{dt}$$

$$= \frac{1}{N\tau_p}\Gamma_0'(N + R_3)R_3 - \frac{2\nu}{N\tau_p}\left[\frac{d}{dt} + \Gamma_2\right]S_0\ . \tag{2.19}$$

It is easily to check that obtained equation coincides with the equation (1.59).

3.2.3. *Superradiance in weakly amplifying media.* Let us now discuss how the processes of the transversal relaxation influence on the dynamics of a superradiance. To demonstrate the role of these processes we should obtain the solution of equations (2.7) accounting the finiteness of the relaxation time T_2 .

The analysis of the SR dynamics can be divided into the three stages. In the first stage the correlated state is formed from the completely uncorrelated initial state. In this stage the condition $N - R_3 \ll N$ holds and the equations (2.7) become linear. The set of solutions of the characteristic equation for this stage is determined by (2.3), where all τ_k is equal to $\tau = L/c$. The equations (2.7) become also linear in the third stage, where $N - |R_3| \ll N$. In this stage, the decay of the correlated states occurs. This decay has the exponential or oscillatory form in depend on the system parameters. Therefore, the parameters of the emitted pulses, such as the delay time t_0 and the frequency of oscillations, can be obtained by analysing the solutions of the linearized equations. For example, using (2.3) we can easily obtain the following equation for the delay time t_0 of a superradiance pulse

$$t_0 = 2\tau \left[\left[\left(1 - \frac{\tau}{T_2}\right)^2 + 4\frac{\tau}{\tau_c}\right]^{1/2} - \left(1 + \frac{\tau}{T_2}\right)\right]^{-1} \ln\left(\frac{N}{\nu}\right) \quad .(2.20)$$

The second – intermediate – stage is stage of superradiance pulse formation. It is clear that this stage should be described by the nonlinear equations, but in this case we can again try to find some ways to simplify the analytic solution of the problem. For example, for systems with $\tau \ll \tau_c$ we have $d^2R_3/dt^2 \ll (1/\tau)dR_3/dt$. In this case the population difference dynamics is described by

$$R_3(t) = \frac{N\tau_c}{T_2} - N\left[\left(1 - \frac{\tau_c}{T_2}\right)^2 + \frac{4\nu}{N}\right]^{1/2}$$

$$\times \tanh\left[\frac{1}{2}\left(\frac{1}{\tau_c} - \frac{1}{T_2}\right)(t - t_0)\right] . \quad (2.21)$$

The intensity of emission is determined by [15]:

$$I(t) = \frac{N\hbar\omega}{4\tau_c}\left(1 - \frac{\tau_c}{T_2}\right)^2 \operatorname{sech}^2\left[\frac{1}{2}\left(\frac{1}{\tau_c} - \frac{1}{T_2}\right)(t - t_0)\right] \quad ,(2.22)$$

where

$$t_0 = \left(\frac{1}{\tau_c} - \frac{1}{T_2}\right)^{-1} \ln\left[\frac{N}{\nu}\left(1 - \frac{\tau_c}{T_2}\right)\right] . \qquad (2.23)$$

It is seen from (2.22) that the peak intensity is given in terms of the gain coefficient μ_0 and medium length L by

$$I_{max} = \frac{N\hbar\omega}{4\tau_c}\left(1 - \frac{1}{\mu_0 L}\right)^2 . \qquad (2.24)$$

The Eq. (2.24) shows that the condition for superradiative decay is

$$\mu_0 L > 1 . \qquad (2.25)$$

From (2.21)–(2.23) we see that, while in strongly amplifying media all N initially excited atoms come into the correlated state, only a certain effective number of atoms do so in a weakly amplifying media. This effective number of atoms is given by

$$N_{eff} = N\left(1 - \frac{1}{\mu_0 L}\right) .$$

Only these atoms involve in the process of the cooperative radiative decay, hence, the population difference immediately after the superradiance pulse emission is

$$R_{3f} = -N\left(1 - \frac{2}{\mu_0 L}\right) . \qquad (2.26)$$

Let us compare the equations (2.20) and (2.23) for the delay time of a superradiance pulse t_0 . The first of them was derived by equating the sum of ν exponents with the increments (2.23) to its extreme value, which is N . The second one is a result of the approximate solution of the nonlinear equations (2.7). A comparison with the results of numerical calculations shows that in the region $\tau \ll \tau_c$ the approximate solution essentially coincides with the exact one. Thus, the analysis above shows that the equation (2.20) becomes the most general formula for a superradiance delay time if we replace N by N_{eff} in the logarithm in the right–hand side of (2.20). It is not difficult to see that cofactor of the logarithm in Eq. (2.23) follows from the corresponding cofactor in (2.20) in the case $\tau \ll \tau_c$ and $\tau \ll T_2$. The resultant formula for decay time t_0 is

$$t_0 = 2\tau \left[\left[\left(1 - \frac{\tau}{T_2} \right)^2 + \frac{4\tau}{\tau_c} \right]^{1/2} - \left(1 + \frac{\tau}{T_2} \right) \right]^{-1} \ln \left[\frac{N}{\nu} \left(1 - \frac{1}{\mu_0 L} \right) \right] .$$

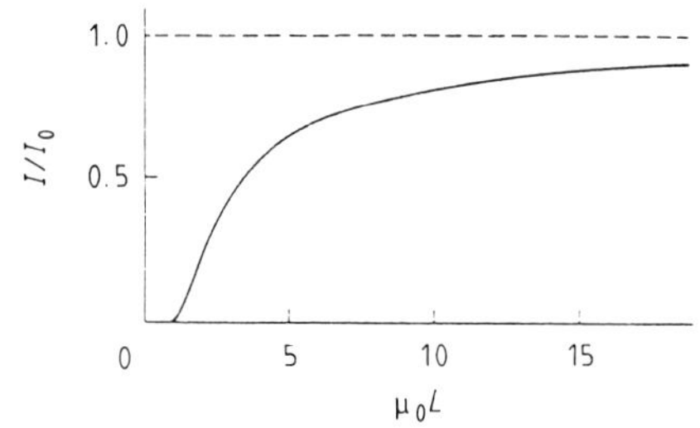

Fig.3.2. Peak intensity as a function of gain. The peak intensity is normalized to the peak intensity of strongly amplifying media ($I = N/4\tau_c$).

Fig. 3.2 shows the peak intensity, normalized to the peak intensity of a superradiance in strongly amplifying media $I_0 = N/4\tau_c$, as a function of gain $\mu_0 L$. We can see from the results presented in Fig. 3.2 that for $\mu_0 L \simeq 3$ the peak intensity is only twice less than the peak intensity for strongly amplifying media. Notice, that plot in Fig. 3.2 is not a real dependence of the peak intensity on the medium gain, because the peak intensity of strongly amplifying media I_0 depends, in its turn, on the gain. The pulse profiles for three values of a gain are shown in Fig. 3.3 (Ref.16).

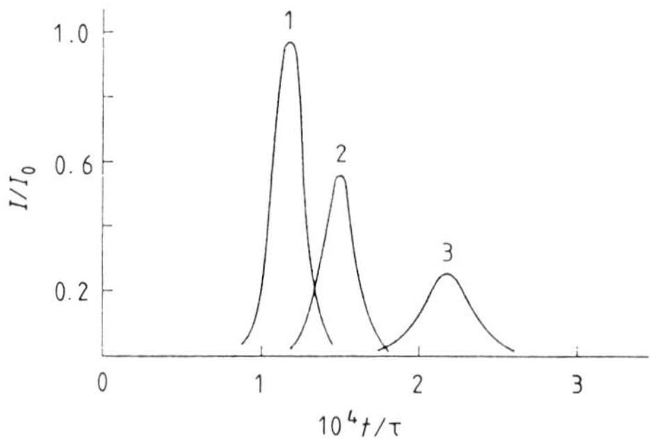

Fig.3.3. Dependence of the SR pulse shape on gain: $\mu_0 L = 200$ (curve 1), 4 (2), 2 (3).

3.2.4. *Oscillatory superradiance.* The successful experiments by Skribanowitz et al. [17], which revealed the basic features of the effect – the high directivity of emission and the quadratic dependence of the peak intensity on the atoms density – induced the enhanced interest in the problem. An interesting feature of the superradiance pulse profile recorded in [17] was its oscillatory structure. Some typical oscilloscope traces with superradiance pulse from Ref. 17 are reproduced in Fig. 3.4.

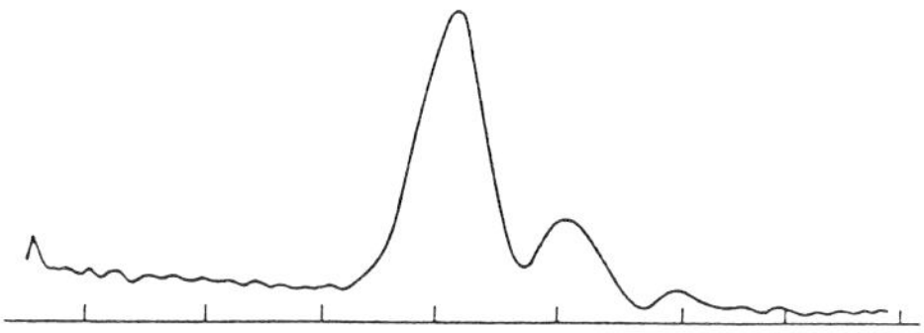

Fig.3.4. Oscilloscope trace of SR pulse (Ref.17).

It is not difficult to understand the reason of the appearance of oscillations. In fact, in the case $\tau_c < \tau$ the radiators can be phased only inside the regions with the dimensions of the order of $l_c = c\tau_c$. As the radiation, emitted in the central regions of active volume, propagates through the adjacent "coherence regions", it may be absorbed there. Since, the field may induce the transitions of the atoms, which have emitted the photons, back to the excited state. It is subsequent decay of these repeatedly excited atoms, which leads to the appearance of the post pulses. The frequency of oscillations was determined for the first time in Refs. 5, 6, 18.

The set of the solutions of the characteristic equation in the third stage is

$$\lambda_1 = -\frac{1}{2}\left[\frac{1}{\tau} + \frac{1}{T_2}\right],$$
$$\lambda_{2,3} = \lambda_1 \pm \left[\frac{1}{4}\left(\frac{1}{\tau} - \frac{1}{T_2}\right)^2 + \frac{1}{T^2}\frac{R_{3f}}{N}\right]^{1/2}. \tag{2.27}$$

We can see that the oscillations will appear when $R_{3f} < 0$. The condition for the oscillatory superradiance is

$$\mu_0 L > \frac{1}{4}\left[6 + \frac{\tau}{T_2} + \frac{T_2}{\tau}\right]. \tag{2.28}$$

The oscillation frequency is defined by [5,6]:

$$\Omega = \left[\frac{1}{T^2} \frac{|R_{3f}|}{N} - \frac{1}{4} \left(\frac{1}{\tau} - \frac{1}{T_2} \right)^2 \right]^{1/2} . \tag{2.29}$$

3.2.5. *Competition between collective and spontaneous processes.* We turn now to the study of the radiative decay in slightly anisotropic systems. In studying the decay of extended systems we summed over end–fire and off–axial modes separately in (2.1). In the case of nearly isotropic systems, all the field modes are equivalent and Eq. (2.1) must now be summed over all modes. The quantity ν in (2.7) is now determined by the number of modes in the active volume V. Using the number of modes in the volume V, we get

$$\nu_0 = 8\pi V\omega^2\Delta\omega/(2\pi c)^3 = V\omega^2/2\pi c^3\tau . \tag{2.30}$$

On the other hand, the probability of a spontaneous decay is determined by

$$T_1^{-1} = (2\pi/\hbar)|g|^2\rho(\hbar\omega) = 2\tau\nu_0/T_0^2 , \tag{2.31}$$

where $\rho(\hbar\omega)$ is a density of the final states. The equations (2.14), (2.30), and (2.31) generate the following relationship

$$\frac{\nu_0}{\tau_c} = \frac{N}{T_1} .$$

Thus, in a completely isotropic active medium, there is absolutely no correlated emission, because $\nu = N$ and hence, $\tau_c = T_1$. On the other hand, if the condition $\nu \leq N$ holds, a slight correlation may be obtained.

Using the Eq. (2.31) as the base we can give now an explicit interpretation of formulas of Sec. 3.2.2. For example, the pulse width τ_p is determined by Eq. (2.18b):

$$\frac{1}{\tau_p} = \frac{2\tau}{T_0^2} (N + \nu) .$$

Comparing this relationship with the Eq. (2.31) we can rewrite it in the form

$$\frac{1}{\tau_p} = \frac{1}{\tau_c} + \frac{1}{T''_1},$$

where

$$\frac{1}{T''_1} = \frac{1}{T_1} - \frac{1}{T'_1}$$

is a rate of spontaneous emission into the end–fire modes. Thus, an atomic system evolves in superradiance if $\tau_c \ll T''_1$. When τ_c approaches to T''_1, the coherent part of emission disappears.

If (2.30) and (2.31) are used in (2.7), we get for the temporal profile of a population difference [19]:

$$R_3(t) = N\alpha - N\beta \tanh[\beta(t - t_0)/2\tau_c], \tag{2.32}$$

where

$$\alpha = \frac{\tau_c}{T_2} + \frac{\tau_c}{T_1}, \quad \beta = \left(\left[1 - \left(\frac{\tau_c}{T_2} - \frac{\tau_c}{T_1}\right)\right]^2 + \frac{4\tau_c^2}{T_2 T_1}\right)^{1/2},$$

$$t_0 = \frac{\tau_c}{\beta} \ln \frac{\beta + (1 - \alpha)}{\beta - (1 - \alpha)}. \tag{2.33}$$

The peak intensity of emission is defined by

$$I = \frac{N}{2T_1}\left(1 + \frac{\tau_c}{T_2} + \frac{\tau_c}{T_1}\right) + \frac{N}{4\tau_c}\left[\left(1 - \frac{\tau_c}{T_2}\right)^2 - \left(\frac{\tau_c}{T_1}\right)^2\right], \tag{2.34}$$

where the first term describes the intensity of noncoherent spontaneous decay, while the second describes the intensity of the coherent decay. Thus, the coherent component of the field disappears, when

$$\frac{1}{T_1} = \frac{1}{\tau_c} - \frac{1}{T_2}.$$

In this case, the intensity of emission is defined by

$$I = \frac{N\hbar\omega}{T_1},$$

and the delay time becomes equal zero. Really, according to (2.33) $t_0 = 0$

when $\alpha = 1$. There is a purely spontaneous decay in a such system.

Notice that the radiative decay of atomic ensembles with the comparable relaxation times $\tau \sim T_1 \sim T_2$ can no longer be referred to as a superradiance, since the intensities of coherent and noncoherent components of the emitted field become comparable, according to (2.34), so that I ceases to vary in proportion to $(N/V)^2$.

To analyse the angular spectra of emission for weakly anisotropic systems we can use the equations (1.56). In isotropic active volumes the parameter $\gamma(\mathbf{k})$ ceases to be dependent on vector $\mathbf{k}$ direction. We can easily check this, if we remember the equation for $P(\mathbf{k}' - \mathbf{k})$:

$$P(\mathbf{k}' - \mathbf{k}) = \frac{1}{V} \int_V \exp[i(\mathbf{k}' - \mathbf{k})\mathbf{r}] \, d\mathbf{r} \ .$$

It is seen that for an isotropic active volume the function $P(\mathbf{k}' - \mathbf{k})$ depends on $\Delta k = |\mathbf{k}' - \mathbf{k}|$ only, i.e., it depends on the angle between the vectors $\mathbf{k}'$ and $\mathbf{k}$ only. Hence, the parameter $\gamma(\mathbf{k})$, which, in its turn, is defined by

$$\gamma(\mathbf{k}) = \frac{1}{4\pi} \int d\Omega_{\mathbf{k}'} \, P(\mathbf{k}' - \mathbf{k}) \ ,$$

does not depend on the vector $\mathbf{k}$ direction. If we assume, for example, that volume V has a form of a ball with the radius R, we get

$$\gamma(\mathbf{k}) = \frac{3}{2} k^2 R^2 \ .$$

Thus, the anisotropy of $S(\mathbf{k})$ due to the anisotropy of $\gamma(\mathbf{k})$ disappears in the case of an isotropic active volume, so that the intensity of emission becomes also isotropic, according to (1.6).

3.2.6. *Markovian limit.* In Sec. 3.2.2 we have shown that the equations (2.7) have transformed to the Markovian equations under an adiabatic exclusion of the field variables. Further insight into the implication of the Markovian type approximations can be obtained by considering the Markov approximation for the operator equations directly. In obtaining Eqs. (2.7) we have used the equations (2.3.13). The Eq. (2.3.13a) can be transformed to the following form

$$a_{\mathbf{k}}(t) = a_{\mathbf{k}}^{(v)}(t) + \frac{i g_{\mathbf{k}}^{*}}{\hbar} \int_0^t R_{\mathbf{k}}^{-}(t') \exp[-i\omega(t - t')] \, dt' \ . \qquad (2.35a)$$

By applying the canonical transformation defined by

$$R^-_{\mathbf{k}}(t) = \mathcal{R}^-_{\mathbf{k}}(t)\exp(-i\omega_0 t)$$

we can transform the equation (2.35a) in slowly varying approximation to the following form

$$a_{\mathbf{k}}(t) = a^{(v)}_{\mathbf{k}}(t) + \frac{ig^*_{\mathbf{k}}}{\hbar}\mathcal{R}^-_{\mathbf{k}}(t)\,\zeta(\omega-\omega_0)\,, \tag{2.35b}$$

where

$$\zeta(\omega-\omega_0) = \lim_{\varepsilon\to 0}\frac{1-\exp[-i(\omega-\omega_0-i\varepsilon)t]}{i(\omega-\omega_0-i\varepsilon)}$$

$$= \pi\delta(\omega-\omega_0) - i\frac{P}{\omega-\omega_0}\,, \tag{2.36}$$

here the symbol "P" means a mean–value integration, and $a^{(v)}_{\mathbf{k}}(t)$ is a free "vacuum" component of the electromagnetic field. The imaginary part of Eq. (2.36) is responsible for the frequency shift effect. Disregarding the latter effect, we get in result of substitution (2.35b) into the equation (2.3.13c) the following equation of motion for population difference

$$\frac{dR_3}{dt} = -\sum_{\mathbf{k}}\frac{4\pi|g_{\mathbf{k}}|^2}{\hbar^2}\langle R^+_{\mathbf{k}}R^-_{\mathbf{k}}\rangle\,\delta(\omega-\omega_0)$$

$$= -\frac{1}{4\pi}\int d\Omega_{\mathbf{k}_0}\frac{4\omega_0|\mathbf{m}|^2}{\hbar c^3}\sin^2\vartheta_{\mathbf{m}}\,\langle R^+_{\mathbf{k}_0}R^-_{\mathbf{k}_0}\rangle$$

$$= -\frac{3\Gamma_0}{4\pi}\int d\Omega_{\mathbf{k}_0}\langle R^+_{\mathbf{k}_0}R^-_{\mathbf{k}_0}\rangle\sin^2\vartheta_{\mathbf{m}}\,. \tag{2.37}$$

The integrand of this equation can be transformed by the following way

$$\langle R^+_{\mathbf{k}_0}R^-_{\mathbf{k}_0}\rangle = \sum_{i=j=1}^{N}\langle\sigma_{i+}\sigma_{i-}\rangle + \sum_{(i}\sum_{j)j}{}'\langle\sigma_{i+}\sigma_{j-}\rangle\exp[i\mathbf{k}_0(\mathbf{r}_i-\mathbf{r}_j)]$$

$$= \frac{N+R_3}{2} + S(\mathbf{k}_0)\,.$$

Using this transformation we can rewrite the equation (2.37):

$$\frac{dR_3}{dt} + \Gamma_0(N + R_3) = -\frac{3\Gamma_0}{4\pi}\int d\Omega_{\mathbf{k}}\ S(\mathbf{k})\ \sin^2\vartheta_{\mathbf{m}}\ . \tag{2.38}$$

If (2.35) is applied to (2.3.13b), the following equation for correlator $S(\mathbf{k})$ is obtained

$$\frac{dS(\mathbf{k})}{dt} + \Gamma_2 S(\mathbf{k})$$

$$= \Gamma_0 S(\mathbf{k})\ R_3\ \frac{3}{8\pi}\int d\Omega_{\mathbf{k}'}\ P(\mathbf{k} - \mathbf{k}')\ \sin^2\vartheta_{\mathbf{m}}$$

$$+ \frac{3}{16\pi}\ \Gamma_0 R_3(N + R_3)\int d\Omega_{\mathbf{k}'}\ P^2(\mathbf{k} - \mathbf{k}')\ \sin^2\vartheta_{\mathbf{m}}\ . \tag{2.39}$$

If we average the equations (2.38), (2.39) over the vector $\mathbf{m}$ directions, they coincide with the equations (1.56) describing superradiance in Markov approximation.

Thus, if we neglect the retardation in arguments of operators $\mathcal{R}_{\mathbf{k}}(t - \tau)$ and ignore the frequency shift effects, the general set of equations is transformed to the set of Markov approximation equations.

3.3. Superradiance under Bragg diffraction conditions

3.3.1. *Equations of motion for two-wave Bragg superradiance.* Here we shall consider a radiative decay in spatially periodic media. Up to now we have supposed that the atoms were situated chaotically and that there was not any regularity in their positions. Let us now turn to the superradiance of an collection of atoms where the atoms are situated in the nodes of a three–dimensionally periodic lattice. The dynamics of superradiance in a spatially periodic medium differs most essentially from that for an irregular medium when the lattice period is comparable with the wavelength of radiation. Indeed, we have assumed earlier that the field modes with the different wave vectors $\mathbf{k}$ do not essentially interact. It follows from the fact that for a needle–shaped extended active volume the function $P(\mathbf{k}' - \mathbf{k})$, defined by relationship $R_{3,\mathbf{k}'-\mathbf{k}} = R_3\ P(\mathbf{k}' - \mathbf{k})$, is a highly–directional function.

In general case the operator $R_{3,\mathbf{k}'-\mathbf{k}}$ is

$$R_{3,\mathbf{k}'-\mathbf{k}} = \sum_{i=1}^{N} \sigma_{i3} \exp[(i(\mathbf{k}' - \mathbf{k})\mathbf{r}_i] \ . \qquad (3.1)$$

The positions of the atoms in a three-dimensional periodic lattice are determined by

$$\mathbf{r}_i = n_1\mathbf{a} + n_2\mathbf{b} + n_3\mathbf{c} \ , \qquad (3.2)$$

where n_1 , n_2, n_3 are the integers, and $\mathbf{a}$, $\mathbf{b}$, $\mathbf{c}$ are the basic lattice vectors. Hence, the spatial harmonics of a population difference $R_{3,\mathbf{k}'-\mathbf{k}}$ become comparable with R_3 , when the wave vectors difference $\mathbf{k}' - \mathbf{k}$ coincides with the reciprocal lattice vector

$$\mathbf{H} = 2\pi(h_1\mathbf{a}^* + h_2\mathbf{b}^* + h_3\mathbf{c}^*) \ , \qquad (3.3)$$

where h_1 , h_2 , h_3 are integers, and $\mathbf{a}^*$, $\mathbf{b}^*$, $\mathbf{c}^*$ are the basic reciprocal lattice vectors. Indeed, if the equality $\mathbf{k}' - \mathbf{k} = \mathbf{H}$ holds, we get

$$R_{3,\mathbf{H}} = \sum_i \sigma_{i3} \exp(i\mathbf{H}\mathbf{r}_i) = \sum_i \sigma_{i3} = R_3 \ . \qquad (3.4)$$

because $\mathbf{H}\,\mathbf{r}_i = 2\pi\,(h_1n_1 + h_2n_2 + h_3n_3) = 2\pi\,M$, where M is an integer. Thus, in the crystal lattice the field modes with the wave vectors $\mathbf{k}$ and $\mathbf{k}' = \mathbf{k} + \mathbf{H}$ interact effectively with each other.

Let us now discuss the specific features of the two-wave Bragg diffraction. In the case of a two-wave diffraction it is supposed that there are only two diffractionally coupled plane waves inside a crystal with the wave vectors $\mathbf{k}_0$ and $\mathbf{k}_0 + \mathbf{H}$. It means that only one set of the crystal planes is in diffraction position for generated radiation. We shall see later, that this model is not simply an illustrative model, and there is a real physical basis for this model realization. As an particular example we can consider the superradiance of a collection of excited nuclei in thin crystal film [23].

The equations of dynamics for the number of photons $n_{\mathbf{k}}$ in the field mode $\mathbf{k}$ are determined by (3.2.1). It is convenient to introduce a new set of field operators in terms of the old ones. We let

$$a_1 = (a_{\mathbf{k}_0} - a_{\mathbf{k}_0+\mathbf{H}})/\sqrt{2}\,, \qquad a_3 = (a_{-\mathbf{k}_0} - a_{-(\mathbf{k}_0+\mathbf{H})})/\sqrt{2}\,,$$
$$\tag{3.5}$$
$$a_2 = (a_{\mathbf{k}_0} + a_{\mathbf{k}_0+\mathbf{H}})/\sqrt{2}\,, \qquad a_4 = (a_{-\mathbf{k}_0} + a_{-(\mathbf{k}_0+\mathbf{H})})/\sqrt{2}\,.$$

We assume that the vector $2\mathbf{k}_0$ does not coincide with any vector $\mathbf{H}$ of the reciprocal lattice, and therefore the waves with the wave vectors $\mathbf{k}_0$ and $-\mathbf{k}_0$ are not coupled diffractionally. The operators a_1 and a_3 in (3.5) are the annihilation operators for photons in the Borrmann modes. The amplitude of the interference field for any Borrmann mode turns into zero at the atomic planes. The operators a_2 and a_4 are the annihilation operators for photons in the anti–Borrmann modes, for which the amplitude of the interference field is a maximum at the atomic planes. In fact, the vector potential of the field inside a crystal can be represented by

$$A = c_{\mathbf{k}_0}\,[a_{\mathbf{k}_0}\exp(i\mathbf{k}_0\mathbf{r}) + a_{\mathbf{k}_0+\mathbf{H}}\exp(i(\mathbf{k}_0+\mathbf{H})\mathbf{r})]\,. \tag{3.6}$$

Substituting the operators $a_{\mathbf{k}}$, and $a_{\mathbf{k}+\mathbf{H}}$ by the operators a_1, and a_2 in (3.6), we can rewrite it:

$$A = -i\sqrt{2}\,c_{\mathbf{k}_0}\,[a_1\sin(\tfrac{1}{2}\mathbf{Hr}) + ia_2\cos(\tfrac{1}{2}\mathbf{Hr})]\,\exp[i(\mathbf{k}_0+\tfrac{\mathbf{H}}{2})\mathbf{r}]\,. \tag{3.7}$$

At the atomic planes the radius–vector $\mathbf{r}$ is defined by (3.2), hence

$$\frac{1}{2}\,\mathbf{Hr}_i = \pi(n_1h_1 + n_2h_2 + n_3h_3) = \pi N\,,$$

where N is an integer. We can see from (3.7) that the amplitude of the field in the Borrmann mode is equal to zero at the atomic planes, because $\sin(\frac{1}{2}\mathbf{Hr}_i) = 0$. While the amplitude in the anti–Borrmann mode reaches a maximal value at the atomic planes, as $|\cos(\frac{1}{2}\mathbf{Hr}_i)| = 1$.

If (3.2) and (3.3) are applied to (2.3.12), we get the following rules for the operators $R^{\pm}_{\mathbf{k}}$:

$$R^{\pm}_{\mathbf{k}_0+\mathbf{H}} = \sum_{i=1}^{N}\sigma_{i\pm}e^{\pm i(\mathbf{k}_0+\mathbf{H})\mathbf{r}_i} = \sum_{i=1}^{N}\sigma_{i\pm}e^{\pm i\mathbf{k}_0\mathbf{r}_i} = R^{\pm}_{\mathbf{k}_0}\,. \tag{3.8}$$

By analogy with the equations (3.5), we introduce the new atomic operators

$$
\begin{aligned}
R_1^+ &= R_{\mathbf{k}_0}^+ (g_{\mathbf{k}_0} - g_{\mathbf{k}_0+\mathbf{H}})/\sqrt{2}\,, \\
R_3^+ &= R_{-\mathbf{k}_0}^+ (g_{-\mathbf{k}_0} - g_{-(\mathbf{k}_0+\mathbf{H})})/\sqrt{2}\,, \\
R_2^+ &= R_{\mathbf{k}_0}^+ (g_{\mathbf{k}_0} + g_{\mathbf{k}_0+\mathbf{H}})/\sqrt{2}\,, \\
R_4^+ &= R_{-\mathbf{k}_0}^+ (g_{-\mathbf{k}_0} + g_{-(\mathbf{k}_0+\mathbf{H})})/\sqrt{2}\,.
\end{aligned}
\tag{3.9}
$$

The numbers of photons in the Borrmann and anti–Borrmann modes are defined by

$$
n_B = \langle a_1^+ a_1 \rangle + \langle a_3^+ a_3 \rangle\,, \quad n_A = \langle a_2^+ a_2 \rangle + \langle a_4^+ a_4 \rangle\,.
$$

The evolution of these variables is defined by the following equations

$$
\begin{aligned}
&\frac{dn_B}{dt} + \frac{n_B}{\tau_B} = F_B\,, \quad \frac{dn_A}{dt} + \frac{n_A}{\tau_A} = F_A\,, \\
&\frac{dF_B}{dt} + \frac{1}{2}\left(\frac{1}{\tau_B} + \frac{1}{T_2}\right)F_B = \frac{1-p}{T^2}(n_B R_3 + S + S_0)\,, \\
&\frac{dF_A}{dt} + \frac{1}{2}\left(\frac{1}{\tau_A} + \frac{1}{T_2}\right)F_A = \frac{1+p}{T^2}(n_A R_3 + S + S_0)\,, \\
&\frac{dS}{dt} + \frac{S}{T_2} = (F_A + F_B)R_3\,, \\
&\frac{dR_3}{dt} + \frac{N + R_3}{T_1} = -2(F_A + F_B)\,,
\end{aligned}
\tag{3.10}
$$

where

$$
\frac{1}{\tau_B} = \frac{c(1 - p_1)}{l_{abs}} + \frac{c}{L}\,, \qquad \frac{1}{\tau_A} = \frac{c(1 + p_1)}{l_{abs}} + \frac{c}{L}\,,
$$

$$p = (g^*_{\mathbf{k}_0} g_{\mathbf{k}_0+\mathbf{H}} + g^*_{\mathbf{k}_0} g_{\mathbf{k}_0+\mathbf{H}})/2|g_{\mathbf{k}_0}|^2 , \qquad T_0^{-2} = 2|g_{\mathbf{k}_0}|^2/\hbar^2 ,$$

l_{abs} is a length of the nonresonant quantum absorption. For the γ – quanta of the Mossbauer energy range the main effect of the quantum absorption is a photoabsorption, it results in the following lengths of the nonresonant absorption for the crystals: $l_{abs} \leq 0.1$ cm. Therefore, for the crystals with the dimensions $L > 0.1$ cm we get

$$\frac{1}{\tau_B} = \frac{c(1 - p_l)}{l_{abs}} , \qquad \frac{1}{\tau_A} = \frac{c(1 + p_l)}{l_{abs}} .$$

The coefficient $(1 - p_l)^{-1}$ determines the degree of suppression of a nonresonant absorption due to the structure of the Borrmann mode. We have mentioned, that the field amplitude for the Borrmann mode tends to zero at the atomic planes of a crystal. In result of a such distribution of the field amplitude the photoabsorption due to the absorption of the γ– quanta by the electrons of the atomic shells is suppressed. However, the same reasons cause the suppression of a resonant interaction of the γ – rays with the nuclei. The factor $(1 - p_l)^{-1}$ in (3.10) determines the degree of suppression of the resonant interaction. To achieve the high values of a net γ – rays amplification we must retain the interaction of γ – rays with nuclei, and remove only their interaction with electrons. It is shown in [24, 25] that this can be achieved by using nuclear transitions of the high multipolarity. In this case the Hamiltonian of the resonant interaction depends on the derivatives of the field amplitude, in contrast to the usual dipole interaction case when it depends on the field amplitude.

The set of equations (3.10) includes the three nonlinear equations, what explains the difficulties of their analyses in general case. Therefore, we shall use the Markov approximation to analyse the evolution of a system at the linear stage. In this case, $n_B(t)$ and $n_A(t)$ are defined by

$$n_B(t) = \frac{4\tau}{\tau_c} \frac{1 - p}{(1 - p_l)^2} \exp\left[\frac{1 - p}{(1 - p_l)^2} \frac{t}{\tau_c}\right] \tag{3.11a}$$

$$n_A(t) = \frac{4\tau}{\tau_c} \frac{1 + p}{(1 + p_l)^2} \exp\left[\frac{1 + p}{(1 + p_l)^2} \frac{t}{\tau_c}\right] \tag{3.11b}$$

From these equations we obtain for the delay times for the Borrmann (t_{0B})

and anti–Borrmann (t_{0A}) modes:

$$t_{0B} = -\tau_c \frac{1 - p_1}{1 - p} \ln N ,$$
$$t_{0A} = \tau_c \frac{1 + p_1}{1 + p} \ln N \simeq \tau_c \ln N . \qquad (3.12)$$

It follows from (3.12), that $t_{0B} \ll t_{0A}$. But the numerical solutions of the complete set of the nonlinear equations (3.10) show that the variables n_B and n_A approach their maxima simultaneously. This is due to the strong nonlinear interaction of the Borrmann and anti–Borrmann modes.

3.3.2. *Superradiance under thermal dynamics of crystal lattice.* We have assumed up to now, that the crystal lattice temperature is not changed during the stage of formation of a superradiance pulse. In this case the parameters p_1 and p , characterizing the degree of interaction of γ – rays with the electrons and nuclei, respectively, are the constants. But these parameters depend significantly on the lattice temperature. In [26] the analysis of the superradiance dynamics under Bragg diffraction conditions was presented, where the thermal dynamics of a crystal lattice was taken into account. The interest in this problem study has been motivated by the investigation of the possibility of generation of the coherent Mossbauer radiation (see [27] for a review on γ – ray laser problem). The threshold values of the pumping particle fluxes are very heavy, therefore the pump can lead to the crystal lattice distortion and even to melting of the active region of crystal. If the energy of pumping is sufficient for melting of the active volume than the subsequent cool leads to the recrystallization of the crystal active region. In this case the parameters p and p_1 become a function of time because they depend on temperature, which, in its turn, depends of time. The temperature dependence of the parameters p and p_1 is defined by

$$1 \pm p = (1 \pm p_0)\, e^{-W(\mathbf{k}-\mathbf{k}',T)} , \quad p_1 = p_{01}\, e^{-W(\mathbf{k},T)} , \qquad (3.13)$$

where $f = \exp[-W(\mathbf{k},T)]$ is a Debye–Waller factor [22], which defines the probability of Mossbauer effect. The cross–section of interaction of the γ – rays with nuclei depends on the wave vector of an incident γ – quantum $(\mathbf{k})$, while the cross–section of their interaction with electrons depends on the deference $\mathbf{k} - \mathbf{k}'$ between the wave vectors of the incident and scattered quanta (**k′**). The temperature dynamics is determined by the thermodiffusion equation

$$\frac{\partial T}{\partial t} = D \frac{\partial^2 T}{\partial x^2} , \qquad (3.14)$$

which describes the heat transfer from the active region of a crystal to the unexcited parts of it.

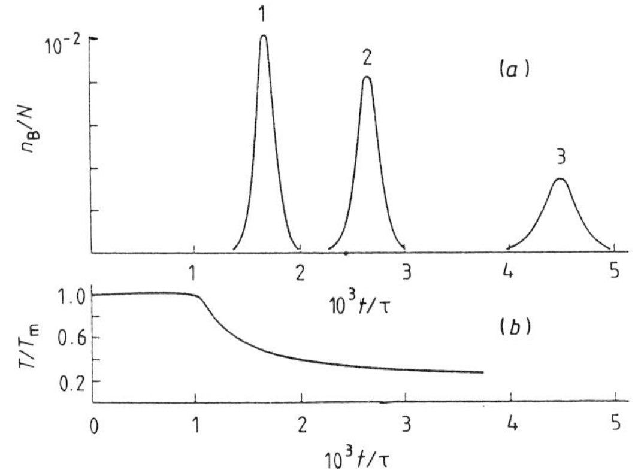

Fig.3.5. (a) intensity profiles for three different values of the ratio $\tau/T_2 = 10^{-3}$ (1), 10^{-2} (2), $2{\cdot}10^{-2}$ (3) and with the same value of the ratio $\tau/\tau_c = 2{\cdot}10^{-2}$; (b) temporal profile of an active layer temperature.

Some results of numerical calculations with the equations (3.10), (3.13), and (3.14) are presented in Fig. 3.5 and 3.6. Fig. 3.5a shows the effect upon the pulse profile of variation in the ratio τ/T_2 with the fixed value of τ/τ_c. The temporal profile of temperature is shown in Fig. 3.5b. We see that the delay time of superradiance pulse decreases with increasing of gain $\mu_0 L = T_2/\tau_c$. This is due to the following reason: the temperature, at which the threshold condition of generation is met, increases with the increase of gain.

In the case of the Bragg–diffraction the threshold condition of generation is

$$\mu_0 L_{eff} > 1 , \qquad (3.15)$$

where

$$L_{\text{eff}} = L \left[\frac{(1 - p)\ \exp(-W)}{1 - p_1 \exp(-W_1)} + \frac{(1 + p)\ \exp(-W)}{1 + p_1 \exp(-W_1)} \right] . \quad (3.16)$$

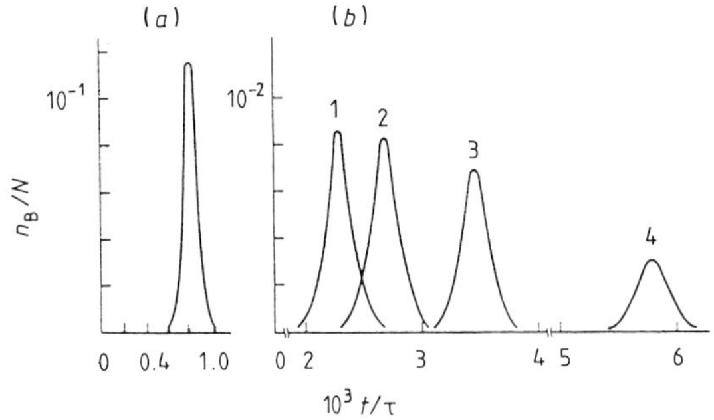

Fig.3.6. Intensity profiles for the different durations of the melting phase: (a) $t_m/\tau = 0$, (b) $t_m/\tau = 500$ (1), 10^3 (2), 2×10^3 (3), 5×10^3 (4).

Fig. 3.6 shows the effect upon superradiance pulse profile of variation in the duration (t_p) of the melting phase. The pulse profile for an ideal crystal is shown in Fig. 3.6a. There is no annealing in this case, and the initial parameters are the same as those for Fig. 3.6b. Fig. 3.6b shows the pulse profile for four values of the melting phase duration, t_p/τ : 500 (curve 1), 1000 (2), 2000 (3), 5000 (4). It is seen that a superradiance takes place even when the melting phase duration exceeds significantly the time of the superradiance pulse formation.

3.3.3. *Localized excitation and superradiance under Bragg-diffraction conditions.* Another model of a superradiance in a spatially periodic medium was studied by Mantsyzov and Kuzmin [28]. They suppose that a discrete one–dimensional periodic resonant medium consists of a set of periodically positioned thin layers of the thickness $a \ll \lambda$ (λ is the wavelength of radiation) containing two–level atoms. To ensure diffraction the period d of the structure should satisfy the condition $d \quad m\lambda/2$. It is assumed that the quantum states of atoms belonging to the one layer are strongly correlated. The full set of layers forms a parallelepiped with the length l (in the direction, perpendicular to the layers). The area of one layer, Z , and the length of structure, l , satisfies the condition $\Phi = Z/l\lambda \quad 1$, that gives a possibility to use the one–dimensional model.

In this case the equations of dynamics for field and matter are

$$\left[\frac{\partial}{\partial t} \pm c\frac{\partial}{\partial x}\right] A^{\pm}(x,t) = p\ ,$$

$$\frac{\partial p}{\partial t} = \frac{2\pi|m|^2}{\hbar\omega} R_3(A^+ + A^-), \tag{3.17}$$

$$\frac{\partial R_3}{\partial t} = -\frac{2k}{\pi\hbar c} p(A^+ + A^-)\ .$$

If we introduce the Bloch angle

$$\vartheta(x,t) = \frac{2\,\mathrm{i}\,\omega m}{\hbar c}\int_{-\infty}^{t}[A^+(x,t') + A^-(x,t')]\,dt' + \pi\ ,$$

the equation (3.17) can be transformed to the following equation

$$\frac{\partial^2\vartheta}{\partial t^2} - c^2\frac{\partial^2\vartheta}{\partial x^2} = \frac{2}{\tau_c^2}\sin\vartheta\ . \tag{3.18}$$

Numerical solution of Eq. (3.18) shows the possibility of the appearance of the localized dynamical excitations in the light–matter interacting system under condition $\tau > \tau_c$ and in the case of strongly coupled Bragg waves $A^+(\mathbf{r},t)$ and $A^-(\mathbf{r},t)$. The spatio–temporal dynamics of a population difference for such type localized excitation is shown in Fig. 3.7. The simplest localized nonstationary solution of Eq. (3.18) is

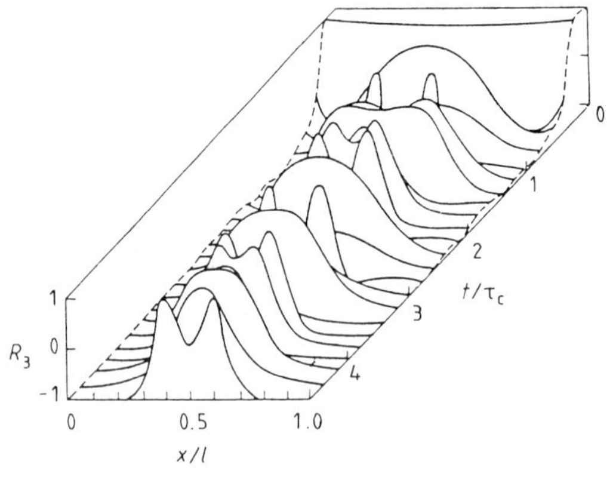

Fig.3.7. Spatio-temporal dynamics of a population difference for the case of appearance of stable localized field-matter excitations.

the breather

$$\vartheta(\tilde{x},\tilde{t}) = 4 \text{ atan}[\tan \nu \sin(\tilde{t} \cos \nu) \operatorname{sech}(\tilde{x} \sin \nu)] ,$$

where $\tilde{x} = x/(l_c/\sqrt{2})$, $\tilde{t} = t/(\tau_c/\sqrt{2})$, and the parameter ν($0 < \nu < 2\pi$) determines the characteristic breather width and amplitude.

3.4. Spatio-temporal dynamics of superradiance

3.4.1. *Account of retardation and quantum fluctuations in boson operator exclusion method.* The relationship between the atomic and field operators is determined, according (2.3.8a), by

$$a_{\mathbf{k}}(t) = a_{0\mathbf{k}}\exp(-i\omega_{\mathbf{k}}t) \tag{4.1}$$

$$+ \frac{i}{\hbar}\int_0^t \sum_i [g_{\mathbf{k}}\sigma_{i+}(t') + g^*_{\mathbf{k}}\sigma_{i-}(t')]\exp[-i\mathbf{k}\mathbf{r}_i - i\omega_{\mathbf{k}}(t - t')]\, dt' .$$

In rotating wave approximation, according to (2.3.1)–(2.3.4), we get

$$A_\lambda(\mathbf{r}_i,t) = A_{0\lambda}(\mathbf{r}_i,t) + i \sum_{\mathbf{k}} \sum_j q_\lambda c_{ij} \int_0^\infty \rho_{j-}(t')G_{\mathbf{k}}(t - t')\, dt' , \tag{4.2}$$

where $A_{0\lambda}(\mathbf{r},t)$ is a vacuum component of electromagnetic field,

$$q_\lambda = \frac{2\pi c m^*_\lambda}{W\omega_{\mathbf{k}}} , \qquad c_{ij}(\mathbf{k}) = \exp[i\mathbf{k}(\mathbf{r}_i - \mathbf{r}_j)] ,$$

$$G_{\mathbf{k}}(t - t') = \vartheta(t - t') \exp[-i(\omega_{\mathbf{k}} - \omega_0)(t - t')] . \tag{4.3}$$

Substituting (4.2), and (4.3) into (2.1.10d) and averaging over the initial quantum state, we get

$$\frac{d\langle\sigma_{i3}\rangle}{dt} + \Gamma_0(1 + \langle\sigma_{i3}\rangle) = -2\, [F^+(\mathbf{r}_i,t) + F^-(\mathbf{r}_i,t)] . \tag{4.4}$$

Here

$$F^+(\mathbf{r}_i,t) = \sum_{\mathbf{k}} F_{\mathbf{k}}(\mathbf{r}_i,t) \,, \quad k_x > 0 \,,$$
$$F^-(\mathbf{r}_i,t) = \sum_{\mathbf{k}} F_{\mathbf{k}}(\mathbf{r}_i,t) \,, \quad k_x < 0 \,, \tag{4.5}$$

$$F_{\mathbf{k}}(\mathbf{r}_i,t) = \beta_{\mathbf{k}} \sum_j \int_0^\infty dt'[c_{ij} <\rho_{i+}(t)\rho_{j-}(t')> G_{\mathbf{k}}(t-t') + \text{c.c.}] \,, \; j \neq i, \tag{4.6}$$

$$\beta_{\mathbf{k}} = \sum_\lambda \frac{|g_{\mathbf{k}}|^2}{\hbar^2} = \sum_\lambda \frac{2\pi |m_\lambda|^2}{W\hbar\omega_{\mathbf{k}}} \,.$$

Differentiating $F_{\mathbf{k}}(\mathbf{r}_i,t)$ with respect to the time, we get ($j \neq i$)

$$\frac{\partial F_{\mathbf{k}}(\mathbf{r}_i,t)}{\partial t} = \beta_{\mathbf{k}} \sum_j \int_0^\infty dt' \Big[c_{ij} < \frac{d\rho_{i+}(t)}{dt} \rho_{i-}(t') > G_{\mathbf{k}}(t-t')$$
$$+ c_{ij} <\rho_{i+}(t)\, \rho_{j-}(t') > \frac{dG_{\mathbf{k}}(t-t')}{dt} + \text{c.c.} \Big] \tag{4.7}$$
$$= \beta_{\mathbf{k}} \sum_j \int_0^\infty dt' c_{ij}(\mathbf{k}) < \frac{d\rho_{i+}(t)}{dt} \rho_{j-}(t') > G_{\mathbf{k}}(t-t')$$
$$+ \beta_{\mathbf{k}} \sum_j \int_0^\infty dt' c_{ij}(\mathbf{k}) < \rho_{i+}(t) \frac{d\rho_{j-}(t')}{dt'} > G_{\mathbf{k}}(t-t') + \text{c.c.}$$

We have changed the differentiation with respect to the time t by the differentiation with respect to the time t' in the second term in right–hand side of (4.7), and we have used the following initial conditions $<\sigma_{j-}(0)> = 0$. If (2.1.10c), (4.2), (4.5), and (4.6) are applied to (4.7), the following equation of motion is obtained for variable $F^+(\mathbf{r}_i,t)$:

$$\frac{\partial F^+(\mathbf{r}_i,t)}{\partial t} = Q_1^+(\mathbf{r}_i,t) + Q_2^+(\mathbf{r}_i,t) \,, \tag{4.8}$$

where

$$Q_1^+(\mathbf{r}_i,t)=\sum_{\mathbf{k}}\sum_{\mathbf{k}'}\beta_{\mathbf{k}}\beta_{\mathbf{k}'}\sum_j\sum_l c_{ij}(\mathbf{k})c_{jl}(\mathbf{k}')\int_0^\infty dt'\int_0^\infty dt''\; G_{\mathbf{k}}(t-t')G_{\mathbf{k}'}(t'-t'')$$

$$\times <\rho_{i+}(t)\rho_{j3}(t')\rho_{l-}(t'')> + \text{c.c.}\,, \quad k_x > 0\,, \quad j \neq i\,, \tag{4.9}$$

$$Q_2^+(\mathbf{r}_i,t)=\sum_{\mathbf{k}}\sum_{\mathbf{k}'}\beta_{\mathbf{k}}\beta_{\mathbf{k}'}\sum_j\sum_l c_{ij}(\mathbf{k})c_{li}(\mathbf{k}')\int_0^\infty dt'\int_0^\infty dt''\; G_{\mathbf{k}}(t-t')G^*_{\mathbf{k}'}(t-t'')$$

$$\times <\rho_{l+}(t'')\rho_{i3}(t)\rho_{j-}(t')> + \text{c.c.}\,, \quad k_x > 0\,, \quad j \neq i\,. \tag{4.10}$$

We shall use the rule (2.9.10) for decoupling of the triple correlations and we assume that the active volume has a form a cylinder with the length L and cross–section area Z, satisfying the condition $\Phi = Z/L\lambda \le 1$, and, secondly, that the operators $\sigma_{i3}(t)$ are slowly varying functions of $\mathbf{r}_i$. Under these two assumptions the variables $Q_1^+(\mathbf{r}_i,t)$ and $Q_2^+(\mathbf{r}_i,t)$ can be written in the following form

$$Q_1^+(x,t) = \frac{\gamma\Gamma_0^2}{2L}\int_0^{\vartheta_F} d\vartheta \sin\vartheta \sum_l \Big[< \rho_{i+}(t)\, \rho_{l-}\Big(t - \frac{x_i - x_l}{c}\Big) >$$

$$\times \exp(-ikr_{il}\cos\vartheta) + \text{c.c.}\Big] \int_{t-(x_i-x_l)/c}^{t} dt'\, R_3(t')$$

$$+ \frac{\Gamma_0}{2}\int_0^{\vartheta_F} d\vartheta \sin\vartheta \sum_j \Big[< \rho_{i+}(t)\, \rho_{j-}\Big(t - \frac{x_i - x_j}{c}\Big) > \tag{4.11}$$

$$\times \exp(-ikr_{ij}\cos\vartheta) + \text{c.c.}\Big] \frac{\partial}{\partial t} < \sigma_{j3}\Big(t - \frac{x_i - x_j}{c}\Big) > ,\; j \neq i\,,$$

$$Q_2^+(x_i,t) = \frac{F^+(x_i,t)}{2}\,\frac{\partial <\sigma_{i3}>}{\partial t} - \frac{\Gamma_0^2}{4}\int_0^{\vartheta_F} d\vartheta \sin\vartheta \int_0^{\vartheta_F} d\vartheta' \sin\vartheta'$$

$$\times \sum_l \sum_j \langle\sigma_{i3}(t)\rangle \left[\left\langle \rho_{l+}\left(t - \frac{x_i - x_l}{c} \right) \rho_{j-}\left(t - \frac{x_i - x_j}{c} \right) \right\rangle \right.$$

$$\left. \times \exp\left[ik(r_{ij}\cos\vartheta' - r_{il}\cos\vartheta)\right] + \text{c.c.} \right] . \qquad (4.12)$$

Averaging (4.11), (4.12) over cylinder cross-section we finally get the following equations

$$\frac{\partial R_3(x,t)}{\partial t} + \Gamma_0(R_3(x,t) + 1) = -2\,[F^+(x,t) + F^-(x,t)] , \qquad (4.13)$$

$$\frac{\partial F^+(x,t)}{\partial t} + \Gamma_2 F^+(x,t) = F^+(x,t)\,\frac{\partial R_3(x,t)}{\partial t}$$

$$+ \frac{\gamma\Gamma_0}{L} \int_0^x dx' \left[NR_3(x,t) + R_3\left(t - \frac{x - x'}{c} \right) \right] \left\{ F^+\left(x', t - \frac{x - x'}{c} \right) \right.$$

$$\left. + \frac{\Gamma_0}{4} \left[R_3\left(x', t - \frac{x - x'}{c} \right) + 1 \right] \right\} \vartheta\left(t - \frac{x - x'}{c} \right) , \qquad (4.14)$$

$$F^-(x,t) = F^+(L - x,t) , \qquad (4.15)$$

where

$$R_3(t) = \frac{1}{L} \int_0^L R_3(x,t)\, dx .$$

The initial conditions for equations (4.13)–(4.15) are

$$F^+(x,0) = F^-(x,0) = 0 , \quad R_3(x,0) = 1 , \quad t - (x - x')/c \geq 0 .$$

The term $F^+\partial R_3/\partial t$ in equation (4.14) is due to the quantum fluctuations, and if we ignore them, then this term disappears. This term is essential in the region of intensity maximum, where the derivative $\partial R_3/\partial t$ is a maximum. This term leads to broadening of a pulse width.

Fig. 3.8 shows the results of the numerical calculations for the variable $F^+(x,t)$ at the center of active volume $(x = L/2)$ and at the right boundary $(x = L)$. The curves labeled "1" are the solutions of equations (4.13)–(4.15) without the term $F^+\partial R_3/\partial t$ in right–hand side of (4.14), and the curves "2" are the solutions of the complete set of equations.

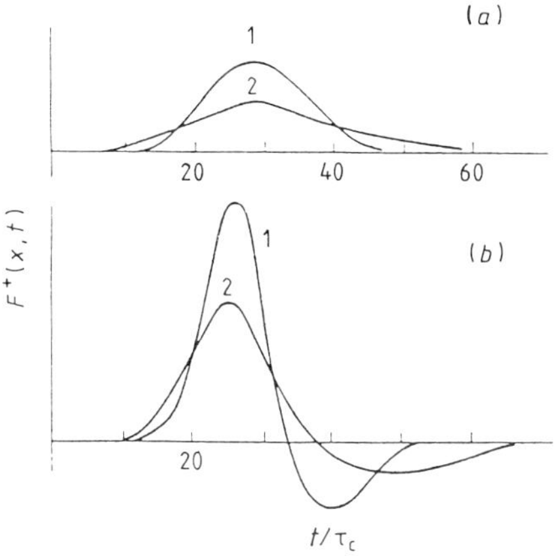

Fig.3.8. Temporal profiles of SR decay rates: (a) at the center of the active medium $(x = L/2)$, (b) at the right boundary $(x = L)$. The profiles indicated by "1" correspond to the Eqs.(4.13)-(4.15) without the term $F^+(x,t)\ \partial R_3/\partial t$ in the right-hand side of the Eq.(4.14), the profiles indicated by "2" correspond to the complete (unreduced) set of equations.

Integrating equations (4.13)–(4.15) with respect to the coordinate x we get in the Markov approximation

$$\frac{\gamma\Gamma_0}{L^2} R_3 \int_0^L dx \left[\int_0^x dx'\ F^+(x',t) + \int_x^L dx'\ F^-(x',t) \right]$$

$$= \frac{\gamma\Gamma_0}{L} R_3(t) \int_0^L dx' \left[\frac{L - x'}{L} F^+(x',t) + \frac{x'}{L} F^-(x',t) \right] . \qquad (4.16)$$

If we substitute $L - x'$ and x' by their average values $L/2$, the resultant equation will coincide with the Eq. (2.9.30).

Fig. 3.9 shows the results of the numerical calculations for the integrated intensity profile

$$I(t) = -\frac{N_0\hbar\omega\gamma}{2L}\int_0^L \frac{\partial R_3(x,t-x/c)}{\partial t}\,dx\ .$$

The intensity profile is indicated by "1" when it is referred to the case when the term $F^{+}\partial R_3/\partial t$ is included in right–hand side of (4.14), and by "2", when it does not. For comparison the results of solution of Eq. (2.9.30) (curve "3") and Eq. (1.59) (curve "4") are shown.

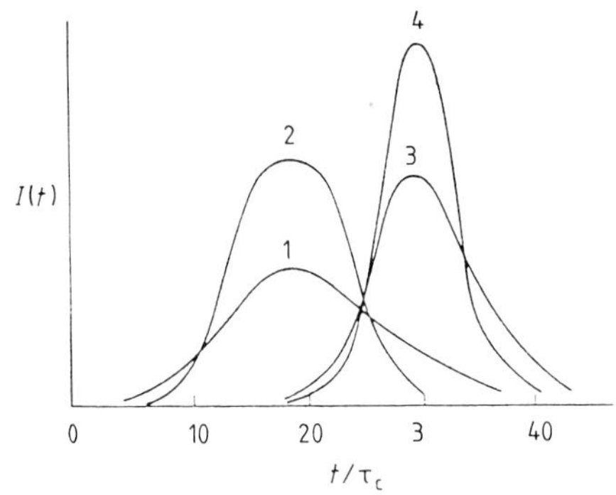

Fig.3.9. The SR intensity profiles given by the Eqs.(4.13)-(4.15) with the term $F^{+}(x,t)\ \partial R_3/\partial t$ (curve 1), and without it (curve 2). For comparison, the SR intensity profiles given by Eq.(2.9.30) (curve 3), and by Eq.(1.59) (curve 4).

3.4.2. *Spatio-temporal dynamics of field.* We turn now to the study of the spatio–temporal dynamics of superradiance avalanche at the linear stage, when the condition $N - |R_3| \ll N$ holds. In the linear stage and under one–wave approximation the equations (2.7.9) are ($R_3 = N_0$)

$$\frac{\partial n}{\partial t} + c\frac{\partial n}{\partial x} = F\ ,$$

$$\frac{\partial F}{\partial t} + \frac{c}{2}\frac{\partial F}{\partial x} + \frac{F}{2T_2} = \frac{1}{{}_0T^2}\,(N_0 n + S + S_0)\ , \qquad (4.17)$$

$$\frac{\partial S}{\partial t} + \frac{S}{T_2} = N_0 F\ .$$

It is convenient to apply the Laplace–Carson transformation

$$n(x,q) = \int_0^\infty n(x,t)\ e^{-qt}\ dt\ . \tag{4.18}$$

If the transformation (4.18) is applied to Eqs. (4.17) the resultant equations of motion for the Laplace transforms $n(x,q)$ and $F(x,q)$ are

$$\frac{d}{dx}\, n(x,q) + \frac{q}{c}\, n(x,q) = \frac{1}{c}\, F(x.q)\ ,$$

$$\frac{d}{dx}\, F(x,q) + \frac{1}{c}\left[2q + \frac{1}{T_2} \right] F(x,q) \tag{4.19}$$

$$= \frac{1}{T^2 c}\left[n(x,q) + F(x,q)\left[\frac{T_2}{1 + qT_2} \right] + 1 \right],$$

where $T^2 = T_0^2/2N_0$. The solution of (4.19) is determined by

$$n(x,q) = \frac{\lambda_2(\exp(\lambda_1 x) - 1) - \lambda_1(\exp(\lambda_2 x) - 1)}{T^2 c^2 q \lambda_1 \lambda_2 (\lambda_1 - \lambda_2)}, \tag{4.20}$$

where

$$\lambda_1 = \frac{q}{c} + \frac{1}{T^2 c}\ \frac{T_2}{1 + qT_2}\ ,$$

$$\lambda_2 = -\frac{1}{c}\left[2q + \frac{1}{T_2} \right]. \tag{4.21}$$

The inverse Laplace–Carson transformation is given by

$$n(x,t) = \frac{1}{2\pi i} \int_{\delta - i\infty}^{\delta + i\infty} dq\ n(x,q)\ e^{qt}\ . \tag{4.22}$$

If (4.22) is applied to (4.20), we get for $t > x/c$

$$n(x,t) = \frac{1}{T^2} \exp\left[-\frac{1}{T_2}\left[t - \frac{x}{c} \right] \right] \sum_n I_n\left\{ \frac{2}{T} \left[\left[t - \frac{x}{c} \right] \frac{x}{c} \right]^{1/2} \right\}$$

$$\times (-1)^{n}\left[T^{2}\left(t - \frac{x}{c}\right)\left(\frac{x}{c}\right)^{-1}\right]^{n/2} \sum_{\alpha} A_{\alpha}[(q_{0} - q_{\alpha})^{n} - q_{0}^{n}] \ , (4.23)$$

$n = 0,...,\infty$; $\alpha = 1,...,4$,
where

$$q_{0} = -\frac{1}{T_{2}} , \quad q_{1,2} = -\frac{1}{T_{2}} \pm \frac{\mathrm{i}}{T} ,$$

$$q_{3,4} = -\frac{1}{2T_{2}} \pm \left(\frac{1}{4T_{2}^{2}} + \frac{1}{T^{2}}\right)^{1/2} ,$$

$$A_{\alpha} = \frac{(q_{\alpha} - q_{0})^{2}}{q_{\alpha}} \frac{1}{\prod_{\alpha'} (q_{\alpha} - q_{\alpha'})} , \quad \alpha' = 1,...,4 \ ; \quad \alpha' \neq \alpha \ .$$

Using the asymptotic expansion of the modified Bessel functions $I_{n}(z)$ for $z \gg 1$, we get

$$n(x,t) = \frac{1}{\sqrt{4\pi T}\ T^{2}} \left[\frac{x/c}{(t - x/c)^{3}} \right]^{1/4} \exp\left[\frac{2}{T} \left[\left(t - \frac{x}{c}\right) \frac{x}{c}\right]^{1/2} - \frac{1}{T_{2}}\left(t - \frac{x}{c}\right)\right] (p - q_{0})^{2}[p(p - q_{1})(p - q_{2})(p - q_{3})(p - q_{4})]^{-1}, \quad (4.24)$$

where

$$p(x,t) = \frac{1}{T} \left(\frac{x/c}{t - x/c} \right)^{1/2} - \frac{1}{T_{2}} .$$

This result generates the following formula for the delay time t_{0} of a superradiance pulse

$$t_{0} = \tau_{c}\left\{ \frac{T_{2}}{\tau_{c}} - \left[\left(\frac{T_{2}}{\tau_{c}}\right)^{2} - \frac{T_{2}}{\tau_{c}} \ln(N \sqrt{\pi}\,) \right]^{1/2}\right\}^{2} . \quad (4.25)$$

This formula holds for the following values of gain: $\mu_{0}L > \ln(N\sqrt{\pi})$. When

$\mu_0 L \gg \ln(N\sqrt{\pi})$, this formula is simplified

$$t_0 = \frac{\tau_c}{4} \ln^2(N\sqrt{\pi}) . \tag{4.26}$$

Thus, if the retardation effect are taken into account the superradiance pulse delay time increases in $\ln(N\sqrt{\pi})$ times in comparison with that for the retardationless models treated above.

The final stage of the radiative process, when $R_3 \simeq -N$, is characterized by the appearance of oscillations. In fact, in this case the modified Bessel functions $I_n(z)$ are substituted by Bessel functions $J_n(z)$, which have the following asymptotics for $z \gg 1$

$$n(x,t) \propto \cos\left[\frac{2}{T}\sqrt{\left(t - \frac{x}{c}\right)\frac{x}{c}}\right] .$$

Particularly, at the active volume boundary $(x = L)$ we get

$$n(L,t) \propto \cos\left[2\sqrt{\frac{t - \tau}{\tau_c}}\right] .$$

It is seen that the period of oscillations increases with time, while in the frames of the retardationless models the frequency of oscillations did not depend on time and was defined by

$$\Omega_0 \simeq \frac{1}{T}\sqrt{1 - \frac{\tau_c}{4\tau}} , \tag{4.27a}$$

or, in the case $\tau_c \ll \tau$, by

$$\Omega_0 \simeq \frac{1}{T} = \sqrt{\tau\tau_c} = \frac{1}{\tau_c}\sqrt{\frac{\tau_c}{\tau}} . \tag{4.27b}$$

Thus, the retardation effects and the nonuniformity of the field amplitude along the active medium length result in substantial changes of the main pulse parameters. The significant increase in the delay time, pulse width, and oscillation period occurs.

Fig. 3.10 shows the intensity profiles calculated from the equations (2.1) [29 – 31] and from the equations (2.7.9), for comparison. The curves in Fig. 3.10a correspond to the systems with $\tau \gg T_2$, while those in Fig. 3.10c correspond to the systems with $\tau \ll T_2$. So, there is a superradiance in the last case. It is seen that the difference between the two profiles is more pronounced in the case of a superradiance. The spatial variations of the field amplitudes bring to increasing of the characteristic times of a superradiance pulse formation. The presence of spatial derivatives in the equations of motion for the systems with $\tau \gg T_2$ does not essentially change the pulse shape, because in this case we have $\partial n/\partial x \propto n/\tau \ll \partial n/\partial t$. The latter follows from the unequalities $\tau \gg T_2 > \tau_c$.

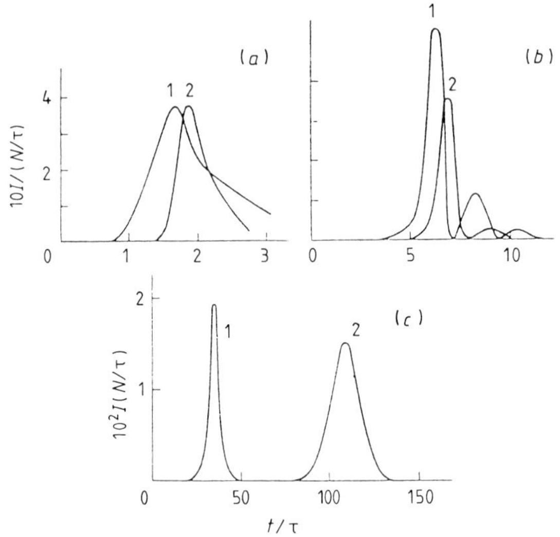

Fig.3.10. Emission intensity profiles for three different values of ratio τ/T_2 = 50 (a), 1 (b), 0.05 (c). The curves indicated by "1" are the solutions of the equations (2.1), and the curves indicated by "2" are the solutions of the equations (2.7.9). In all cases the gain is $\mu_0 L$ = 25.

3.5. Superradiance dynamics of multilevel systems

We have considered above the superradiance for a collection of the two-level atoms. In experiments this phenomenon is sometimes complicated by the presence of the additional radiative transitions from the levels involved in superradiance. The dynamics of generation will depend on the specific structure of the atomic levels.

In the model of three-level atoms which is the subsequent simplest case after the model of two-level atom, the following possibilities can be realized:

a) cascade superradiance, when the transitions $3 \longrightarrow 2$ and $2 \longrightarrow 1$ are allowed, and the transition $3 \longrightarrow 1$ is forbidden (Fig. 3.11a);

b) two color superradiance, when the transitions $3 \longrightarrow 2$ and $3 \longrightarrow 1$ are allowed, and the transition $2 \longrightarrow 1$ is forbidden (Fig. 3.11b);

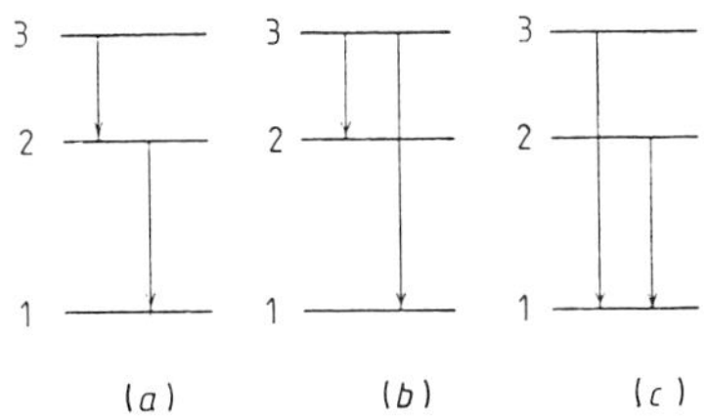

Fig.3.11. SR regimes for three-level systems: (a) cascade SR, (b) two-coloured SR, (c) V-type SR.

c) *V*–type superradiance, when the transition $3 \longrightarrow 2$ is forbidden (Fig. 3.11c);

d) superradiance between the degenerated levels.

The method, developed above for the analyses of a superradiance from collections of the two–level atoms, can be generalized to treat the radiative decay of the multilevel atoms [37]. In close analogy with the theory for the two–level atoms we can obtain the equation of motion and then apply the numerical or asymptotical methods of their analyses. In the case of the pure superradiance, when the characteristic times of a superradiance are much shorter then those for an uncorrelated decay, we can easily generalize the equations of a superradiance dynamics to the case of multilevel systems.

The case of a cascade superradiance has been treated with the help of quasiclassical approximation in Sec.1.4 for the simplest model of a three–level atom. More detail discussion of these problems is given in Ref. 38. As we have mentioned, the dynamics of superradiance depends essentially on the ratio of the characteristic times of a collective decay τ_{c1} and τ_{c2} for two adjacent transitions. Under condition $\tau_{c1} < \tau_{c2}$ (where τ_{c1} is a collective decay time for transition $3 \longrightarrow 2$, and τ_{c2} is a collective decay time for transition $2 \longrightarrow 1$), the superradiance pulses for the upper and bottom transitions are not overlapped and the radiative decay for each of these transitions can be described with the help of the two-level atom model. In fact, the initial population of the upper level for the second transition $(2 \longrightarrow 1)$ coincides with the final population of the lower level for the first transition $(3 \longrightarrow 2)$. For $\tau_{c1} > \tau_{c2}$ the superradiance pulses in these two transitions are overlapped and the dynamics of emission in the second transition depends essentially on the appropriate parameters for the first transition.

Such mutual influence of the collective processes in the two coupled transitions is a characteristic feature of a radiative decay in other types of three-level systems. For example, there appears a strong mutual influence of the emission process in the two transitions in the case of two-color superradiance (see Fig. 3.11b). Now we discuss, following the Ref. [37], the dynamics of formation of the pulses in the case of a two-color superradiance.

To get a self-consistent set of equations for correlators of atomic operators we will use the method of exclusion of the boson operators, which has been widely used above for analyses of a superradiance in the case of the two-level systems. The Hamiltonian, which describes the field-matter interactions for this system comprising N atoms, is

$$H = \sum_{\mathbf{k}} \hbar\omega_{\mathbf{k}}\, a^{+}_{\mathbf{k}} a_{\mathbf{k}} + \sum_{l} \sum_{s} \hbar\omega_{s} \sigma^{(l)}_{ss}$$

$$- \frac{1}{c} \sum_{l} \sum_{s'} \sum_{s} \{\mathbf{m}_{s's} \mathbf{A}^{+}(\mathbf{r}_l) \sigma^{(l)}_{ss'} + \mathbf{m}^{*}_{s's} \sigma^{(l)}_{s's} \mathbf{A}^{-}(\mathbf{r}_l)\}, \tag{5.1}$$

$$l = 1,\ldots,N\,, \qquad s, s' = 1,\ldots,3\,, \qquad s' > s\,,$$

where $\hbar\omega_s = E_s$ is the energy-level spacing such that $E_3 > E_2 > E_1$, $\omega_{ss'} = \omega_{s'} - \omega_s$, $\mathbf{m}_{s's}$ are the matrix elements of the transition current density operators, $\sigma^{(l)}_{s's}$ are the operators defined by the following matrix elements

$$\langle r' | \sigma^{(l)}_{s's} | r \rangle = \delta_{r's'} \delta_{rs}\,, \qquad r, s = 1, 2, 3, \qquad l = 1,\ldots,N\,.$$

The operators $\sigma^{(l)}_{s's}$ are closely related to the generators of the $U(3)$ and $SU(3)$ groups. They obey the following commutation rules

$$[\sigma^{(l)}_{s's}\,,\,\sigma^{(l')}_{r'r}] = \delta_{ll'}(\sigma^{(l)}_{s'r}\delta_{sr'} - \sigma^{(l)}_{r's}\delta_{s'r})\,. \tag{5.2}$$

The vector–potential operators $\mathbf{A}^{\pm}(\mathbf{r}_l,t)$ are expressed in terms of the photon creation $a^{+}_{\mathbf{k}}$ and annihilation $a_{\mathbf{k}}$ operators for a field mode with frequency $\omega_{\mathbf{k}}$, wave vector $\mathbf{k}$, and polarization $\mathbf{e}_{\lambda}$:

$$\mathbf{A}^{+}(\mathbf{r}_l,t) = \sum_{\mathbf{k}} \sqrt{\frac{2\pi\hbar c^2}{W\omega_{\mathbf{k}}}}\; a^{+}_{\mathbf{k}}\, \mathbf{e}_{\lambda} \exp(-i\mathbf{k}\mathbf{r}_l)\,.$$

By analogy with the case of the two–level atoms (see Sec.2.4), the operators $\mathbf{A}^{\pm}(\mathbf{r},t)$ can be expressed in terms of the atomic operators and operators associating with the vacuum field

$$\mathbf{A}^{+}(\mathbf{r}_m,t) = \mathbf{A}^{+}_{v}(\mathbf{r}_m,t) \tag{5.3}$$

$$+\, i \sum_{\mathbf{k}} \sum_{l} \mathbf{e}_{\lambda}\, q_{\mathbf{k}} \exp[i\mathbf{k}(\mathbf{r}_m - \mathbf{r}_l)] \int_0^t dt' \exp[-i\omega_{\mathbf{k}}(t - t')]\sigma^{(l)}_{s's}(t')\,.$$

If the commutation rules (5.2) are applied to the Heisenberg equations with Hamiltonian (5.1), the following equations of motion are obtained for the atomic operators

$$\frac{d\sigma^{(l)}_{s's}}{dt} = i(\omega_{s'} - \omega_{s})\sigma^{(l)}_{s's}$$

$$+\, \frac{i}{\hbar c} \sum_{r} \mathbf{m}_{r'r}\mathbf{A}\, (\sigma^{(l)}_{r's}\delta_{s'r} - \sigma^{(l)}_{rs}\delta_{r's'} - \sigma^{(l)}_{s'r'}\delta_{sr})\,. \tag{5.4}$$

Substituting (5.3) into (5.4), and then integrating it with respect to the wave vector $\mathbf{k}$ and time, we get the following hierarchy of coupled nonlinear equations of motion for mean values and second–order correlations of the atomic operators

$$\frac{d\langle\sigma^{m}_{ss}\rangle}{dt} = \frac{1}{2\tau_{os}} \sum_{l} c_{lm}(\omega_{3s}) D_{3s}(m,l)\,,$$

$$\frac{dD_{3s}(m,l)}{dt} = -\frac{(1-\delta_{ml})}{T_{2s}} D_{3s}(m,l)$$

$$+\frac{1}{2\tau_{0s}} \sum_n [c_{mn}(\omega_{3s})E_{3s}(n,m,l) + c_{ln}(\omega_{3s})E_{3s}(m,l,n)] \qquad (5.5)$$

$$-\frac{1}{2\tau_{0r}} \sum_n [c_{mn}(\omega_{3r})F(n,l,m) + c_{ln}(\omega_{3r})F(m,l,n)] \; .$$

It is to be noted that the following relationship holds:

$$<\sigma_{11}^{(n)}> + <\sigma_{22}^{(n)}> + <\sigma_{33}^{(n)}> = 1 \; ,$$

$$s\; , r = 1\; ,2, \qquad s \neq r \; , \qquad l\; , \mathrm{n} = 1,...,N \; .$$

In (5.5) we have introduced the following notations:

$$D_{3s}(m,l) = <\sigma_{3s}^{(m)}(t)\; \sigma_{3s}^{(l)}(t)> + \text{c.c.} \; ,$$

$$E_{3s}(n,m,l) = (<\sigma_{33}^{(m)}(t)> - <\sigma_{ss}^{(m)}(t)>)\; <\sigma_{3s}^{(m)}(t)\; \sigma_{s3}^{(l)}(t)>$$

$$-\; 2\; <\sigma_{3s}^{(n)}(t)\; \sigma_{s3}^{(l)}(t)>\; <\sigma_{3s}^{(l)}(t)\; \sigma_{s3}^{(m)}(t)> + \text{c.c.} \; ,$$

$$F(n,l,m) = <\sigma_{31}^{(n)}(t)\; \sigma_{13}^{(m)}(t)>\; <\sigma_{32}^{(m)}(t)\; \sigma_{23}^{(l)}(t)> + \text{c.c.},$$

$$c_{lm}(\omega_{3s}) = \frac{\sin(\omega_{3s} r_{lm}/c)}{\omega_{3s} r_{lm}/c} \; , \quad r_{lm} = |\mathbf{r}_l - \mathbf{r}_m| , \quad \frac{1}{\tau_{0s}} = \frac{4\omega_{3s}|\mathbf{m}|^2}{3\hbar c^3} .$$

τ_{0s} is a spontaneous decay time in transition $3 \longrightarrow s$, and T_{2s} is a transversal relaxation time for this transition.

In obtaining Eqs. (5.5) we have used the decoupling rules for the triple correlations, which takes into account the fluctuations of populations difference,

$$<(\Sigma_{33} - \Sigma_{ss})^2> - (<\Sigma_{33}> - <\Sigma_{ss}>)^2 ,$$

where

$$\Sigma_{rr} = \sum_{l} \sigma_{rr}^{(l)} .$$

To get this, the decoupling rules analogous to (4.10) have been used. To decouple the correlation $<\sigma_{31}^{(n)}\sigma_{12}^{(m)}\sigma_{23}^{(l)}>$ the following rule has been used

$$<\sigma_{31}^{(n)}\sigma_{12}^{(m)}\sigma_{23}^{(l)}> = <\sigma_{31}^{(n)}\sigma_{13}^{(m)}\sigma_{32}^{(m)}\sigma_{23}^{(l)}> \simeq <\sigma_{31}^{(n)}\sigma_{13}^{(m)}><\sigma_{32}^{(m)}\sigma_{23}^{(l)}> .$$

For the case of the N identical three-level atoms comprised into the needle-shaped active volume the equations of motion for the level populations ($x = <\Sigma_{11}>/N$ and $y = <\Sigma_{22}>/N$) are

$$\frac{d^2x}{dt^2} = \frac{1}{\tau_{c1}}(1 - 2x - y)\left[\left(1 + \frac{\tau_{c1}}{\tau_{01}}\right)\frac{dx}{dt} - \frac{1 - x - y}{\tau_{01}}\right]$$

$$- 2\frac{dx}{dt}\frac{dx}{dt} - \frac{dy}{dt}\frac{dx}{dt} - \frac{1}{T_{21}}\left(\frac{dx}{dt} - \frac{1 - x - y}{\tau_{01}}\right) , \qquad (5.6)$$

$$\frac{d^2y}{dt^2} = \frac{1}{\tau_{c2}}(1 - 2y - x)\left[\left(1 + \frac{\tau_{c2}}{\tau_{02}}\right)\frac{dy}{dt} - \frac{1 - x - y}{\tau_{02}}\right]$$

$$- 2\frac{dy}{dt}\frac{dy}{dt} - \frac{dx}{dt}\frac{dy}{dt} - \frac{1}{T_{22}}\left(\frac{dy}{dt} - \frac{1 - x - y}{\tau_{02}}\right) .$$

In these equations, the characteristic time,

$$\tau_{cr}^{-1} = \frac{\lambda_{3r}^{2} N}{4\pi Z \tau_{0r}},$$

is a collective decay time in transition $3 \longrightarrow r$, and Z is a cross–section area of the active volume.

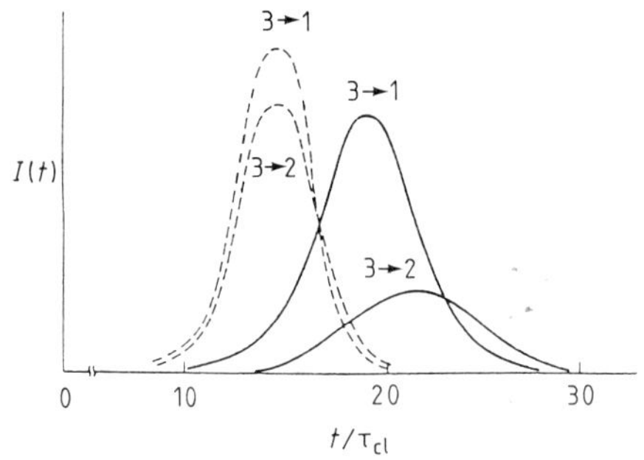

Fig.3.12. Intensity profiles for two-coloured SR. The parameters are $T_{21} = T_{22} = 10\ \tau_{c1},\ \tau_{01} = \tau_{02} = 10\ \tau_{c1},\ \tau_{c2} = 1.05\ \tau_{c1}$. The solid curves are the solutions of the equations (6.6), the dashed curves are the solutions of the equations (6.7).

Fig. 3.12 shows the results of the computer calculations for the intensity profiles at the carrier frequencies ω_{31} and ω_{32}. The dashed lines in this figure shows for comparison the results of the numerical calculations with the equations based upon the application of semiclassical decoupling (see (3.2.6)). These equations are

$$\frac{d^2x}{dt^2} = -\frac{1}{\tau_{01}}\left(\frac{dx}{dt} + \frac{dy}{dt}\right)$$

$$+ \frac{1}{\tau_{c1}}(1 - 2x - y)\left[\left(1 + \frac{\tau_{c1}}{\tau_{01}}\right)\frac{dx}{dt} - \frac{1 - x - y}{\tau_{01}}\right],$$

$$\frac{d^2y}{dt^2} = -\frac{1}{\tau_{02}}\left(\frac{dx}{dt} + \frac{dy}{dt}\right)$$

$$+\frac{1}{\tau_{c2}}(1-2y-x)\left[\left(1+\frac{\tau_{c2}}{\tau_{02}}\right)\frac{dy}{dt}-\frac{1-x-y}{\tau_{02}}\right].$$

It follows from this figure, that there appears a strong mutual influence of emission processes in these two transitions, when parameters τ_{c1} and τ_{c2} approach to one another. These results agree qualitatively with the experimental investigations of the two-color superradiance [39].

3.6. Experimental investigations of superradiance

The superradiance has been observed in the first time by Skribanowitz et al [17], in which they observed the effect in *HF* vapor. To achieve complete initial inversion and absence of initial polarization in SR transition the experiment was carried out through the three-level scheme. It means that the transition between the two excited states was chosen as a superradiant transition, and the upper superradiant level was excited from the ground state. The energy-level spacing for molecule *HF* is shown schematically in Fig. 3.13. The transition between the two adjacent rotational sublevels of the vibrationally excited state (v = 1) was chosen as the superradiant transition in Ref. 17 (see Fig. 3.13). Similarly, the superradiance pulses could be produced by making use of other rotational sublevels of the v = 1 level. The corresponding transition wavelength at which superradiance was observed in Ref. 17 are summarized in Table 3.1.

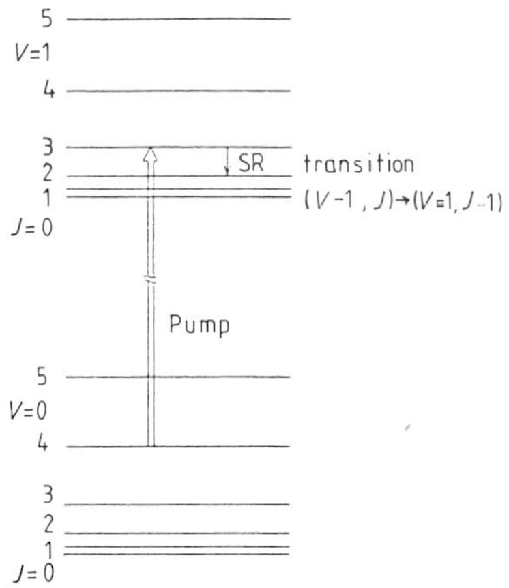

Fig.3.13. Energy-level diagram of the *HF* molecules.

A cell, containing HF vapor at room temperature, had the length from 30 to 100 cm and inner diameter from 12 to 28 mm. The pressure of HF vapors could vary over a wide range, that provided a possibility for obtaining a concentrational dependence. Typical values of pressure were in the range 1–20 mTorr. Below 10 mTorr the superradiance pulses were delayed 500–2000 ns with respect to the pump pulse. As the pressure increased, the delay time and pulse width increased, while the intensity decreased. At these pressures, the emitted pulse had an oscillatory structure, i.e. there appeared a train of pulses of decreasing height. At pressures above 10 mTorr, a single superradiance pulse was observed.

In experiments on superradiance there are two different possible pumping scheme: along the sample axis and across it. In the first case all atoms in the active volume are excited simultaneously. In the second case, the atoms are excited by the pulse, which propagates with the speed of light along the sample axis. Therefore the initial population difference depend on the retarded time $t - x/c$. If the transit time, $\tau = L/c$, is smaller than the coherence time, i.e. $\tau_c \gg \tau$, then there is no any essential difference between these two types of pumping. If the condition $\tau \approx \tau_c$ holds then there appears the difference between these cases. For example, in the first case, when pumping is homogeneous, the atoms at the right boundary of sample begin to emit earlier than the radiation emitted by the atoms at the left boundary reaches them, and vice versa. In this case, the intensity of emission is due to the independent decay of parts of active volume with the length $L_c = c\tau_c$, as we have mentioned above. The shape of superradiance pulses emitted from the both sides of active volume will be the same. In the systems with $\tau \approx \tau_c$, but in the case of travelling pump, the conditions for the right wave copropagating with pump pulse will be different than those for the left wave. And the shapes of the pulses leaving the two ends of the cell become different.

Such asymmetry of the right and left superradiance pulses have been observed by Rosenberger et al [40], in which they observed the superradiance in rotational transitions of CH_3F molecule. The radiation of CO_2 laser was used for pumping. The pump pulse propagated along the axis of cell, containing gas of CH_3F molecules, and the length of which varied from 2.3 m to 9.7 m . For the length above 6.3 m the left and right superradiance pulses became different. The difference in peak intensity was as great as of two order of magnitude.

The vapors of alkali metals has found wide use as the active media in experiments on superradiance (see Table 3.2). The emission lines of these materials lie, as a rule, in visible spectra. The specific feature of these group of experiments was the appearance of superradiance pulse beating. The beats in SR are due to the fact that the superradiance conditions could be satisfied simultaneously for a number of transitions with the close frequencies, for example, for the different sublevels of superfine atomic structure (SFS). Vrehen et al [46] observed the beats due to the SFS of $7P_{1/2}$ level in *Cs* (see Fig. 3.14). Gross et al [56] observed the beats due to the interaction of groups of atoms with the

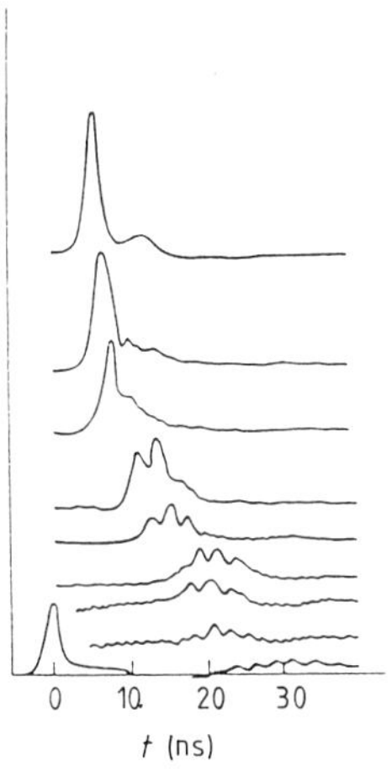

Fig.3.14. Quantum beats in SR intensity due to the hyperfine structure of the $7P_{1/2}$ level of *Cs* [56].

different velocities lying inside the Doppler line (500 MGz) of *Cs* vapor. In contrast to [46], the excitation of sublevels of $7P_{1/2}$ state has been done here by the pulse of N_2 – laser with the bandwidth 100 MGz, which is less than the SFS sublevels energy spacing for $7P_{1/2}$ state. As a result, the state $7P_{1/2}$ is populated by the atoms, velocities of which obey the conditions

$$\nu_{gF} = \nu_L - k_L v_F, \qquad \nu_{gF'} = \nu_L - k_L v_{F'},$$

where ν_L and k_L are the mean frequency and mean wave vector of excited pulse, ν_{gF} ($\nu_{gF'}$) is a frequency difference between the upper (lower) sublevel of $7P_{1/2}$ state and one of the sublevels of $6S_{1/2}$ state. Thus, there are excited the two groups of atoms with the different velocities and, hence emitted the pulses with the different frequencies $\nu_{fF} + k_{SR}V_F$ and $\nu_{fF'} + k_{SR}V_{F'}$ respectively. ν_{fF} ($\nu_{fF'}$) ia a frequency difference between the upper (lower) sublevels of $7S_{1/2}$ state, k_{SR} is the mean wave vector of superradiant pulse. Thus, as one should expect the superradiant pulses to be modulated at the frequency

$$\nu = \nu_{FF'}\left(1 - \frac{k_{SR}}{k_L}\right), \quad \nu_{FF'} = \nu_{gF} - \nu_{gF'} = \nu_{fF} - \nu_{fF'} .$$

For radiation emitted along the direction of exciting pulse ($k_{SR}/k_L > 0$), the frequency will be red-shifted; if k_{SR} and k_L have opposite signs, the frequency will be blue-shifted. Such shift was observed experimentally in Ref. [56]. An important feature of the experiments of Ref. 56 was in that the beats arose, not a result of the formation of a coherent superposition of states of one atom, but because of a coherent superposition of states of physically distinct atoms.

It should be stressed that the beats in superradiance and the oscillatory superradiance are the two different effects. The beats are due to the hyperfine structure of levels, while the oscillations can occur even in nondegenerate pure two-level system and they are the result of energy exchange process in the resonant field interaction with matter.

The multicoloured superradiance was observed in vapors of *Na* [45], *Rb* [48], and *Sr* [55]. The experiments on superradiance in the Rydberg levels of sodium atoms [42, 43] extended the region of wavelength to the *mm*–waves. In this case the dimensions of active volume were comparable with the emitted radiation wavelength. The superradiance in solids was observed in [39, 57–59]. In [39,57] the crystals of *KCl* , activated by molecular O_2^- ions, and in [58, 59] the molecular pyrene–doped diphenyl crystals were used.

In experiments, mentioned above, the resonance one– or two–photon pumping was applied. In contrast to that, in [53] the superradiant level of *Bi* vapor was pumped by the two–photon nonresonant manner. In [60] the experiments on superradiance were performed, where the *Ba* vapors were pumped by nonresonant manner and the energy deficit was covered by collisions of the *Ba* atoms. Jaroszynski D.A., King T.A. [61] showed the possibility of the superradiant excitation by the photodissociation of molecules, when the dissociation products were found to be in excited states. This offers considerable scope for applying traditionally laser scheme of pumping to superradiant media.

A theory of SR has grown out of an analysis of the problem of radiative decay of a poliatomic system of dimensions much less than the radiation wavelength. For a long time it has been assumed that this problem is of purely model nature and its attractiveness is simply due to the feasibility of providing an exact description of its dynamics. However, recent experiments [62, 63] have shown that the model has a physical realization, which is undoubtedly of interest in the theory of SR because it makes it possible to test the fundamentals of this theory.

CHAPTER 4

SEMICLASSICAL THEORY OF SUPERRADIANCE

Optical superradiance is one of the most striking examples of the cooperative behavior of multiatomic systems. The difference of this phenomena from other collective processes is in the fact, that atomic system itself via the internal interactions among the comprising atoms comes from the uncorrelated state to the correlated one, when the radius of correlation of atomic transition moments takes the macroscopic values. We have already discussed that a superradiance and lasing are the two limiting cases of the same process. Any process of generation is based on the interaction of the two subsystems, one of which is an active medium, which accumulates energy in the initial stage, and another is a field, which accepts the energy stored up initially in the active medium. The emitted radiation is coherent, if at least one of the interacting subsystems is in consistent state, i.e., it should have the least possible number of the degrees of freedom between which the energy can be distributed. In the ideal case, the field and radiators should have the same resonant frequency with the extremely small widths $\Delta\omega_a$ and $\Delta\omega_f$, and, besides, the width of the angular spectrum of emission should have the smallest possible values. In lasers this is accomplished via the field modes discrimination. Whereas, narrowness of the field line $\Delta\omega_f$ is achieved by the use of th high–Q resonators. Therefore, the field must be in optical resonator for a sufficiently long time to make at least a few round trips. We are interested here in the regime of free running generation, when initially inverted medium emits the pulse of radiation in result of a spontaneous decay. The width of a field line is determined in this case by the active medium length *L*. In the absence of reflection at the active medium boundaries, the width of field line is $\Delta\omega_f = 1/\tau$. If the field reflection at the active volume boundaries is essential, then the width of a field line is

$$\Delta\omega_f = \frac{1}{\tau} \ln \frac{1}{r_1 r_2} ,$$

where r_1 and r_2 are the amplitude reflection coefficients. If the active medium length is smaller than the resonator length l, then τ is determined by $\tau = l/c$. When $l >> L$, we have to excite the medium synchronously with the arrival of the pulse, therefore this case is, in principle, the case of amplification of an external pulse, which should be considered separately. The width of an atomic transition is $\Delta\omega_a = 1/T_2$. If $\Delta\omega_f \ll \Delta\omega_a$, the ordering is due to the field. If $\Delta\omega_f \ll \Delta\omega_a$, it is due to the medium. The first case is close to lasing, and in the case of mirrorless systems it is called *superluminescence* (SL) or amplified spontaneous emission. The second case, when the ordering is due to the narrowness of the atomic transition lines, is called *superradiance*. The field mode discrimination is achieved in both cases by the shape of the active volume, which has usually the needle–shaped form. In this case, the waves propagating along the active volume axis are amplified much more strongly than the waves propagating in any other directions.

An ensemble of atoms interacting with resonant field can be characterized by the following parameters: N/V is the atom number density, $\Delta\omega_a = 1/T_2$ is the atomic transition width, L is an active volume length determining the width of the field line $\Delta\omega_f = c/L$, T_1 is a spontaneous decay time for an individual atom, r_1 and r_2 are the reflection coefficients at the active volume boundaries. In the case of inhomogeneous broadening, the additional parameter $\Delta\omega^* = 1/T_2^*$ appears. In this chapter we try to answer the following questions. How do the parameters of the free running generation pulses depend on the parameters listed above ? How can the parameters of generated pulses be controlled and in what limits ? Which regime – superradiance or superluminescence – is more profitable from the point of view of the highest rate of energy transfer from an excited medium to the field of radiation ? The answers to these questions enable us to determine the possibilities of pulse controlling and suggest the methods for optimization of the processes of pulse generation.

The optimization discussed above does not exhaust all the possibilities of controlling of the pulse parameters. In fact, the general problem on the resonant light–matter interaction depends not only on the parameters mentioned above, but also on the initial and boundary conditions. The boundary conditions we have already taken into account by including the reflection coefficients r_1 and r_2 in the parameters of the problem. The initial conditions could essentially influence on the nature of a radiative decay of the atomic system. For example in Ref.[1], the supersymmetric solutions of the general equations describing the interaction of a system of the two–level atoms with the resonant radiation were found. It was shown that the selection of the initial conditions corresponding to this solutions results in generation of pulses of maximal intensity. The high rate of a superradiative decay is a result of the fact, that the atomic subsystem during the radiative decay comes into the state which is completely symmetric under the rearrangements of the atoms. This situation can only be realized into the two following cases, when the atoms are confined into the volume with the dimensions much smaller than the wavelength of radiation and when the atomic subsystem interacts with the one mode of the field. In the real extended systems the inhomogeneity of the field along the active volume length breaks the symmetry of the atomic subsystem. This results in increasing of the generated pulse width. In Ref.[1] it was shown that there exist the states accounting the symmetry of the whole atomic– field system and not only atomic subsystem. The process of the radiative decay in the extended multiatomic system occurs in this case as if in the point–source system and is characterized by the maximum intensity and minimum width of the emitted pulse for a given population inversion density.

The problems discussed above are very closely related to the problems of a pulse propagation in the resonant media under coherent interaction conditions (see Ref.[2–11]). One of the most striking effects here is the effect of a self–induced transparency. It has been investigated thoroughly both experimentally [2, 3, 12, 13] and theoretically [4–9]. Simultaneously with the development of the theory of propagation of the 2π– pulses in the resonantly absorbing media the theory of their propagation in resonantly amplifying media has been developed (see Ref.[11, 16] as a review). The theory of the soliton propagation in the semiinfinite amplifying media under certain values of the problem parameters leads to the seemingly unphysical results such as the pulse propagation with the velocities exceeding the velocity of the light, and so on. In this chapter we shall obtain the soliton solutions for the spatially inhomogeneous media (including the spatially bounded ones), which provide the clear interpretation all of the specific features of soliton propagation in resonantly amplifying media. We shall demonstrate the possibilities of the medium state control, which offer considerable scope for generation of pulses with a given profile.

4.1. Semiclassical equations for resonant light-matter interaction

4.1.1. *Wave equation.* The semiclassical equations for resonant light–matter interaction have been obtained in Chap.2 with the help of semiclassical decoupling of operators:

$$\langle\sigma_+ A\rangle = \langle\sigma_+\rangle\langle A\rangle, \quad \langle\sigma_- A\rangle = \langle\sigma_-\rangle\langle A\rangle, \quad \langle\sigma_3 A\rangle = \langle\sigma_3\rangle\langle A\rangle \qquad (1.1)$$

If the atoms interact via transversal electromagnetic field, the equations are

$$\left(\Delta - \frac{1}{c^2}\frac{\partial^2}{\partial t^2}\right) A(\mathbf{r},t) = -\frac{4\pi}{c}(j^+ + j^-),$$

$$\frac{\partial j^+}{\partial t} - i\omega_0 j^+ + \frac{j^+}{2T_2} = \frac{i|m|^2}{\hbar c} A\rho,$$

$$\frac{\partial j^-}{\partial t} - i\omega_0 j^- + \frac{j^-}{2T_2} = -\frac{i|m|^2}{\hbar c} A\rho, \tag{1.2}$$

$$\frac{\partial \rho}{\partial t} + \frac{\rho - \rho_e}{T_1} = \frac{2i}{\hbar c}(j^+ - j^-)A,$$

where

$$j^+(\mathbf{r},t) = m\sum_{i=1}^{N} \langle\sigma_{i+}(t)\rangle\,\delta(\mathbf{r} - \mathbf{r}_i),$$

$$\rho(\mathbf{r},t) = \sum_{i=1}^{N} \langle\sigma_{i3}(t)\rangle\,\delta(\mathbf{r} - \mathbf{r}_i), \qquad m = \mathbf{e}m,$$

e is a unit vector of the wave polarization, T_2 is a characteristic time of the homogeneous transversal relaxation, T_1 is the longitudinal relaxation time, ρ_e is an equilibrium value of the population difference. In deriving (1.2), we have neglected by the inhomogeneous broadening effects and assumed that $m_i = m$, which is true in the absence of the level degeneracy, i.e., in the real two-level system. In the absence of relaxation, when $T_2 = T_1 = \infty$, the Eqs.(1.2) have the very well known integral of motion

$$j^+(\mathbf{r},t)\, j^-(\mathbf{r},t) + \frac{|m|^2}{4}\rho^2(\mathbf{r},t) = \text{const.} \tag{1.3}$$

Basing on the analogy between the system of the classical oscillators and the ensemble of the two-level atoms, the Eqs.(1.2) often are written in the following equivalent form

$$\left[\Delta - \frac{1}{c^2}\frac{\partial^2}{\partial t^2}\right] A\,(\mathbf{r}\,,t) = -\frac{4\pi}{c}\,j(\mathbf{r}\,,t)\,,$$

$$\frac{\partial^2 j}{\partial t^2} + \frac{1}{T_2}\frac{\partial j}{\partial t} + \left[\frac{1}{4T_2} + \omega_0^2\right] j = -\frac{2\,\omega_0 |m|^2}{\hbar c} A\,\rho, \tag{1.4}$$

$$\frac{\partial \rho}{\partial t} + \frac{\rho - \rho_e}{T_1} = \frac{2}{\hbar\omega_0 c} A \frac{\partial j}{\partial t}\,,$$

where $j = j^+ + j^-$. In this case, if $T_2 = T_1 = \infty$, the integral of motion (1.3) takes the form

$$\left[\frac{\partial j}{\partial t}\right]^2 + \omega_0^2 j^2 + |m|^2\,\omega_0^2\,\rho^2 = \text{const.} \tag{1.5}$$

Thus, the conservation laws (1.3) and (1.5) follow from the equations (1.2) and (1.4), respectively, in the case $T_2 = T_1 = \infty$. These conservation laws consist the basis for treatment of the processes of the coherent light–matter interactions. It should be noted that the equations (1.3) and (1.5) include the variables oscillating at the optical frequency. Traditionally, the conservation laws are derived from the equations for slowly varying amplitudes. Notice, that if in (1.2) j^+ and j^- are the complex variables, then the variables j in (1.4) is a real function.

Let us now turn to the equations for slowly varying amplitudes. The slowly varying amplitudes $A_{1,2}(x,t)$ for right and left waves are introduced by

$$A(\mathbf{r},t) = A_1(x,t)\,\sin\,(\omega t - kx) + A_2(x,t)\,\sin\,(\omega t + kx + \phi)\,, \tag{1.6}$$

the wave vector k is determined by the conditions $k = 2\pi n/L$, where L is the active medium length, $\omega = kc$. In the case when $\omega = \omega_0$, the equations for slowly varying amplitudes are #

See Ref. 16 for equations in general case, when $\omega \neq \omega_0$.

$$\frac{\partial A_1}{\partial t} + c\,\frac{\partial A_1}{\partial x} = \frac{2\pi}{\kappa}\,j_1\,, \qquad \frac{\partial A_2}{\partial t} - c\,\frac{\partial A_2}{\partial x} = \frac{2\pi}{\kappa}\,j_2\,,$$

$$\frac{\partial j_1}{\partial t} + \frac{j_1}{2T_2} = \frac{|m|^2}{\hbar c}\,(A_1\rho_0 + A_2\rho_1)\,,$$

$$\frac{\partial j_2}{\partial t} + \frac{j_2}{2T_2} = \frac{|m|^2}{\hbar c}\,(A_2\rho_0 + A_1\rho_1)\,, \tag{1.7}$$

$$\frac{\partial \rho_0}{\partial t} + \frac{\rho_0 - \rho_e}{T_1} = -\frac{1}{\hbar c}\,(A_1 j_1 + A_2 j_2),$$

$$\frac{\partial \rho_1}{\partial t} + \frac{\rho_1}{T_1} = -\frac{1}{2\hbar c}\,(A_1 j_2 + A_2 j_1)\,,$$

where

$$j(\mathbf{r},t) = j_1(x,t)\cos(\omega t - kx) + j_2(x,t)\cos(\omega t + kx + \phi), \tag{1.8}$$

$$\rho(\mathbf{r},t) = \rho_0(x,t) + 2\,\rho_1(x,t)\cos(2kx + \phi). \tag{1.9}$$

It is clear from this equations that we have taken into account the effects of light induced population inversion grating at the spatial frequency $K = 2k$, which are due to the processes of the four–wave light scattering. These effects can influence on the dynamics of the radiative decay of a multiatomic system under certain conditions.

4.1.2. *Normalized equations and main dimensionless parameters.* It is convenient to introduce the dimensionless coordinate $x' = x/L$ and time $t' = t/\tau$, where $\tau = L/c$, and also the new set of normalized variables in terms of the old ones

$$a_{1,2}(x,t) = A_{1,2}(x,t)\left[\frac{8\pi\hbar c^2}{\omega}\frac{N}{V}\right]^{-1/2},$$

$$p_{1,2}(x,t) = \frac{2\pi\tau}{\kappa}\,j_{1,2}(x,t)\left[\frac{8\pi\hbar c^2}{\omega}\frac{N}{V}\right]^{-1/2}, \tag{1.10}$$

$$R(x,t) = \rho(x,t) \left[\frac{N}{V} \right]^{-1}.$$

If the equations (1.10) are applied to (1.7), we get the following equations of motion for new variables

$$\frac{\partial a_1}{\partial t} + \frac{\partial a_1}{\partial x} = p_1, \qquad \frac{\partial a_2}{\partial t} - \frac{\partial a_2}{\partial x} = p_2,$$

$$\frac{\partial p_1}{\partial t} + \alpha p_1 -= \beta (a_1 R_0 + a_2 R_1),$$

$$\frac{\partial p_2}{\partial t} + \alpha p_2 -= \beta (a_2 R_0 + a_1 R_1), \tag{1.11}$$

$$\frac{\partial R_0}{\partial t} + \alpha_1 (1 + R_0) = - 4 - (a_1 p_1 + a_2 p_2),$$

$$\frac{\partial R_1}{\partial t} + \alpha_1 R_1 = - 2 - (a_1 p_1 + a_2 p_2),$$

where

$$\beta = \frac{2\pi |m|^2}{\hbar\omega} \frac{N}{V} \tau, \quad \alpha = \frac{\tau}{2T_2}, \quad \alpha_1 = \frac{\tau}{T_1}. \tag{1.12}$$

In (1.11) we have omitted the primes of the dimensionless variables x' and t'. The ratio of the dimensionless parameters β and α, according to Eq.(1.11) determine the steady–state gain. Really, if

$$\frac{\partial a_n}{\partial t} = \frac{\partial p_n}{\partial t} = 0,$$

it follows from (1.11) that

$$\frac{\partial a_1}{\partial x} = \frac{\beta}{\alpha} R_0(0)\, a_1 = \mu_0 L\, a_1.$$

Consequently, in the case of the complete inversion, when $R_0(0) = 1$, we get

$$\mu_0 L = \frac{\beta}{\alpha} = \frac{4\pi |m|^2}{\hbar\omega} \frac{N}{V} \tau T_2 = \frac{3\lambda^2}{4\pi} \frac{N}{V} \frac{T_2}{T_1} L . \quad (1.13)$$

In accordance with the normalization (1.10) the variable $n(x,t) = |a(x,t)|^2$ is the photon number density normalized to the atomic number density. In particular, the field energy density is expressed in terms of n_1 and n_2 by

$$W(x,t) = \hbar\omega_0 \frac{N}{V} (n_1(x,t) + n_2(x,t)), \quad (1.14)$$

and the energy flux density (Poynting vector) is

$$\mathbf{S}(x,t) = \mathbf{e}_x \hbar\omega_0 \frac{N}{V} (n_1(x,t) - n_2(x,t)). \quad (1.15)$$

4.1.3. *Initial and boundary conditions.* In the case of a spontaneous decay of multiatomic systems the initial conditions for field and polarization should be homogeneous:

$$a_1(x,0) = a_2(x,0) = p_1(x,0) = p_2(x,0) = 0. \quad (1.16)$$

However, the set of equations (1.11) with the initial conditions (1.16) has the identically zero solution. This is due to the fact that semiclassical operator decoupling, determined by Eq.(1.1), leads to the loss of the sources of spontaneous emission, as we have already discussed in Chap.3. Therefore, we either have to supplement the equations (1.11) by the effective sources of the spontaneous polarization or we have to use the nonzero initial conditions. For example, we can use the following ones

$$a_n(x,0) = a_{0n}(x), \qquad p_n(x,0) = p_{0n}(x), \quad (1.17)$$

where either a_{0n} or p_{0n} can be equal to zero. The value of p_{0n} can be defined if in the initial stage we shall use the two–particle correlations instead the decoupling (1.1) (see Chap.3 for details):

$$p_{01}^2 + p_{02}^2 = \frac{\beta}{2N^2} < R^+R^- + R^-R^+ > |_{t=0} = \frac{\beta}{2N}.$$

Consequently, we have

$$p_{01} = p_{02} = \frac{1}{2}\left(\frac{\beta}{N}\right)^{1/2} \tag{1.18}$$

In the frames of more rigorous theories the initial conditions are specified in the form of the correlation matrices (see, for example, Sec. 2.8 and Ref.[14])

$$K_{nm}(x,x') = \langle a_n(x,0)\, a_m(x',0)\rangle, \tag{1.19}$$

where the angular brackets mean the quantum averaging in the frames of quantum theories and averaging over the realizations in the frames of the phenomenological theories.

The boundary conditions for the slowly varying amplitudes $a_n(x,t)$ are

$$a_1(-1/2,\ t) = r_1\, a_2(-1/2,\ t), \tag{1.20a}$$

$$a_2(1/2,\ t) = r_2\, a_1(1/2,\ t), \tag{1.20b}$$

where $r_{1,2}$ are the amplitude reflection coefficients at the left ($x = -1/2$) and right ($x = 1/2$) boundaries. Notice, that the introduction of the phase ϕ into the equations (1.6), (1.8) enables us to account for the phase difference of the reflection coefficients $r_{1,2} = |r_{1,2}| \exp(i\phi_{1,2})$, and $\phi = \phi_2 - \phi_1$.

4.1.4. *Integrals of motion.* The equations (1.11) generate the following relationship, which has the meaning of the local energy change law:

$$\frac{\partial}{\partial t}(W_F + W_A) + \operatorname{div} \mathbf{S} = 0, \tag{1.21}$$

where

$$W_A = \frac{\hbar\omega_0}{2} R_0. \tag{1.22}$$

Integrating equation (1.21) over the time and along the length of the sample and accounting for the boundary conditions (1.20a), and (1.20b), we get

$$\bar{n}_1 + \bar{n}_2 + (1 - |r_2|^2) \int_0^t n_1(1/2,t')\, dt' + (1 - |r_1|^2) \int_0^t n_2(-1/2,t')\, dt'$$

$$= (1 - \bar{R}_0)/2 \qquad (1.23)$$

where

$$\bar{n}_{1,2}(t) = \int_{-1/2}^{1/2} n_{1,2}(x,t)\, dx, \qquad \bar{R}_0(t) = \int_{-1/2}^{1/2} R_0(x,t)\, dx.$$

This equation defines the energy conservation law in the integral form.

The second integral expression can be deduced from the last three equations of set of equations (1.11):

$$\sum_{n=1}^{2} |p_n(x,t)|^2 + \frac{\beta}{4} (R_0^2(x,t) + 2\, R_1^2(x,t)) +$$

$$+ 2 \int_0^t \{\alpha \sum_{n=1}^{2} |p_n(x,t')|^2 + \alpha_1 \frac{\beta}{4} [(1 + R_0)R_0 + 2\, R_1^2]\}\, dt' = \text{const.} \quad (1.24)$$

In the case $\alpha = \alpha_1 = 0$, this expression is transformed into the generalized Bloch integral of motion

$$p_1^2(x,t) + p_2^2(x,t) + \frac{\beta}{4} [R_0^2(x,t) + 2\, R_1^2(x,t)] = \text{const.} \qquad (1.25)$$

This equation follows as dircctly from thc sct of cquations (1.11), so from the substitution of (1.8), and (1.10) into (1.5) and subsequent averaging of the latter over the period $T = 2\pi/\omega_0$ and over the length $\lambda = 2\pi / k$.

4.1.5. *Equations for slowly varying amplitudes.* The processes of light–induced population inversion grating can have a significant influence on the dynamics of the radiative decay only under specific conditions, which will be discussed later. It is clear from the last equation in (1.11) that R_1 will be closer to R_0, if $|a_1p_1| \simeq |a_1p_2|$. As we shall show in the next paragraph, under spontaneous decay of the open systems (r_1, $r_2 < 0.9$) we usually have $|a_1p_1| >> |a_1p_2|$. In this case the set of equations (1.11) assumes a simpler form

$$\frac{\partial a_1}{\partial t} + \frac{\partial a_1}{\partial x} = p_1, \qquad \frac{\partial a_2}{\partial t} - \frac{\partial a_2}{\partial x} = p_2,$$

$$\frac{\partial p_1}{\partial t} + \alpha\, p_1 = \beta\, a_1 R,$$

$$\frac{\partial p_2}{\partial t} + \alpha\, p_2 = \beta\, a_2 R,$$

$$\frac{\partial R}{\partial t} + \alpha(1 + R) = -\,4\,(a_1 p_1 + a_2 p_2)\,. \qquad (1.26)$$

4.2. Linear Stage of Pulse Generation

We shall consider here the linear stage of the emission dynamics, when the condition $1 - R(t) \ll R(0) = 1$ holds.

4.2.1. *Homogeneous boundary conditions.* If we neglect the field reflection at the active volume boundaries, then the boundary conditions (3.6) can be rewritten in the form

$$a_1\left(-\frac{1}{2}, t\right) = a_2\left(\frac{1}{2}, t\right) = 0. \qquad (2.1)$$

In this case the right and left waves are evolved independently during the linear stage of a system dynamics. If we assume that $R_0(x) = 1$, then the evolution of the right wave will be described by the following equations

$$\frac{\partial a}{\partial t} + \frac{\partial a}{\partial x} = p\,, \qquad \frac{\partial p}{\partial t} + \alpha p = \beta a\,. \qquad (2.2)$$

Applying the Laplace–Carson transformation

$$a(x,u) = \int_0^\infty a(x,t)\, e^{-ut} dt \qquad (2.3)$$

we get for the transform $a(x,u)$ the equation

$$\frac{\partial a(x,u)}{\partial x} + \left[u - \frac{\beta}{u + \alpha}\right] a(x,u) = a(x,0) + \frac{p(x,0)}{u + \alpha} . \qquad (2.4)$$

If $p(x,0) = 0$, then, applying the inverse Laplace–Carson transformation, we obtain for the amplitude $a(x,t)$ the following equation

$$a(x,t) = \int_{-1/2}^{x} a(x',0)\, G_a(x - x',t)\, dx' , \qquad (2.5)$$

where the Green's function $G_a(x - x',t)$ is determined by

$$G_a(x - x',t) = \frac{1}{2\pi i} \int_{\delta - i\infty}^{\delta + i\infty} \exp\left\{u\,[t - (x - x')] + \frac{\beta(x - x')}{u + \alpha}\right\} du$$

$$= \left[\frac{\beta(x - x')}{t - (x - x')}\right]^{1/2} I_1(2\{\beta(x - x')\,[t - (x - x')]\}^{1/2})$$

$$\times \exp\{-\alpha[t - (x - x')]\}\ \vartheta(t - (x - x')) ; \qquad (2.6)$$

here $\vartheta(z)$ is the Heaviside step function

$$\vartheta(z) = 1 , \ z \geq 0 ,$$
$$= 0 , \ z < 0 . \qquad (2.7)$$

If in the initial state $a(x,0) = 0$ and there is a small but nonzero initial polarization in the SR transition, then the solution of the equations (2.2) is

$$a(x,t) = \int_{-1/2}^{x} p(x',0)\, G_p(x - x',t)\, dx' , (2.8)$$

where

$$G_p(x - x',t) = I_0(2\{\beta(x - x')[t - (x - x')]\}^{1/2})$$

$$\times \exp\{-\alpha\,[t - (x - x')]\}\ \vartheta(t - (x - x')) . \qquad (2.9)$$

4.2.2. *Delay time.* Let us determine the effect on the characteristic times of pulse evolution of the main parameters of the problem. Let us assume, for the sake of simplicity, that

$$a(x,0) = 0\ , \quad p(x,0) = \left(\frac{\beta}{2N}\right)^{1/2} \delta(x)\ . \tag{2.10}$$

In this case the amplitude of the right wave at the right boundary ($x = 1/2$) is determined, according to (2.8) and (2.9), by

$$a\left(\frac{1}{2},t\right) = \left(\frac{\beta}{2N}\right)^{1/2} I_0\left(\left[2\beta\left(t-\frac{1}{2}\right)\right]^{1/2}\right)$$

$$\times \exp\left[-\alpha\left(t-\frac{1}{2}\right)\right]\vartheta\left(t-\frac{1}{2}\right)\ . \tag{2.11}$$

For $\beta t \gg 1$ we can use the asymptotics of the modified Bessel function, and the last equation becomes

$$a\left(\frac{1}{2},t\right) = \left(\frac{\beta}{2N}\right)^{1/2} \frac{\exp[\{2\beta[t-(1/2)]\}^{1/2} - \alpha\,[t-(1/2)]]}{[2\pi\{2\beta[t-(1/2)]\}^{1/2}]^{1/2}}\ . \tag{2.12}$$

It follows from the conservation law (1.23), that

$$\int_{1/2}^{\infty} n\left(\frac{1}{2},t\right)\, dt \le \frac{1}{2}\ .$$

This means that under symmetric initial conditions the number of quanta emitted from each boundary of active volume is not greater than the half of the initially excited atoms. According to (2.12) we get

$$\int_{1/2}^{t_0+1/2} n\left(\frac{1}{2},t\right) dt = \frac{1}{8\sqrt{\pi}\ N}\, G^{1/2}\, e^{G}[\Phi(G^{1/2}) - \Phi(G^{1/2} - (2t^{*}\alpha)^{1/2})]$$

$$\frac{1}{8\pi N}\left[\frac{\exp\left\{2\alpha\left[\left(\frac{2G}{\alpha}\, t_0^{*}\right)^{1/2} - t_0^{*}\right]\right\}}{1-(2\alpha t_0^{*}/G)^{1/2}} - 1\right] = \frac{1}{4}\ , \tag{2.13}$$

where

$$t_0^* = t_0 + 1/2 \ .$$

The delay time t_0 in (2.13) is determined by the moment when the intensity reaches its peak value. For the symmetric pulse shape, the half of the total number of quanta will be emitted up to that moment. That is why we have substituted the one forth in the right–hand side of the Eq. (2.13). We have used in (2.13) the following notation

$$G(\Gamma) = \frac{\beta}{\alpha} = \frac{3\lambda^2}{8\pi}\frac{N}{V}\frac{1}{\Gamma T_1}L \ , \tag{2.14}$$

where $\Gamma = 1/2T_2$ is a width of the homogeneously broaden line of the SR transition. From Eq. (2.13) we get the following expression for the delay time t_0 :

$$t_0 = \frac{1}{2} + \left\{\left(\frac{G}{2\alpha}\right)^{1/2} - \left[\frac{G}{2\alpha} - \frac{1}{2\alpha}\ln(2\pi N)\right]^{1/2}\right\}^2 . \tag{2.15}$$

On the other hand, the right–hand side of the Eq. (2.12) reaches its maximum at the moment

$$t_m = \frac{1}{2} + \frac{2G}{\alpha} \ . \tag{2.16}$$

Thus, as far as $G > \ln(2\pi N)$ the decay time is determined by (2.15). Notice, that $G = \mu_0 L$. Hence, the delay time t_0 is determined by the Eq. (2.15) for the case of strongly amplifying media. Particularly, when

$$G(\Gamma) \gg \ln(2\pi N) \ ,$$

we obtain from Eq. (2.15)

$$t_0 = \frac{1}{2} + \frac{1}{8\beta}\ln^2(4\pi N) \ . \tag{2.17}$$

For the case $G < \ln(2\pi N)$ the left–hand side of the Eq. (2.13) does not reach $1/4$, i.e , the number of the emitted quanta becomes less than the number of the initially excited atoms and we are dealing with the regime of the superradiance in the weakly amplifying media [15]. The delay time is determined in this case by (2.16). In the case $G \simeq \ln(2\pi N)$ we should directly use the Eq. (2.13) for the determination of the delay time. It should be noted that if $G < \ln(2\pi N)$ the Eq. (2.16) can be written in the form

$$t_0 = \frac{1}{2} + \frac{2}{\alpha} \ln(2\pi N_{\mathrm{eff}}) , \qquad (2.18)$$

where N_{eff} is an effective number of the emitted atoms (see Sec.3.2.3 and Ref.15).

4.2.3. *Superradiance in a resonator.* In contrast to the boundary conditions (2.1), the boundary conditions (1.20a), (1.20b) connect the amplitudes of the counterpropagating waves even in the linear stage of the pulse evolution. In the case $p_n(x,0) = 0$, following the procedure described in Sec. 4.2.1 we can easily show that the amplitude of the right wave is determined by

$$a_1(x,t) = \int_{-1/2}^{x} a_1(x',0)\, G_a(x - x',t)\, dx'$$

$$+ \int_{-1/2}^{1/2} a_1(x',0) \sum_{k=1}^{\infty} (r_1 r_2)^k\, G_a(x - x' + 2k,t)\, dx'$$

$$+ \int_{-1/2}^{1/2} a_2(x'.0)\, r_1 \sum_{k=0}^{\infty} (r_1 r_2)^k\, G_a(x + x' + 2k + 1,t)\, dx'. \qquad (2.19)$$

The meaning of the additional terms in (2.19) in comparison with that in Eq. (2.5) is obvious. The second term describes the wave which is a result of the k–fold reflections of the right wave from both boundaries of the active volume. The third term describes the wave formed in result of the left wave reflection from the left boundary and subsequent k–fold reflections from the both boundaries. The upper limit of the sums in the right–hand side of (2.19) is determined, according to (2.6), by the number of reflections, which occur by the moment t. In particular case $r_1 = r_2 = r$ and $a_2(x,0) = a_1(-x,0)$ the Eq. (2.19) takes the more simple form

$$a_1(x,t) = \int_{-1/2}^{x} a(x',0)\, G_a(x - x',t)\, dx'$$

$$+ \int_{-1/2}^{1/2} a(x',0) \sum_{k=1}^{\infty} r^k G_a(x - x' + k,t)\, dx' . \tag{2.20}$$

In the case $a_1(x,0) = a_2(x,0) = 0$ and $p_1(x,0) \neq 0$, and $p_2(x,0) \neq 0$, we should substitute in the Eqs. (2.19), and (2.20) $a_n(x,0)$ by $p_n(x,0)$, and $G_a(x - x',t)$ by $G_p(x - x',t)$, where $G_p(x - x',t)$ is determined by (2.9).

4.2.4. *Correlation properties of a superradiance pulse.* Let the correlation properties of the initial spontaneous field $a_n(x,0)$ are determined by the correlation matrix

$$K_{nm}(x,x',0) = \langle a_n(x,0)\, a_m(x',0)\rangle = A(x - x')\delta_{nm}. \tag{2.21}$$

For the sake of simplicity, we assume that all the amplitudes are real. The generalization to the case of the complex fields can be easily done.

The evolution of correlation function is given by

$$K_{nm}(x,x',t) = \int_{-1/2}^{x} d\xi \int_{-1/2}^{x'} d\eta\; A(\xi - \eta)\, G(x - \xi,t)\, G(x' - \eta,t).$$

It is seen, that with increasing of the time, t, the initial correlation properties of the field are "forgotten", and the correlation radius begins to depend on the value of parameters α and β.

The main difference between SR in free space $(r_1 = r_2 = 0)$ and SR in a resonator $(r_1 \neq 0,\ r_2 \neq 0)$ is that in the first case $K_{12}^{(1)}(1/2,-1/2,t)$ is always equal to zero, i.e., the SR pulses emitted from the two ends of the active volume remain always uncorrelated, if they are due to the uncorrelated sources of the initial spontaneous field of polarization. In the case $r_1 \neq 0$, $r_2 \neq 0$, it follows from (2.19) that

$$K_{12}^{(2)}\left(\frac{1}{2}, -\frac{1}{2}, t\right) = \int_{-1/2}^{1/2} dx \int_{-1/2}^{1/2} dx'\; A(x - x') \sum_{k=0}^{\infty} \sum_{l=0}^{\infty} (r_1 r_2)^{k+l}$$

$$\times \left[r_2 G_a\left(\frac{1}{2} - x + 2k,t\right) G_a\left(-\frac{1}{2} + x' + 2l + 1,t\right)\right.$$

$$+ r_1 G_a\left[\frac{1}{2} + x + 2k + 1, t\right] G_a\left[-\frac{1}{2} - x' + 2l, t\right]\bigg] \ . \quad (2.22)$$

It is seen, that for $t > 1$ the correlation is established between the waves emitted from the opposite boundaries of the active volume. The degree of the mutual correlation increases with increasing of the parameter β and the reflection coefficients $r_{1,2}$, and decreases with increasing of the parameter α.

4.2.5. *Inhomogeneous broadening.* Up to now we have not took into account the effects of the inhomogeneous broadening of the atomic transitions. If we allow for such broadening, then the equations (1.11) become

$$\frac{\partial a_1}{\partial t} + \frac{\partial a_1}{\partial x} = \int p_1(\Delta)\, f(\Delta)\, d\Delta\ , \qquad \frac{\partial a_2}{\partial t} - \frac{\partial a_2}{\partial x} = \int p_2(\Delta)\, f(\Delta)\, d\Delta$$

$$\frac{\partial p_1(\Delta)}{\partial t} + (\alpha + i\alpha^*\Delta)p_1(\Delta) = \beta(a_1 R_0(\Delta) + a_2 R_1(\Delta))$$

$$\frac{\partial p_2(\Delta)}{\partial t} + (\alpha + i\alpha^*\Delta)p_2(\Delta) = \beta(a_2 R_0(\Delta) + a_1 R_1(\Delta)) \qquad (2.23)$$

$$\frac{\partial R_0(\Delta)}{\partial t} + \alpha_1(1 + R_0(\Delta)) = -\,2(a_1 p_1^*(\Delta) + a_2 p_2^*(\Delta) + \text{c.c.})\ ,$$

$$\frac{\partial R_1(\Delta)}{\partial t} + \alpha_1 R_1(\Delta) = -(a_1 p_2^*(\Delta) + a_2^* p_1(\Delta) + \text{c.c.})\ ,$$

where $f(\Delta)$ is a profile of the inhomogeneously broadened line, which obeys the following normalization condition

$$\int_{-\infty}^{\infty} f(\Delta)\, d\Delta = 1\ . \qquad (2.24)$$

The parameter α^* in (2.23) is determined by $\alpha^* = \tau\Delta_0$, and Δ_0 is a characteristic width of the function $f(\Delta)$.

Let us estimate the main variations of the pulse parameters in the linear stage of generation, which are due to the inhomogeneous broadening effect. Following the procedure described in Sec. 4.1.1 we find that the Laplace–Carson transform of the amplitude of the right wave is governed by the following equation

$$\frac{da_1(x,u)}{dx} + \left[u - \beta \int \frac{f(\Delta)\ R_0(\Delta)\ d\Delta}{u + \alpha + i\alpha^* \Delta} \right] a_1(x,u)$$

$$= a_1(x,0) + p_1(x,0) \int \frac{f(\Delta)\ d\Delta}{u + \alpha + i\alpha^* \Delta} , \tag{2.25}$$

where $R_0(\Delta) = R(\Delta,x,0)$. Let $p_1(x,0) = 0$, the form of the Green function, $G_a(x - x',t)$, was determined in the case of the homogeneously broadening systems by the singular point $u_0 = -\alpha$ (see Eq. (2.6)). In the case of the inhomogeneously broaden media the form of the Green function will be determined by the singular points of the following function

$$F(u) = \int \frac{f(\Delta)\ R_0(\Delta)\ d\Delta}{u + \alpha + i\alpha^* \Delta} . \tag{2.26a}$$

Let $R_0(\Delta) = 1$, i.e., all the atoms within the inhomogeneous line width of the transition are excited uniformly.

For the convenience of the interpretation, let us return in this section to the dimensional variables, then α is replaced by $\Gamma = 1/2T_2$, and the function $F(u)$ takes the form

$$F(u) = \int \frac{f(\Delta)\ d\Delta}{u + \Gamma + i\Delta} . \tag{2.26b}$$

Let us assume that the function $f(\Delta)$ is

$$f(\Delta) = \frac{\Delta_0}{\pi} \frac{1}{\Delta^2 + \Delta_0^2} , \tag{2.27}$$

where Δ_0 is a half-width of the function $f(\Delta)$, and the function $f(\Delta)$ still obeys the condition (2.24). If (3.27) is applied to (3.26a), we get

$$F(u) = \frac{1}{u + \Gamma + \Delta_0} . \tag{2.28}$$

Hence, the amplitude $a_1(x,t)$ is again described by the Eqs. (3.5), (3.6), if we substitute in them the homogeneous line width Γ by

$$\Gamma + \Delta_0 = \frac{1}{2T_2} + \Delta_0 = \frac{1}{2T_2^*} .$$

Thus, when the whole inhomogeneous spectrum is excited uniformly, the increment of the field growth is determined not by the ratio β/α, but by β/α^*, which is

$$G(\Gamma + \Delta_0) = \frac{\beta}{\alpha}\frac{\Gamma}{\Gamma + \Delta_0} = \frac{3\lambda^2}{4\pi}\frac{N}{V}\frac{T_2^*}{T_1}L . \tag{2.29}$$

Let us consider now the case when only a part of the inhomogeneous spectrum, with the width Δ_1, is excited in the initial state:

$$R_0(\Delta) = \frac{\Delta_1^2}{\Delta^2 + \Delta_1^2} . \tag{2.30}$$

In this case the function $F(u)$ takes the form

$$F(u) = \frac{\Delta_1}{\Delta_0^2 - \Delta_1^2}\left(\frac{\Delta_0}{u + \Gamma + \Delta_1} - \frac{\Delta_1}{u + \Gamma + \Delta_0}\right) . \tag{2.31}$$

Particularly, if $\Delta_0 \gg \Delta_1$, we get

$$F(u) = \frac{\Delta_1}{\Delta_0}\frac{1}{u + \Gamma + \Delta_1} . \tag{2.32}$$

The nature of the field growth is determined now by the parameter

$$G(\Gamma + \Delta_1,\Delta_0) = \frac{\beta}{\alpha}\frac{\Delta_1}{\Delta_0}\frac{\Gamma}{\Gamma + \Delta_1} = \frac{3\lambda^2}{4\pi}\frac{N}{V}\frac{\Delta_1}{\Delta_0}\frac{T_{21}^*}{T_1}L , \tag{2.33}$$

where $2T_{21}^* = (\Gamma + \Delta_1)^{-1}$. The atoms with the frequencies lying inside the band $\Gamma + \Delta_1$ near the resonance frequency are only involved now in the process of the SR decay. The number of these atoms is determined by $N(\Delta_1/\Delta_0)$.

If the part of the inhomogeneous spectrum of the width Δ_1 is excited now, but the center of this band does not coincide with the central line of the atomic inhomogeneous spectrum and is shifted by Δ_2 from the central line, then

$$R_0(\Delta) = \frac{\Delta_1^2}{(\Delta - \Delta_2)^2 + \Delta_1^2} . \tag{2.34}$$

If $\Delta_0 \gg \Delta_1$, we get

$$F(u) = \frac{\Delta_0 \Delta_1}{\Delta_0^2 + \Delta_2^2} \frac{1}{u + \Gamma + \Delta_1 - i\Delta_2} . \tag{2.35}$$

Consequently, the wave amplitude will be the oscillating function at the frequency Δ_2 in time and at the wave vector Δ_2/c in space. The increment of the field growth in the linear stage is determined by

$$G(\Gamma + \Delta_1, \Delta_0, \Delta_2) = \frac{3\lambda^2}{4\pi} \frac{N}{V} \frac{\Delta_0 \Delta_1}{\Delta_0^2 + \Delta_2^2} \frac{T_{21}^*}{T_1} L . \tag{2.36}$$

In conclusion, comparing the Eqs. (2.11) – (2.17) and the Eqs. (2.29), (2.33), and (2.36) we see that the pulse evolution during the linear stage of its dynamics is determined by Eqs. (2.11)–(2.17) when the substitutions $\beta \longrightarrow \beta_{eff}$ and $\alpha \longrightarrow \alpha_{eff}$ are made, where the parameters α_{eff} and β_{eff} are determined by the following expressions

a) for $R_0(\Delta) = 1$:

$$\beta_{eff} = \beta , \quad \alpha_{eff} = \alpha \frac{\Gamma + \Delta_0}{\Gamma} ;$$

b) for $R_0(\Delta) = \Delta_1^2/(\Delta^2 + \Delta_1^2)$:

$$\beta_{eff} = \beta \frac{\Delta_1}{\Delta_0} , \quad \alpha_{eff} = \alpha \frac{\Gamma + \Delta_1}{\Gamma} ;$$

c) for $R_0(\Delta) = \dfrac{\Delta_1^2}{(\Delta - \Delta_2)^2 + \Delta_1^2}$:

$$\beta_{eff} = \beta \frac{\Delta_0 \Delta_1}{\Delta_0^2 + \Delta_2^2} , \quad \alpha_{eff} = \alpha \frac{\Gamma + \Delta_1}{\Gamma} .$$

Decreasing of β corresponds to decrease in the density of the number of the excited atoms, and increasing of α corresponds to increase in the transition line width.

4.3. One - wave problems

We have shown in the previous paragraph that in absence of the wave reflections on the active volume boundaries counterpropagating waves evolve independently of one another, at least in the linear stage of the system dynamics. In this case the pulse parameters can be determined with the help of the one-wave models. Here we discuss the nonlinear effects which can be described in the frame of the one-mode model. They include the problem of the pulse amplification and the coherent propagation.

4.3.1. *Interaction of counterpropagating waves.* In the one-wave case the equations (1.26) become

$$\frac{\partial a}{\partial t} + \frac{\partial a}{\partial x} = p , \quad \frac{\partial p}{\partial t} + \alpha p = \beta a R ,$$

$$\frac{\partial R}{\partial t} + \alpha_1 (1 + R) = -4ap, \tag{3.1}$$

the initial conditions are

$$a(x,0) = a\ (x) , \quad p(x,0) = 0 \tag{3.2a}$$

or

$$a(x,0) = 0 , \quad p(x,0) = p_0(x) \tag{3.2b}$$

and the boundary condition is

$$a\left(\frac{1}{2} , t \right) = A(t) . \tag{3.3}$$

In the case of a spontaneous decay $A(t) = 0$. The solution of the equations (3.1) coincides exactly with the solution of the equations (1.26), if we choose in the latter the following initial and boundary conditions for the left wave:

$$a_2(x,0) = p_2(x,0) = 0 , \quad a_2\left(\frac{1}{2} ,t \right) = 0 . \tag{3.4}$$

Fig.4.1 enables us to compare the emitted pulse profiles described by the equation (1.26) for the two types of the initial conditions:

$$p_1(x,0) = p_0 \, , \quad p_2(x,0) = 0 \, , \tag{3.5a}$$

$$p_1(x,0) = p_2(x,0) = p_0 \, , \tag{3.5b}$$

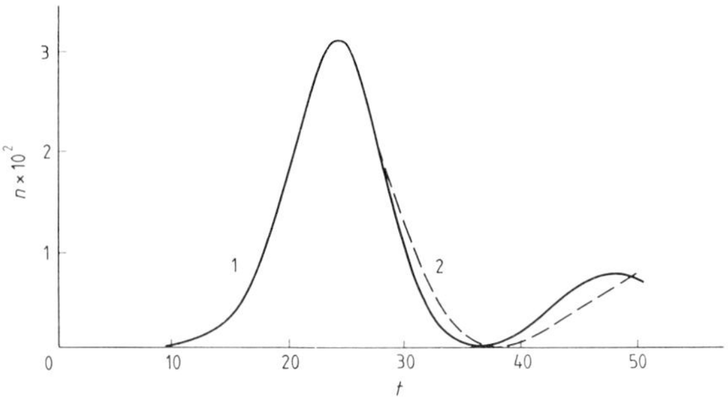

Fig.4.1. Influence of the initial conditions on the emission intensity profiles. Curve "1" corresponds to the initial conditions (3.5a), and curve "2" corresponds to the initial conditions (3.5b).

and for the zero initial and boundary conditions for field in both cases. The computer calculations show that $a_2(x,t) = 0$ in the case (3.5a), and $a_2(x,t) = a_1(-x,t)$ in the case (3.5b). We can see that the counterpropagating waves do not essentially interact up to the moment of formation of the first maximum of the emission intensity, and they begin to interact only in process of formation of the subsequent maxima. It follows from the fact that, as we can see from the equations (1.26), the interaction of the counterpropagating waves occurs only via the mutual depletion of the population inversion. It is seen from the Fig. 4.2a, that the main part of the right wave energy up to its maximum is concentrated in the interval $0 \leq x \leq 1/2$, and only after the depletion of the population inversion in the region near the right end of the active volume the maximum of the right wave amplitude begins to shift to the center of the active volume (see Fig. 4.2b). Starting from this moment the counterpropagating pulses begin to overlap inside the active volume, and they begin to interact via the mutual depletion of the population inversion.

Fig.4.2 demonstrates also another specific feature of the spontaneous decay of the open systems. As for the symmetric initial conditions (see Eq. (3.5b)) the conditions $a_2(x,0) = a_1(-x,0)$, and $p_2(x,0) = p_1(-x,0)$ hold, then, as it follows from Fig.4.2, during the

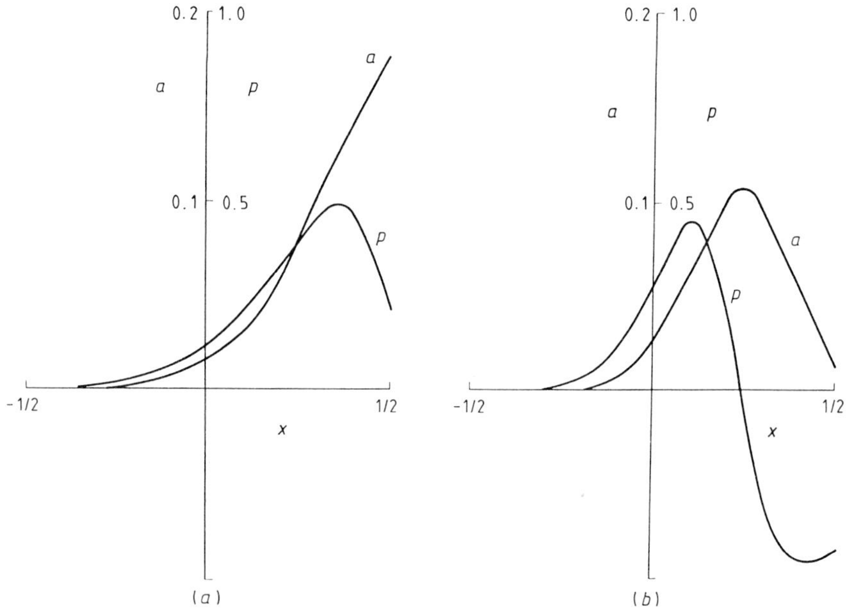

Fig.4.2. Spatial distributions of the right-wave field amplitude and of the polarization associated with the right-wave for two different moments of time: t = 24 (a), 35 (b).

formation of the first maximum of emission we get $|a_1(x,t)\ p_1(x,t)| \gg |a_1(x,t)\ p_2(x,t)|$. Consequently, the effects of the light induced formation of the population inversion grating do not influence the radiative decay during the first peak of emission and can manifest themselves only in the afterglow.

4.3.2. *Soliton propagation in unbounded media.* One of the most interesting effect described by Eq. (3.1) is the appearance of self-similar solutions or solitons. The self-similar solutions appear in the coherent regime, when $\alpha = \alpha_1 = 0$. In this case the Bloch integral of motion begins to play an important role. It has the form

$$p^2(x,t) + \frac{\beta}{4} R^2(x,t) = p^2(x,0) + \frac{\beta}{4} R^2(x,0) \, . \tag{3.6}$$

Introducing the Bloch angle ϑ :

$$R = R_0 \cos \vartheta \, , \tag{3.7}$$

we get

$$p = \frac{\beta^{1/2}}{2} R_0 \sin \vartheta \, , \quad a = \frac{1}{2\beta^{1/2}} \frac{\partial \vartheta}{\partial t} \, , \tag{3.8}$$

and the following equation of motion for the Bloch angle

$$\frac{\partial^2 \vartheta}{\partial t^2} + \frac{\partial^2 \vartheta}{\partial t \, \partial x} = \beta R_0 \sin \vartheta \, . \tag{3.9}$$

The one-soliton solution of the Eq. (3.9) in the case of the unexcited medium, when $R_0 = -1$, has the very well known form [2-8]:

$$\vartheta(x,t) = 4 \text{ atan } [\exp \frac{t - (x/v)}{\tau_0}], \tag{3.10}$$

where the soliton velocity v and its width τ_0 are related by condition (see Fig. 4.3a)

$$\tau_0 = \left(\frac{1}{\beta} \frac{1 - v}{v} \right)^{1/2} , \quad \text{or} \quad v = \frac{1}{1 + \beta\tau_0^2} \, . \tag{3.11}$$

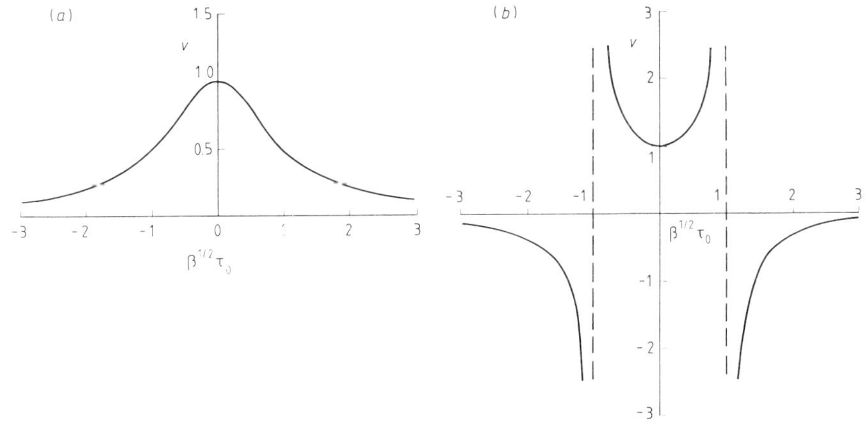

Fig.4.3. Soliton velocity v versus its temporal width τ_0: (a) for the resonantly absorbing media, $R_0 = -1$, (b) for resonantly amplifying media, $R_0 = 1$.

In the case of an excited medium, the soliton-like solutions are also known (see Ref. [11] for a review), but in this case their propagation is characterized by the number of specific features. For example, the relationship between the soliton velocity and its width now become

$$\tau_0 = \left[\frac{1}{\beta} \left(1 - \frac{1}{v} \right) \right]^{1/2}, \quad \text{or} \quad v = \frac{1}{1 - \beta\tau_0^2} . \qquad (3.12)$$

Thus, as it follows from the Fig. 4.3b, if in the case $R_0 = -1$ the velocity is $v \leq 1$, for all possible values of τ_0, in the case $R_0 = 1$ the velocity $v > 1$ for $\beta\tau_0^2 < 1$ and $v < 0$ for $\beta\tau_0^2 > 1$. We can see, that the velocity of the maximum pulse movement is either higher then the light velocity (in dimensionless units, introduced in Sec. 4.1.2, the light velocity is $v_2 = 1$) or is negative. In the latter case, we have $|v| > 1$ if $1 < \beta\tau_0^2 < 2$, and $|v| \leq 1$ if $\beta\tau_0^2 \geq 2$, i.e., the negative velocity can also be higher than the light velocity.

The field amplitude in both cases $(R_0 = \pm 1)$ is described by

$$a(x,t) = \frac{1}{\beta^{1/2}\tau_0} \cosh^{-1}\left[\frac{1}{\tau_0} \left(t - \frac{x}{v} \right) \right] . \qquad (3.13)$$

4.3.3. *Exponential and lethargic amplification.* The general solution of the Eq. (2.4) for $a(x,u)$ is

$$a(x,u) = A(u)\, e^{-\lambda x} + \int_0^x \left(a(x',0) + \frac{p(x',0)}{u + \alpha} \right) \exp[-\lambda(x-x')]dx' \qquad (3.14)$$

where

$$\lambda = u - \frac{\beta}{u + \alpha} .$$

Let now $a(x,0) = 0$ and $p(x,0) = 0$, then

$$a(x,u) = e^{-\lambda x} \int_0^\infty e^{-ut} a(0,t)\, dt . \qquad (3.15)$$

Consequently, we get

$$a(x,t) = \int_0^\infty a(0,t')\, G_b(t - t',x)\, dt' + a(0,t) , \qquad (3.16)$$

where

$$G_b(t - t',x) = \left[\frac{\beta x}{t - t' - x} \right]^{1/2} I_1(2[\beta x(t - t' - x)]^{1/2})$$

$$\times \exp[-\alpha(t - t' - x)]\ \vartheta(t - t' - x)\ . \tag{3.17}$$

In this section we shall consider the semi infinite medium, which is in the interval $[0,\infty]$ and not in the interval $[-1/2,1/2]$, as we have used so far. Hence, $a(0,t)$ determines the form of a pulse incident on the boundary of the active volume.

Let the boundary condition has the form

$$a(0,t) = a_0 e^{\delta t}\ , \tag{3.18}$$

then, according to (3.16), and (3.17), we get

$$a(x,t) = \Bigg\{ a_0 \exp\left[\delta(t - x) + \frac{\beta x}{\alpha + \delta} \right] \tag{3.19}$$

$$+ a_0 \exp[-\alpha(t-x)] \sum_{n=0}^{\infty} \left[(\alpha+\delta) \left(\frac{t-x}{\beta x} \right)^{1/2} \right]^n I_n(2[\beta x(t-x)]^{1/2}) \Bigg\}\ \vartheta(t-x).$$

If the incident pulse is a pulse of a constant amplitude ($\delta = 0$), then from (3.19) we have

$$a(x,t) = \Bigg\{ a_0 \exp\left(\frac{\beta x}{\alpha} \right) + a_0 \exp[-\alpha(t - x)] \sum_{n=0}^{\infty} \left[\alpha \left(\frac{t - x}{\beta x} \right)^{1/2} \right]^n$$

$$\times I_n(2[\beta x(t - x)]^{1/2})\ \vartheta(t - x) \Bigg\}\ . \tag{3.20}$$

In the case when $\alpha = 0$, the Eq. (3.19) is transformed to

$$a(x,t) = \Bigg\{ a_0 \exp\left[\frac{\beta x}{\delta} + \delta(t - x) \right] + a_0 \sum_{n=0}^{\infty} \left[\delta \left(\frac{t - x}{\beta x} \right)^{1/2} \right]^n$$

$$\times I_n(2[\beta x(t - x)]^{1/2}) \Bigg\}\ \vartheta(t - x)\ . \tag{3.21}$$

If $\alpha = \delta = 0$, then

$$a(x,t) = a_0\ I_0(2[\beta x(t - x)]^{1/2})\ \vartheta(t - x)\ . \tag{3.22}$$

It follows from the above expression, that if at least one of the parameters α and δ differs from zero, then for $\beta x \ll 1$ we have the usual exponential amplification of the incident pulse characterized by the gain

$$\mu_0 L = \frac{\beta}{\alpha} .$$

With increasing of product $\beta x(t - x)$ such regime of the amplification is substituted by the regime with the lower rise of amplitude along x:

$$a(x,t) \quad \exp\{2[\beta x(t - x)]^{1/2}\} ,$$

which is called by the lethargic amplification regime. As it follows from (3.22), in the case $\alpha = \delta = 0$ we have the lethargic amplification regime just from the beginning.

4.3.4. *Self-similar solutions in spatially inhomogeneous media.* The results discussed in the previous two sections allow us to put forward the following problems which require the further insight into the implication of them. Can the soliton velocity exceed the light velocity as shown in Fig.4.3. What does the divergence of v at $\beta\tau_0^2 = 1$ mean (see Fig.4.3b) ? The soliton-like solution is an eigenfunction of Eq. (3.9) in the infinite medium. What happens if the pulse of radiation of the type (3.13) is transmitted by an amplifying medium, which has the finite length along the axis x ? Can we obtain again the soliton as an output pulse, or is its profile deformed by the boundaries of the active volume ?

We shall show that the answer to the last question gives us the answer to the previous questions. Let us consider the propagation of the radiation pulses in the spatially inhomogeneous media in the regime of coherent amplification ($\alpha = 0$) . The Bloch integrals of motion (1.24), (1.25), and (3.6) display the property that they hold in each spatial point x .Therefore, assuming that there is some profile of the atomic number density inside the active volume

$$R(x,0) = R_0(x) , \tag{3.23}$$

we can rewrite Eqs. (3.7) and (3.8) in the form

$$R(x,t) = R_0(x) \cos \vartheta(x,t) , \quad p(x,t) = \frac{\beta^{1/2}}{2} R_0(x) \sin \vartheta(x,t) ,$$

$$a(x,t) = \frac{1}{2\beta^{1/2}} \frac{\partial \vartheta(x,t)}{\partial t} . \tag{3.24}$$

If (3.24) is applied to (3.1), we get in the case $\alpha = \alpha_1 = 0$ the following equation for the Bloch angle $\vartheta(x,t)$:

$$\frac{\partial^2 \vartheta(x,t)}{\partial t^2} + \frac{\partial^2 \vartheta(x,t)}{\partial x\, \partial t} = \beta R_0(x) \sin \vartheta(x,t) \,. \tag{3.25}$$

The solution of (3.25) shall be sought in the form

$$\vartheta(x,t) = 4 \text{ atan } e^{\Phi(x,t)} \,. \tag{3.26}$$

Then, for the variable $\Phi(x,t)$ we get the equations

$$\frac{\partial \Phi}{\partial t}\left[\frac{\partial \Phi}{\partial t} + \frac{\partial \Phi}{\partial x}\right] = \beta R_0(x) \,, \quad \frac{\partial}{\partial t}\left[\frac{\partial \Phi}{\partial t} + \frac{\partial \Phi}{\partial x}\right] = 0 \,. \tag{3.27}$$

It follows from the last equation that

$$\frac{\partial \Phi}{\partial t} + \frac{\partial \Phi}{\partial x} = C(x) \,, \tag{3.28}$$

then from the first equation of (3.27) we get

$$\frac{\partial \Phi}{\partial t} = \frac{\beta R_0(x)}{C(x)} \,,$$

or

$$\Phi(x,t) = \frac{\beta R_0(x)}{C(x)}\, t + \Phi_0(x) \,. \tag{3.29}$$

Substituting Eq. (3.29) into Eq.(3.28) and integrating the latter we obtain

$$\Phi(x,t) = \int_{-\infty}^{x} \left[C(x') - \frac{\beta R_0(x')}{C(x')} \right] dx' + \Phi_1(t) \,. \tag{3.30}$$

Comparing Eq. (3.29) and Eq. (3.30), we can see that $C(x) = \beta R_0(x)\tau_0$, where τ_0 is a constant. Finally, the variable $\Phi(x,t)$ is determined by the following expression

$$\Phi(x,t) = \frac{t - x}{\tau_0} + \beta\tau_0 \int_{-\infty}^{x} R_0(x')\, dx' \,. \tag{3.31}$$

It is seen from this equation that the constant τ_0 is a pulse width outside the active volume, where $R_0(x) = 0$.

By way of illustration, we write down the form of $\Phi(x,t)$ for some profiles $R_0(x)$:

a) $R_0(x) = R_0/\cosh^2\alpha_0 x$:

$$\Phi(x,t) = \frac{t - x}{\tau_0} + \frac{\beta\tau_0 R_0}{\alpha_0} \tanh \alpha_0 x ; \tag{3.32a}$$

b) $R_0(x) = R_0[\tanh \alpha_0 x + \tanh \alpha_0(L - x)]/2$:

$$\Phi(x,t) = \frac{t - x}{\tau_0} + \frac{\beta\tau_0 R_0}{2\alpha_0} \ln \frac{\cosh \alpha_0 x}{\cosh \alpha_0 (L - x)} ; \tag{3.32b}$$

c) $R_0(x) = R_0[\mathrm{sgn}\, x + \mathrm{sgn}(L - x)]/2$:

$$\Phi(x,t) = \frac{t - x}{\tau_0} + \frac{\beta\tau_0 R_0}{2} x\, [\mathrm{sgn}\, x + \mathrm{sgn}(L - x)] + \frac{\beta\tau_0 R_0 L}{2} [1 + \mathrm{sgn}(x - L)] ; \tag{3.32c}$$

d) $R_0(x) = R_0 + R_1 \cos \alpha_0 x$:

$$\Phi(x,t) = \frac{t - x}{\tau_0} + \beta\tau_0 \left[R_0 x + \frac{R_1}{\alpha_0} \sin \alpha_0 x \right] . \tag{3.32d}$$

Considering that the equation (3.25) is resulted from the equations for slowly varying amplitudes, we can neglect the field reflection at the medium boundaries. The medium length is designated as L in (3.32b), and (3.32c), and $L = 1$ according to normalization introduced above.

If (3.26) is applied to (3.24), we get the relationships between the variables $a(x,t)$, and $R(x,t)$ and the variable $\Phi(x,t)$:

$$a(x,t) = \frac{1}{\beta^{1/2}\tau} \frac{1}{\cosh \Phi(x,t)} , \quad R(x,t) = R_0 \left[1 - \frac{2}{\cosh^2 \Phi(x,t)} \right] . \tag{3.33}$$

We can see that the amplitude $a(x,t)$ is a maximum when $\Phi(x,t_0) = 0$, the population difference $R(x,t_0) = -R_0$ at this moment. We can represent $\Phi(x,t)$ in the form

$$\Phi(x,t) = \frac{1}{\tau_0} (t - \phi(x)) . \tag{3.34}$$

The dependencies $\phi(x)$, determined by (3.32b), and (3.32d), in the case of the initially excited media $(R_0 > 0)$ are shown in Fig.4.4. It is seen from Fig.4.4a, that at a moment t_1 we have a solitary pulse in the

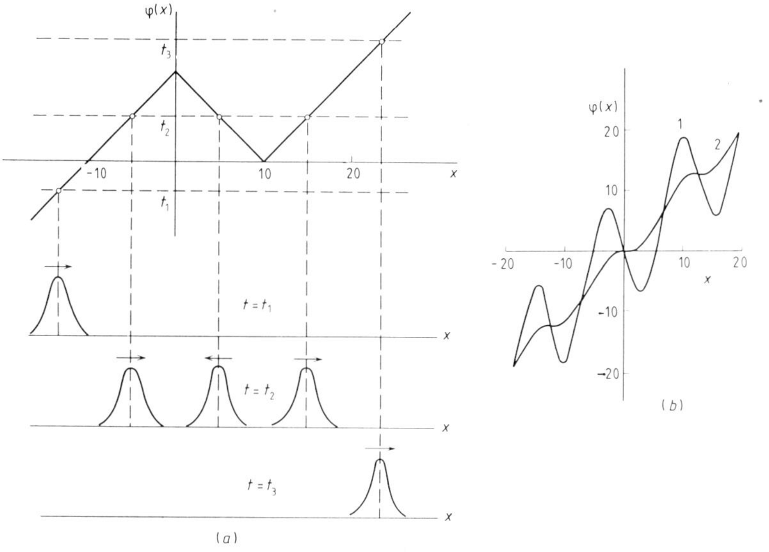

Fig.4.4. To the interpretation of Eqs.(4.32)-(4.34): (a) function $\phi(x)$ described by Eq.(4.32b) for $\alpha_0 = 10$, $L = 10$, $\beta\tau_0^2 R_0 = 2$, and spatial distribution of the field amplitude at three different moments of time; (b) function $\phi(x)$ described by Eq.(4.32d) for the following values of parameters: $\alpha_0 = 0.5$, $R_0 = 0$, $\beta\tau_0^2 R_1 = 5$ (curve 1), and $\alpha = 0.5$, $R_0 = 0$, $\beta\tau_0^2 R_1 = 1$ (curve 2).

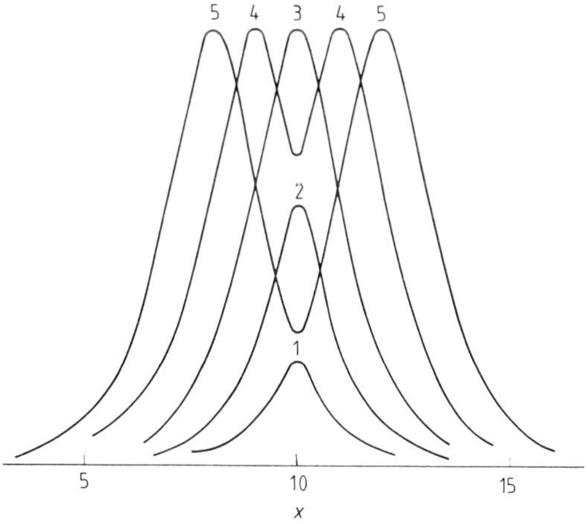

Fig.4.5. Dynamics of output pulse formation at the exit face of the active medium. The curves are referred to the following moments of time: t = -2 (1), -1 (2), 0 (3), 1 (4), 2 (5).

region $-\infty < x < \infty$. This pulse is a peak value at a point x_1 , determined by the condition $\phi(x_1) = t_1$, and it moves in the positive direction of the x axis. At a moment $t = t_2$ we have three pulses, two of them, which are out of the amplifying medium, move in the positive direction of the x axis, while the third pulse, which is inside the amplifying medium, moves in the opposite direction. At a moment $t = t_3$ we again have the solitary pulse, propagating in the direction of the incident pulse. The appearance of the three maxima at a moment $t = t_2$ is due to the fact that as a result of the amplification of the incident pulse the population difference at the right end of the active medium at a moment t_0 is $R(L,t_0) = -R_0$. In the case $\beta\tau_0^2 R_0 = 2$ and $\alpha_0 = 10$, the moment t_0 is $t_0 = 0$, as is shown in Fig.4.4. Fig.4.5 shows the pulse evolution at the right end of the active medium. We can see that at $t > 0$ the pulse splits into two parts, one of which leaves the active medium, whereas the other counterpropagates with the incident pulse. When $\beta\tau_0^2 R_0 = 2$, the two parts are symmetric relative to the point $x = 1$. This is explained by the fact that the rising and falling parts of the curve $\phi(x)$ have the same inclination with respect to the coordinate axis. Therefore, the projections onto the x axis of the $\phi(x)$ curve segments, which are cut off by the lines t_2 and $t_2 + \tau_0$, have the same value. If $\beta\tau_0^2 R_0 > 2$, the linear dimension of the pulse inside the medium is shortened in comparison with its dimension outside the medium, whereas, if $\beta\tau_0^2 R_0 < 2$, the pulse becomes elongated.

It should be noted, that at $t = t_0$ the amplitude of the incident pulse at the entrance to the active volume is

$$a_{in}(0,t_0) = \frac{1}{\tau_0\beta^{1/2}} \frac{1}{\cosh[(1/\tau_0)(\beta\tau_0^2 R_0 - 1)]} \frac{1}{\tau_0\beta^{1/2}} \exp\left(-\frac{\beta\tau_0^2 R_0 - 1}{\tau_0}\right),$$

while at the exit it is

$$a_{in}(L,t_0) = \frac{1}{\tau_0\beta^{1/2}} \frac{1}{\cosh(\beta\tau_0 R_0)} \frac{1}{\tau_0\beta^{1/2}} e^{-\beta\tau_0 R_0} .$$

Since the amplitude of the output pulse at $t = t_0$ is

$$a_{out}(L,t_0) = \frac{1}{\tau_0 \beta^{1/2}} ,$$

then we can determine the gain from the condition

$$a_{out}(L,t_0) = e^{GL} a_{in}(L,t_0) ,$$

hence,

$$G = \beta\tau_0 R_0 - \ln 2 \simeq \beta\tau_0 R_0 .$$

Comparing this formula with Eq. (3.21), we can see that the gain corresponds to the first term in Eq. (3.21). In fact, according to (3.18) $\delta = 1/\tau_0$, therefore assuming that $R_0 = 1$ we get $\beta/\delta = \beta\tau_0$. Thus, the amplification of the exponential incident pulse of the type $a(0,t) = a_0 \exp(t/\tau_0)$ results in formation at the medium exit face of the soliton-like pulse of the type (3.33a) with the width τ_0 determined by the increment of the input signal.

When $\beta\tau_0^2 R_0 = 1$ the curve $\phi(x)$ has a form shown in Fig.4.6. In this case the population inversion evolves identically along the whole length of the sample. Such regime of a pulse propagation is in interest for the pulse generation and amplification in the three-level systems. Under certain conditions, we can realize the uniform population inversion for the adjacent transition along the whole length of the sample.

Further opportunities for controlling of the pulse shape in medium can be realized in the case of the medium with the periodic profile of excitation or density of the active atoms. It is clear from Fig.4.4b , that variation of the parameters R_0 , R_1 , τ_0 , and α_0 (see Eq. (3.32d) enables to produce the periodic profile of excitation in adjacent transition in the three-dimensional medium. This profile can be realized as over the whole length of the active volume, or in the separate local region of the active volume.

Finally, the above analysis shows that soliton pulse can pass through the resonantly amplifying or absorbing media of the finite length without change in its profile. The pulse shape can be essentially modified inside the medium, but it is described everywhere by the same equation (3.33a). The pulse of the type (3.33a) can have a few maxima (see, for example, Fig.4.5) or it can even reach a maximum in the macroscopical region along the x axis (see, for example, Fig.4.6). We can now explain why the soliton velocity becomes equal to infinity in the frames of theories based on the analysis of the soliton propagation in the unbounded media (see Fig.4.3). It is explained by the appearance

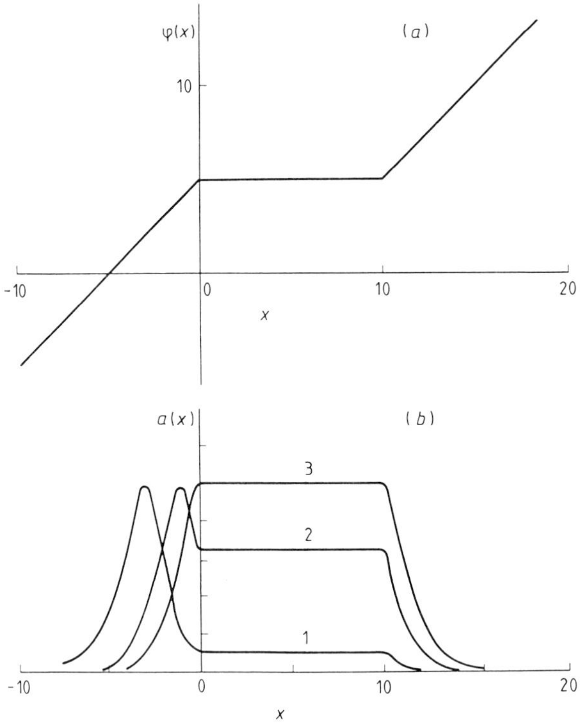

Fig.4.6. (a) function $\phi(x)$ described by Eq.(4.23b) for the following values of parameters: $\alpha_0 = 10$, $L = 10$, $\beta\tau_0^2 R_0 = 1$; (b) spatio-temporal dynamics of pulse for the same values of parameters as those for (a); the moments of time are t = 2 (1), 4 (2), 5 (3).

of the amplitude maxima in the pulse front resulted from its amplification in the active medium. In fact, the velocity of the pulse maximum movement is determined, according to (3.34), by

$$v_{max} = \frac{dx}{dt} = \left(\frac{d\phi}{dt} \right)^{-1} .$$

At a point x_0, where

$$\frac{d\phi}{dx}(x_0) = 0 ,$$

at a moment $t_0 = \phi(x_0)$ the pulse of the type (3.33a) has a maximum equal to $a(x_0,t_0) = 1/\tau_0\beta^{1/2}$. At this moment the pulse maxima appears infinitely fast in the new position. However, we can see from Fig.4.5 that time of formation of the pulse maxima at the exit face of the medium has a finite duration. The limitation of the output pulse amplitude by the value $1/\tau_0\beta^{1/2}$ is due to the fact that the peak value of the polarization amplitude in the resonant transition is restricted by the Bloch integral of motion (see Eq.(3.6)).

4.4. Two - wave problems

In this section we shall study the dependence of the nature of generation and amplification of pulses on the matter parameters in the frames of the models accounting for the interaction of the counterpropagating waves. The equations describing the dynamics of the field and polarization amplitudes, and population inversion for the case of the inhomogeneously broadened media are

$$\frac{\partial a_2}{\partial t} - \frac{\partial a_2}{\partial x} = \int p_2(\Delta)\, f(\Delta)\, d\Delta,$$

$$\frac{\partial a_1}{\partial t} + \frac{\partial a_1}{\partial x} = -\int p_1(\Delta)\, f(\Delta)\, d\Delta,$$

$$\frac{\partial p_{1,2}(\Delta)}{\partial t} + (\alpha + i\,\alpha_1^{*}\Delta)p_{1,2}(\Delta) = \beta\, a_{1,2}R(\Delta), \tag{4.1}$$

$$\frac{\partial R(\Delta)}{\partial t} + \alpha_1(1 + R(\Delta)) = -2\sum_{n=1}^{2}(a_n^{*}p_n(\Delta) + a_n p_n^{*}(\Delta)),$$

In the case of a spontaneous decay of multiatomic systems the initial conditions for the field and polarization can be chosen in the form

$$a_1(x,0) = a_2(x,0) = 0,$$

$$p_n(x,0,\Delta) = p_{0n}(x,\Delta),$$

$$R(x,0,\Delta) = R_0(x,\Delta), \tag{4.2}$$

where $p_{0n}(x,\Delta)$ determine the sources of the initial polarization. The boundary conditions for the slowly varying amplitudes $a_n(x,t)$, in general case, are

$$a_1(-1/2,\ t) = r_1\ a_2(-1/2,\ t)$$

$$a_2(1/2,\ t) = r_2\ a_1(1/2,\ t) \tag{4.3}$$

It is clear from (4.1) that the main parameters determining the nature of a decay dynamics are α, α^*, and β, which are the ratios of the active volume transit time, $\tau = L/c$, to the homogeneous broadening relaxation time T_2 ($\alpha = \tau/2T_2$) and to the inhomogeneous broadening relaxation time T_2^* ($\alpha^* = \tau/2T_2^*$), and the ratio of τ to the coherent time τ_c ($\beta = \tau/\tau_c$). We shall study the influence of these parameters on the intensity and shape of the generated pulses.

4.4.1. *Superradiance and superluminescence.* Let us consider first the open (mirrorless) systems. In this case

$$r_1 = r_2 = 0, \tag{4.4}$$

and the boundary conditions (4.3) become homogeneous. The profile of the emission intensity for the case of the media, which were excited completely and uniformly in the initial moment

$$R_0(x,0) = 1, \tag{4.5}$$

and in the absence of the inhomogeneous broadening, when

$$f(\Delta) = \delta(\Delta), \tag{4.6}$$

are shown in Fig.4.7. The ratio β/α is the same for all three cases and equal to $\beta/\alpha = 100$, whereas the parameter β varies as follows: β = 0.1(a), 1(b), 10(c). It is seen that the pulse profile does not almost affected by the variation of β, however the intensity peak value I_0 and

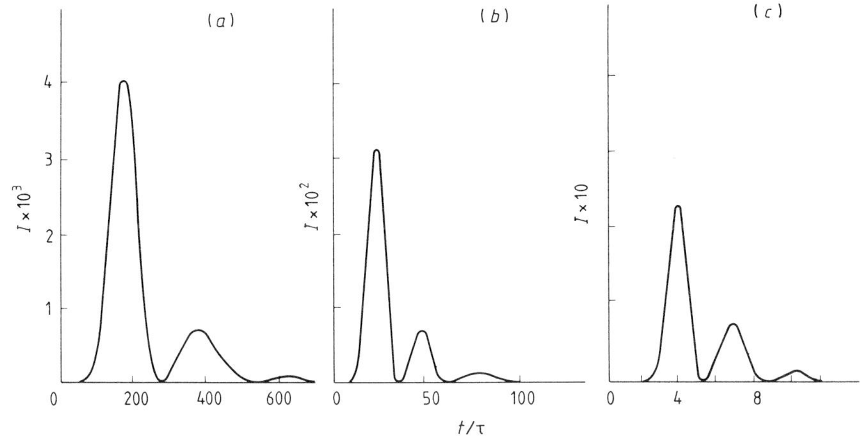

Fig.4.7. Emission intensity profile for three different values of parameter β and fixed value of the gain $\mu_0L = \beta/\alpha$. The parameters α and β are: (a) $\alpha = 10^{-3}$, $\beta = 10^{-1}$, (b) $\alpha = 10^{-2}$, $\beta = 1$, (c) $\alpha = 10^{-1}$, $\beta = 10$.

the delay time t_0 depend essentially on β. The intensity increases with increasing β, while the delay time decreases (approximately as β^{-1}). Fig.4.8 shows the effect upon the peak intensity of variation in β for a fixed α (Fig.4.8a,b) and in α for a fixed β (Fig.4.8c). We can see that there is approximately linear relationship between I_0 and β in the region $\beta>\alpha$, which then reaches the saturation, when I_0 approaches to $I_{0max} = 1$. It will be remind that the field amplitude in the equations (4.1) is normalized in such a way that

$$n(x,t) = |a(x,t)|^2$$

is a photon number density normalized to the atomic number density. In its turn, the output intensity is determined by

$$I(t) = n(1/2,t).$$

Therefore, if $I(t)$ is equal to unity it means that the photon number density at the active volume boundary approaches to the atomic number density. It is clear from Fig.4.8c that in the region $\alpha<<\beta$ the value of the peak pulse intensity does not depend on the parameter α, but the value of the intensity in the region of "saturation" depends on the value of the parameter β. The peak intensity falls when α approaches to β and becomes close to the spontaneous background level when $\alpha \cong \beta$. The dashed curve in Fig.4.8b corresponds to the curve of Fig.4.8a. Therefore, we can see that in the region $\beta >> \alpha$ the peak intensity I_0 does not depend on α

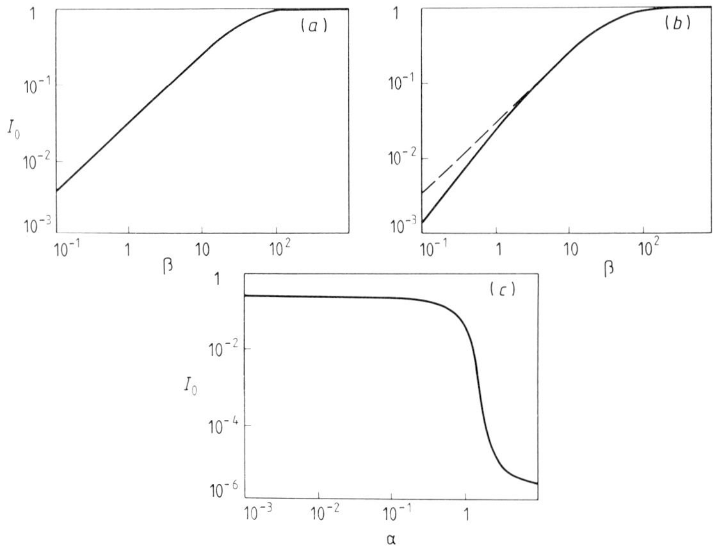

Fig.4.8. The peak intensity versus the parameter β for a fixed value of the parameter $\alpha = 10^{-3}$ (a) and $\alpha = 10^{-2}$ (b), and versus the parameter α for a fixed value of the parameter $\beta = 10$ (c).

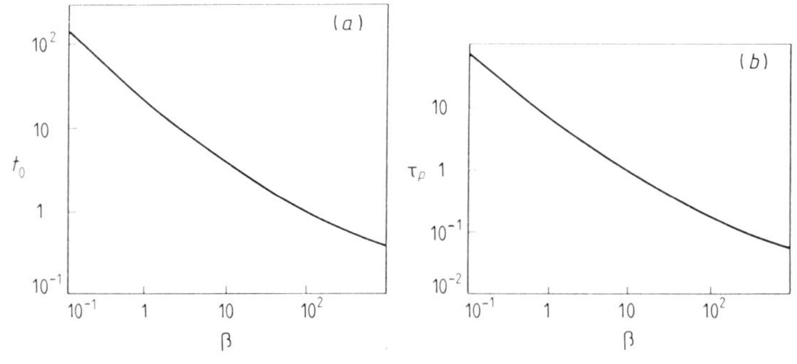

Fig.4.9. Delay time (a) and pulse width (b) versus the parameter β for a fixed value of the parameter $\alpha = 10^{-3}$.

and is determined by the value of the parameter β. When β approaches to α the gain $\mu_0 L = \beta/\alpha$ falls and the intensity of the emission decreases. In Fig.4.9a the delay time is plotted versus the parameter β for a fixed value of the parameter α. The relationship between the pulse width and the parameter β is shown in Fig.4.9b. It is clear from these figures that in the region $\beta >> \alpha$ we have $t \propto \beta^{-1}$ and $\tau \propto \beta^{-1}$. Fig.4.10 shows the behavior of the peak intensity in the plane of the parameters (α,β). We can see that the peak intensity exceeds the spontaneous emission background level when $\beta > \alpha$. The pulse intensity increases inside the region $\beta > \alpha$ along the straight lines parallel to the line $\lg \alpha = - \lg \beta$. The intensity is approximately constant along the straight lines parallel to the line $\lg \alpha = \lg \beta$.

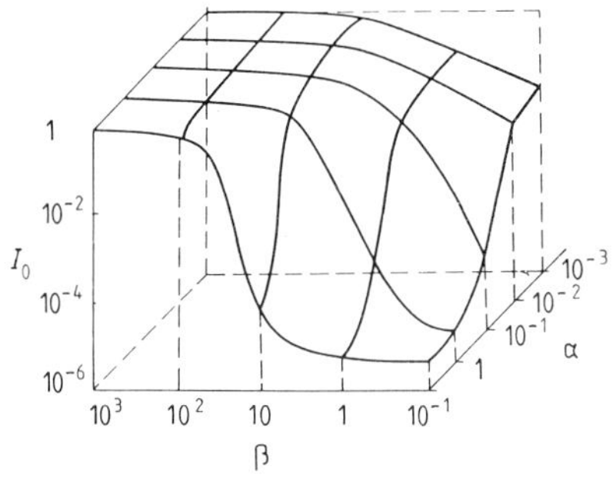

Fig.4.10. Peak pulse intensity as a function of parameters α and β.

Figs.4.11 and 4.12 illustrate the effect upon the intensity and shape of the pulse of variation in the width of the inhomogeneous broadening spectrum. It is clear from these figures that the generation occurs if $\beta > \alpha^*$, that is in full agreement with analysis given in Sec.4.2.5. When $\beta \propto \alpha^*$ the peak intensity is close to the spontaneous emission background level.

Finally, the results presented above enable us to make the following conclusions. First, in the case of homogeneously broaden system the coherent generation threshold is determined by the condition $\beta > \alpha$. Secondly, in the case of the inhomogeneously broaden systems and for the uniform excitation of the whole inhomogeneous spectrum (Figs.4.11 and 4.12 refer especially to this case) the coherent generation condition becomes $\beta > \alpha + \alpha^*$. Thirdly, the nature of the peak intensity dependence on the value of the parameter β allows us to distinguish the two qualitatively different regions. It is follows from Fig.4.8 that for $\beta <$ 10 there is the quadratic dependence of the peak intensity on the initial population inversion density. In fact, in the region $\beta < 10$ we have $I_0 \propto \beta$. Considering the normalization used above (see Eqs.(1.10), and (1.14)) and Eq.(1.12), we get

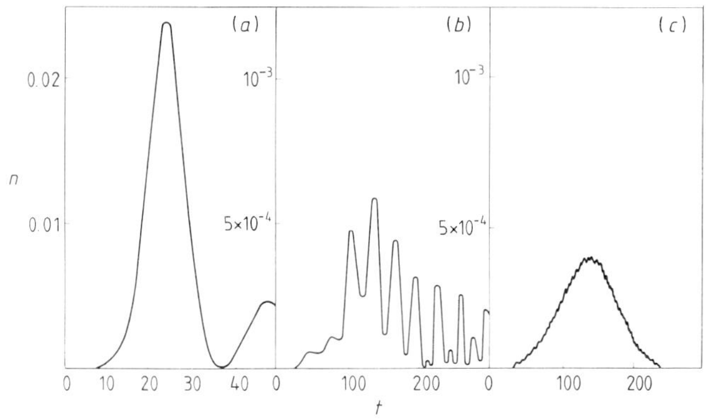

Fig.4.11. Influence of inhomogeneous broadening on the emission intensity profile. The parameters α and β are given as $\alpha = 10^{-2}$, $\beta = 1$. The normalized half-width of the inhomogeneously broadened spectrum is $\alpha^* = 0.1$ (a), 1 (b), 5 (c).

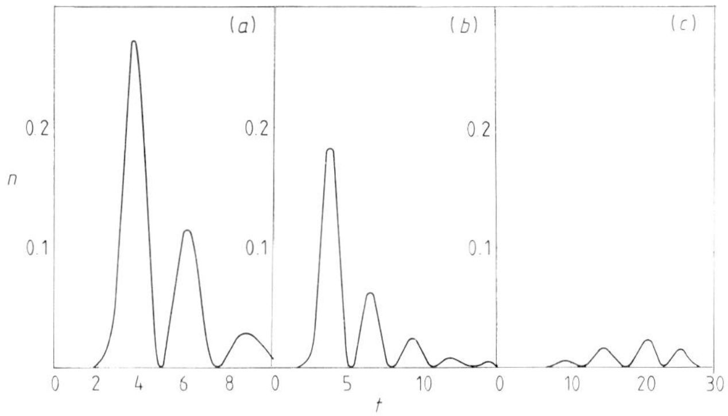

Fig.4.12. Influence of inhomogeneous broadening on the emission intensity profile. The parameters α and β are given as $\alpha = 10^{-2}$, $\beta = 10$. The normalized half-width of the inhomogeneously broadened spectrum is $\alpha^* = 0.1$ (a), 1 (b), 5 (c).

$$I \propto cW = c\hbar\omega \frac{N}{V} I_0 \propto \frac{N}{V} \beta \propto \left[\frac{N}{V}\right]^2 . \tag{4.7}$$

The regime of generation with the quadratic dependence of the peak intensity on the population inversion density is usually called by superradiance. It is also seen from Fig.4.8 that in the region $\beta > 10$ the peak intensity increases more slowly with increasing of β and it approaches in the limit to the linear dependence $I_{max} \propto N/V$. This is usually called by a superluminescence or amplified spontaneous emission.

4.4.2. *Superradiance, superluminescence and selfexcitation in an optical cavity.* We now turn to the study of the main characteristics of emission, when generation occurs in an optical cavity. It should be noted that we shall not consider the case when the active medium is placed into the high–Q resonator, but we shall treat the more general problem concerning to the influence of the field reflection at the boundaries of the active volume on the nature of the generation dynamics. We shall assume that the value of the reflection coefficients $r_{1,2}$ in the boundary conditions (4.3) varies from 0 to 1. Thus, the case of a generation in a high–Q resonator is one of the particular cases of the general problem. The necessity of a field reflection accounting is due to the fact that in all practically important cases the reflection coefficient is different from zero, because the dielectric permittivity of the active medium always differs from the dielectric permittivity of surrounding. This difference exists even in the case of the low–pressure gases and can essentially change the nature of the generation dynamics in the case of the solid–state active media. Thus, the decay of the open systems ($r_1 = r_2 = 0$) and the decay in the high–Q optical resonator are the limiting cases of the general problem under consideration.

Fig.4.13 shows the effect upon the pulse profile of variation in the value of the reflection coefficients. The resultant curves for the dependence of the peak intensity on the reflection coefficients are plotted in Fig.4.14 for the various values of the parameters α and β. Figs.4.14a – 4.14d show the dependence of the peak field intensity inside the cavity $I_0 = |a_1(1/2,t_0)|^2$ and outside it $I = I_0(1 - r_2^2)$ for the two limiting cases, when both mirrors have the same reflection coefficients ($r_1 = r_2 = r$) and when the left mirror is totally reflecting $r_1 = 1$, and the reflection coefficient of the second mirror is varied from 0 to 1. It is clear from the figures that in the case of the superluminescent media ($\alpha << \beta < 1$) there is a quadratic rise of the peak intensity with the reflection coefficient r. Whereas in the case of the superluminescent media ($1 < \alpha << \beta$), the peak intensity decreases with increasing r. Fig.4.15 clearly demonstrates the difference in the field dynamics for the cases of the superradiant and superluminescent decay. It is seen from Fig.4.15a that in the case of the superradiant media when $r = 1$ the field is regularly absorbed and reemitted by the resonant atoms. So, there is a continuous energy exchange between the atomic and field subsystems. The peak value of I_0 in this case is $I_0 = 1/2$, i.e., the net peak number of the photons in two counterpropagating waves is equal to the number of the resonant atoms. On the other hand, in the case of the superluminescent

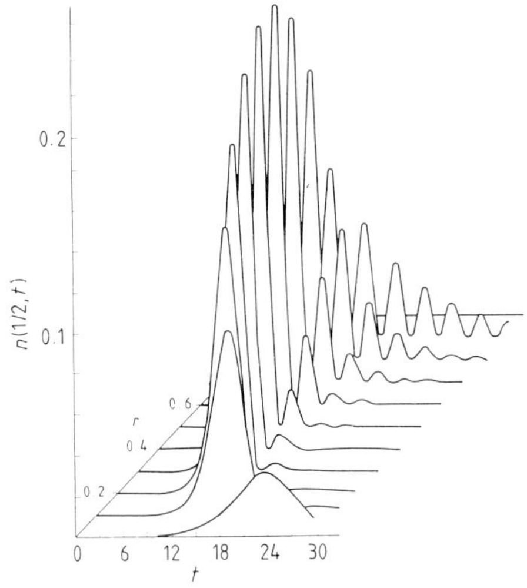

Fig.4.13. Dependence of the emission intensity profile on the reflection coefficients $r_1 = r_2 = r$. The parameters α and β are given as $\alpha = 10^{-2}$, $\beta = 1$.

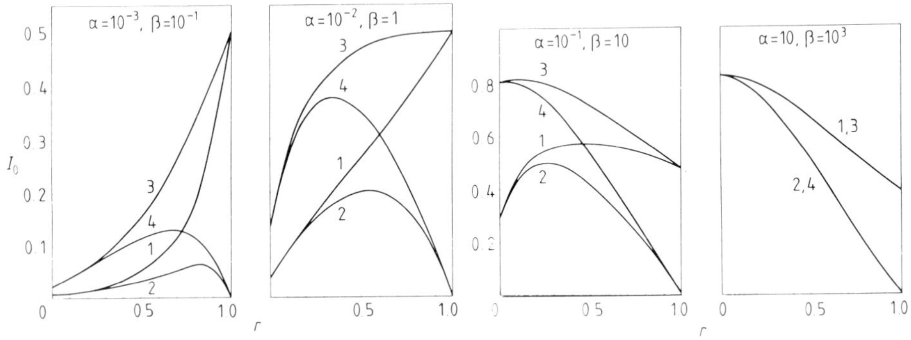

Fig.4.14. Peak intensity inside (curves 1,3) and outside (curves 2,4) the active volume versus the reflectivity coefficients. The curves 1,2 correspond to the case when $r_1 = r_2 = r$, the curves 3,4 correspond to the case when $r_1 = 1$, $r_2 = r$.

media when $r = 1$ the field intensity for a short time reaches value $I_0 = 1/4$ and does not change further. It means that the resonant atoms emit the $N/2$ photons and then the population difference becomes equal to zero,

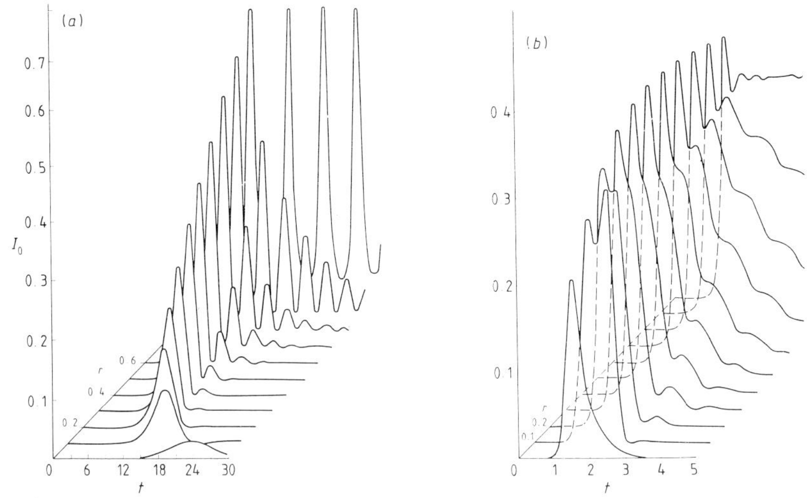

Fig.4.15. Influence of the value of the reflection coefficients on the intensity profile $I_0(t) = |a(1/2,t)|^2$ inside the active volume. The parameters α and β are: (a) $\alpha = 10^{-2}$, $\beta = 1$, (b) $\alpha = 10$, $\beta = 100$.

following which the atomic subsystem no longer interacts with the field in a resonator.

In Fig.4.16 the delay time and the pulse width are plotted versus the reflection coefficient for the same values of the parameters α and β as in Fig.4.14.. It is clear from these figures that if in the case of

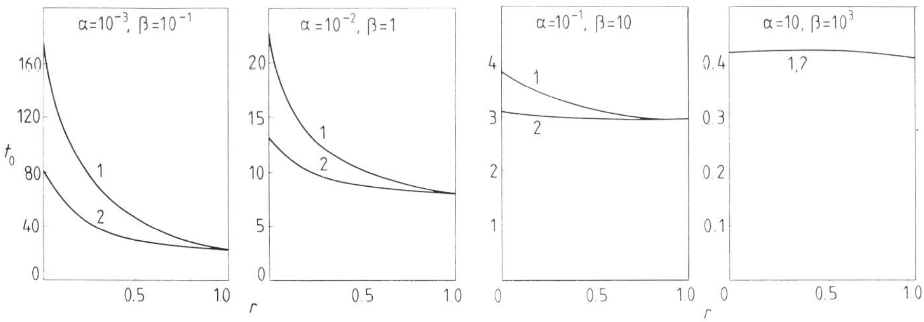

Fig.4.16. Delay time versus reflection coefficients. The curves indicated "1" correspond to the case when $r_1 = r_2 = r$, and the curves indicated by "2" correspond to the case when $r_1' = 1$, $r_2 = r$.

the superradiant media the delay time and the pulse width decrease approximately inversely with r, in the case of the superluminescent media they are practically independent of r.

It is customary to assume that the superradiance and superluminescence can be identified on the value of the parameter $\alpha = \tau/2T_2$. It is believed that the atomic ensemble will decay via superradiance, if the cavity transit time $\tau = L/c$ is less than the characteristic relaxation time of the homogeneous broadening T_2. When $\alpha > 1$, i.e., when $T_2 < \tau$ it is assumed that the superluminescence takes place. It is clear from Figs.4.14 – 4.16 that the separation of these two regimes on the basis of the value of the parameter α only is sufficiently rough. The parameter β proved to be quite important. Fig.4.17a – 4.17c illustrates the influence of the parameter β on the form of the dependencies $I_0 = I_0(r)$ and $I = I(r)$ for a fixed value of the parameter α. We can see that in the case $\beta < 1$ the peak intensity I_0 increases with r, while in the case $\beta > 1$ it decreases with increasing r. Therefore the self-consistent criteria of the type of the generation dynamics regime should be based on the use of the qualitative differences in the dependencies of the characteristic pulse parameters on the main matter parameters. As it follows from Figs.4.14, 4.15, and 4.17, the suitable criterion can be based on the sign of the derivatives $\left.\frac{\partial I_0}{\partial r}\right|_{r=0}$ and $\left.\frac{\partial I_0}{\partial r}\right|_{r=1}$:

$$\left.\frac{\partial I_0}{\partial t}\right|_{r=0} > 0, \qquad \left.\frac{\partial I_0}{\partial r}\right|_{r=1} \geq 0 \text{ – superradiance,}$$

$$\left.\frac{\partial I_0}{\partial t}\right|_{r=0} \leq 0, \qquad \left.\frac{\partial I_0}{\partial r}\right|_{r=1} < 0 \text{ – superluminescence,}$$

$$\left.\frac{\partial I_0}{\partial t}\right|_{r=0} > 0, \qquad \left.\frac{\partial I_0}{\partial r}\right|_{r=1} < 0 \text{ – intermediate regime.}$$

The rise of the peak intensity I_0 with r in the case of the superradiant media can be explained by the following way. The line width of a field mode $\Delta\omega_f \cong 1/\tau$ in the case of superradiance in open system is greater than the line width of an atomic transition $\Delta\omega_a \cong 1/T_2$. The line width of the field mode in the optical cavity is

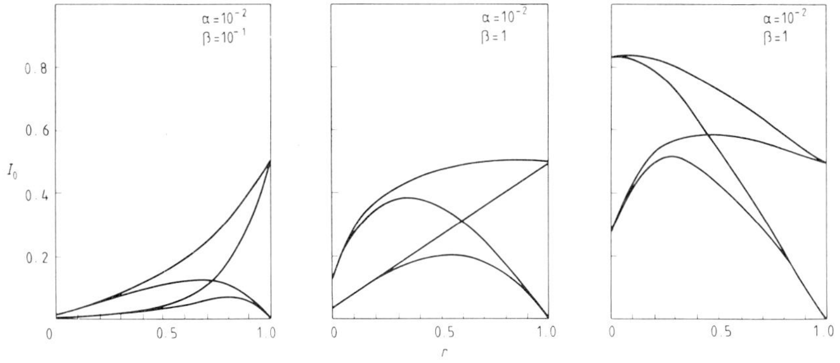

Fig.4.17. Peak pulse intensities inside $I_0(r)$ and outside $I(r)$ the active volume versus reflectivity coefficients for three different values of the parameter β: $\beta = 10^{-1}$ (a), 1 (b), 10 (c).

$$\Delta\omega_f = \frac{1}{\tau} \ln \frac{1}{r} .$$

Therefore, the width of the field mode $\Delta\omega_f$ decreases with increasing of r, which results in an enhancement of the efficiency of interaction of the atomic and field subsystems. In the case of the superluminescent media we have $\Delta\omega_f < \Delta\omega_a$ even at $r = 0$. Therefore, further reduction of $\Delta\omega_f$ with the increase of the cavity q–factor results in the fall of the efficiency of interaction of the atomic and field subsystems.

Let us now study the influence of inhomogeneous broadening on the parameters of the emitted pulses. Fig.4.18 shows the effect upon the pulse intensity of variation in the width of the inhomogeneous broadening spectrum α^* (it will be recalled that $\alpha^* = \tau\,\Delta_0$, where Δ_0 is a characteristic line width of the inhomogeneous spectrum). We can see that for $\alpha^* >> \alpha$ the peak intensity decreases with increasing of α^*. However, the type of the dependence $I_0(r)$ does not change qualitatively. In particular, the sign of the derivative $\left.\frac{\partial I_0}{\partial r}\right|_{r=1}$ does not change. The fall of the intensity with increasing α^* is due to the reduction in the density of resonantly interacting atoms.

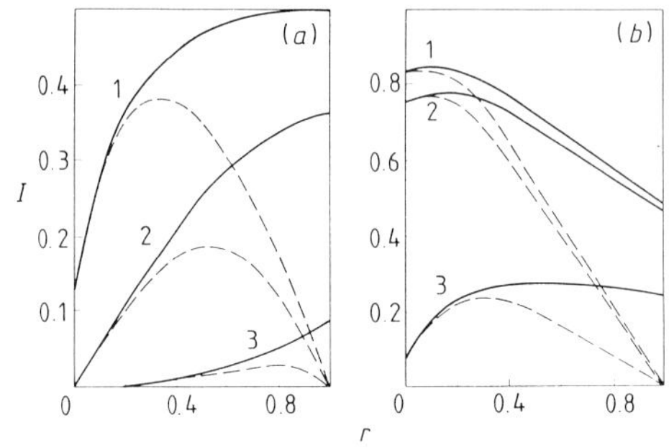

Fig.4.18. Influence of inhomogeneous broadening on the shape of the functions $I_0(r)$ and $I(r)$. The parameters are: (a) $\alpha = 10^{-2}$, $\beta t = 1$, $r_1 = 1$, $\alpha^* = 0.1$ (1), 1 (2), 5 (3); (b) $\alpha = 10^{-2}$, $\beta = 10$, $r_1 = 1$, $\alpha^* =$ 0.1 (1), 1 (2), 5 (3).

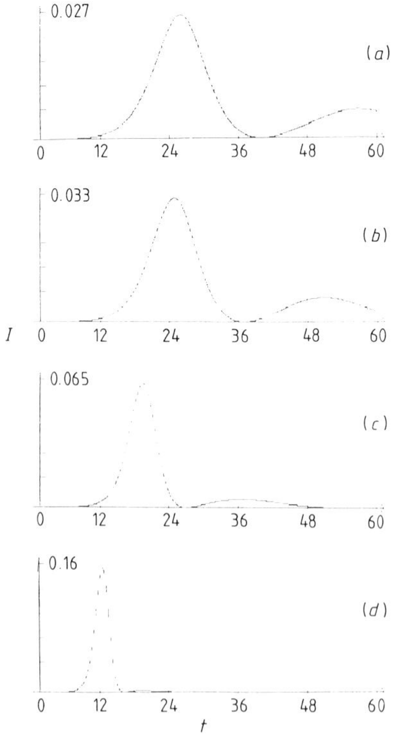

Fig.4.19. Intensity profile for four different values of the parameter D_1: $D_1 = 0$ (a), 0.01 (b), 0.1 (c), 0.5 (d). The reflection coefficients are given as $r_1 = r_2 = 0$.

4.4.3. *Generation dynamics under light-induced grating of population inversion.* Here we discuss the specific features of a generation dynamics under conditions of appearance of a light–induced grating of the population inversion. The equations of motion in this case are

$$\frac{\partial a_1}{\partial t} + \frac{\partial a_1}{\partial x} = p_1, \qquad \frac{\partial a_2}{\partial t} - \frac{\partial a_2}{\partial x} = p_2,$$

$$\frac{\partial p_1}{\partial t} + \alpha p_1 = \beta\,(a_1 R_0 + a_2 R_1),$$

$$\frac{\partial p_2}{\partial t} + \alpha p_2 = \beta\,(a_2 R_0 + a_1 R_1), \tag{4.8}$$

$$\frac{\partial R_0}{\partial t} + \alpha_1 (1 + R_0) = -4\,(a_1 p_1 + a_2 p_2),$$

$$\frac{\partial R_1}{\partial t} + \alpha_1 R_1 = -2\,(a_1 p_1 + a_2 p_2),$$

The boundary conditions have again the form

$$a_1(-1/2,\, t) = r_1\, a_2(-1/2,\, t),$$

$$a_2(1/2,\, t) = r_2\, a_1(1/2,\, t), \tag{4.9}$$

and we choose the initial conditions in the form

$$a_n(x,0) = 0, \quad p_n(x,0) = p_{0n}(x), \quad R_0(x,0) = 1, \quad R_1(x,0) = D_1. \tag{4.10}$$

Fig.4.19 shows the effect upon the superradiance pulse profile of variation in the value of the parameter D_1 for the case of the homogeneous boundary conditions: $r_1 = r_2 = 0$. We can see that the pulse peak intensity increases approximately for order–of–magnitude when D_1 increases from 0 to 0.5. In the case $D_1 = 0$ the dependence of the pulse profile on the value of the reflection coefficients repeats practically the dependence shown in Fig.4.15 and obtained from the solution of the equations (4.1) with $f(\Delta) = \delta(\Delta)$. The only difference is in the value of the peak intensity which becomes smaller when r_1 and r_2 approach to unity, as it is shown in Fig 4.20. . This is due to the fact that the Bloch integral of motion is substituted now by the generalized Bloch integral of motion defined by the Eq.(1.25). According to this generalized integral of motion the unexcited ground state of the atomic subsystem no longer is a stable state of the atom–field system, and the new metastable states are appeared. In fact, it is seen from (1.25) that $R_0(x,t_1) \neq -1$ even when $p_1(x,t_1) = p_2(x,t_1) = 0$, and it is determined in this moment by

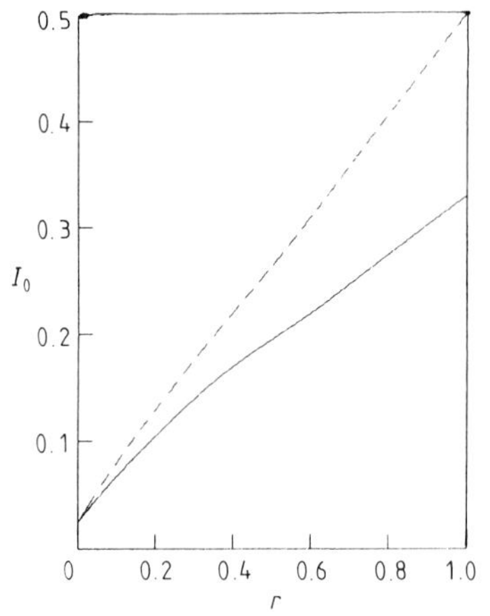

Fig.4.20. Peak pulse intensity versus reflection coefficients. The solid curve has been obtained from the numerical calculations using the equations (4.8), and dashed curve is the curve "1" in Fig.4.14b.

$$R_0(x,t_1) = \pm\,[1 - 2\,R_1^2(x,t_1)]^{1/2}. \tag{4.11}$$

Consequently, in spite of the fact that the emitting component of the polarization is equal to zero, the atomic subsystem does not reaches its ground state when $R_1(x,t) \neq 0$. The appearance of such non-radiative states in the atom–field system is due to the processes of the destructive interference of the counterpropagating waves with the waves reflected on the light–induced grating. This phenomenon is illustrated by Fig.4.21a where the time profiles of the field intensity, polarization amplitude and variables R_0 and R_1 are shown in the case when $r_1 = r_2 = 1$. In this specific case, the decay rate has the same value in all points of the active medium, therefore, the equations (4.8) can be solved analytically. For example, the field amplitudes are described in this case by

$$a_1(x,t) = a_2(x,t) = a(t) = \frac{[(1 + D_1)/3]^{1/2}}{\cosh\{[\beta(1 + D_1)]^{1/2}(t - t_0)\}}, \qquad (4.12)$$

where the delay time t_0 is determined by the initial value of the Bloch angle.

Fig.4.21b shows the time dynamics of the field intensity, polarization amplitude and the variables R_0 and R_1 in the case of a superluminescent medium placed into the high–Q optical cavity. We can see from these figures that the polarization at the resonance frequency disappears for a time equal to the few cavity transit times, but the intensity at the right end of the active volume oscillates even after disappearance of the polarization. These oscillations are due to the fact that in this case, in contrast to the previous one, the linear dimensions of the steady–state pulses are smaller than the active volume length. And the intensity oscillations are explained by the subsequent reflections of these pulses at the active volume boundaries. We can also see that, as in the case of the superradiant media, the equilibrium state of the atomic subsystem differs from the zero inversion state, which is the stable state for superluminescent media in the case of the absence of the light–induced grating effects (see, for example, Fig.4.15b). It should be recalled that we assumed everywhere that $T_1 >> \tau$.

4.4.4. *Superradiance dynamics of bulk media.* One of the most complex problems of a superradiance theory is the analysis of the specific features of a superradiance from the bulk media. The need for such analysis is due to the fact that the one–dimensional models of SR give the results which are essentially different in the main pulse parameters from the results following from the point–source models of SR (cf. Sec.3.1 and Sec.4.4). The natural question arises, whether the variations in the transversal dimensions of the active media will change the dynamics of the emission. This question is as important from the point of view of the general theory of SR, as from the point of view of its practical applications, because the answer to this question enables us to withdraw the rigid requirements to the Fresnel number: $\Phi \cong 1$, which determines the area of applications of the one–dimensional models. The analysis of SR in bulk media should be based on the equations including the second–order transversal derivatives. Thus, we have to solve the multidimensional equations, which complicates the analysis essentially, and another complication concerns the boundary conditions in the transversal plane, since, in the frames of a rigorous analysis we have to

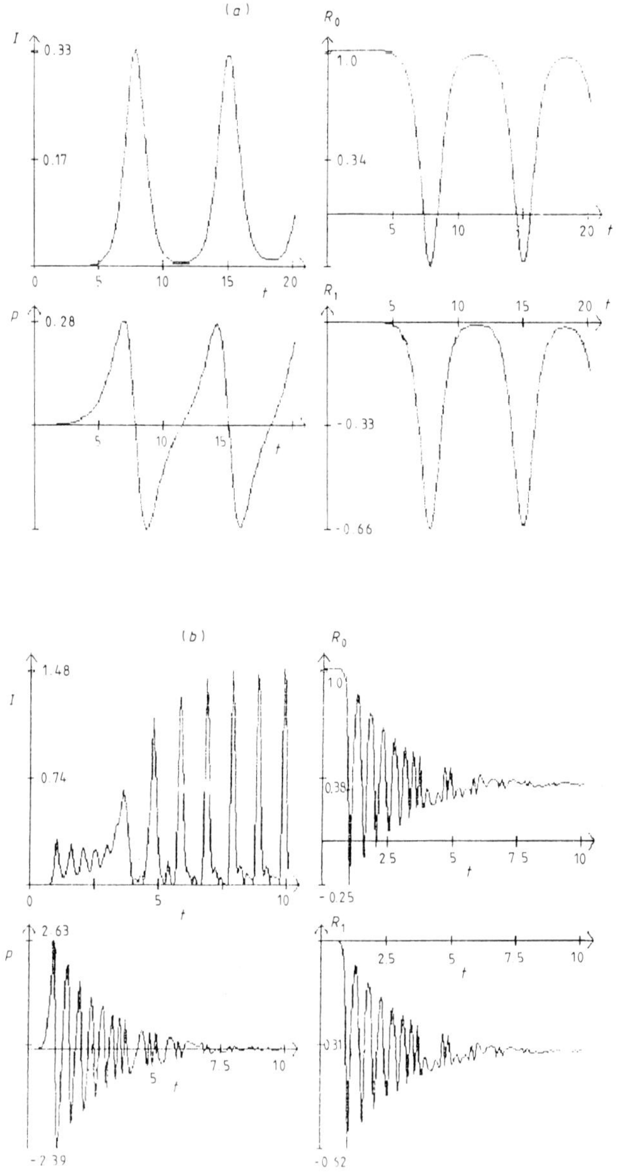

Fig.4.21. Profiles of intensity I, polarization amplitude p, and population difference components R_0 and R_1. The parameters are: (a) $\alpha = 0.01$, $\beta = 1$, $r_1 = r_2 = 1$, (b) $\alpha = 1$, $\beta = 100$, $r_1 = r_2 = 1$.

use the boundary conditions at infinity.

Here, we present the results of the computer calculations for the case $\Phi < 1$. The dependence of the main pulse parameters on Φ are discussed. If $\Phi < 1$, we can use the parabolic approximation of the wave equation and the equation of motion becomes

$$\frac{\partial a_n}{\partial t} - (-1)^n \frac{\partial a_n}{\partial x} + i\,\gamma\,\Delta_\perp a_n = p_n\,,$$

$$\frac{\partial p_n}{\partial t} + \alpha\,p_n = \beta\,a_n\,R\,, \qquad (4.13)$$

$$\frac{\partial R}{\partial t} = -2\sum_{n=1}^{2}(a_n^* p_n + a_n p_n^*)\,,$$

In the case of the active volumes with a cylinder symmetry the transversal Laplacian is

$$\Delta_\perp = \frac{\partial^2}{\partial \rho^2} + \frac{1}{\rho}\frac{\partial}{\partial \rho}\,, \qquad (4.14a)$$

and the parameter γ is determined by

$$\gamma = \frac{1}{4\Phi} = \frac{\lambda L}{4\pi R^2}\,, \qquad (4.14b)$$

where R is a cylinder radius. In the case of the two–dimensional media, they are

$$\Delta_\perp = \frac{\partial^2}{\partial z^2}\,, \qquad (4.15a)$$

$$\gamma = \frac{1}{2\pi\Phi} = \frac{\lambda L}{4\pi d^2}\,, \qquad (4.15b)$$

where d is a thickness of the active medium. The initial conditions are

$$a_n(x,\rho,0) = 0, \quad p_n(x,\rho,0) = p_{0n}(x,\rho), \quad R\,(x,\rho,0) = R_0(x,\rho). \qquad (4.16)$$

The boundary conditions are chosen in the form

$$a_1(-1/2,\rho,t) = a_2(1/2,\rho,t) = 0, \qquad (4.17a)$$

$$a_n(x,\rho,t)\Big|_{\rho=D} = 0, \qquad \frac{\partial a}{\partial \rho}\Big|_{\rho=0} = 0. \qquad (4.17b)$$

We shall restrict ourselves here by considering of the open (mirrorless) systems, therefore, we use the homogeneous boundary conditions (4.17a). In the case when the initial population inversion density $R_0(x,\rho)$ and the sources of the initial spontaneous polarization $p_{0n}(x,\rho)$ have a cylindrical symmetry, the solutions of the equations (4.13) are also symmetric, that is why we use the condition $\left.\frac{\partial a}{\partial \rho}\right|_{\rho=0} = 0$. The value D is chosen in a such a way that its further rise does not affect on the emission dynamics. In fact, the numerical calculations showed that in the specified below region of variation in Φ it was adequate to assume, that $D = 6R$. Fig.4.22 shows the results of computer calculations for the intensity profile in the cases of a cylindrical medium (Fig.4.22a) and a two–dimensional medium (Fig.4.22b). It is clear from these figures that the qualitative character of dependence of the pulse profile on the variation in the parameter γ is the same in both cases. Fig.4.23 shows

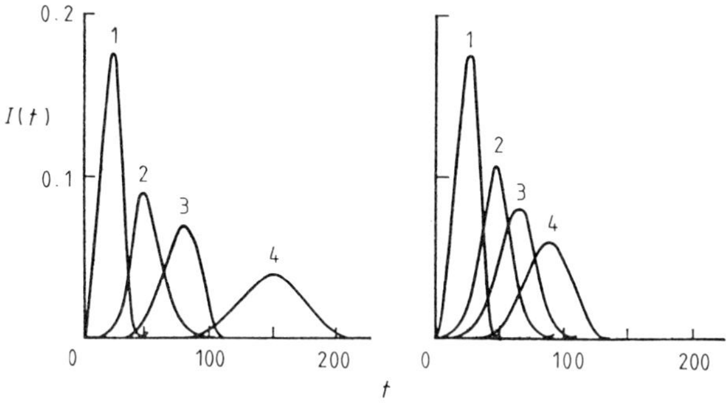

Fig.4.22. Intensity profiles for a cylindrical (a) and two-dimensional (b) active volumes for four different values of a parameter γ: (a) $\gamma = 0$ (1), 1 (2), 2 (3), 5 (4); (b) $\gamma = 0$ (1), 2 (2), 5 (3), 10 (4). The parameter α and β are given as $\alpha = 0.01$ and $\beta = 1$, respectively.

the transversal profile of the field amplitude. It is seen that for the moments of time exceeding the delay time the transversal amplitude distribution becomes non-monotonous. Consequently, the time dynamics of the field is different at the various points in the transversal plane. Fig.4.24 shows the effect upon the peak intensity and delay time of variation in the value of the parameter γ. It is clear from this figure that for the cylindrical media the fall of the peak intensity and the rise of the delay time are more intensive than in the case of the two–dimensional media. The inclusion of the effects of the field transversal inhomogeneity changes the pulse parameters essentially in comparison with the predictions of the one–mode models. We can see that for cylindrical media even in the case $\gamma = 2$ we have three-fold increase in the delay time and two-fold decrease in the peak intensity. These results should be taken into account under comparison of the theoretical and experimental data.

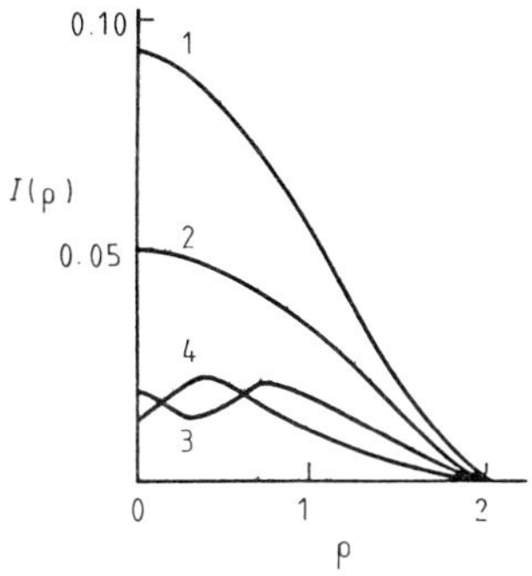

Fig.4.23. Transversal distribution of the field amplitude for a cylindrical active volume at four moments of time: (1) $t = t_0$ (where t_0 is a delay time), (2,3,4) $t > t_0$.

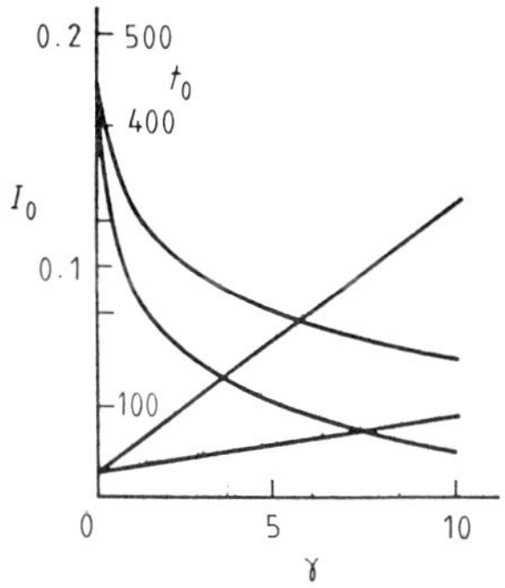

Fig.4.24. Delay time t_0 (curves 1,3) and peak intensity (curves 2,4) versus parameter γ for a cylindrical (1,4) and two-dimensional (2,3) active volume.

4.4.5. *Three-dimensional model of superradiance.* The main distinctive features of a superradiance from bulk media are due to the reflection of the waves, came into being inside the active volume, on its boundaries. It is especially important for the glancing angles of incidence because the reflection coefficient approaches to unity for such waves. In the frames of the approaches based on the plane wave expansion, the refraction effects are not taken into account. The time dynamics of a dielectric permittivity due to the atomic subsystem dynamics and the dynamics of the frequency spectrum due to the frequency modulation result in the time dynamics of the emission angular spectra.

The semiclassical equations in a three-dimensional case in rotating wave approximation and in the absence of inhomogeneous broadening effects are

$$\left[\Delta - \frac{1}{c^2}\frac{\partial^2}{\partial t^2}\right] A_\lambda^+(\mathbf{r},t) = -\frac{4\pi}{c}\,(j_\lambda^+ + j_{\lambda 0}^+),$$

$$\frac{\partial j_\lambda^+}{\partial t} - \mathrm{i}\omega_0 j_\lambda^+ + \frac{j_\lambda^+}{2T_2} = \frac{\mathrm{i}\,m_\lambda}{\hbar c}\sum_{\lambda'} m_{\lambda'}^* A_{\lambda'}^+ \rho,$$

$$\frac{\partial j_\lambda^-}{\partial t} + \mathrm{i}\omega_0 j_\lambda^- + \frac{j_\lambda^-}{2T_2} = -\frac{\mathrm{i}\,m_\lambda^*}{\hbar c}\sum_{\lambda'} m_{\lambda'} A_{\lambda'}^- \rho, \qquad (4.18)$$

$$\frac{\partial \rho}{\partial t} + \frac{\rho - \rho_e}{T_1} = \frac{2\mathrm{i}}{\hbar c}\sum_{\lambda} (j_\lambda^+ A_\lambda^- - A_\lambda^- j_\lambda^+),$$

We have introduced in the right-hand side of the first equation the source of a spontaneous polarization j_0, therefore, we can use now the homogeneous initial conditions both for the field and polarization:

$$A(\mathbf{r},0) = 0, \quad j(\mathbf{r},0) = 0, \quad \rho(\mathbf{r},0) = \rho_0(\mathbf{r}). \qquad (4.19)$$

The indexes "+" will be omitted further. We shall investigate the dynamics of the decay in the linear stage of a system evolution. Let us introduce the following notations:

$$A = \sum_\lambda m_\lambda^* A_\lambda\,, \quad j_\lambda = m_\lambda\, p\,.$$

In the initial stage, when $\rho(\mathbf{r},t) = \rho_0(\mathbf{r})$, Eqs.(4.18) become

$$\left[\Delta - \frac{1}{c^2}\frac{\partial^2}{\partial t^2}\right] A = -\frac{4\pi|m|^2}{c}(p + p_0), \tag{4.20}$$

$$\frac{\partial p}{\partial t} - \mathrm{i}\omega_0 p + \frac{p}{2T_2} = \frac{\mathrm{i}\,\rho_0(\mathbf{r})}{\hbar c} A .$$

Let us assume that $\rho_0(\mathbf{r})$ has the form

$$\rho_0(\mathbf{r}) = \begin{cases} \rho_0, \; \mathbf{r} \in V, \\ 0, \; \mathbf{r} \in V, \end{cases} \tag{4.21}$$

where V is an active medium volume. Then, expending the variables into the Fourier integrals:

$$A(\mathbf{r},t) = \int \mathrm{d}\omega \; A(\mathbf{r},\omega) \exp(\mathrm{i}\omega t),$$

$$p(\mathbf{r},t) = \int \mathrm{d}\omega \; p(\mathbf{r},\omega) \exp(\mathrm{i}\omega t), \tag{4.22}$$

we can transform the instantaneous equations (4.20) into the one equation of the following type

$$[\Delta + k^2(\mathbf{r})] \; A(\mathbf{r},\omega) = -\frac{4\pi|m|^2}{c} p_0(\mathbf{r},\omega), \tag{4.23}$$

where

$$k^2(\mathbf{r}) = \begin{cases} k^2 = \kappa^2 + \dfrac{4\pi|m|^2 n_0}{\hbar c^2}\left[\omega - \omega - \dfrac{\mathrm{i}}{2T_2}\right]^{-1}, \; \mathbf{r} \in V, \\ \kappa^2 = \omega^2/c^2, \; \mathbf{r} \notin V. \end{cases} \tag{4.24}$$

First, we shall define the eigensolutions of the homogeneous equation (4.23), when the right–hand side of this equation is equal to zero. Expending $A(\mathbf{r},\omega)$ in the Fourier integral:

$$A(\mathbf{r},\omega) = \int \mathrm{d}q \; A(\rho,q,\omega) \exp(-\mathrm{i}qz), \tag{4.25}$$

and substituting this expansion into the left–hand side of (4.23), we get

$$[\Delta + \kappa^2 - q^2]\, A(\rho,q) +$$

$$+ \frac{4\pi|m|^2 n_0}{\hbar c^2} \left(\omega - \omega - \frac{i}{2T_2}\right)^{-1} \frac{1}{2\pi} \int_{-L/2}^{L/2} A(\mathbf{r}) \exp(iqz)\, dz = 0. \quad (4.26)$$

For $qL >> 1$ the integral in (4.26) can be replaced by the $A(\rho,q)$, then this equation becomes

$$[\Delta + \beta^2]\, A(\rho,q) = 0, \quad \rho \leq R, \qquad (4.27a)$$

$$[\Delta + \mu^2]\, A(\rho,q) = 0, \quad \rho \geq R, \qquad (4.27b)$$

where

$$\beta^2 = \mu^2 + \frac{4\pi|m|^2 n_0}{\hbar c^2} \left(\omega - \omega_0 - \frac{i}{2T_2}\right)^{-1}, \quad \mu^2 = \kappa^2 - q^2. \quad (4.28)$$

The solution of the equation (4.27a) is

$$A_1(\rho,q) = C\, J_0(\beta\rho), \qquad (4.29a)$$

as the solution of the equation (4.27b), satisfying the boundary conditions at the infinity ($\rho \longrightarrow \infty$), has the form

$$A_2(\rho,q) = B\, H_0^{(1)}(\mu\rho). \qquad (4.29b)$$

The solutions (4.29a), (4.29b) have to satisfy to the conditions of continuity of the functions and their derivatives at the boundary:

$$C\, J_0(\beta R) = B\, H_0^{(1)}(\mu R), \qquad C\, \beta\, J_1(\beta R) = B\, \mu\, H_1^{(1)}(\mu R), \quad (4.30)$$

Consequently, according to the condition of an existence of the non-trivial solutions we get

$$\mu\, J_0(\beta R)\, H_1^{(1)}(\mu R) - \beta\, J_1(\beta R)\, H_0^{(1)}(\mu R) = 0. \qquad (4.31)$$

Thus, the common solution of the equations (4.27a) – (4.27b) can be written in the form

$$\phi_\mu(\rho) = \begin{cases} C\, H_0^{(1)}(\mu R)\; J_0(\beta\rho),\ \rho \le R, \\ C\, J_0(\beta R)\; H_0^{(1)}(\mu\rho),\ \rho > R. \end{cases} \tag{4.32}$$

Using the asymptotic representation of the Bessel functions, we can easily show that the functions $\phi_\mu(\rho)$ satisfy the following orthogonality condition

$$\int_0^\infty \phi_\mu^*(\rho)\, \phi_{\mu'}(\rho)\, \rho d\rho = |C|^2 J_0(\beta R)\, J_0(\beta' R)\, \frac{2}{\pi(\mu\mu')^{1/2}} \left.\frac{\exp[i(\mu-\mu')\rho]}{i(\mu - \mu')}\right|_{\rho\to\infty}$$

$$= 2\, |C|^2\, J_0^2(\beta R)\, \mu^{-1}\, \delta(\mu - \mu')\,. \tag{4.33}$$

Therefore, assuming

$$|C|^{-2} = J_0^2(\beta R)/\pi\mu, \tag{4.34}$$

we finally get

$$\int_0^\infty \phi_\mu^*(\rho)\, \phi_{\mu'}(\rho)\, \rho d\rho = 2\pi\, \delta(\mu - \mu'). \tag{4.35}$$

Thus, eigenfunctions of the homogeneous equation (4.23) are

$$u_{\mu,q}(\mathbf{r}) = \phi_\mu(\rho)\, \exp(-iqz). \tag{4.36}$$

The functions $u_{\mu,q}(\mathbf{r})$ satisfy the following orthogonality condition

$$(1/2\pi)^3 \int u_{\mu,q}^*(\mathbf{r})\, u_{\mu',q'}(\mathbf{r})\, d\mathbf{r} = \delta(\mu - \mu')\, \delta(q - q'). \tag{4.37}$$

Using (4.36), (4.37), the solution of the equation (4.23) can be represented in the form

$$A(\mathbf{r},\omega) = -\frac{4\,\pi|m|^2}{c} \int d\mu \int dq\, \frac{p_0(\mu,q)}{\kappa^2 - q^2 - \mu^2}\, u_{\mu,q}(\mathbf{r})\ , \tag{4.38}$$

where

$$p_0(\mu,q) = (2\pi)^{-3}\int d\mathbf{r}\ u^*_{\mu,q}(\mathbf{r})\ p_0(\mathbf{r},\omega).$$

We can expand the density of spontaneous polarization sources $p_0(\mathbf{r})$ in the integral on eigenfunctions $u_{\mu,q}(\mathbf{r})$:

$$p\ (\mathbf{r}) = \begin{cases} dq_0\ p_{q_0}(\omega)\ u_{\mu_0 q_0}(\mathbf{r}), & \mathbf{r} \in V, \\ 0, & \mathbf{r} \notin V. \end{cases} \tag{4.39}$$

Consequently,

$$p_0(\mu,q) = \frac{1}{2\pi}\int dq_0\ p_{q_0} R\ C^* C_0\ H_0^{(2)}(\mu R)\ H_0^{(1)}(\mu_0 R) \times \tag{4.40}$$

$$\times \frac{2\sin[(q - q_0)L/2]}{(q - q_0)}\ \frac{\beta\ J_1(\beta R)\ J_0(\beta_0 R) - \beta_0\ J_0(\beta R)\ J_1(\beta_0 R)}{\beta^2 - \beta_0^2}$$

and the equation (4.38) becomes

$$A(\mathbf{r},\omega) = -\ \frac{|m|^2 R}{\pi c}\int dq_0\ p_{q_0}(\omega)\int dq\ C^* C_0\ H_0^{(2)}(\kappa_\perp R)\ H_0^{(1)}(\mu R)\ \kappa_\perp^{-1}$$

$$\times \frac{\sin[(q - q_0)L/2]}{(q - q_0)}\ \frac{k_\perp J_1(k_\perp R)\ J_0(\beta_0 R) - \beta_0\ J_0(k_\perp R)\ J_1(\beta_0 R)}{k_\perp^2 - \beta_0^2}$$

$$\times\ u_{\kappa_\perp,q}(\mathbf{r}),$$

where $\kappa_\perp = (\kappa^2 - q^2)^{1/2}$, $k_\perp = (k^2 - q^2)^{1/2}$. Introducing the notation $q = k\cos\theta$ and comparing then the last equation with the equation (3.1.50), finally, we get

$$A(\mathbf{r},\omega) = -\ \frac{\omega\ |m|^2 V}{2\pi^2 c^2}\int dq\ p_q(\omega)\ |C|^2\ |H_0^{(1)}(\kappa_\perp R)|^2\ \kappa_\perp^{-1}\ u_{\kappa_\perp,q}(\mathbf{r})\ \gamma(\mathbf{k}),$$

where $\mathbf{k} = \{\mathbf{k}_\perp, q\}$.

Thus, the angular spectrum of the emitted intensity differs from the angular spectra obtained in Sec.3.1. This difference is due to the fact that we have used here the expansion into the integral on the eigenfunctions (4.36), whereas in Sec.3.1 the expansion into the integral on the plane waves has been used. However, we can see from the last equation, that the angular spectrum of emission is mainly determined by the term $\gamma(\mathbf{k})$.

4.5. Collective Superradiance

We have investigated above the dependence of the pulse intensity and shape on the main dimensionless parameters of the emitting system: α, β, α^*, r_1, r_2. The first three parameters are the internal energy parameters, because we have to increase the pump power to rise the value of the parameter β, or to supply energy to establish more ordered state in the atomic subsystem and, hence, to reduce the value of the parameters α and α^*. The parameters r_1 and r_2 are external. We can realize the pulse parameters control by varying these parameters in certain available for us region, i.e., we can to some extent optimize the process of the extraction of energy stored up initially in the atomic subsystem.

An optimization over the parameters mentioned above does not exhaust all the possibilities for optimization of the process of the radiation emission by the extended inverted multiatomic systems. One additional opportunity is in the possibility of the control by the initial conditions described by Eq.(4.2). But realization of this opportunity brings to the more complex problem than previous ones, because to determine the dependence of a solution on the parameters of the problem we can simply vary these parameters in certain region, as it was done above. Thus, while the peak intensity I_0 is a function of the parameters α, β, α^*, r_1, r_2, it is, at the same time, the functional in the space of the initial conditions. It is practically impossible to determine the form of this functional simply varying the form of the initial conditions. However, we are interested primarily in the extremal regions of this functional. To solve this particular problem we can apply the analysis of the symmetrical properties of the investigated physical or mathematical problem. For example, in the case of the point-source system, i.e., the system with the dimensions much smaller than the wavelength of radiation, the symmetry of the atomic subsystem manifesting itself in the superradiant state leads to the maximum decay rate proportional to N^2. In the case of the extended multiatomic system, as we have seen above, this symmetry breaks down. This is due to the fact that in the point–source system the symmetry of the atomic subsystem determines the symmetry of the complete system, because in this case the field amplitude is the same in all the points inside the active volume and the photons leave the active volume immediately after their emission. In the extended systems the inhomogeneity of the field along the active volume axis will break down the symmetry of the atomic subsystem, because the rate of the induced emission is proportional to the field intensity. Therefore, the states symmetric only relative to the variables of the atomic subsystem will not be characterized by the highest rates of the radiative decay. The highest rates of the decay will be appropriate to the states, which include the symmetry of the complete atom–field system.

4.5.1. *Supersymmetric solutions.* In Ref.[1] the high–symmetry solutions of the equations (4.1) with the boundary conditions (4.3) were found. They are

$$a_1(x,t) = \frac{1}{2\beta^{1/2}} \frac{\partial \theta(x,t)}{\partial t} \cos \phi(x),$$

$$a_2(x,t) = \frac{1}{2\beta^{1/2}} \frac{\partial \theta(x,t)}{\partial t} \sin \phi(x),$$

$$p_1(x,t) = \frac{\beta^{1/2}}{2} R_0 \sin \theta(x,t) \cos \phi(x),$$

$$p_2(x,t) = \frac{\beta^{1/2}}{2} R_0 \sin \theta(x,t) \sin \phi(x),$$

$$R(x,t) = R_0 \cos \theta(x,t), \tag{5.1}$$

where the Bloch angle $\theta(x,t)$ is determined by

$$\theta(x,t) = A\ (\cosh \Phi(x))^{1/2}\ u(t); \tag{5.2}$$

here

$$A = \frac{4 \ln [1/(r_1 r_2)]^{1/2}}{[(1 - r_1^2)/r_1] + [(1 - r_2^2)/r_2]}, \tag{5.3}$$

$$\Phi(x) = x \ln \frac{1}{r_1 r_2} - \ln \left(\frac{r_2}{r_1}\right)^{1/2}, \tag{5.4}$$

$$\sin 2\phi(x) = \frac{1}{\cosh \Phi(x)}. \tag{5.5}$$

It is clear from the Eqs. (4.8) – (4.12) that the spatial dependence of the population difference, and the field and polarization amplitudes is uniquely determined by the boundary conditions. The time dynamics all of the five unknown functions $a_1(x,t)$, $a_2(x,t)$, $p_1(x,t)$, $p_2(x,t)$, $R(x,t)$ is determined by one equation

$$\frac{d^2 u}{dt^2} + \gamma \frac{du}{dt} = \beta R_0 \sin u \,, \tag{5.6}$$

where

$$\gamma = \ln \frac{1}{(r_1 r_2)^{1/2}} \,.$$

In Ref.[1] the inhomogeneously broaden systems were investigated also. In this case the Bloch angle is

$$\theta(x,t,\Delta) = A \,(\cosh \Phi(x))^{1/2} \, u(t,\Delta),$$

and the functions $u(t,\Delta)$ are determined by the equations

$$\frac{d^2 u(t,\Delta)}{dt^2} + (\gamma - i\,\alpha^* \Delta) \frac{du(t,\Delta)}{dt} =$$

$$= \beta \int \exp[-\, i(\,\Delta - \Delta')\, \alpha^* t] \sin u(t,\Delta') \; f(\Delta')\, d\Delta'. \tag{5.7}$$

The field and polarization amplitudes and the population inversion now become

$$a_1(x,t) = \frac{1}{2\beta^{1/2}} \frac{\partial \theta(x,t,\Delta)}{\partial t} \exp\,(-\, i\,\Delta\,\alpha^* t)\, \cos\, \phi(x),$$

$$a_2(x,t) = \frac{1}{2\beta^{1/2}} \frac{\partial \theta(x,t,\Delta)}{\partial t} \exp\,(-\, i\,\Delta\,\alpha^* t)\, \sin\, \phi(x),$$

$$p_1(x,t,\Delta) = \frac{\beta^{1/2}}{2} R_0 \exp\,(-\, i\,\Delta\,\alpha^* t)\, \sin\, \theta(x,t,\Delta) \;\; \cos\, \phi(x),$$

$$p_2(x,t,\Delta) = \frac{\beta^{1/2}}{2} R_0 \exp\,(-\, i\,\Delta\,\alpha^* t)\, \sin\, \theta(x,t,\Delta) \;\; \sin\, \phi(x),$$

$$R(x,t,\Delta) = R_0 \cos\, \theta(x,t,\Delta), \tag{5.8}$$

It follows from (5.1) that $\phi(x)$ is the azimuthal angle in the energy pseudospin space (p_1, p_2, R), where the Bloch angle $\theta(x,t)$ is the polar angle. These two angles uniquely determine the position of the Bloch vector end on the sphere of the radius R_0. The difference between the supersymmetric solutions defined by (5.1), and (5.8) and any arbitrary solution of the equations (4.1) for the case $\alpha = 0$ is that in general case the angle ϕ is dependent on time $\phi(x,t)$. Therefore we can say that in general case the end of the Bloch vector winds itself on the Bloch sphere in the processes of a radiative decay of the atomic system (see Fig.4.25). It is this feature which explains the rise of the time of radiative decay and oscillatory form of the emitted pulses. As a result the peak intensity decreases in comparison with the possible case of a monoimpulse emission with the same net energy. In the case of the decay determined by the equations (5.1) or (5.8) the end of the Bloch vector moves from the uppermost point on the Bloch sphere to the lowest one along the shortest way, i.e., along the azimuth, that results in the highest decay rate at a given population inversion density.

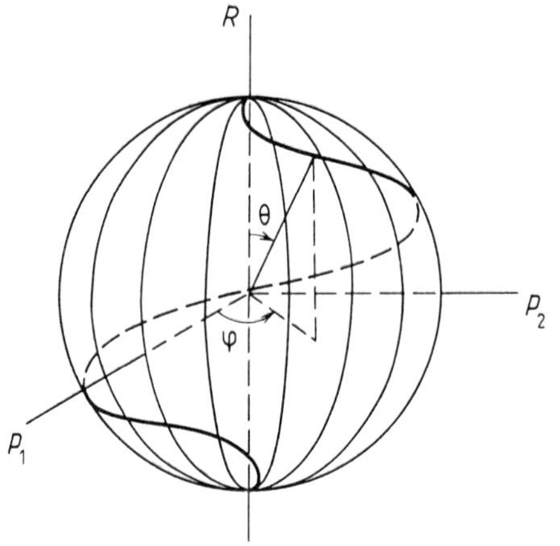

Fig.4.25. Angles $\theta(x,t)$ and $\phi(x,t)$ specify uniquely the position of the Bloch vector in the pseudospin space (p_1, p_2, R). In general case of a radiative decay, the end of the Bloch vector winds itself on the Bloch sphere, that results in increase of a radiative decay time. The highest decay rate corresponds to the states, which are associated with the motion of the end of the Bloch vector along the azimuth.

Using the Eqs.(5.1) – (5.5) we can easily show that the intensity of the emission from the right end of the active medium, where the reflection coefficient is equal to r_2, is determined by

$$I_1(t) = |a_1(1/2,t)|^2(1 - r_2^2) = 2\gamma a^2(t) \frac{(1 - r_2^2)r_1}{(1 - r_1^2)r_2 + (1 - r_2^2)r_1}, \tag{5.9}$$

where

$$a(t) = \frac{1}{2\beta^{1/2}} \frac{du}{dt}.$$

The net intensity from both ends of the active volume is

$$I(t) = I_1(t) + I_2(t) = 2\,\gamma\, a^2(t). \tag{5.10}$$

The field intensity inside the active volume is

$$n(x,t) = n_1(x,t) + n_2(x,t) = A^2\, a^2(t) \cosh \Phi(x). \tag{5.11}$$

The integral intensity of the field inside the active volume is determined by

$$I_0(t) = \int_{-1/2}^{1/2} n(x,t)\, dx = a^2(t). \tag{5.12}$$

4.5.2. *Scaling properties of collective superradiance.* Let us assume for simplicity that in the initial moment the active medium is in the state of the complete inversion, so that $R_0 = 1$. We shall introduce the new dimensionless time

$$\tau = \beta^{1/2} t,$$

then the Eq.(5.6) can be rewritten in the form

$$\frac{d^2u}{d\tau^2} + \delta \frac{du}{d\tau} = \sin u\,, \tag{5.13}$$

where

$$\delta = \gamma/\beta^{1/2}. \tag{5.14}$$

Therefore, the intensity profile is now determined by one parameter only. Fig.4.26 shows the pulse shape for one of the values of the parameter β, as well as the dependence of the normalized delay time τ_0 and the square of the peak amplitude $a^2(\tau_0)$ on the value of the parameter δ. The quantity $2\delta a^2(\tau_0)$, according to (5.10), determines the net intensity of the emission from both boundaries of the active volume. We can see that it has an extremum, where the peak intensity is a maximum.

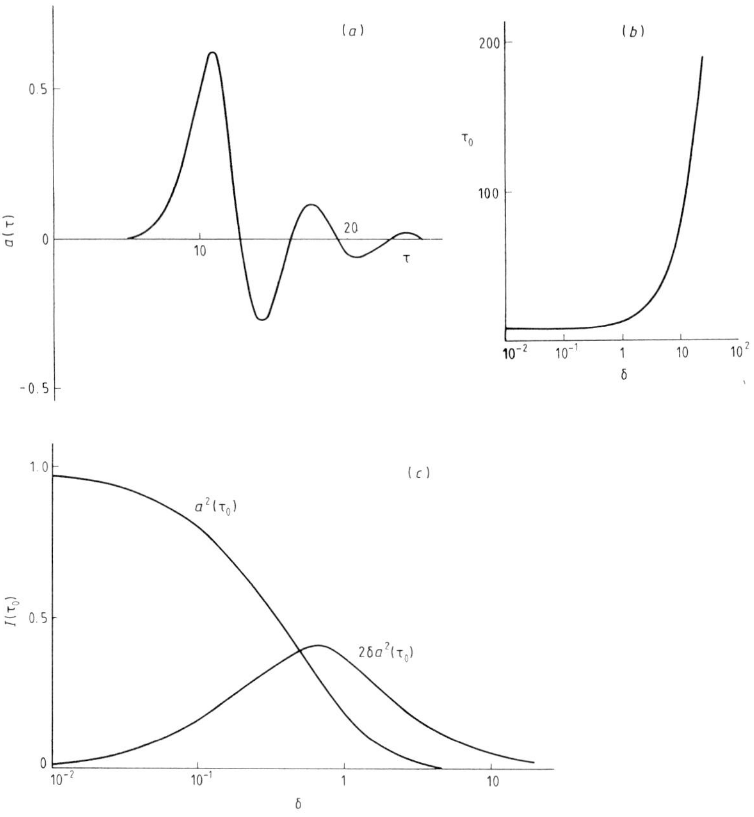

Fig.4.26. (a) amplitude profile for collective superradiance, the parameter δ is $\delta = 0.5$, (b) delay time versus parameter δ, (c) peak intensities $I_0(\tau_0) = a^2(\tau_0)$ and $I(\tau_0) = 2\delta a^2(\tau_0)$ versus parameter δ.

It should be noted that when $r_1 = r_2 = 1$ the solution of the equation (5.6) has the familiar form

$$a(t) = \frac{1}{\cosh\,[\beta^{1/2}\,(t - t\,)]},$$

where

$$t_0 = \frac{1}{\beta^{1/2}} \ln \frac{1}{\theta_0},$$

here θ_0 is an initial value of the Bloch angle. Using this solution we can write down the expression for the field amplitudes $a_{1,2}(x,t)$ in the interval $(0, 2t_0)$:

$$a_1(x,t) = a_2(x,t) = \frac{1}{2^{1/2} \cosh [\beta^{1/2} (t - t_0)]},$$

and then they are repeated periodically in the intervals $(2nt_0, 2(n + 1)t_0)$.

4.5.3. *Initial condition for collective superradiance.* It is clear from the equations (5.1) that the amplitudes $a_1(x,t)$, $a_2(x,t)$, $p_1(x,t)$, $p_2(x,t)$ can be represented in the form

$$a_1(x,t) = a(x,t) \cos \phi(x,t), \quad a_2(x,t) = a(x,t) \sin \phi(x,t),$$

$$p_1(x,t) = p(x,t) \cos \phi(x,t), \quad p_2(x,t) = p(x,t) \sin \phi(x,t).$$

The high symmetry of the solution (5.1) and the corresponding states of the atomic–field system is clearly demonstrated in the fact that for this solution, in addition to the Bloch integral of motion

$$p^2(x,t) + \frac{\beta}{4} R^2(x,t) = p^2(x,0) + \frac{\beta}{4} R^2(x,0),$$

there appear the two new integrals of motion:

$$\phi(x,t) = C_1(x), \quad a(x,t) [\sin (2 \phi(x,t))]^{1/2} = C_2(t).$$

In contrast to the Bloch integral of motion depending on the variables of the atomic subsystem only, these two new integrals of motion depend on the variables referring to the whole atomic – field system.

Now, when we have found the supersymmetric solutions, we can answer the question what should be done to realize them, i.e., what kind of the initial condition should we choose. According to the equation (5.6), we can choose the two following types of the initial conditions:

$$(1) \quad u(0) = u_0, \quad \frac{du}{dt}(0) = 0,$$

$$(2) \quad u(0) = 0, \quad \frac{du}{dt}(0) = \frac{du_0}{dt}.$$

In the first case, the initial conditions for $a_1(x,t)$, $a_2(x,t)$, $p_1(x,t)$, $p_2(x,t)$, and $R(x,t)$ are

$$a_1(x,t) = a_2(x,t) = 0,$$

$$p_1(x,0) = \frac{\beta^{1/2}}{2} R_0 \frac{\sin\,[A\ (\cosh\,\Phi(x))^{1/2}\ u_0]}{(2\,e^{-\Phi}\ \cosh\,\Phi)^{1/2}} \cong \frac{\beta^{1/2}}{2} R_0 \frac{A\ u_0}{2^{1/2}}\ e^{\Phi/2},$$

$$p_2(x,0) = \frac{\beta^{1/2}}{2} R_0 \frac{e^{-\Phi} \sin\,[A\ (\cosh\,\Phi(x))^{1/2}\ u_0]}{(2\,e^{-\Phi}\ \cosh\,\Phi)^{1/2}} \cong \frac{\beta^{1/2}}{2} R_0 \frac{A\ u_0}{2^{1/2}}\ e^{-\Phi/2}$$

$$R(x,0) = R_0\ \cos\,[A\ (\cosh\,\Phi(x))^{1/2}\ u_0]. \tag{5.15}$$

In the second case the initial conditions are

$$a_1(x,0) \cong \frac{A}{2\beta^{1/2}}\ \frac{1}{2^{1/2}}\ \frac{du_0}{dt}\ e^{\Phi/2},$$

$$a_2(x,0) \cong \frac{A}{2\beta^{1/2}}\ \frac{1}{2^{1/2}}\ \frac{du_0}{dt}\ e^{-\Phi/2},$$

$$p_1(x,0) = p_2(x,0) = 0,$$

$$R(x,0) = R_0. \tag{5.16}$$

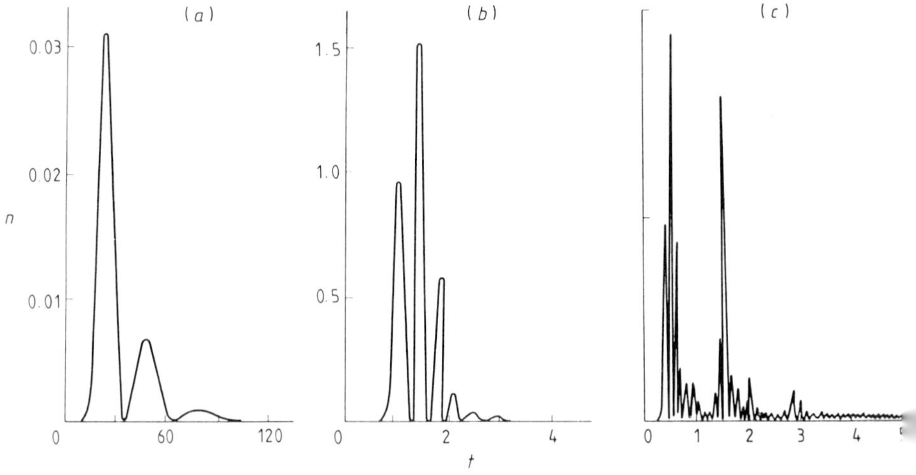

Fig.4.27. Intensity profiles for three different values of the parameter $\beta = 1$ (a), 100 (b), 10^3 (c) and for a fixed value of parameter $\alpha = 10^{-3}$. The initial conditions are $a_1(x,0) = a_2(x,0) = a_0$, $p_n(x,0) = 0$, $R(x,0) = R_0$.

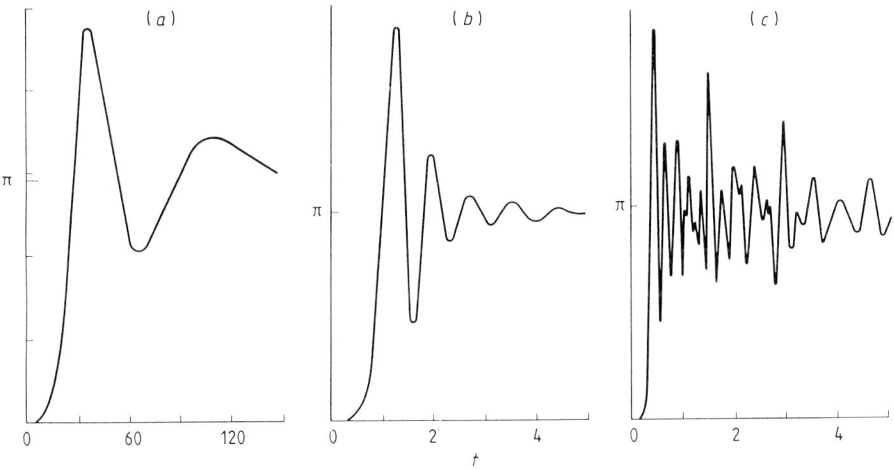

Fig.4.28. Pulse area profiles corresponding to intensity profiles shown in Fig.4.27.

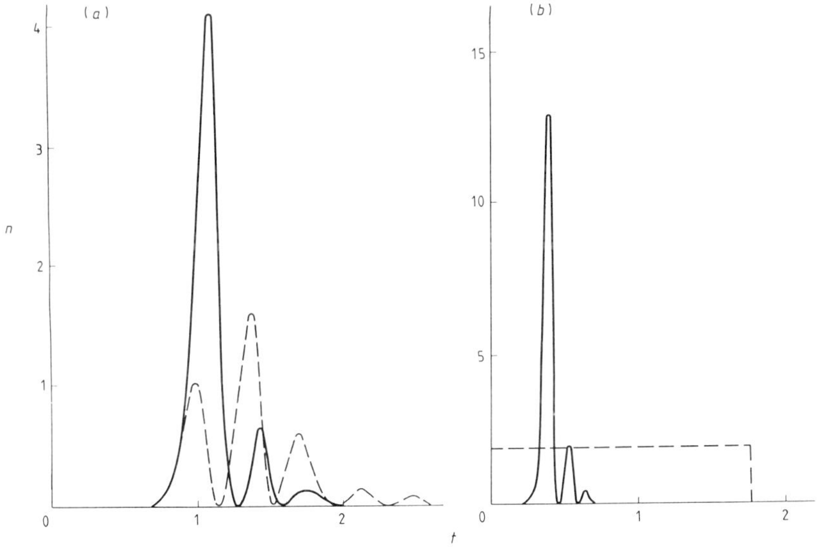

Fig.4.29. Comparison of intensity profiles for collective superradiance (solid curves) and usual superradiance (see Fig.4.27b,c) for the same parameters: (a) $\alpha = 10^{-3}$, $\beta = 10^{2}$, (b) $\alpha = 10^{-3}$, $\beta = 10^{3}$. In the case *b* the height of the dashed rectangle is equal to the maximum peak value of the intensity, whereas the width is equal to the characteristic duration of a train of pulses shown in Fig.4.27c.

In the first case, we have to create a certain spatial distribution of the initial population inversion and polarization density. In the latter case, we have to create the initial field amplitude distribution determined by the equations (5.16).

Fig.4.27 – 4.29 clearly demonstrate the opportunities for the optimization of the process of the energy extraction from the excited medium in the regime of the collective spontaneous emission. Fig.4.27 shows the effect upon the pulse profile of the variation in the value of the parameter β (for example, the variation in the initial population inversion density) for a fixed value of the parameter $\alpha = \tau/T_2$ and for the following form of the initial conditions:

$$a_n(x,0) = 0,\ p_n(x,0) = p_{0n},\ R(x,0) = 1. \tag{5.17}$$

We can see that, as the population inversion density increases, the emitted pulse is transformed into a sequence of pulses with the irregular variation in the peak value of the subsequent pulses. In all the cases, we have the coherent regime of generation, which is confirmed by the pulse area time dependence shown in Fig.4.28. We can see from these graphs that

$$\theta(1/2,t)\Big|_{t\to\infty} \longrightarrow \pi.$$

Fig.4.29 shows in comparison the pulse profiles corresponding to the initial conditions (5.15) (or (5.16)) and (5.17). We can see that the peak intensity of the pulse corresponding to the initial conditions (5.15) exceeds considerably the peak intensity of any pulse in train corresponding to the initial conditions (5.17).

Chapter 5
LIGHT-INDUCED RESONANCE PHASE TRANSITIONS DUE TO INTERATOMIC INTERACTIONS THROUGH RADIATION FIELD

5.1. Classification and general characteristics of resonance phase transitions

The studies of nonequilibrium phase transitions in the systems of quantum electronics have begun with the establishing of analogy between laser generation threshold and equilibrium phase transitions [1, 2]. The subsequent experimental and theoretical investigations [3, 4] have revealed, that this analogy goes very deeply. In the last time it was shown, that in the typical systems of quantum electronics - ensembles of two-level atoms, interacting through the radiation field, electrostatic field, dipole electrostatic field, or lattice displacement field a whole number of novel light-induced phase transitions is possible. The chapters 5 and 6 of this book are devoted to the exposition of the results, obtained in studies of models of these phase transitions.

As it is known [5], phase transitions are characterized by spontaneous appearance (disappearance), or jumplike change of certain quantity, called order parameter. This change occurs, when some externally controlled parameter - temperature, number density of atoms or the intensity of external field reaches a certain critical value. In the equilibrium case the appearance of order parameter minimizes the free energy of the system, so that thermodynamically the ordered state is more preferable.

The typical example of the equilibrium phase transition is the spontaneous appearance of magnetization in spin system, when the temperature is decreased down to critical value $T = T_c$ [5]. This phase transition occurs due to the exchange interactions between atoms. The static dipole-dipole interactions leads to the equilibrium ferro-electric transition [6], the effective interaction through the exchange of phonons - to the superconducting phase transition [5].

The phase transitions in the systems of quantum electronics, considered in this chapter, occur due to the interaction through the dipole radiation field. Thus, they are linked with the effect of superradiance (SR), studied in previous chapters. Moreover, we shall show in Sec.5.2, that superradiance is, in essence, one of the possible phase transitions in the system considered and establish the analogy of this effect with ferromagnetism and superconductivity.

The interaction through radiation field is the most universal type of interaction in the present system. However, in some practically interesting cases other types of interactions can play the dominant role. Thus, in the medium of two-level atoms, possessing constant dipole moments, collectivization of atomic ensemble can be effected by the static dipole-dipole interaction, which can be described in Lorentz field approximation (see Sec.6.2). In crystals there appears also extra interaction through the lattice displacement field. Light-induced phase transitions, occurring due to these last two types of interactions, are considered in chapter 6.

In the effects considered below the role of externally varied parameter is , as a rule, played by the intensity of the light field, acting on cooperative atomic system, or in general case - by the pump intensity. As it will be shown, under the particular conditions, when the pump intensity reaches a certain critical value, a nonzero order parameter spontaneously appears in the system. Depending on the situation considered an order parameter can serve move a static polarization, electrostatic field, a nonzero value of macroscopic polarization, oscillating with atomic transition frequency, quadratic macroscopic susceptibility, jump-like change of atomic energy levels population difference etc. The appearance of order parameter, may reveal itself through the change of the optical material spectroscopic characteristics (e.g. the transition frequency shift), lead to a jumplike change of output intensity and other optical manifestations. The light wave, leading to these threshold changes, can be generated in the working medium itself. Thus, the light induced phase transitions can essentially influence the work of quantum electronics devices.

One of the essential features of phase transitions, considered here, is their nonequilibrium character: they take place either in the system, inverted by an external pump, or occur under the action of the external light wave, which changes the populations of the working energy levels, i.e. these phase transitions occur due to the effect of the saturation of two-level atoms. This imposes the resonance condition on the frequency of the exciting wave: this frequency, or specific combination of frequencies of interacting light waves must lie in the vicinity of atomic transition frequency. Thus, phase transitions considered here can be called the resonance light-induced transitions.

For the classification of light-induced phase transitions it is convenient as in the case of equilibrium phase transitions, to consider the corresponding soft modes. In equilibrium systems the soft modes are the vibration, or relaxation variables, the frequencies of which cancel at $T = T_c$. In nonequilibrium systems of quantum electronics the frequencies of soft modes depend on pump intensity and at critical value of pump intensity E_c^2 they cancel, signalling that the initial disordered phase becomes unstable and the system tends to make a transition to a new stable state.

The typical system considered here consists of an ensemble of N ($N >> 1$)

two-level atoms with the upper level energy ε_a and lower level energy ε_b. Each atom is characterized by two relaxation frequencies $\gamma_{ab} \equiv T_2^{-1}$ and $\gamma \equiv T_1^{-1}$ (T_1 and T_2 are correspondingly the relaxation times of energy levels populations and atomic polarization) and by transition frequency $\omega_0 = (\varepsilon_a - \varepsilon_b) / \hbar$. If the system is placed inside resonator, one extra frequency appears - the decay constant of light field τ^{-1}. The interatomic interaction in the presence of external light wave leads to the renormalization of these frequencies. The renormalized frequencies γ^*_{ab}, γ^*, ω^*_0, τ^{*-1} are now referred to as the atomic ensemble and depend on the number density of atoms in the system n and intensity of external pump wave E^2. One may envisage, that under certain conditions some of the renormalized frequencies γ^*_{ab}, γ^*, ω^*_0, or τ^{*-1} will tend to zero, when E^2 tends to E_c^2, or T tends to T_c. The corresponding variable is then the nonequilibrium soft mode.From the soft mode concept of phase transitions it follows then, that in the system of two-level atoms at least four different types of light-induced phase transitions are possible.

The first one is the transition to the state of laser generation. Because this question has been thoroughly studied [3, 4] we shall mention it only for the comparison and illustration of general principles. This phase transition occurs in the inverted atomic system, when the difference of populations of the upper (a) and lower (b) levels $D = \sigma_{aa} - \sigma_{bb} > 0$ and it is linked with the field relaxation soft mode

$$\tau^{*\,-1} = \tau^{-1} \left[1 - \frac{2\pi n d_{ab}^2 \omega_0 D}{\hbar \gamma_{ab}\, \tau^{-1}}\right], \qquad (1.1)$$

where d_{ab} is the transition dipole moment matrix element and $n\,D$ is the inversion density.As an order parameter serves the light field amplitude in the resonator (or the photon number, see Fig. 5.1). Since $\tau^{-1} = c\,(1 - R)\,l^{-1}$, where c is the light velocity, l is the medium length, R is the optical reflection coefficient of resonator mirrors, one has from (1) the following necessary condition of phase transition

$$\frac{2\pi\, n\, d_{ab}^{2}\, \omega_0}{\gamma_{ab}\, c\, (1 - R)\, l^{-1}} > 1 \qquad (1.2)$$

This example illustrates the important principle of order parameter enslaving [3], which points out the way to discovering other possible light-induced phase transitions in the considered system, described by modal variables D, non-diagonal element of the density matrix $\sigma_{ba} \equiv \sigma_{+}$ and and the light field amplitude E.

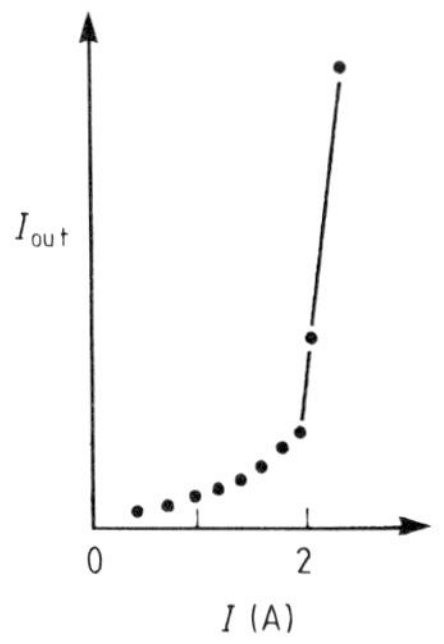

Fig.5.1. The typical critical dependence of output intensity on the pump intensity in a laser.

Let us assume, that under the variation of the external control parameter some of these modes can become soft, in the case considered now it is the field amplitude E. Then photon relaxation frequency τ^{*-1} tends to zero, when the pump intensity tends to its critical value, whereas γ^{*}, γ^{*}_{ab} and ω^{*}_{0} undergo only slight variations (hard modes). The variables, describing the hard modes - D and σ_{+} can be adiabatically eliminated, and as a result one obtains a single equation for the order parameter E with the soft relaxation frequency τ^{*-1}. This procedure of adiabatic elimination of fast relaxing variables contains the essence of principle of the enslaving by slow relaxing order parameter.We shall use this principle for the description of other types of phase transitions in the considered system. One must note, that the enslaving principle is closely connected with the idea of the hierarchy of relaxation times and reduced description of multicomponent system, formulated in the book [7].

If the same system of two-level atoms, placed inside resonator, is non-inverted and is excited by an external light wave with intensity E^2, then at a certain critical value E_c^2 the jumplike increase of the transmission of exciting radiation occurs; under subsequent decrease of the pump intensity at the value $E^2 = E_{c1}^2 < E_c^2$ the jumplike decrease of the transmission occurs (bistability of transmission). Resonator bistability is well investigated theoretically and experimentally [8,9]. It is linked with soft relaxation mode $\gamma_* \longrightarrow 0$ and its necessary condition coincides with condition of laser generation (1.2) with the difference, that now $n\,D < 0$.

In this chapter we shall be interested in the light-induced phase transitions in the open, resonatorless medium of two-level atoms. In contradistinction to case of medium placed inside resonator, the studies of the resonatorless phase transitions are now only in the beginning. Conditions of their observation are more severe than the condition (1.2), but are quite realistic. Moreover, as we shall show, the mechanisms of nonequilibrium phase transitions, considered here, are able to give a good qualitative and in some cases quantitative description of light-induced critical phenomena observed in the system of quantum electronics.

In Sec.5.2 the phase transition to the state of polariton (polarization) generation in the open system of effective two-level atoms, interacting through the Stokes field in Raman scattering (RS) of light is studied. Simultaneously the considered model describes the system of stationary inverted two-level atoms with the transition frequency ω_0, interacting through radiation field . We consider this phase transition on the example of RS, because in this case a vast experimental data is available, which can be used to verify the theory. The phase transition manifest itself in a sharp jump-like (6 - 11 orders of magnitude) increase of RS intensity, when the pump intensity exceeds the critical value E_c^2, accompanied by the critical narrowing of RS-line in the vicinity of E_c^2 and subsequent splitting of it into separate components. In Sec.5.2 this critical light-induced phenomenon is considered to be the consequence of the critical behavior of the collective spontaneous radiation (CSR). As an order parameter serves the Fourier-component of the pair correlator of atomic polarizations and transition itself is linked with the soft relaxation mode $\gamma_{ab}^* \longrightarrow 0$. The necessary condition of phase transition for the case of two-level atoms has the following form

$$(\tau_c \gamma_{ab})^{-1} \equiv \frac{2\pi n d_{ab}^2 \omega_0}{\hbar \gamma_{ab}^2} > 1 , \qquad (1.3)$$

(nD is the inversion density). In the case of RS this condition has the same form with the substitution being made: $\omega_0 \longrightarrow \omega_s$, $d_{ab} \longrightarrow d_s$,where d_s and ω_s are the Stokes dipole moment and frequency correspondingly. From the comparison of (1.2), (1.3) it is seen, that the condition of the phase transition to polariton generation state in open system is more severe, then the condition of laser generation in resonator, because usually $\gamma_{ab} \gg c\,(1 - R)\,l^{-1}$. At strong exceeding of the threshold , when $D\,(\tau_c \gamma_{ab})^{-1} \gg 1$, the stimulated polariton generation in nonstationary regime is nothing else as Dicke superradiance (SR). In Sec.5.3 the dynamics of multimode SR in Raman scattering (SRRS) in the media with arbitrary geometrical form is studied and the influence of the geometry of the medium on the characteristics of SRRS is revealed.

As in the case of resonator system, in the open system the bistability of transmission can take place under the action of the light with the intensity $E^2 > E_c^2$ on the atomic system in ground state. This effect occurs due to the interatomic interaction through the radiation field, which qualitatively modifies the dependency of atomic transitions saturation on the pump intensity. In Sec.5.4 bistability of transmission in the open atomic system with two-photon pumping is considered. The results obtained are also applicable to the case of one photon pumping. In open system the field adiabatically rapidly adjust itself to the atomic polarization and population difference, so that the field variable can be eliminated by solving Maxwell equation. The necessary condition of bistability of transmission in open system is given by Eq. (1.3) with $D\,n < 0$.

Phase transitions, considered in Secs.5.5 - 5.6, which can be called quasiequilibrium phase transitions, occur due to the interatomic interaction through radiation field appearing under the action of laser field. They take place under the condition, that during the action of the light pump pulse τ_{pump} the quasiequilibrium state is installed, described by Boltzmann distribution function for the order parameter. Under this condition for the description of light-induced phase transitions one can use free energy minimum principle.

In Sec.5.5 the phenomenon of the inversion filamentation in laser media due to the effective interatomic interactions through the radiation field is considered. It is shown, that the state of spatially uniform inversion is energetically unfavorable and the system minimizes its energy making transition to the state, in which inversion forms the thin threads along the direction of the working medium extension. The theory yields the expressions for radius of these threads and their number in the beam cross section. The theoretical results are compared with experimental data of the studies of the grained structure of the radiation of open inverted atomic systems.

In Sec.5.6 the ferromagnetic phase transition in paramagnetic gas, induced by the coherent light wave with an arbitrary (in particular linear) polarization is considered. It is shown, that in the presence of the effective interatomic interaction through the radiation field the appearance of the asymmetry of populations of magnetic sublevels minimizes the system free energy, that leads at $E^2 > E^2_c$ to appearance of nonzero order parameter-the macroscopic magnetization. The theory of this phase transition explains the main experimental results , obtained in studies of laser-induced generation of electromotive force (EMF) in the coil, placed inside the cavity with paramagnetic gas.

In Sec.5.7 it is shown, that the analogous phase transition can take place under the action of coherent light wave with the frequency ω in the system of anisotropic molecules, interacting with each other through the second harmonic (SH) radiation field with frequency 2ω. At $E^2 > E^2_c$ the transition dipole moments of the molecules are arrayed in one direction. The order parameter of this orientational phase transition is the macroscopic quadratic susceptibility X^2. In chapter 6 the analogous phase transition in the system of doping centers in solids with mutual interactions through the strong Lorentz field, lattice displacement field and relatively weak SH field is considered and the results obtained are applied to the interpretation of the phenomenon of threshold SH generation (SHG) in centrosymmetric glass fibers.

The consideration of phase transitions, occurring due to the radiation field, is concluded with the equilibrium ferroelectric transition in the system of two-level atoms (Sec.5.8). This phase transition leads to the appearance at $T < T_c$ of macroscopic polarization on zero frequency and it is linked with equilibrium polariton soft mode($\omega^*_0 \longrightarrow 0$). Its necessary condition has the form

$$\frac{2\pi\ n\ d^2_{ab}}{\hbar\ \omega_0} > 1\ . \tag{1.4}$$

This condition on the density of atoms is the most strong of all above listed conditions of phase transitions in the system of two-level atoms.

5.2.Phase transition to active polariton generation state and SRS anomalies

Light amplification in resonatorless active media, in particular in RS, is usually considered as a process of stimulated amplification of spontaneous emission of the individual atoms, independent of each other. The notion of the independence of the spontaneous radiation of individual atoms is used for the description as of stationary stimulated RS (SRS), when the pump pulse duration $\tau_{pump} \gg T_2$ [10], as of non-stationary SRS ($\tau_{pump} \ll T_2$) [11,12]. A number of aspects of these theories have obtained solid experimental evidences (see, for example [13-15]).

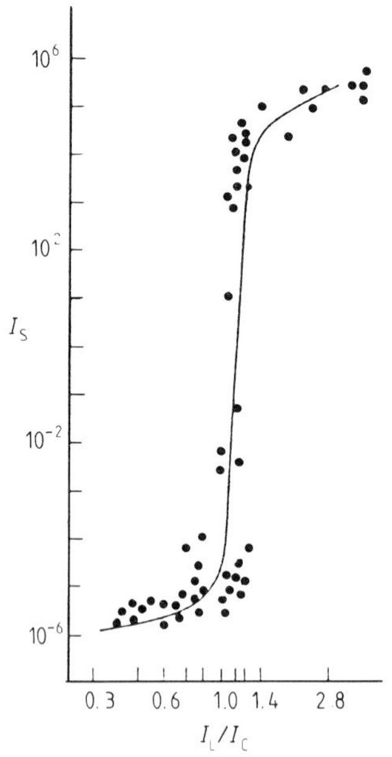

Fig.5.2. The critical dependence of the SRS intensity on the pump intensity in compressed hydrogen ($l < l_2$): the solid line-the theory (according to Eq.(5.2.41), see also sec.5.2.8), the experimental points are from Ref.[19].

It is known, at the same time, that at high pump intensities n_L SRS manifests a combination of properties, that one must consider as anomalous from the viepoint of the concept of stimulated amplification of spontaneous emission of individual atoms.To these anomalies of stationary SRS belong the following effects.

1. At a certain critical value of the pump intensity $n_L = n_c$ the Stokes signal intensity n_s undergoes giant jumplike increase by 5 - 10 orders of magnitude (see Fig. 5.2) with subsequent saturation.

Regime of exponential spontaneous emission gain takes place only in the region $n_L < n_c$ and in many cases the SRS signal jump occurs imediately from the spontaneous noise level (Fig. 5.2) [14-21]. This phenomenon can't be explained from the viewpoint of the concept of gain of spontaneous emission of independent atoms[1]

2. With approaching the threshold value n_C from below the SRS-line width $\delta\omega_{st}$ decreases approximately two orders of magnitude relative to the spontaneous RS line width $\delta\omega_{sp}$: $\delta\omega_{st} \cong \delta\omega_{sp} \cdot 10^{-2}$. At the same time the theory of the exponential gain yields $\delta\omega_{st} \cong \delta\omega_{sp} \cdot (g_0\, n_L\, l)^{-1/2} \cong 10^{-1} \cdot \delta\omega_{sp}$, i.e. it does not describe this critical SRS-line narrowing [21].

3. In the region above threshold value n_C, after SRS-intensity jump, SRS-line consists of separate narrow lines ,lying inside spontaneous line

---

[1] SRS intensity jump was treated in [16-18] as a result of the critical enhancement of the SRS gain coefficient g_0. One can however show, that this critical g_0 enhancement is the consequence of the unjustified assumption about the mutual cancelling of the two terms, taking into account of the effects of photon propagation (see [16] after the Eq.(2.8)). The taking into account of these terms restores the usual expression for g_0. Other aspects of this question are discussed in [22-24].

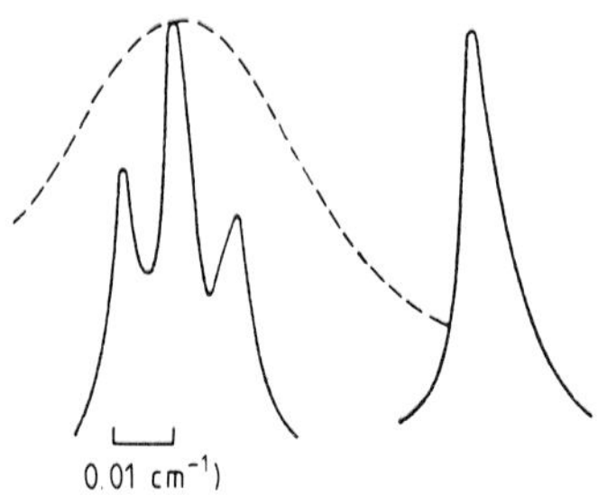

Fig.5.3. The spectral distribution of the first Stokes SRS component in liquid nitrogen. the dashed line is the spontaneous RS line. In the right shown excitation laser line [21].

width, the number of which increases with increase of n_L.The distance between these components and the width of each component do not depend on pump intensity (Fig.5.3) [21].

4. The SRS pulse duration τ_s decreases sharply several times from the value $\tau_s=\tau_{pump}$, and increases afterwards with the increase of n_L up to the values of the order of τ_{pump} (Fig.5.4) [25].

A giant jump of n_s at a certain critical value of pump intensity is characteristic also for the non-stationary SRS [26].

The above anomalies bear the strikingly critical character and suggest the idea about the threshold transition of SRS at $n_L = n_c$ to some new

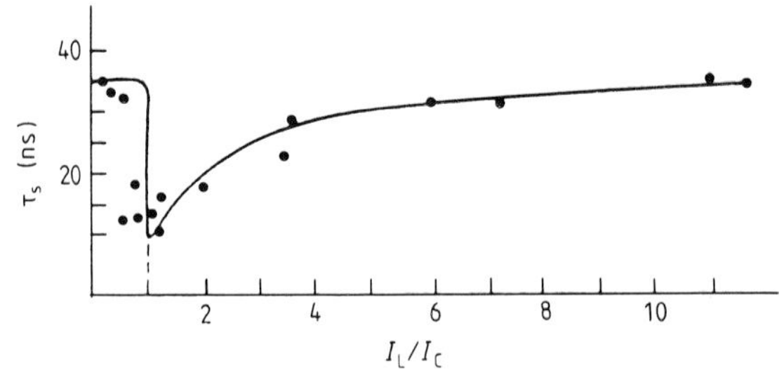

Fig.5.4 The critical dependence of the SRS pulse duration τ_s on the pump intensity: the experimental points are from Ref.[25], the theoretical curve corresponds to Eq.(5.2.57).

scattering regime.

In the present section it is shown, that the above SRS anomalies can be explained from the unified viewpoint, if one besides the individual spontaneous emission in the spontaneous radiation source takes into account also the collective spontaneous radiation(CSR).The SRS is considered then as a process of stimulated gain of individual spontaneous emission and CSR.

At $n_L < n_c$ the spontaneous radiation is determined, in the main, by the individual contribution and at $n_L > n_c$ it consists in the main of CSR. The region of validity of the theory of the exponential gain of the spontaneous emission of independent atoms, from this viewpoint, is limited by the inequality $n_L \ll n_c$.

The process of collective spontaneous radiation in RS without taking into account of the stimulated gain was considered previously in Ref. [27, 28] (see also [29] and [30] as an essentially non-stationary process (Dicke superradiance in RS), duration of which is determined by collective time $\tau_c = (2\pi\, n\, d_s^2 \omega_s / \hbar\, \gamma_{eff})^{-1} \ll T_2$, where n is the number density of atoms, ω_s is the Stokes frequency, $\gamma_{eff} = c\, /\, l$, if the length of the medium $l < c\, T_2$, $d_s = r\, E_L$ is the Stokes dipole moment ($n_L \sim E_L^2$). The new conclusion made in this chapter is the assertion, that for $n_L < n_c$ (i.e. for $\tau_c > T_2$) the stationary regime of collective spontaneous emission is possible [31]. At $n_L = n_c$ ($\tau_c = T_2$) this regime looses its stability and for $n_L \gg n_c$ ($\tau_c \ll T_2$) it transforms into the pulsed Dicke superradiance. Such stationary regimes of CSR are possible as in the case $l < c\, T_2$, as in the case $l > c\, T_2$ (in the last case $\gamma_{eff} = T_2^{-1}$) .

Let us at first consider the general picture of the CSR process from the viewpoint of the suggested conception.

To establish the connection between SRS and CSR we use the kinetic equation for the number density of the Stokes quanta in the derivation of which we distinguish between the processes of the stimulated emission and spontaneous emission of independent atoms. This method consists in the following. One finds the rate of Stokes photon generation by some distinguished (j-th) atom in the ensemble of atoms,which serves as a local source of radiation in the equation for the Stokes quanta. This radiation source consists of three terms. The first term is determined by the work of the field induced part of the polarization of j-th atom over the same field in the point $\mathbf{r}_j$. This term is responsible for the stimulated emission and determines the SRS gain coefficient. The second term is determined by the work of the spontaneous, independent of the field part of polarization $P_j^{sp}(t)$over that part of the field in the point $\mathbf{r}_j$, which correlates with $P_j^{sp}(t)$. This part of the field arises due to the previous action of the spontaneous radiation of j-th atom on others atoms, distributed in the nearby region with radius of order $l_2=cT_2$, with the subsequent reemission by these atoms backward to the j-th atom.This part of the field is essentially the sum of the retarded values of the spontaneous polarizations of j-th atom, i.e. the retarded reaction field of a given atom. The radiation of j-th atom under the action of this field determines the collective spontaneous emission. At last the third term-the individual spontaneous emission is determined by the work of $P_j^{sp}(t)$ over the radiation field of a given atom, the latter being expressed through the same polarization $P_j^{sp}(t)$, taken without retardation.

The local rate of the Stokes quanta production, occurring as a source in the equation for the number of Stokes quanta consists thus of the stimulated term, spontaneous term and collective spontaneous term. The latter is expressed through the correlator of the non-diagonal elements of the density matrix of different atoms

$$S_{ij} = < \sigma_{i-} \sigma_{j+} >, \quad i \neq j \ .$$

The solution of this equation associates the intensity and the spectrum of the output Stokes signal with the intensity of spontaneous radiation and CSR, amplified by the stimulated emission of the photons.

The correlator S_{ij} obeys the equation, describing the processes of spontaneous and stimulated creation of correlations due to the effective interatomic interaction through the exchange of photons inside the region with the correlation radius $l_{cor} = l_{cor}(T_2, \Gamma_L, n_L) \geq l_2$, where Γ_L^{-1} the correlation time of the pump field, and the process of the decay of these correlations with the characteristic time T_2. The interplay of these rival processes determines the magnitude of the correlations of the spontaneous polarizations of different atoms($S_{ij} \neq 0$). Thus, even in the stationary conditions the spontaneous radiation of the separate parts of the atomic ensemble with the dimensions of order of l_{cor} have the collective components - the stationary CSR, which are small for sufficiently small values of n_L.

However, when n_L approaches the certain critical value n_c, defined by the condition $\tau_c = T_2$, the CSR intensity radiated into the central axial mode with the wave vector $\mathbf{k} = \mathbf{k}_s$ is critically increased.In this process the intensity of spontaneous radiation, integrated over the spectrum of the wave vectors remains unchanged, i.e. the redistribution of the spontaneous radiation over the different modes takes place. The geometry of the active medium, in particular the extension of the cylindrical working volume, plays the essential role in this process. The critical value of the pump intensity in the case of long cylinder with the length $l > c\, T_2$ is formally defined by formula, following from the condition $\tau_c = T_2$

$$g_0\, n_c\, l_2 = 1\ , \quad l_2 = c\, T_2 \qquad (T_2 \ll \Gamma_L^{-1})\ , \tag{2.1}$$

where g_0 is the SRS gain coefficient.For the medium of short length $l < l_2$ one must substitute $l_2 \longrightarrow l$. We must note that the occurrence of the factor g_0 in the critical condition (2.1) bears the formal character and in no way means, that the process of the stimulated emission of photons plays any role in the critical behavior of CSR. In the general case one must substitute in (2.1) $l^{-1}_2 \longrightarrow (T_2 + \Gamma_L^{-1}) / c$.

The nonlinear equation for S_{ij}, describing such critical behavior of CSR, is in linear approximation analogous to well known Ornstein-Zernike equation for the pair correlation function, which for the case of the system of spins with exchange interaction describes the ferromagnetic phase transition, and in the system of Cooper electron pairs with the effective exchange interaction through the phonons - the superconducting phase transition [5]. The CSR instability thus bears the character of the non-equilibrium phase transition in the active cooperative atomic system with the effective interaction through the radiation field.

The Fourier-transform of the pair correlation function S_k can be considered as an intensity of the polarization wave - the collective molecular excitation (polariton). The equation for S_k, following from the nonlinear Ornstein-Zernike equation, describes the processes of spontaneous and stimulated creation of polaritons. The latter process for $n_L \geq n_c$ leads to the absolute instability of these non-propagating excitations. The stationary or non-stationary CSR, thus, can be considered as a consequence of self excitation and stimulated generation of standing polaritons in an active medium (active polaritons). At $n_L =$

n_c the initial stationary regime loses its stability. The CSR starts to grow exponentially in time, the resulting change of the population difference leads to the subsequent decrease of S_k and the systems forms the CSR-pulse with the duration $\tau_p \sim T_2[(n_L / n_c) - 1]^{-1}$, the intensity of which $S_k \sim [1 - (n_c / n_L)]^2$ increases in threshold manner with the increase of n_L. Correspondingly the Stokes signal n_s is critically increased. Only a limited cooperative fraction of number of atoms $n_{max} < 10^{18}$ cm^{-3} takes part in the formation of CSR-pulse. After the end of the pulse they found themselves at the upper level, because the time of population relaxation $T_1 \gg \tau_{pump}$.

The delay time of the CSR-pulse t_0, that is the time lag between the switching on the pump and the maximum of CSR-intensity, is of order of magnitude greater then τ_p. At high exceeding of the threshold

$\tau_p \sim T_2 (n_L / n_c)^{-1/2}$ and at $t_0 \ll \tau_{pump}$ several CSR pulses can be formed.

The transition to the state of non-stationary active polariton generation in the open medium, considered here, has a number of features characteristic for phase transitions.At $n_L \longrightarrow n_c$ the CSR line critically narrows, the correletion length of the polarizations of different atoms $l_{cor} \rightarrow \infty$, the order parameter critically slows down, polariton decay constant $T_k^{-1} \longrightarrow 0$, the polariton intensity undergoes the jumplike increase. If $\tau_{pump} > T_1$ (this case can be realized in the system of two-level atoms), then a new regime can be installed-the regime of stationary polariton generation, or regime of coherent CSR. The stationary value of the polarization wave amplitude in this regime obeys the square root dependency on the pump intensity (proportional to $[1 - (n_c/n_L)]^{1/2}$) analogously in some extent to the dependency of the amplitude of light field on the pump intensity in lasers.

As the non-stationary(transient), as new stationary regime of CSR are the regimes of the stimulated generation of polaritons - the non-propagating collective polarization excitations, appearing in active open medium due to the effective interatomic interactions. This process is basically different from the stimulated photon generation.

In this section all the above listed SRS-anomalies are treated as the consequences of the phase transition at $n_L = n_c$ into the state of non-stationary polariton generation in the open system. For $n_L \ll n_c$ the present theory yields the usual expression, describing the gain of the spontaneous emission of independent atoms. We note, that although the theory of the phase transition is developed here for the case of instant switching on the pump, it can describe the phenomenon of SRS-intensity jump in the case of the pump pulses with nanosecond fronts, because in the vicinity of $n_L = n_c$ the duration of transient process is of order of $\tau_p \sim T_2[(n_L/n)_c - 1]^{-1}$ and can exceed the duration of the leading pump pulse front . Some peculiarities arising due to the slow switching on the pump are considered in Sec. 5.2.7.

The results, obtained for the threshold and the magnitude of the jump, line width, period of fine structure are compared with the experimental results. There is agreement between the theory of SRS behavior under the high levels of excitations, developed here, and the experimental results. The results obtained are applicable in the main also to the case of light amplification in the open system of inverted two-level atoms with the stationary maitained inversion. One must however note, that in this case the inequality $\Gamma_L \gg T_2^{-1}$ is realized, which means, that the threshold condition on inversion (2.1) is much more severe.

Let us now pass over to the detailed mathematical presentation of the above described picture of the CSR instability.

5.2.1. *Initial field and atomic equations. Local rates of induced, collective spontaneous and spontaneous creation of quanta.* Let us consider the medium of N multilevel atoms (molecules) confined in cylindrical volume $V = A\ l$, extended in z-direction ($l \gg A^{1/2}$). The system is acted upon by the light pump wave

$$\mathbf{E}_L(\mathrm{r}, t) = \mathbf{E}_L^-(\mathrm{r}, t) + \mathbf{E}_L^+(\mathrm{r}, t)$$

$$= \mathbf{e}_L \{ E_L(t) \exp [- i (\omega t - k_L z)] + \mathrm{c.c.} \} \qquad (2.2)$$

with the correlation function $\overline{E_L(t)\, E_L^*(t')} = |E_L|^2 \exp[-\Gamma_L (t - t')]$, where $|E_L|^2 = \mathrm{const}$ for $0 \le z \le l$ in the constant pump intensity approximation, $z = 0$ is the coordinate of entry face of the medium, $|E_L|^2 = 0$ for $t < 0$ and $t > \tau_{pump}$ and $|E_L|^2 \neq 0$ for $0 \le t \le t_{pump}$.

Due to the RS, the Stokes field E_s with the frequency $\omega_s = \omega - \omega_{ab}$, where ω_{ab} is the transition frequency of the working pair of levels,arises in the medium

$$\mathbf{E}_s(\mathbf{r}_j, t) = \mathbf{E}_s^-(\mathbf{r}_j, t) + \mathbf{E}_s^+(\mathbf{r}_j, t)$$

$$= \mathbf{E}_s^- \exp(-i\,\omega_s t) + \mathbf{E}_s^+ \exp(i\,\omega_s t)\,. \tag{2.3}$$

The Stokes field is found from the solution of Maxwell equation

$$\frac{\partial^2 \mathbf{E}_s}{\partial t^2} - c^2 \Delta\, \mathbf{E}_s = -\frac{4\pi\, \partial^2\, \mathbf{P}_s}{\partial t^2}\,, \tag{2.4}$$

where the nonlinear polarization

$$\mathbf{P}_s(\mathbf{r}_i, t) = \sum_i (\mathbf{P}_{si}^+ + \mathbf{P}_{si}^-)\,\delta(\mathbf{r}_i - \mathbf{r}_j)$$

$$= -\sum_i \mathrm{r}\, E_L^+(\mathbf{r}_j, t)\,\sigma_{i-}\,\delta(\mathbf{r}_i - \mathbf{r}_j) + \text{c.c.} \tag{2.5}$$

Here σ_{i-} are the operators of the atomic density matrix,

$$\mathrm{r} = \hbar^{-1} \sum_\alpha [(\mathbf{d}_{b\alpha}\, \mathbf{e}_L)\, \mathbf{d}_{\alpha a}\, (\omega_{b\alpha} + \omega)$$

$$+ \mathbf{d}_{b\alpha}\, (\mathbf{d}_{\alpha a}\, \mathbf{e}_L)\, (\omega_{b\alpha} - \omega_s)]. \tag{2.5a}$$

The Eqs. for the non-diagonal element of density matrix σ_{j-} and the atomic energy levels populations difference σ_{j3} have the form of the effective two-level atom equations [32]:

$$\frac{d\sigma_{j-}}{\partial t} + i\,(\omega_{ab} - i\,T_2^{-1})\,\sigma_{j-} =$$

$$= \frac{i\,\mathbf{d}_{sj}\,(t)}{\hbar}\,\mathbf{E}_s^+\,(\mathbf{r}_j,\,t)\,\sigma_{j3}\,, \qquad (2.6)$$

$$\frac{\partial\sigma_{j3}}{\partial t} + \frac{\sigma_{j3} - \sigma_{j3}^0}{T_1} = -\,\frac{2\,i\,\mathbf{d}_{sj}\,(t)}{\hbar}\,\mathbf{E}_s^+(\mathbf{r}_j,t)\,\sigma_{j+} + \text{c.c.}\,, \qquad (2.7)$$

where the effective Stokes dipole moment

$$\mathbf{d}_{sj}\,(t) = \mathbf{r}^*\,E_L^-\,(\mathbf{r}_j,\,t) \equiv \mathbf{d}_s\,(t)\,\exp\,[-\,i\,(\omega t - k_L\,z_j)]\,.$$

The equation (2.7) can be represented in the form

$$\frac{\partial\sigma_{j3}}{\partial t} = \left(\frac{\partial\sigma_{j3}}{\partial t}\right)_{\text{rel}} + \left(\frac{\partial\sigma_{j3}}{\partial t}\right)_{\text{rad}}. \qquad (2.8)$$

Here the first term corresponds to the relaxation with the time T_1, The quantity $\hbar\omega_s(\partial\sigma_{j3}\;/\;\partial t)_{\text{rad}}\;/\;2$ describes the change of the energy of j-th atom per unit time due to the work of the polarization of j-th atom over the Stokes field in the point $\mathbf{r}_j$.

Using the solution of the Eq. (2.4) in the far field, we express the field variable through the retarded polarizations of atoms

$$\mathbf{E}_s\,(\mathbf{r}_j,\,t) = c^{-2}\,\sum_i \frac{\partial^2}{\partial t^2}\,\frac{1}{r_{ij}}\,[\,[\,\mathbf{P}_{si}\,(t - \tau_{ij})\,\mathbf{n}_{ij}\,]\,\mathbf{n}_{ij}]\,, \qquad (2.9)$$

where $\mathbf{n}_{ij} = \mathbf{r}_{ij}\,/|\,\mathbf{r}_{ij}|$, $\quad \mathbf{r}_{ij} = \mathbf{r}_i - \mathbf{r}_j$, $\quad \tau_{ij} = |\mathbf{r}_{ij}|\;/\;c$. Substituting (2.9) into (2.7), we perform averaging over the atomic ensemble and the pump. Then, taking into account (2.8), we obtain, substituting $c^{-2}\;\partial^2\;/\;\partial t^2 \to -\,k_s^2\,,$

$$\frac{1}{2}\frac{\partial}{\partial t}\langle\sigma_{j3}\rangle_{\mathrm{rad}}$$

$$= (i\,k_s^2/\hbar)\sum_{i\neq j} a_{ij} < \sigma_{i-}(t-\tau_{ij})\,\sigma_{j+}(t) > + \text{ c.c.} \qquad (2.10)$$

Here

$$a_{ij} = (a^0_{ij}/r_{ij})\exp\left[-i\,k_s r_{ij} - ik_L(z_i - z_j)\right],$$

$$a^0_{ij} = \overline{\mathbf{d}_s(t)\,[\,[\mathbf{d}^*_s}\,\overline{(t-\tau_{ij})\,\mathbf{n}_{ij}]\,\mathbf{n}_{ij}}\,],$$

where the bar denotes the averaging over the pump(see Eq. (2.2)). The mean value of a^0_{ij}, averaged over the directions of the vector $\mathbf{n}_{ij}$, is equal to $2\,|d_s|^2/3$, $k_s = \omega_s/c = 2\pi/\lambda_s$.

Thus, the rate of creation of Stokes quanta is determined by the two-time correlation function $\langle\sigma_{i-}(t-\tau_{ij})\,\sigma_{j+}(t)\rangle$. To evaluate it we write down the equations for the operators $\sigma_{i-}(t-\tau_{ij})$ and $\sigma_{j+}(t)$, following from (2.6) with taking into acount of (2.9), multiply the first one on the right by $\sigma_{j+}(t)$ and the second one-on the left by $\sigma_{i-}(t-\tau_{ij})$, then take the sum of both and average. Then we obtain

$$\left(\frac{\partial}{\partial t} + \frac{2}{T_2}\right)\langle\sigma_{i-}(t-\tau_{ij})\,\sigma_{j+}(t)\rangle$$

$$= -(i\,k_s^2/\hbar)\,\langle\sigma_{j2}(t)\rangle\sum_{l(l\neq j)} a^*_{jl}\,\langle\sigma_{i-}(t-\tau_{ij})\,\sigma_{l+}(t-\tau_{jl})\rangle$$

$$+ (i k_s^2 / \hbar) <\sigma_{i3} (t - \tau_{ij})>$$

$$\times \sum_{l(l \neq i)} a_{li} <\sigma_{l-} (t - \tau_{ij} - \tau_{il}) \sigma_{j+} (t)> . \qquad (2.11)$$

For clarity of interpretation of (2.11) let us consider in more details the stationary case,putting $\partial / \partial t = 0$. As one can see from 2.11 and (2.10) (see also (2.7)), the work of the polarization of j–th atom over the field consists of two comntributions. The first contribution(the first term in the right hand side of (2.11)) is the work performed over the field by the induced part of the polarization of j–th atom, which itself is determined by the field, created in the point $\mathbf{r}_j$ by all other atoms of atomic ensemble. This first contribution is proportional to $<\sigma_{j3} (t)>$ and thus it describes the induced transitions of j-th atom. For this contribution we have from (2.11)

$$<\sigma_{i-} (t - \tau_{ij}) \sigma_{j+} (t)>_{\text{ind}}$$

$$= - (i T_2 k_s^2 / 2\hbar)<\sigma_{j3}(t)> \sum_{l \neq j)} a_{jl}^* <\sigma_{i-}(t - \tau_{ij}) \sigma_{l+}(t - \tau_{jl}) \qquad (2.12)$$

The second term in the right hand side of (2.11) corresponds to the work of the spontaneous part of the polarization σ_{j+}^{sp} of j-th atom , performed over that part of the field in the point $\mathbf{r}_j$, which correlates with σ_{j+}^{sp}. From (2.11) we have for this second contribution

$$<\sigma_{i-} (t - \tau_{ij}) \sigma_{j+} (t)>_{\text{sp}} = (i T_2 k_s^2 / 2\hbar) <\sigma_{i3} (t - \tau_{ij})$$

$$\times \{a_{ji} <\sigma_{j-} (t - 2\tau_{ij}) \sigma_{j+} (t)>_{\text{sp}}$$

$$+ \sum_{l \neq i,j)} a_{li} <\sigma_{l-} (t - \tau_{ij} - \tau_{il}) \sigma_{j+} (t)>_{\text{sp}} . \qquad (2.13)$$

Extracted from the sum term with $l = j$ is expressed through the two-time correlator of the spontaneous polarization of j-th atom. It is determined by the work of the spontaneous part of the polarization of j-th atom over the field, determined by the value of spontaneous polarization $\sigma^{sp}_{j+}(t')$, taken at the time moment $t' = t - 2\, r_{ij} / c$. The double time retardation means, that a photon has been spontaneously emitted by j-th atom at the time moment $t - 2\, r_{ij} / c$, propagated during the time interval r_{ij} / c to the i-th atom and backward after the reradiating by i-th atom and at time moment t acted upon j-th atom. The term with $j=l$, thus, is responsible for the appearance and stationary maintenance of the correlations between the polarizations of the individual atoms due to the exchange of photons. The above process of photon exchange is, thus, responsible for spontaneous creation of interatomic correlations.

The terms with $l \neq i, j$ in (2.13) describe the processes of induced creation of correlations: if there has already been established correlation between j-th and l-th atoms, that is $<\sigma_{l-}\, \sigma_{j+}>_{sp} \neq 0$, then the action of l-th atom on i-th atom (the propagation time τ_{il}) and subsequent action of i-th atom on j-th atom (propagation time τ_{ij}) lead to the appearance of correlation between the i-th and j-th atoms. These processes are responsible for the critical behavior of the CSR, i.e. for the phase transition considered. Both types of terms (with $l = j$ and $l \neq j$) take into acount in essence, the retarded reaction field of j-th atom: the spontaneous polarization having existed in j-th atom in the preceding time moments $t' < t$, as a result of the emission of j-th atom and reemission of other atoms, acts with retardation on the spontaneous polarization of j-th atom and stimulates it to radiate. This part of the radiation of j-th atom comprises the CSR.

The spontaneous part of the radiation of j-th atom is determined by the correlator $<\sigma_{j-}(t)\, \sigma_{j+}(t)>$, i.e. by the reaction field of j-th atom, taken without the retardation.

The above analysis shows, that the correlation function in (2.10) consists of the two parts

$$\sigma_{i-}(t - \tau_{ij})\, \sigma_{j+}(t)>$$

$$= <\sigma_{i-}(t - \sigma_{ij})\, \sigma_{j+}(t)>_{ind} + <\sigma_{i-}(t - \tau_{ij})\, \sigma_{j+}(t)>_{sp} .$$

Let us substitute this expression into (2.10). Then, taking into account of (2.12), going over in (2.12) to the field variable $\mathbf{E}_s$, using for the correlation function of the pump expression (2.2) and substituting $a^0_{ij} \rightarrow \langle a^0_{ij} \rangle$, we obtain the expression for the rate of j-th atom energy change

$$\left[\frac{1}{2}\,\frac{\partial \sigma_{j3}}{\partial t}\right]_{\rm rad} = -\frac{T_2}{\hbar^2}\,|\mathbf{d}_s\,\mathbf{E}_s|^2\,\langle \sigma_{j3} \rangle$$

$$+\frac{2\,i\,k_s^2\,d_s^2}{3\,\hbar}\sum_i\{\,r_{ij}^{-1}\exp\left[\,-(k_s\,r_{ij} - k_{\rm L}\,(z_j - z_i) - \Gamma_{\rm L}\,\tau_{ij})\right]$$

$$\times \langle \sigma_{i-}(t-\tau_{ij})\,\sigma_{j+}(t) \rangle_{\rm sp} + {\rm c.c.}\,\}\,. \qquad (2.14)$$

The first term in the right hand side of (2.14) determines the rate of the induced creation of the Stokes quanta $1\,/\,2 \langle \partial\,\sigma_{j3}\,/\,\partial t \rangle_{\rm ind}$. This is the case under the condition $t > T_2$, i.e. for the stationary regime. In the non-stationary case, when $t < T_2$ the rate of the induced creation of quanta is obtained by integrating the equation (2.11), where only the first term in the right hand side is retained, expressed through $\mathbf{E}(t)_s$. The second term in (2.14) determines the rate of the spontaneous creation of the Stokes quanta. To take into account the reaction field of j-th atom, we have extended the summation in (2.14) also over the value $i = j$.

Let us express the rate of the spontaneous creation of quanta: the two-time correlator $\langle \sigma_{i-}\,(t - \tau_{ij})\,\sigma_{j+}\,(t) \rangle_{\rm sp}$ in (2.14) through the single time correlator. We note, that since $\sigma_{j+}\,(t)$ corresponds to the spontaneous part of the polarization, its temporal evolution is determined by the solution of the homogeneous equation (2.6)

$$\sigma_{j+}\,(t) = \sigma_{j+}\,(t - r_{ij}\,/\,c)\,\exp\left[i\,(\omega_{ab}\,/\,c)\,r_{ij} - r_{ij}\,l_2^{-1}\right], \qquad (2.15)$$

where $\sigma_{j+}(t-r_{ij}/c)$ depends slowly on time.We use this formula in (2.14) and introduce the operators

$$\rho_{i-}(t) = \sigma_{i-} \exp(-i k_L z_i) . \tag{2.15a}$$

In stationary regime, when $n_L < n_c$ and $\tau_{pump} > T_2$, or $n_L > n_c$, $\tau_{pump} > T_1$, the single time correlators do not depend on the time, therefore

$$<\rho_{i-}(t - \tau_{ij}) \rho_{j+}(t - \tau_{ij}) = <\rho_{i-}(t) \rho_{j+}(t)> . \tag{2.16}$$

In the non-stationary regime of CSR generation, when $n_L > n_c$ and $\tau_{pump} > \tau_p$, where τ_p is the CSR pulse duration, this equality takes place for the pairs of atoms, separated by the distances smaller then the cooperative length $l_c = c\,\tau_p$, which effectively restricts the dimension of region of collective behavior of atoms in the non-stationary regime of CSR.

Using the equation (2.16) in (2.14), we obtain for the rate of the energy change of j-th atom occurring due the Stokes quanta emission

$$\frac{1}{2} \frac{\partial}{\partial t} <\sigma_{j3}>_{rad} = \frac{1}{2} \frac{\partial}{\partial t} <\sigma_{j3}>_{ind}$$

$$+ \frac{\Gamma_s}{2} \sum_{i\neq j} b_{ij} <\rho_{i-}(t) \rho_{j+}(t)> + \Gamma_s <\rho_{j-}(t) \rho_{j+}(t)> . \tag{2.17}$$

where $\Gamma_s = 4 k_s^3 |d_s^2| / 3\hbar$ is the constant of the spontaneous Stokes decay and b_{ij} is the matrix of the effective interatomic interaction

$$b_{ij} = \frac{1}{-ik_s r_{ij}} \exp[-i k_s r_{ij} - r_{ij} l_2^{-1} - r_{ij} \Gamma_L c^{-1}] . \tag{2.18}$$

As one can see from (2.18) the effective interatomic interaction is cutted down at the distance of order of $(l_2^{-1} + \Gamma_L c^{-1})^{-1}$. Because in stationary CSR $l_2^{-1} \gg \Gamma_L c^{-1}$, we in the following shall neglect the pump line width Γ_L.

The formula (2.17) shows, that the Stokes photons are created as a result of three processes: the induced creation, collective spontaneous creation ($\sum_{i\neq j}$) and spontaneous creation of quanta with the rate

$$\Gamma_s \, \langle\rho_{j-}\, \rho_{j+}\rangle = \Gamma_s \, \langle\sigma_{jbb}\rangle.$$

Now we shall derive the equation for the number density of the Stokes quanta $n_s = |E_s|^2 \, / \, (2\pi \, \hbar \, \omega_s)$. For this we obtain at first from (2.4) and (2.3) the equations for the amplitudes of the Stokes field $E_s^{\pm}$. From these equations, together with (2.7) and (2.8) one can easily obtain the local conservation law, which can be represented in the form

$$\frac{\partial n_s}{\partial t} + \frac{i}{4\pi \, \hbar \, k_s^2} \{(\Delta \, \mathbf{E}_s^+) \, \mathbf{E}_s^- - \mathbf{E}_s^+ \, \Delta \, \mathbf{E}_s^-\}$$

$$= \sum_j \delta \, (\mathbf{r} - \mathbf{r}_j) \frac{1}{2} \, \left\langle \frac{\partial \sigma_{j3}}{\partial t} \right\rangle_{\text{rad}} , \tag{2.19}$$

where $(1 \, / \, 2) \, \partial \, \langle\sigma_{j3}\rangle_{\text{rad}} \, / \, \partial t$ is given by (2.17).

Let us expand the field in the Fourier-series

$$\mathbf{E}_s^+ = \sum_k \mathbf{E}_k^+ \exp \, (- \, i\mathbf{k}\mathbf{r}) \; , \quad n_s = \sum_k n_k \; , \quad n_k = |E_k|^2 \, / \, (2\pi \, \hbar \, \omega_s) \; .$$

Here $\mathbf{E}_k^{\pm} = \mathbf{E}_k^{\pm} \, (z,t)$ are the slow varying amplitudes, $\langle E_k^+ E_{k'}^-\rangle = \delta_{kk'} \cdot \langle E_k^+ E_k^-\rangle$. Let us now obtain from (2.19) the equation for the number density of quanta in a mode. For this aim we represent the right hand side of (2.19), using (2.17), as a sum over all **k**-modes.

We introduce the spectral function

$$S_k \equiv \langle\rho_- \rho_+\rangle_k = \sum_{i(i\neq j)} \langle\rho_{i-} \rho_{j+}\rangle \exp \, [i \; \mathbf{k} \; (\mathbf{r}_i - \mathbf{r}_j)]$$

$$= \langle \sigma_- \sigma_+ \rangle_{-q} , \qquad \mathbf{q} = \mathbf{k}_L - \mathbf{k}_s . \tag{2.20}$$

Then the term, describing CSR in (2.17), assumes the form

$$\frac{\Gamma_s}{2} \sum_{i(i \neq j)} b_{ij} \langle \rho_{i-}(t)\, \rho_{j+}(t) + \text{c.c.}$$

$$= \frac{\Gamma_s}{N} \operatorname{Re} \sum_k b_k S_k , \tag{2.17a}$$

where b_k is the Fourier-transform of the matrix of interatomic interaction, evaluated for the case of long cylinder

$$b_k = b_{-k} = \sum_i b_{ij} \exp\,[i\mathbf{k}\,(\mathbf{r}_i - \mathbf{r}_j)]$$

$$= \frac{2\pi\, n}{k_s^2} \, \frac{l_{eff}}{1 + i\,(k - k_s)\, l_{eff}} , \tag{2.21}$$

where

$$l_{eff} = \begin{cases} l_2 & \text{for } l > l_2 , \\ l & \text{for } l < l_2 . \end{cases}$$

The rate of Stokes spontaneous radiation of individual atom can be represented in the form

$$\Gamma_s \langle \rho_{j-} \rho_{j+} \rangle = \frac{\Gamma_s}{|d_s|^2} \sum_{\mathbf{k}} (\mathbf{P}^*_{sj}\, \mathbf{P}_{sj})_{\mathbf{k}} ,$$

where $\mathbf{P}_{sj}$ is defined in (2.5)and the spectral function of polarization is given by

$$(\mathbf{P}^*_{sj}\,\mathbf{P}_{sj})_{\mathbf{k}} = \frac{2\pi <\sigma_{jbb}> |d_s|^2\, l_2^{-1}}{V k_s^2\, [l_2^{-2} + (k - k_s)^2\,]} \,. \qquad (2.17b)$$

Representing now the left and right hand side of (2.19) as the sums over **k** and using in (2.17) the obtained expressions for the spectral functions (2.17a) and (2.17b), we have

$$\frac{1}{c}\,\frac{\partial n_{\mathbf{k}}}{\partial t} + \frac{\partial n_{\mathbf{k}}}{\partial z} - g\, n_L\, n_{\mathbf{k}} = I_{\mathbf{k}}^{sp} \,. \qquad (2.22)$$

Here we have introduced the SRS gain coefficient g

$$g\, n_L = \frac{2\pi\, d_s^2\, n\, \omega_s\, |\sigma_3|}{3\,\hbar\, c\, T_2^{-1}} \equiv g_0\, n_L\, |\sigma_3| \,, \qquad (2.23)$$

where n is the number density of atoms.

The intensity of the full source of spontaneous radiation consists of two parts:

$$I_{\mathbf{k}}^{sp} = \frac{\Gamma_s}{cV}\,[\mathrm{Re}\,(\,b_k S_k) + \mathrm{Re}\,(b_k <\sigma_{bb}>)\,] \,. \qquad (2.24)$$

In (2.24) the CSR contribution is expressed through the spectral function S_k and the second term in right hand side describes the contribution from the spontaneous radiation of independent atoms. The source I_k^{sp} depends on z through the pump intensity $E_L^2\,(z)$. In the constant intensity pump approximation E_L^2 = const and I_k^{sp} = const.

The spectral distribution of intensity of the spontaneous radiation of the independent atoms is formally expressed through $b_k = b_k\,(l_2)$, since the exponential spatial dependency of b_{ij} in (2.18) in the considered case $l_2 \ll c\,/\,\Gamma_L$, is determined by the temporal dynamics of the decay of spontaneous polarization of an atom, described by Eq.(2.15). If in (2.24) one neglects CSR, putting $S_k = 0$, then, using for Re b_{k} the formula (2.21), one has for the spectral distribution of the spontaneous radiation of the independent atoms (compare (2.30)) the following expression

$$I_k^{\mathrm{sp}} = \frac{1}{cV} \frac{2\pi\, n\, \Gamma_{\mathrm{s}} <\sigma_{bb}>}{k_{\mathrm{s}}^2} \frac{l_2^{-1}}{l_2^{-2} + (k - k_{\mathrm{s}})^2}. \qquad (2.25)$$

The intensity of spontaneous radiation of independent atoms , integrated over the **k**-spectrum is

$$I^{\mathrm{sp}} = \frac{V}{(2\pi)^3} \int d\,\mathbf{k}\, I_k^{\mathrm{sp}} = \frac{\Gamma_{\mathrm{s}}}{c}\, n <\sigma_{bb}> .$$

We note, that the gain coefficient g in (2.23) must have the same Lorentz shape as in Eq. (2.25). Here, however we neglect the effect of the line narrowing in the spatial gain and use g in the form (2.23), which correspond to the maximum gain coefficient value.

5.2.2. *Nonstationary nonlinear Ornstein-Zernike equation. Balance equations for spectral function S_k and population difference.* To obtain the equation for the spectral function S_k we use the equation (2.11), where in accordance with the discussion of (2.12), (2.13) only the second term is retained in the right hand side.Going over in (2.11) to one time correlators of the operators ρ_{i+} , ρ_{i-}, using (2.15), (2.15a), we obtain the following equation for pair correlation function of nondiagonal elements of atomic density matrices

$$\frac{S_{ij}}{\partial t} + \frac{2}{T_2} S_{ij} = - \frac{\Gamma_{\mathrm{s}}}{2} b_{ij} <\sigma_{i3}> <\sigma_{jbb}>$$

$$- \frac{\Gamma_{\mathrm{s}}}{2} <\sigma_{i3}> \sum_{l \neq i,j} b_{il} S_{lj} . \qquad (2.26)$$

The equation (2.26), as it was discussed, describes the process of inducing and maintaining of interatomic correlations due to the exchange of virtual photons. We note, that in stationary case the equation (2.26) is analogous to the well known equation for pair correlation function - the equation of Ornstein-Zernike type [5], describing the phase transition into ferroelectric and superconducting states. On the other hand in the nonstationary case the equation of the type (2.26) describes the transition of initially incoherent system with $S_{ij} = 0$ to the superradiant state in RS with $(\sum_i S_{ij} \sim N^2)$ (see Sec.5.3). With the substitution $\langle\sigma_3\rangle \longrightarrow - \langle\sigma_3\rangle$, $\langle\sigma_{bb}\rangle \longrightarrow \langle\sigma_{aa}\rangle$, $k_s \longrightarrow \omega_{ab} / c$ the equation (2.26) describes the superradiance in the system of two-level atoms.

Let us go over in (3.26) to the Fourier-representation, using (2.21). Then we obtain, defining $\langle\sigma_{i3}\rangle \equiv D$

$$\frac{\partial S_k}{\partial t} + \left(\frac{2}{T_2} + \frac{\Gamma_s b_k D}{2}\right) S_k = - \frac{\Gamma_s}{2} \langle\sigma_{bb}\rangle D\, b_k. \qquad (2.27)$$

The quantity S_k can be considered as an intensity of the collective Stokes polariton excitation with the wave vector $\mathbf{q} = \mathbf{k}_L - \mathbf{k}_s$ (compare (2.20)) and the frequency $\omega_{qs} \sim \omega_{ab}$ (polariton dispersion relation is discussed in Sec.5.2.9). Then the equation (2.27) describes the competition of the processes of spontaneous creation of polaritons, taken into account by the right hand side, induced creation of polaritons $(\Gamma_s b_k D\, S_k / 2)$ and the process of dissipative decay of polaritons on the time T_2.

The equation for the population difference D we obtain from (2.7), (2.8) and (2.14), taking into account of (2.16). Going over to the Fourier-representation, using (2.20), we have

$$\frac{\partial D}{\partial t} + \frac{D}{T_{1\,\mathrm{eff}}} - \frac{D_0}{T_1}$$

$$= \frac{2\,\Gamma_s}{N} \sum_k \mathrm{Re}\,(b_k S_k) + 2\,\Gamma_s \langle\sigma_{bb}\rangle\,,$$

where

$$\frac{1}{T_{1eff}} = \frac{1}{T_1} + \frac{2\,d_s^2\,E_s^2}{3\,\hbar^2 T_2^{-1}} \simeq \frac{1}{T_1}\,. \tag{2.28}$$

The first term in the right hand side of (2.28) describes the CSR contribution and the second corresponds to the individual spontaneous radiation.

5.2.3. *Stationary subthreshold regime of CSR. Critical behavior of stationary CSR ($n_L < n_c$).* Let the duration of the pump pulse $\tau_{pump} > T_k$, where T_k is the polariton lifetime (see (2.33)). Then from (2.27) we have for the intensity of the stationary polariton excitation

$$S_k = -\frac{\Gamma_s T_2 D \langle\sigma_{bb}\rangle\, b_k}{4\,(1 + \Gamma_s T_2 D\, b_k / 4)}\,. \tag{2.29}$$

Let us consider the case $l > l_2$, for which in (2.21) $l_{eff} = l_2$, and substitute (2.29) into (2.24). Then the second term in the right hand side of (2.24) is exactly cancelled and, taking into account (2.23), we have for the full intensity of spontaneous radiation into the **k**-mode

$$I_k^{sp} = \frac{1}{Vc}\,\frac{2\pi\, n\, \Gamma_s \langle\sigma_{bb}\rangle}{k_s^2}\,\frac{l_{cor}^{-1}}{l_{cor}^{-2} + (k - k_s)^2}\,. \tag{2.30}$$

Here we introduced the correlation length of polarizations of different atoms

$$l_{cor} = \frac{l_2}{1 - |D|\, n_L/n_c}\ , \quad n_c = \frac{1}{g_0 l_2}\ . \tag{2.31}$$

From (2.30) we obtain for the integrated intensity of the full source of spontaneous radiation the following expression

$$I^{sp} = \sum_k I_k^{sp} = \frac{1}{c}\, n\, \Gamma_s <\sigma_{bb}> ,$$

which coincides with the integral intensity of spontaneous radiation of the system of independent atoms. Thus, the interatomic interactions in a volume of particular geometry do not enhance the spontaneous radiation, but only redistribute it over different modes **k**.

As one can see from (2.31), at weak pump intensities, when $n_L \ll n_c$, the correlation length $l_{cor} \simeq l_2$ and the formula (2.30) coincides with (2.25). Thus, at weak pump the collective effects are unimportant and the spontaneous radiation is largely of individual nature.

However, when $n_L \rightarrow n_c$ the sharp increase of CSR-intensity into the $k = k_s$ mode occurs, accompanied by sharp narrowing of CSR-line, which tends to δ-function.As one can see from (2.31) the size of the region of correlated behavior of polarizations of different atoms critically increases, i.e, $l_{cor} \rightarrow \infty$ for $n_L \rightarrow n_c$.

Simultaneously and the RS-line must undergo the corresponding changes.In fact, the stationary solution of (2.22) yields for the output SRS-intensity ($z = l$) :

$$n_{\mathbf{k}}(l) = I_{\mathbf{k}}^{sp}\, l\, \frac{\exp\,(g\, n_L l) - 1}{g\, n_L\, l}\ , \tag{2.32}$$

where I_k^{sp} for $n_L < n_c$ is given by Eq. (2.30). At $n_L \ll n_c$ the CSR-contribution is unimportant, $l_{cor} \simeq l_2$ and Eq.(2.32) together with Eq.(2.30) describe the exponential gain of the spontaneous radiation of individual atoms. Remind, that we have neglected the dispersion of the gain coefficient and, consequently neglect the line narrowing in the process of spatial gain. For $n_L \rightarrow n_c$ the SRS-line, described by (2.32), critically narrows and tends to δ-function. Such critical behavior occurs due to cancelling of decay constant of collective polarization excitation(polariton), the expression for which follows from (2.27)

$$T_k^{-1} = \frac{1}{T_2}\left(1 - \frac{n_L}{n_c}\right) \qquad (2.33)$$

At $n_L > n_c$ polariton decay constant $T_k^{-1} < 0$ and the polariton instability develops in the system.

5.2.4.*Nonstationary stimulated polariton generation and the SRS-intensity jump.* We assume, that the transient process occurs on times $t \ll T_1$, i.e. the CSR delay time $t_0 \ll T_1$. Then from (2.28) and (2.27) we have, confining ourselves by the consideration of generation on the central mode with $\mathbf{k} = \mathbf{k}_s$,

$$\frac{\partial D}{\partial t} = \frac{8\alpha(t)}{T_2 N} S_{\mathbf{k}_s} + \Gamma_s(1-D). \qquad (2.34)$$

$$\frac{\partial S_{\mathbf{k}_s}}{\partial t} + \frac{2\alpha(t)}{T_2}[\alpha^{-1}(t) + D\,)\,S_{\mathbf{k}_s} = -\frac{2\alpha(t)}{T_2} D <\sigma_{bb}> \qquad (2.35)$$

where $\alpha(t) \equiv n_L(t) / n_c$.

In the derivation of the equations (3.34) and (2.35) we assume, that the condition (2.16) is valid for the region with dimension cT_2.Thus, these equations are valid under the condition $l_c = c\tau_p \geq l_2$. Well above threshold, when $n_L/n_c \gg 1$, in (2.34)and (2.35) one must substitute $l_2 \longrightarrow l$ (see below).

We suppose here that the pump is switched on instantly:

$$n_L(t) = \begin{cases} n_L > n_c & \text{for } 0 < t < \tau_{pump}, \quad \alpha \equiv n_L / n_c \\ 0 & \text{for } t<0 \text{ and } t>0 \ . \end{cases} \tag{2.35a}$$

The Eqs.(2.34) and (2.35) are analogous to the equations of Q-switched laser in giant pulse regime[33]. However, instead of the number of photons, occurring in laser equations, in (2.34) and (2.35) the intensity of polariton wave S_k occurs. Besides, the equation for D, in contradistinction to the case of laser, is linear. One may note, that the Eq.(2.34) and (2.35) coincide in form with maser equations [33].

From (2.34) and (2.35), neglecting the right hand side in (2.35) (this approximation gives rise to decrease of CSR pulse delay, see Sec.5.3.2)and introducing the notation

$$\xi = \alpha^{-1} + D,$$

we obtain the following equation

$$\frac{\partial^2 \xi}{\partial t^2} + \frac{\alpha}{T_2} \frac{\partial \xi}{\partial t} + \Gamma_s \frac{\partial \xi}{\partial t} = 0$$

with the initial conditions

$$\xi(0) = \alpha^{-1} - 1, \quad (\partial \xi / \partial t)_{t=0} = 2\Gamma_s \ .$$

The solution of this equation under the condition $\alpha > 1$, with accuracy up to the terms of order of $\tau_p \Gamma_s \ll 1$, has the form

$$D(t) = -\alpha^{-1} + (1 - \alpha^{-1}) \text{th} \ [(t - t_o)/\tau_p] \ . \tag{2.36}$$

Here

$$\tau_p = T_2/(\alpha - 1), \quad t_0 = \frac{\tau_p}{2} \ln [N_c(1 - \alpha^{-1})],$$

where the cooperative number of particles $N_c = nAl_2(\Delta\Omega_g/4\pi)$,$\Delta\Omega_g$ is the diffraction angle $\Delta\Omega_g = \lambda_s^2/A$. At $\alpha \gg 1$ the CSR pulse duration $\tau_p = \tau_c$.In the case of short length medium, when $l < l_2$ in (2.36) one must substitute $l_2 \longrightarrow l$.

The nonstationary correlator of polarizations in the approximation adopted is given by the expression

$$S_{k_s} = \begin{cases} 0, \text{ for } \alpha < 1 \\ \dfrac{N}{8} (1 - \alpha^{-1}) \cosh^{-2} \dfrac{t - t_0}{\tau_p} \quad , \text{ for } \alpha > 1. \end{cases} \tag{2.37}$$

The CSR pulse formation, i.e. the growth of S_k, begins after the exceeding of the pump threshold :$n_L > n_c$, where the critical pump intensity is determined by the condition (2.1).Near the threshold, where $\tau_p \geq T_2$, the condition of validity of (2.34) and (2.35) holds.However at large exceeding, where $\tau_p \ll T_2$, the condition (2.16) holds only inside the cooperative region with dimensions of order of l_c, with the value of l_c being determined from the self consistency condition

$$l_c \equiv c\tau_p = cT_2/g_0 n_L l_c$$

whence

$$l_c = l_2 (n_c / n_L)^{1/2} \tag{2.38}$$

The solutions (2.36) and (2.37) remain valid in this case, if one makes there the substitution

$$\tau_p \longrightarrow T_2 (n_c/n_L)^{1/2} \tag{2.36a}$$

and in the expression for N_c substitute $l_2 \longrightarrow l_c$.

In the process of Stokes radiation molecules are excited from the ground to the upper state, the dynamics of this transition being described by the formula (2.36).After the end of the pulse the number density of excited molecules, as it follows from (2.36), is equal to

$$n_a = n < \sigma_{aa}(t = \infty) > = n\,[1- (n_c/n_L)]$$

Without taking into account of the pump depletion this formula yields unrealistically high value of number density n_a.To take into account of the depletion of the pump, we note, that the pump intensity remains approximately constant, if overall energy, inputted during the radiation time τ_p, exceeds the output Stokes pulse energy, i.e.

$$n_L \hbar\omega c\tau_p A > \hbar\omega_s A l_2 n\,[1- (n_c/n_L)]. \tag{2.39}$$

Near the threshold, where $n_L \geq n_c$, using for τ_p the formula from (2.36) and for g_0-formula (2.23), we obtain, that the pump depletion limitsfrom above the value of n_a:

$$n_a = n[1- (n_c/n_L)] < 1\,/\,(\pi|r|\omega_s T_2) \equiv n_a^{max} \tag{2.40}$$

For $T \sim 10^{-10}$ s, $r \sim 10^{-24}$ ESU, $\omega \sim 3\ 10^{15}\ s^{-1}$, we obtain

$$n_a < n_a^{max} \sim 10^{18}\ \text{cm}^{-3}$$

At high pump intensities $n_L \gg n_c$, using for τ_p the formula (2.36a), we have from (2.39)

$$n_a < n_a^{max} = (n_L n_c n)^{1/3}$$

For $n_L \sim 10^{15}\text{cm}^{-3}$, $n \sim 10^{22}\text{cm}^{-3}$, $\alpha \sim 10$, the maximum number density of cooperatively excited atoms $n_a^{max} \sim 10^{17}\text{cm}^{-3}$ and it weakly increases with increase of n_L.

To calculate the Stokes signal intensity we use the stationary solution for the number of Stokes quanta (2.32) in a **k**-mode. Substituting (2.37) into (2.24) and using the obtained expressions for I_k^s

in (2.32), we have for the number density of the Stokes quanta, radiated from the medium exit end into the solid angle $\Delta\Omega$

$$n_s(t) = \sum_{\mathbf{k} \in \Delta\Omega} n_{\mathbf{k}} = n_{s0} \left\{ \frac{\exp(g_L n l)}{g n_L l} \right\}$$

$$\times \left[1 + \delta \left(1 - \frac{n_c}{n_L} \right) \cosh^{-2} \frac{(t - t_0)}{\tau_p} \right], \qquad (2.41)$$

whereδ is the Stokes intensity jump parameter

$$\delta = n_a^{max} A\, l_2\, \Delta\Omega_g / \Delta\Omega \quad ,$$

and n_{s0} is the intensity of spontaneous radiation of the independent atoms into the solid angle $\Delta\Omega$:

$$n_{s0} = \frac{\Delta\Omega}{4\pi} \quad \frac{n\Gamma_s l}{c} \quad .$$

In the derivation of Eq.(2.41) we have substituted $n(1-n_c/n_L) \longrightarrow n_a^{max}$ and besides we have neglected the change of levels populations during the radiation of CSR pulse in the term, describing the spontaneous radiation of independent atoms, since $n_a^{max} \ll n$.

At $n_L \gg n_c$, when t_0 becomes smaller then τ_{pump}, the additive CSR pulses can be formed so that the Stokes pulse is the envelope of the enhanced CSR pulses(see sec.5.2.7).

The expression (2.41) is valid under the condition $n_L \geq n_c$, $l_2 < l$. (if $l_2 > l$, then one must substitute in (2.41) $l_2 \longrightarrow l$).

The first term in the square brackets of (2.41) describes the contribution of the spontaneous radiation of independent atoms, the second term corresponds to CSR, which is formed as a pulse with the maximum value, reached at the time moment t_0($t_0$is the delay time of the Stokes CSR pulse). The factor in parenthesis describes the exponential gain of the individual spontaneous radiation as well as the CSR in the process of stationary SRS.

In the case of short medium, when $l<l_2$ the onset of the instability of the stationary CSR occurs before the stimulated gain becomes important, since when the critical condition (2.1) is met, still $g_0 n_L l < 1$. Thus at the threshold the stimulated gain is unimportant and the intensity jump occurs from the spontaneous noise level. The formula (2.41) in this case describes directly the Stokes CSR pulse. For a long medium ($l>l_2$) one must take into account the stimulated gain.We remind, that (2.41) has been obtained for the stationary gain case.The condition of the stationary gain of the pulsed CSR is the inequality $\tau_p > T_2$. Using for τ_p the formula from (2.36), we obtain, that this condition is fulfilled for $n_L/n_c < 2$ .Thus the region of intensity jump is described by the stationary expression (2.41).At high exceeding the threshold one must consider the process of nonstationary gain.

The parameter δ shows how strong the maximum value of the CSR signal at $t=t_0$ exceeds the level of spontaneous radiation of individual atoms and so determines the magnitude of the SRS intensity jump at $n_L=n_c$ (for $n_L<n_c$ one must, as it seen from (2.37), put in (2.41) $\delta=0$).If the output radiation is gathered in a diffraction solid angle $\Delta\Omega=\Delta\Omega_g$, then from (2.41) one has

$$\delta = N_{max} = n_a^{max} A\, l_2$$

i.e. the Stokes intensity jump magnitude is determined by the maximum number of molecules in cooperative volume Al_2, making transition to the excited state during the CSR duration τ_p.

5.2.5. *CSR line width in the regime of nonstationary polariton generation.* Let us consider now the question about the Stokes radiation line width in the regime with $n_L>n_c$. Simultaneously with the central component $k=k_s$ other polariton waves are generated.To describe generation of the polariton spectrum, we go back to the equation (2.27), using for Re b the expression (2.21), where we substitute

$$l_2 \longrightarrow \tilde{l}_2 = \begin{cases} l_2 & \text{for} \quad \alpha \geq 1, \\ l_2/\alpha^{1/2} & \text{for} \quad \alpha \gg 1. \end{cases} \tag{2.42}$$

under the condition, that l_2, $\tilde{l}_2 < l$ and expand Re$b_{\mathbf{k}}$ into a power series of $(\tilde{l}_2 \Delta k)^2$, where $\Delta k = k - k_s$.Then, neglecting the spontaneous right hand side, we write down the solution of (2.27) for the time moments $t \leq t_0$, when one may put $D = -1$:

$$S_{\mathbf{k}}(t) = S_{\mathbf{k}_s}(t) \exp\left[-(2g_0 n_L \tilde{l}_2 / T_2) \Delta k^2 \, \tilde{l}_2^2 \, t \right],$$

where $S_{\mathbf{k}_s}(t)$ is given by (2.37).

As a line width one can take that value of Δk, at which the magnitude of $S_{\mathbf{k}}(t)$ is e times smaller then the central component intensity value $S_{\mathbf{k}_s}$.We have

$$\Delta k = \tilde{l}_2^{-1} \left(\frac{t \, 2 \, g_0 n_L l_2}{T_2} \right)^{-1/2} .$$

Thus, owing to the difference of the growth rates of polariton Fourier-amplitudes with different **k**, the line narrowing occurs during the generation process.

The maximum intensity is reached at the time moment t_0, thus for the effective CSR line width one can take the quantaty $\Delta k_{eff} = \Delta \mathbf{k}(\, t = t_0)$:

$$\Delta k_{eff} = \frac{\tilde{l}_2^{-1}}{(\, 2g_0 n_L \tilde{l}_2 \, t_0 \, / \, T_2 \,)^{1/2}}$$

$$= \frac{\tilde{l}_2^{-1} (1 - n_c/n_L)^{1/2}}{[\ln N_c (\, 1 - n_c \, / \, n_L)\,]^{1/2}} .$$

Near the threshold, where $n_L \longrightarrow n_c$(and $n_L > n_c$)the CSR line critically narrows, i.e. $\Delta k_{eff} \longrightarrow 0$.Well above threshold ($n_L \gg n_c$)the CSR line and thus SRS line may be several times narrower then the spontaneous line width.

Along with the line narrowing in the process of avalanche-like growth of the intensity of active polaritons $S_{\mathbf{k}}(t)$, the radius of correlation of polarizations of different atoms increases:

$$l_{cor} = \frac{1}{\Delta k_{eff}} \leq l_{cor}^{max} \sim l_2 \frac{[\ \ln \tilde{N}_c\ (1 - n_c/n_L\ )\ ]^{1/2}}{(\ 1 - n_c/n_L\ )^{1/2}} \ . \tag{2.43}$$

The maximum correlation length is reached at time moment $t=t_0$.Near the threshold the correlation length critically increases: $l_{cor} \longrightarrow \infty$.

Well above the threshold ($n_L \gg n_c$) the correlation length $l_{cor} \tilde{=} l_2(\ln \tilde{N}_c)^{1/2} \geq l_2$. If $l_2 < l$, then the whole length l is broken into the separate"domains" with length l_{cor}, the polariton generation in which occurs independently, however due to the pump depletion the first domain at the volume entrance must play the dominant role.

The above consideration shows, that the transition to the regime of nonstationary polariton generation, i.e. into regime of the CSR pulses generation bears the character of the nonstationary phase transition with such characteristic critical features as order parameter critical slowing down($t_0 \longrightarrow \infty$ for $n_L \longrightarrow n_c$, i.e.the CSR pulse maximum goes to infinity), the correlation radius critical increasing and generation line narrowing.

5.2.6. *Stationary polariton generation regime (regime of coherent CSR).* Let us consider now the system on times $t > T_1$, when the stationary polariton regime sets in.Note, that this condition may be realized in the inverted system of two-level atoms with stationary sustained inversion in the case, when Γ_L is not too large, so that the condition (2.1) is fulfilled.

In the regime of stationary polariton generation on axial mode

$\mathbf{k}_s = \mathbf{z}\, k_s$($\mathbf{z}$ is the unit vector along z-axis) we have

$$\rho_{-}^{i} = \sigma \exp\,(-i\omega_{ab}t - i\mathbf{k}_{s}\mathbf{r}_{i}). \tag{2.44}$$

Now the initial operator of nondiagonal element of density matrix (see (2.15)) is

$$\sigma_{i-} = \sigma \exp\,(-i\omega_{ab}t + i\mathbf{q}\,\mathbf{r}_{i})$$

$$\mathbf{q} = z\,(\,k_{L} - k_{s}). \tag{2.45}$$

Using (2.44), from the definition of $S_{\mathbf{k}}$ - (2.20) we have

$$S_{\mathbf{k}} = |\sigma|^{2}\, N\, \delta_{\mathbf{k},\mathbf{k}_{s}} \tag{2.46}$$

The regime of stationary polariton generation is described by equations (2.27) and (2.28), where now $b_{\mathbf{k}} = b_{\mathbf{k}_{s}}$ and one can neglect the spontaneous terms in the right hand side of (2.27). Using (2.46), we have

$$\frac{\partial|\sigma|^{2}}{\partial t} + \frac{2}{T_{2}}(\,1 + \alpha\, D\,)|\sigma|^{2} = 0, \tag{2.47}$$

$$\frac{\partial D}{\partial t} + \frac{D}{T_{1ef}} - \frac{D_{0}}{T_{1}} = \frac{8\,\alpha}{T_{2}}|\sigma|^{2} \tag{2.48}$$

where

$$\alpha = n_{L} / n_{c}$$

Near the threshold of generation the order parameter critically slows down, because of that one can adiabatically exclude the variable D. From (2.48) we have

$$D = D_0 \frac{T_{1eff}}{T_1} + \frac{T_{1eff}}{T_1} 8 \alpha |\sigma|^2. \qquad (2.49)$$

Substituting (2.49) into (2.47), we obtain the Van-Der-Pol equation for order parameter- the amplitude of the polarization wave (analogue of laser equation for the field amplitude E [3]):

$$\frac{\partial |\sigma|}{\partial t} - \beta (d - |\sigma|^2) |\sigma| = 0 , \qquad (2.50)$$

where

$$\beta = 8 \frac{T_{1eff}}{T_1} \alpha^2 , \qquad d = \frac{|D_0| - T_1 T_{1eff}^{-1} \alpha^{-1}}{8 T_1 T_2 \alpha^{-1}} .$$

From (2.50) we obtain, that in the stationary case the order parameter is given by the formula

$$|\sigma| = \left(\frac{|D_0| - T_1 T_{1eff}^{-1} \alpha^{-1}}{8 T_1 T_2^{-1} \alpha} \right)^{1/2} . \qquad (2.51)$$

The magnitude of T_{1eff} in (2.28) is decreased in the process of pola - riton instability development.In its beginning $E_s \ll E_L$ and $T_{1eff} = T_{1,}$ thus the formula for the critical intensity n_c of the onset of instability, determined from (2.51) , coincides with the expression for n_c, following from(2.37), i.e.(2.1).At $T_1/ T_{1eff} > \alpha$ the stationary regime is impossible.

5.2.7. *Slow switching on the pump. Critical behavior of SRS pulse duration.* Let us define now in eqs.(2.34) and (2.35) the pump intensity in the shape of triangle pulse

$$n_L(t) = \begin{cases} n_L t/\tau_{pump} & \text{for } t \le \tau_{pump}, \\ n_L(2 - t/\tau_{pump}) & \text{for } t \ge \tau_{pump}, \end{cases} \tag{2.52}$$

with the condition $n_L^{-1}(\partial n_L/\partial t) = \tau_{pump}^{-1} \ll T_2^{-1}$.

The CSR pulses generation begins at the time moment

$$t_1 = \tau_{pump}/\alpha, \tag{2.53}$$

which can be found from the condition of polariton decay constant cancelling in Eq.(2.35) with allowence for Eq.(2.52). Similarly we find the time moment of the end of polariton pulses generation

$$t_2 = 2\tau_{pump} - (\tau_{pump}/\alpha). \tag{2.54}$$

From (2.52) and (2.53) we find that the polariton pulses generation occurs during the time interval

$$\tau_{csr} = t_2 - t_1 = 2\tau_{pump}(1 - \alpha^{-1}) \tag{2.55}$$

In the region $n_L < n_c$ the SRS intensity is determined by the gained individual spontaneous radiation and SRS pulse duration $\tau_{srs} = \tau_{pump}$. In the region $n_L > n_c$ the SRS intensity is determined by the gained CSR pulses and τ_{srs} is determined by the polariton generation duration. Assuming $\tau_{srs} = \tau_{csr}/2$, from (2.56), we, thus, obtain the SRS duration

$$\tau_{srs} = \begin{cases} \tau_{pump}, & \alpha < 1, \\ \tau_{pump}[1 - \alpha^{-1}] \text{ for } & \alpha \geq 1. \end{cases} \qquad (2.56)$$

From Fig.5.4 it is seen, that the dependency of τ_{srs} on the pump intensity according to formula (2.56) qualitatively describes the experimental data of Ref.[24].

5.2.8. *Comparison of theoretical and experimental results. Conclusions.* In Fig.5.2 the theoretical dependence of I_s on I_L for compressed hydrogen gas is given, evaluated with formula (2.4) under the condition $g_0 n_L l \ll 1$, in comparison with the experimental results of Ref. 19 (pressure 10 atm, $l<l_2$=1.5cm, the focused beam). The curve is given by the formula $I_s=C\ \alpha\ [1+\delta(1-\alpha^{-1})]$, where $C=5\times10^{-6}$, $\delta=2.4\times10^{10}$ for $\alpha>1$, $\alpha=I_L/I_c$; the parameters C and δ were chosen, using two point of the experimental curve. Estimating now δ with the help of (2.41), substituting with account of (2.40) $n_a^{max}\sim10^{17}\text{cm}^{-3}$, $\Delta\Omega\sim10^{-2}$, $T_2^{-1}\sim 2\ 10^{10}\text{s}^{-1}$, we obtain the theoretical value $\delta\sim10^{10}$ in accordance with value, used in calculation.

As it follows from (2.41) in the case of short length medium ($l<l_2$) the jump occurs from the spontaneous radiation level, so that one can expect in this case a particular great magnitude of the jump. It is seen from fig.5.2 , that in hydrogen at $I_L=I_c$ the Stokes intensity increases about 10 orders of magnitude and that this theory can reproduce this giant SRS intensity jump.

In the case of long medium ($l>l_2$), as it follows from (2.41), the jump occurs from the level of gained spontaneous radiation and one can expect in this case the less magnitude of the intensity jump. In liquid nitrogen, for example (l=25 cm, parallel beam, l_2=2,3 cm, $l/l_2 \cong$ 10,8 [34])the jump occurs from the level of the gained spontaneous noise [35, 34] and the magnitude of the jump is about 10^4 times less then in hydrogen [35,34].

Let us now estimate I_c, rewriting (2.1) in the form $\beta_0 I_c l_2=1$.For liquid nitrogen [34] β_0=2,2 10^{-2}cm/MW (the pump wavelength λ_L=0,53 μm), l_2=2,3 cm and I_c^{theor}= 18 MW/cm^2(the experimental value I_c^{exp} = 16 MW/cm^2).In high pressure hydrogen gas $l_2>l\sim$ 1 cm, $\beta_0\sim$ 1,5 10^{-3}cm/MW [18].Correspondingly I_c^{theor}= 400 MW/cm^2(the experimental value I_c^{exp} ~ 250 MW/cm^2 [18]).

In the case, $l<l_2$, the critical intensity is determined from the condition $\beta_0 I_c l$=1, whence

$$I_c \approx \text{const}\ T_2^{-1} / n \approx T_2^{-1} / P ,$$

where P is the gas pressure.In accordance with the dependence of T_2^{-1} on P in hydrogen gas, the stationary SRS intensity jump threshold must decrease with the increase of P up to the values of several atmospheres, where the maximum value of T_2^{-1} is reached and then remains constant.This prediction qualitatively corresponds to the pressure dependence of the jump threshold, observed in experiments [26,36].

Qualitatively different dependency of jump threshold may be expected in the case of nonstationary SRS, when $\tau_{pump} \ll T_2$. In this case in the region $n_L < n_c$ the cooperative length is $c\tau_{pump}$ and the jump threshold is determined by the condition that delay time of SRS Stokes pulse in (2.36) becomes equal to τ_{pump}, i.e. $t_0 = \tau_{pump}$, whence with logarithmic accuracy one has

$$n_c^{nonst} \cong T_2 \ln N_c / 2\, g_0 c\, \tau_{pump}^2 \qquad (2.57)$$

Here N_c is the cooperative number of the molecules in volume $Ac\tau_{pump}$: N_c = $nA\tau_{pump}\ \Delta\Omega_g/4\pi$.From(2.57) one has

$$n_c^{nonst} = \text{const}/\ P$$

for τ_{pump}=const. The dependency P^{-1} of the jump threshold in fact is observed in experiments on nonstationary SRS in compressed hydrogen gas [26]. Note, that for the numerical estimates of the threshold intensity in the nonstationary case it is convenient to express the formula (2.58) through the stationary expression

$$n_{pump}^{nonst} \cong \left[\frac{T_2^2}{\tau^2} \ln N \right] n^{st} \tag{2.58}$$

At $T_2/\tau_{pulse} \sim 3$, $\ln N_c \sim 3$, we have from (2.58) $n_c^{nonst} \sim 3 \times 10^2 n_c^{st} \sim 10^2 \text{GW/cm}^2$, that correspond to the experimental values [26].

Let us now discuss the question about the fine spectral structure of the SRS signal. As it follows from the results of Secs.5.2.4 and 5.2.7, at $n_L \gg n_c$ the SRS pulse, from adopted here point of view, is the envelope of the gained CSR pulses each of which has the overall time duration of order of t_0 - (2.36) where τ_p is given by the Eq.(2.36a). With allowence for the expression for $n_c = (g\ l)^{-1}$(for $l < l_2$) ,the overall SRS pulse duration can be written in the form

$$t_0 = \frac{T_2}{2} \ln N_c\ /\ (g\ n_L\ l)^{1/2}$$

If one consider the CSR generation as a random process of generation of pulses with equal amplitudes and duration t_0 at random time moments, then the spectral function of this process is

$$S_\omega = \text{const}\ n_L \frac{1 - \cos(\omega - \omega_{ab})t_0}{(\omega - \omega_{ab})^2}. \tag{2.59}$$

The SRS spectrum then consists of separate equidistant lines, lying within the spontaneous width T_2^{-1} provided $t_0 > T_2$, the width of these lines being equal to the distance between them and determined by the formula

$$\Delta\omega = 2\pi / t_0 = \frac{4\pi(gn_L l)^{1/2}}{T_2 \ln N_c} . \tag{2.60}$$

The same spectrum must have the output SRS signal. If one takes into account of the gain coefficient dispersion, then the nonequidistancy and stochasticity of the line positions appears.

The spectral dependency of the type (2.59)) corresponds to the experimental dependency of Ref.21. The magnitude of the splitting practically doesn't depend on the pump intensity and for different substances is proportional to T_2^{-1} (for $(gn_L l)^{1/2} \cong$ const) .From the formula (2.60) we have the relation

$$\ln \Delta\omega = \ln T_2^{-1} - \ln \left[\frac{\ln N_c}{4\pi(gn_L l)^{1/2}} \right] = \ln T_2^{-1} - \text{const}$$

for $(gn_L l)^{1/2} \cong$ const (in SRS experiments usually $gn_L l \sim 20$).

The dependency of the experimentally observed splitting $\Delta\omega$ on T_2^{-1} for different media [21] is shown in Fig.5.5. It is seen, that there is in fact the linear connection between $\ln \Delta\omega$ and $\ln T_2^{-1}$.

Thus one can see, that the conception of SRS at high level of excitation, developed in this section, based on the notion of the sharp threshold transition from the regime of gain of spontaneous radiation of the independent atoms to the regime of the gain of the unstable CSR, i.e. to the generation of polaritons, can explain a whole number of experimental results seeming to be anomalous from the viewpoint of usual theory of exponential gain of spontaneous radiation of independent atoms.

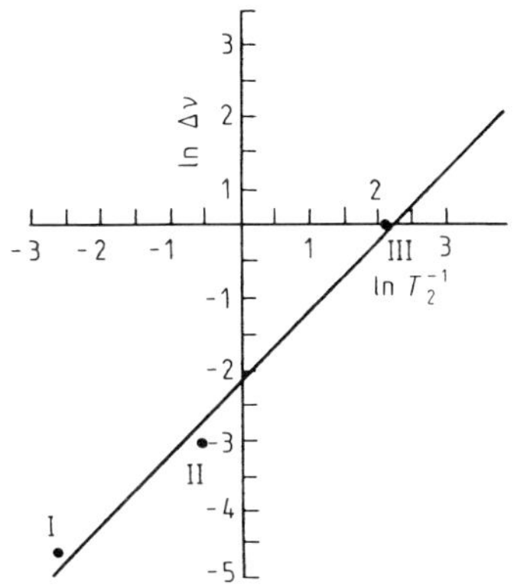

Fig.5.5 The dependence of the distance between the fine structure components Δν on T_2^{-1} (cm^{-1}) : I- the liquid NO_2, II - CS_2 ,III - $C_6H_5NO_2$. The experimental points are taken from Ref.21.

The account of the collective component in the source of spontaneous radiation is, apparently, the necessary element of the SRS theory at high level of excitation.

In the framework of the conception of the gain of CSR the notion of the stationary of SRS is changed.As one can see, even under the condition $\tau_{pump} > T_2$ for $n_L > n_c$ the process of CSR generation is nonstationary.The treatment of this section describes the process of gain of nonstationary CSR for $n_L \gg n_c$ only in case $l \ll l_2$.For $l \gg l_2$ this description is valid only in the vicinity of threshold, where $n_L \geq n_c$.The process of nonstationary gain of CSR can be considered on the basis of eq.(2.11).

In the above treatment we did not take into account the Stokes and antiStokes higher order components. The account of the first antiStokes component, for example, is important for the detailed description of the polarization avalanche on the Stokes frequency(see Ref.37).

We note in conclusion of this section, that the above considered nonequilibrium phase transition is associated with the softnening of the polariton decay constant ($T_k^{-1} \longrightarrow 0$, see (2.33)). In the case of resonance RS on optical phonons in semiconductors the analogous effect of the optical pump induced phonon decay constant softening was considered in earlear works [66-68]. In particular in Ref.68 it was shown that this effect leads to the critical increase of the RS scattered intensity and critical SR line narrowing.

5.2.9. *Active polariton dispersion relation and threshold appearance of cooperative shift of antistokes Raman scattering frequency.* In Sec.5.2.2 we have introduced the conception of active polariton, appearing in an active medium due to the exchange interatomic interactions through the Stokes (or antistokes) field and characterized it by its intensity S_k- (2.20) and lifetime T_k - (2.33). To conclude the characterization of an active polariton we shall find in this section the polariton dispersion law and study some consequences, following from this relation.

The dependence of the active polariton frequency $\omega_{q\alpha}$ ($\alpha = s$ for the Stokes collective spontaneous Raman scattering and $\alpha = a$ for the antistokes , see Fig.5.6) on the modulus of the wave vector $q = |\mathbf{q}|$, ($\mathbf{q} = q\,\mathbf{z}$ is the polariton wave vector in the cylindrical volume extended along z - direction) can be obtained from the expression for the spectral function

$$(\sigma_- \sigma_+)_{\omega,\,\mathbf{q}} = \int d\,\tau\, d\,\mathbf{r}_{ij} \exp\,(\,i\,\omega\,\tau - i\,\bar{\mathbf{q}}\,\bar{\mathbf{r}}_{ij}) < \sigma(t)_{i-} \sigma_{j+}(t') >, \tag{2.61}$$

where the pair correlator $<\sigma_{i-}\sigma_{j+}>$ depends on the distance between the two points $r_{ij} = |\,\mathbf{r}_{ij}| \equiv |\mathbf{r}_i - \mathbf{r}_j\,|$ and the time moments difference $\tau = t - t'$.

Let us obtain at first the equation for the pair correlator $<\sigma_{i-}\sigma_{j+}>_\tau$. For this aim we use Eq.(2.6) for $\sigma_{i-}(t)$, eliminating from it the field $E^+_s(t)$ with the help of (2.9), then multiply it on the right by $\sigma_{j+}(t')$ and perform the averaging of the equation thus obtained. Then we obtain the following equation for the pair correlators

$$\frac{\partial}{\partial t} <\sigma_{i-}(t)\ \sigma_{j+}(t')> + i\omega_{ab}<\sigma_{i\_}\sigma_{j+}> \ = \frac{1}{\hbar} \sum_{l \neq j} \frac{k_s^2}{i\,r_{lj}}\, a^0_{lj}$$

$$\times \exp\,\{\ i\ [k_L(z_j - z_l) - k_{so}\, r_{lj}) \ D <\sigma_{l-}(t - \tau_{lj})\sigma_{j+}(t')> \tag{2.62}$$

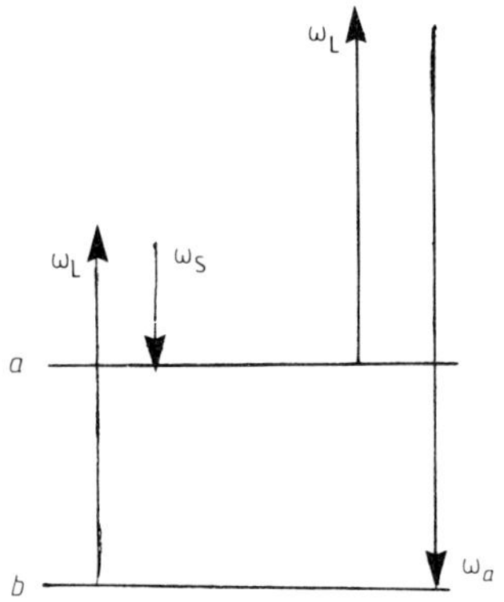

Fig.5.6. Stokes and antistokes RS.

We use then the temporal Laplace transformation

$$(\sigma_{i-}\sigma_{j+})^{+}_{\omega} = \int_{0}^{\infty} \exp\,(\, i\omega\tau\,) < \sigma_{i-}(t)\sigma_{j+}(t') > d\tau \qquad (2.63)$$

and spatial Fourier transformation of the both sides of Eq.(2.62), using Eq.(2.21). Then with the help of the formula , expressing the spectral function (2.61) through one sided spectral function

$$(\sigma_{i-}\sigma_{j+})_{\omega} = 2\,\mathrm{Re}(\sigma_{i-}\sigma_{j+})^{+}_{\omega}\,,$$

we obtain the polariton spectral function, from which follows the polariton dispersion relation in the form [38, 31, 39]:

$$\omega_{q\alpha} = \omega_{ab} + 2\, g_{\alpha} n_{L} D\, l_{eff} \;\; \frac{(q_0 - q)\, D}{1 + (q_0 - q)^2\, l_{eff}^2} \tag{2.64}$$

The formula (2.64) is derived for the case of the Stokes polaritons (α = s), $q_0 = \omega_{ab} / c$, $g_s n_L$ is the Stokes stimulated RS gain coefficient, D is as usual the mean difference of the populations of the upper (a) and the lower (b) working levels.The formula (2.64) describes also the case of antistokes polaritons, for which in (2.64) α = a and $g_a n_L$ is the antistokes stimulated RS gain coefficient.

The dependence of the polariton frequency on the polariton wave number according to formula (2.64) is shown in Fig.(5.7) .

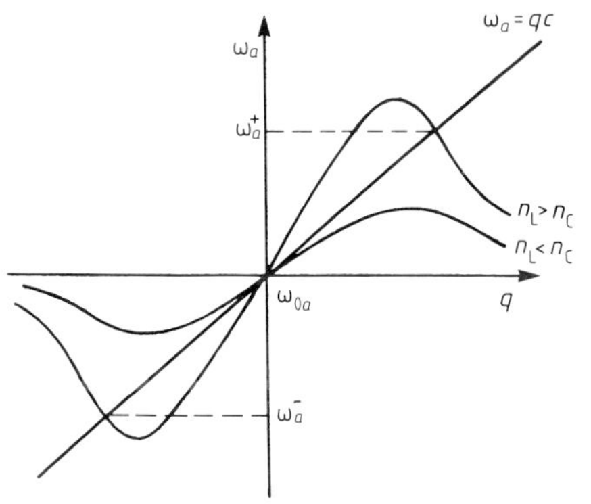

Fig.5.7. The polariton dispersion dependence and the threshold appearance of cooperative antistokes frequency shift: $\omega_a = \omega_{0a}$ for $n_L < n_c$ and $\omega_a = \omega_a^{\pm}$ for $n_L > n_c$.

To find the frequencies ω_α(α= s,a) of the light, scattered on polaritons, one must add to the polariton dispersion relation (2.64) the photon dispersion relation

$$\omega_\alpha = k_\alpha c , \tag{2.65}$$

the momentum conservation law

$$\mathbf{k}_\alpha = \mathbf{k}_L \mp \mathbf{q} , \tag{2.66a}$$

and energy conservation law

$$\omega_\alpha = \omega_L \mp \omega_q , \tag{2.66b}$$

where the sign (-) corresponds to the Stokes and (+) to the antistokes scattering.

From (2.64)-(2.66) one obtains the equation , determining ω_α

$$\Delta_\alpha = 2\, g_\alpha\, n_L\, l_{eff}\, D \,\frac{\Delta_\alpha\, \tau\, T_2^{-1}}{1 + \Delta_\alpha^2\, \tau^2} \tag{2.67}$$

where l_{eff} is given by (2.21), $\Delta_\alpha = \omega_\alpha - \omega_{0\alpha}$ is the cooperative frequency shift from the value $\omega_{0\alpha} = \omega_L \mp \omega_{ab}$, characterizing the individual scattering, $\tau = l_{eff}/ c$.

For $D < 0$ the Eq.(2.67) has the single solution $\Delta = 0$, i.e.the frequency of scattered light $\omega_\alpha = \omega_{0\alpha}$. However for $D > 0$, i.e. in the case of the *a,b*-transition inversion, the Eq.(2.67) has two additive solutions, corresponding to the appearance of cooperative frequency shift

$$\omega_a = \omega_{0\alpha} \pm \tau^{-1} \left(2\, g_\alpha\, n_L\, l_{eff}\, D \,\frac{l_{eff}}{c\, T_2} - 1 \right)^{1/2} \tag{2.68}$$

From (2.68) it is seen, that the appearance of the cooperative frequency shift in RS occurs as the phase transition of the second kind under the exceeding of the threshold pump intensity(see Fig.5.7):

$$n_c = \frac{cT_2}{2g_\alpha \, l_{eff}^2 \, D} \tag{2.69}$$

The necessary condition of this phase transition $D > 0$ can be fulfilled in antistokes CSR due to the inversion of the working transition a,b by the preceding powerful CSR Stokes pulse (see fig.5.6 and[27,28,37]). In the Stokes CSR the reaching of sufficiently high positive values of D is possible only on the fall of the Stokes pulse intensity, because of that one can't expect the appearance of the appreciate cooperative shift of the Stokes component of CSRS.

One can interpret the formula (2.68) as a threshold splitting of the upper level into two sublevels with the energies

$$\varepsilon^{\pm} = \varepsilon_a \pm |\Delta| \hbar ,$$

to which the two components of antistokes CSRS correspond with the frequencies

$$\omega_a^{\pm} = \omega_{a0} \pm \Delta .$$

Since the lower level ε^- is more strong populated, then the upper level ε^+, it should be that $g_a(\omega_a^-) > g_a(\omega_a^+)$ and due the the competition between the gains only one component with the frequency ω_a^- must be observed, which in fact corresponds to the experimental results.

If to consider the field E_a in its turn as a pump field one can draw the conclusion on the basis of the above consideration, that the frequency shift must be observed also for the second antistokes component. This shift, determined by the summed splitting of the levels, must exceed Δ in magnitude .

Experimentally the antistokes shift in RS of light was observed in Ref. [39]. In these experiments the dependency of the frequencies of the first and the second antistokes components on the energy of the exciting pulse in the SRS in compressed hydrogen gas was studied. The SRS was excited with the help of single mode ruby laser with pulse duration 30 ns and the pulse energy variation in the limits 0.05 - 0.25 J. The frequencies shifts were measured with the use of the spectrograms, photographed from Fabry-Perot interferometer.

In experiments the frequency shift of the antistokes components was observed, the magnitude of which depended in threshold manner on the exciting pulse energy. This dependency for the first antistokes component is shown in Fig. 5.8. The analogous dependency takes place also for the higher order antistokes near axis components. The frequency shift for the second antistokes component exceed that for the first antistokes component.The analysis of the spectrogram shows, that the wavelength of the antistokes components is red shifted with the pump energy increase. As one might expected, the delay of the antistokes pulse relative to the maximum of the pump pulse takes place.

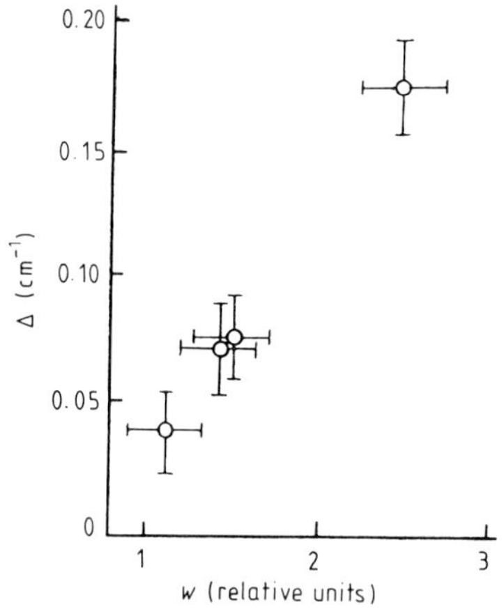

Fig. 5.8. The experimental dependence of the frequency shift Δ of the first antistokes component on the excitation pulse energy [39].

The experimental results obtained are in accordance with the polariton conception of the CSR, developed in this Chapter. The critical condition of the cooperative frequency shift appearance according to formula(2.69) is fulfilled under

$$g_a n_L l > g_a n_c l \sim cT_2 / l \sim \mathbf{1.5}$$

for $T_2 \sim 3{\cdot}10^{-10}$ s^{-1}, l = 3 cm. At $n_L > n_c$ (this condition was met in the experiment of Ref.39) one has from (2.69) for the cooperative frequency shift of the antistokes CSR component $\Delta \sim$ 0.1 cm^{-1}, which lies close to the experimentally measured value. The frequency shift sign also coincides with that predicted by the theory. The frequency shift increases with the exciting pulse energy increase(see Fig.5.8) , that qualitatively corresponds to the dependency, determined by the formula (2.68). Finally the observed increase of the frequency shifts for the higher order antistokes components is also in accordance with the theoretical model.

5.3. Dynamics of multimode superradiance in Raman scattering of light

In Sec.5.2 we have considered the CSR dynamics in the framework of single mode approximation (Sec.5.2.4), focusing the attention on the study of radiation intensity. In this section we shall study in the framework of multimode model the regime of well developed non-stationary CSR ($\tau_c \ll T_2$) - the superradiance in RS (SRRS) under the condition, that the medium length $l \ll l_2$. It gives the possibility to consider explicitly the dynamics of transition from the spontaneous to collective spontaneous emission, accompanied by the radiation angular structure change. Such spatial-temporal dynamics of spontaneous radiation is essentially determined by the geometry of the medium. In sec. 5.3.3 the influence of the pump depletion on the SRSR is investigated in more details, then it was done in Sec. 5.2.4 .

5.3.1. *Initial equations.Collective variables.* We start from the equations for the full population difference of the system

$$D \equiv R_3 = \sum_i \langle\sigma_{3i}\rangle ,$$

that follows from (2.17), and the equation for the single time correlator $S_{ij} = \langle\rho_{i-}\rho_{j+}\rangle$, $i \neq j$ - (2.26) . Neglecting of the induced terms in (2.17), we have from (2.17), (2.7) and (2.26)

$$\frac{\partial D}{\partial t} = \Gamma_s \sum_{i \neq j} \sum b_{ij} S_{ij} + \Gamma_s (N - D) , \qquad (3.1)$$

$$\frac{\partial S_{ij}}{\partial t} = - \frac{\Gamma_s}{2} \langle\sigma_{jbb}\rangle \langle\sigma_{i3}\rangle b_{ij} - \frac{\Gamma_s}{2} \langle\sigma_3\rangle \sum_{l \neq i,j} b_{il} S_{lj} . \qquad (3.2)$$

We assume, that the condition $c\tau_p > l$ holds.Then the population differences in all atoms may be considered to be equal to each other and one can put $<\sigma_{i3}> = D / N$.

To find the solution of the equations (3.1) and (3.2) we introduce the eigenfunctions and eigenvalues of the interatomic interaction matrix $\mathrm{Re}\, b_{ij}$, defined by (2.18), where now max $|r_{ij}| \ll l_2, c / \Gamma_L$:

$$\sum_j C_{pj} \psi_q (\mathbf{r}_j) = q \psi_q (\mathbf{r}_p) , \tag{3.3}$$

$$C_{pj} \equiv \sin k_s \mathbf{r}_{pj} / k_s \mathbf{r}_{pj} =$$

$$= (1 / 4\pi) \int \exp (i\mathbf{k}\mathbf{r}_{pj}) \, d\Omega, \ |\mathbf{k}| = k_s , \tag{3.4}$$

Here $d\,\Omega$ is the element of solid angle.

The eigenfunctions form the orthonormalized and complete set of functions :

$$\sum_j \psi_q (\mathbf{r}_j) \psi_{q'} (\mathbf{r}_j) = \delta_{qq'} , \quad \sum_q \psi_q (\mathbf{r}_j) \psi_q (\mathbf{r}_p) = \delta_{jp} . \tag{3.5}$$

From (3.3) -(3.5) it follows, that

$$C_{ij} = \sum_q q \psi_q (\mathbf{r}_i) \psi_q (\mathbf{r}_j). \tag{3.6}$$

We introduce the collective variable

$$S (q, t) = \sum \sum_{i \neq j} S_{ij} \psi_q (\mathbf{r}_i) \psi_q (\mathbf{r}_j) . \tag{3.7}$$

The variable $S\ (q,t)$ - the analog of (2.20) - is the intensity of the collective polariton excitation.Using (3.3) - (3.6), from the Eqs. (3.1) and (3.2) , we obtain the system of equations for the collective variables :

$$\frac{\partial D}{\partial t} = \Gamma_s \sum_q q\, S (q, t) + \Gamma_s (N - D) , \tag{3.8}$$

$$\frac{\partial S\ (q,\ t)}{\partial t} = -\ \Gamma_s (N - D)\ D\ /\ N^2 - (\Gamma_s D\ /\ N)\ (q - 1)\ S\ (q,t)$$

$$+\ (\Gamma_s D\ /\ N) \sum_{k \neq j} \sum C_{kj}\ \psi_q\ (\mathbf{r}_j)\ \psi_q\ (\mathbf{r}_k)\ S_{kj}(t)\ . \qquad (3.9)$$

Note, that due the definition (3.7) $\sum_q S\ (q,t) = 0$. Having summed both sides of (3.9) over q, one may check out, that both parts of Eq. (3.9) cancel. Note, that eqs. (3.8) and (3.9) are valid for the arbitrary geometry of the scattering medium, in contradistinction from previously used Eqs. (2.34) and (2.35), valid for the medium in the form of a long cylinder.

5.3.2. *Dynammics of population difference in SRSR. Dependence of SRSR pulse delay time on geometry of medium.* We shall analytically investigate the Eqs. (3.8) and (3.9) in the case of a cylindrical volume. The characteristic geometric parameters in this case are

$$H = k_s\ l, \quad h = k_s\ R \quad (H \gg h \gg 1)\ ,$$

where l is the length of cylinder, R and A are the radius and the cross section area.In the limit of large and small Fresnel numbers $F = A\ /\ \lambda_s\ l$ all the largest eigenvalues and corresponding eigenfunctions are known explicitly [40]. For $F \ll 1$ and $F \gg 1$ all the largest eigenvalues are ν-fold degenerated and equal to

$$q = q_0 = N\ \pi\ /\ 2\ H \gg 1, \quad \nu = 1\ /\ F \gg 1, \quad F \ll 1\ ; \qquad (3.10)$$

$$q = q_0 = N\ /\ h^2 \gg 1, \quad \nu = 2F \gg 1, \quad F \gg 1\ . \qquad (3.11)$$

All other $q \ll q_0$. The eigenfunctions, corresponding to the largest eigenvalue are [40]

$$\psi_{q0}(\mathbf{r}_j) = N^{-1/2}(\cos \mathbf{k}_s\mathbf{r}_j + \sin \mathbf{r}_s\mathbf{r}_j) , \quad \mathbf{k}_s = k_s \mathbf{z} , \tag{3.12}$$

where **z** is the unit vector along the cylinder axis.

Let us find the solution of the system (3.8), (3.9) for the cases (3.10), (3.11). At $q = q_0$ the last term in (3.9) is of order of $\nu / N \ll 1$ in comparison with the second term in the right hand side and thus may be neglected. It follows then from (3.9), that all functions $S(q_0,t)$ are equal to each other and in (3.8)

$$\sum_q q\, S(q, t) \cong \nu\, q_0\, S(q_0, t) .$$

With taking this into account one can write Eqs. (3.8) and (3.9) in the form

$$\frac{\partial D}{\partial t} = \Gamma_s q_0 \nu\, S(q_0,t) + \Gamma_s (N - D) \tag{3.13}$$

$$\frac{\partial S(q_0, t)}{\partial t} = - \Gamma_s (N - D)\, D q_0 / N^2$$

$$- (\Gamma_s D\, q_0 / N)\, S(q_0, t) . \tag{3.14}$$

With the substitution $D \rightarrow - R_3$, $\Gamma_s \rightarrow 4\, d_{ab}^2\, \omega_0^3 / 3\, c^3$ the Eqs. (3.13), (3.14) describe the process of CSR of the initially inverted system of two level atoms.

Excluding $S(q_0,t)$ from (3.13), (3.14), we obtain the closed equation for D :

$$\frac{\partial^2 D}{\partial t^2} + (\Gamma_s q_0 / 2N) \frac{\partial D^2}{\partial t} - \frac{\Gamma_s \partial (N - D)}{\partial t}$$

$$= q_0 (\Gamma_s^2 D (N - D) / N) (1 - q_0 \nu / N) \tag{3.15}$$

with the initial conditions

$$D(0) = - N, \qquad \frac{\partial D}{\partial t}\Big|_{t=0} = 2 \Gamma_s N .$$

If one neglects the terms in the right hand side of (3.15) (compare (2.35), (2.36)) then the solution of Eq. (3.15) is written in the form, analogous to (2.36)

$$D (t) = N [(1 + \eta^{-1}) \text{ th } [(t - t_0^*) / \tau] - \eta^{-1}] , \tag{3.16}$$

$$\tau_p = 2 / G_s q_0 , \quad t_0^* = (\tau_p / 2) \ln q_0 , \quad \eta \equiv q_0 , \tag{3.17}$$

where q_0 is given by the formula (3.11) for F » 1 and (3.10) for F « 1.From (3.13) and (3.14), neglecting the spontaneous terms, we obtain the conservation law

$$D^2 + 2 N \nu S (q_0, t) \cong N^2 , \tag{3.18}$$

whence, taking into account (3.16)

$$S (q_0, t) = (N / 2 \nu) \{1 - [(1 + \eta^{-1}) \text{ th } [(t - t_0^*) / \tau_p] - \eta^{-1}]^2\}$$

$$\cong (N / 2q_0) \cosh^{-2} [(t - t_0^*) / \tau_p\} . \tag{3.19}$$

At the time moment $t = t_0^*$ we have $D = 0$, $S(q_0, t) = \max$ and the system is in the superradiant state.

It seems impossible to obtain the analytical solution of full equation (3.15). However from its general form it follows, that the delay time t_0 in fact must be larger, then it is predicted by the formulas (3.16), (3.19) and (3.17), because the right hand side of (3.15) cancels, when the spontaneous term in (3.14) is increased in $N / q_0 \nu \gg 1$ times.

One can find the approximate expression for t_0 in the following way. Assuming that for $0 \leq t \leq t_0$ the population difference $D = -N = \text{const}$, we find from (3.14) and (3.18)

$$S(q_0, t) \cong 2(\exp(\Gamma_s q_0 t_0) - 1) \cong N / 2\nu .$$

whence

$$t_0 \cong (1 / \Gamma_s q_0) \ln N / \nu = \tau_p \ln (N / \nu) / 2 . \qquad (3.20)$$

One can represent the expression for t_0 with taking into account of (3.10), (3.11) in the form

$$t_0 = (\ln \lambda_0 + \ln H) / \Gamma_s q_0 \quad \text{for} \quad F \gg 1 ,$$

$$t_0 = (\ln \lambda_0 + \ln h^2 / \pi) / \Gamma_s q_0 \quad \text{for} \quad F \ll 1 , \qquad (3.21)$$

i.e $t_0 > t_0^*$.

In Fig. 5.9a,b the dependencies $S(q_0,t)$ and $D(t)$, obtained by the numerical solution of (3.13), (3.14), in comparison with the graphs of S and D, corresponding to the formulas (3.19) and (3.16) are shown. It is seen that the numerical solution of the system (3.13), (3.14) corroborates the estimate (3.20).

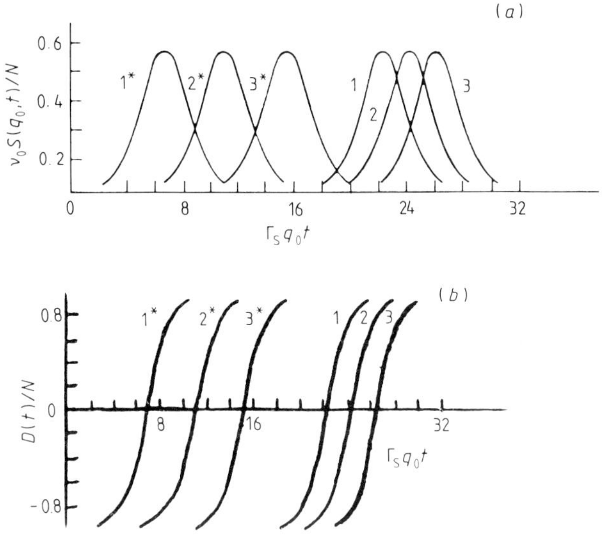

Fig.5.9. a) The SRRS pulse shape : the comparison of the delay times t_0 and t_0^* .The curves 1 - 3 are the numerical solutions of Eqs.(5.3.13) and (5.3.14), 1^* - 3^* are calculated with the help of Eq.(5.3.19) for l = 10 cm, A = 1 cm^2, $n = 10^{14}$ cm^{-3}. For the cases 1 and 1^*, $\lambda_s = 10^{-5}$ cm; for 2 and 2^*, $\lambda_s = 10^{-4}$ cm; for 3 and 3^*, $\lambda_s = 10^{-3}$ cm. b) The dynamics of the population difference in SRRS The values of parameters and designations are the same as in a).

5.3.3. *SRRS pulse shape and spatial-temporal dynamics of SRRS.* Let us pass over to the consideration of the SRRS pulse shape and the angular distribution of radiation. The energy of the scattered field at the observation point $\mathbf{r}_p$ is

$$W_s(\mathbf{r}_p) = |\mathbf{E}_s^+(\mathbf{r}_p)|^2 / (2\pi) . \qquad (3.22)$$

Substituting here the expression for $\mathbf{E}_s(\mathbf{r}_p)$ from (2.9), we have

$$W_s(\mathbf{r}_p) = \hbar\, k_s\, \Gamma_s <\sigma_{bb}> / (2\pi\, r_p^2) +$$

$$+ (\hbar k_s \Gamma_s / 4\pi r_p^2) \sum_{i \neq j} \sum S_{ij} \exp (i\mathbf{k r}_{ij}) . \tag{3.23}$$

Let us introduce the collective polariton variable

$$S (q, q', t) = \sum_{i \neq j} \sum S_{ij} \psi_q (\mathbf{r}_j) \psi_{q'} (\mathbf{r}_i) . \tag{3.24}$$

The inverse transformation yields

$$S_{ij} (t) = \sum_q \sum_{q'} S (q, q', t) \psi_{q'}(\mathbf{r}_i) \psi_q (\mathbf{r}_j) . \tag{3.25}$$

The equation for $S (q, q', t)$ is obtained from (3.2). It has the form, analogous to (3.14), where one must substitute $S(q,t) \longrightarrow S(q,q',t)$, and multiply the first terms in the right- hand- sides of (3.9) and (3.14) by $\delta_{qq'}$. Since $S (q, q', t) = 0$ at $t = 0$, then $S (q, q', t) = \delta_{qq'} S (q, q, t) = \delta_{qq'} S (q, t)$, where $S (q, t)$ is determined by the formula (3.17). Using with account of this the formula (3.25), we obtain from (3.23) the expression for the energy, scattered per unit of time into the unit solid angle in the direction of a unit vector $\mathbf{k}_s$,

$$I_{s,\mathbf{k}} (t) = (\hbar\omega_s \Gamma_s / 4\pi) \{(N - D (t)) / 2 +$$

$$+ (1 / 2) \sum_q S (q, t) \sum_{i \neq j} \sum \psi_q (\mathbf{r}_i) \psi_q (\mathbf{r}_j)\} . \tag{3.26}$$

In the limiting cases F » 1 and F « 1 with the use of (3.22), we find the expression for the intensity of SRRS in explicit form [28] :

$$I_{s,\mathbf{k}} = \frac{\hbar\,\omega_s\,\Gamma_s}{4\pi}\,\frac{N - D}{2} + \frac{N\,\nu\,S\,(q_0,\,t)\,\gamma\,(\mathbf{k})}{2}\,, \qquad (3.27)$$

where $D\ (t)$, $S\ (q_0,\ t)$ are given by the formulas (3.16) and (3.19) with the substitution $t_0 \rightarrow t^*_0$, where t_0 is given by (3.20) and the SRRS angular energy distribution factor is [28]:

$$\gamma\,(\mathbf{k}) = \frac{1}{2N^2}\sum_{i \neq j}\sum\,[\exp\,(i\,(\mathbf{k}_s + \mathbf{k})\,(\mathbf{r}_i - \mathbf{r}_j)) +$$

$$+ \exp\,(i\,(\mathbf{k}_s - \mathbf{k})\,(\mathbf{r}_i - \mathbf{r}_j))]$$

$$= 2\sum_{(+)}^{(-)}\left[\frac{\sin\,[(1/2)\,H\,(1 \mp \cos\,\theta)]}{(1/2)\,H\,(1 \mp \cos\,\theta)}\right]\frac{J_1^2\,(h\,\sin\,\theta)}{h^2\,\sin^2\,\theta}\,.(3.28)$$

Here $\cos\,\theta = \mathbf{k}\,\mathbf{k}_s\,/\,|\mathbf{k}|^2$ (recall that $\mathbf{k}_s$ is directed along the cylinder axis: $|\mathbf{k}| = |\mathbf{k}_s|$), J_1 is the Bessel function of the first kind.

The first term in (3.27) describes the isotropic spontaneous RS, arising at the moment $t = 0$, the second term describes the collective spontaneous radiation, being formed as a pulse with the maximum value reached at the time moment $t = t_0$. As it follows from (3.27) and (3.28), the SRRS occurs into the small solid angles in the opposite directions along the cylinder axis. From (3.27) and (3.19) it is seen, that at $t = t_0$ the intensity $I_{s,k} \sim N^2$. Using for D and $S\ (q_0,\ t)$ the formulas (3.16), and (3.19) with the substitution $\eta \rightarrow N\,/\,\nu$, $t_0 \rightarrow t^*_0$, we obtain for this case

$$I_{s,\mathbf{k}} = \frac{(\hbar\,\omega_s N\,\Gamma_s\,/\,4\pi)\,(1 + \eta)}{\exp\,(2\,t\,/\,\tau_p) + \eta}$$

$$\times \left[1 + \gamma (\mathbf{k}) N \frac{\exp (2 t / \tau_p) - 1}{\exp (2 t / \tau_p) + \eta} \right] \tag{3.29}$$

The formula (3.29) clearly shows, how the SRRS intensity and its angular distribution change in time (we note, that with the substitution $\omega_s \rightarrow \omega_0$, $\Gamma_s \rightarrow (4 / 3)\, \omega_0^3\, d_{ab}^2 / \hbar c^3$, $D \rightarrow -R_3$ the formulas (3.28) and (3.29) describe the superradiant relaxation of the initially inverted two-level medium). At time moment $t = 0$

$$I_{s,\mathbf{k}} (0) = \hbar\, \omega_s\, \Gamma_s\, N / 4\pi ,$$

which corresponds to isotropic spontaneous decay. At $t = t_0 = (\tau_p / 2) \ln (N / \nu)$, using $\eta = N \nu^{-1} \gg 1$, we have

$$I_{s,\mathbf{k}} (t_0) = \frac{\hbar\, \omega_s \Gamma_s}{4\pi} \frac{1 + \eta}{2\eta} \left[1 + N (\eta - 1) \frac{\gamma(\mathbf{k})}{2\eta} \right]$$

$$\cong \frac{\hbar\omega_s \Gamma_s}{8\pi} N^2 \gamma (\mathbf{k}) . \tag{3.30}$$

Thus, at time moment t_0 the system radiates in two opposite directions along cylinder axis, for which $\gamma (\mathbf{k}) \cong 1$, with N^2 times higher rate , then the rate of spontaneous emission of a single atom.The numeric calculations, made with the help of Eq.(3.29) [30], show, that in the typical conditions the intensity, radiated in the SRRS along the cylinder axis, is approximately 10^6 times greater then that in the perpendicular to the axis direction. Let us write down the condition of SRRS observation

$$\frac{l}{c} \ll \frac{T_2}{g_0 n_L l} \ll T_2 . \tag{3.31}$$

The left inequality is necessary for the condition (2.16) holds, i.e. the condition of cooperative behavior of atomic polarization on the length l. The right hand side inequality guarantees the strong exceeding over the threshold of polariton generation, i.e. the fulfillment of the condition (2.11). The right inequality, as it follows from the discussion in Sec. 2.4, may be weakened till the condition $g_0 n_L l = 1$ with gradual disappearing of CSR.

In the regime of non-stationary SRS, occurring as well as SRRS on the time interval shorter then T_2, the amplitude of the Stokes wave gained on the distance l is proportional to the exponential gain factor $\exp(g n_L l^2 / T_2 c)^{1/2}$. Due to the left hand side inequality (3.31) in the SRRS regime we have $(g n_L l^2 / T_2 c)^{1/2} \ll 1$ and thus one can neglect the processes of stimulated gain.

In conclusion of this section we note, that the basic equations of this method of the theory of superradiance - (3.1) and (3.2) were derived initially in consideration of SRRS in Ref. [28]. There the main original results of this method - the closed Eqs. of the SRRS dynamics - (3.15) , the delay time in the form (3.20), the formula (3.27), describing the angular-temporal dynamics of SRRS intensity were obtained. Later this method was used and extended for the description of superradiance in two-level systems (see Chapter 4). Naturally, in similar approximations the results for the cases of SRRS and the two-level superradiance (with corresponding changes of the parameters in the equations) turns out to be analogous.

5.3.4. *SRRS process as a mechanism of collisionless excitation and relaxation of molecules.* As we have seen in Sec.5.3.2, when the SRRS threshold is strongly exceeded, i.e. $n_L \gg n_c$ the SRRS pulse is formed during the time interval $t_0 < T_{min}$, where T_{min} is the minimum molecular relaxation time. After the end of Stokes SRRS pulse the molecular system, initially resting in the ground state, finds itself in the excited state.In multilevel system, for example, in a molecule with the vibration-rotation spectrum, such a process of pairs of levels inversion can occur gradually, step by step, leading to the excitation of higher and higher levels (see Fig. 5.10). The first step - the transition (v = 0, J) $\longrightarrow$ (v = 1, J) occurs due to the resonance absorption of the pump quantum (v, J are the vibration and rotation quantum numbers). Then, due to the Stokes (ω_s) and antistokes (ω_s) SRRS the transitions to the levels (v = 1, J + 2) and (v = 1, J - 2) occur. The bended arrows in Fig.5.10 show the refilling of the level v = 0, J due to the SRRS. Thus, the SRRS can be one of the possible mechanisms of the collisionless excitation (due to Stokes SRRS) and relaxation (anti stokes SRRS) of molecules over

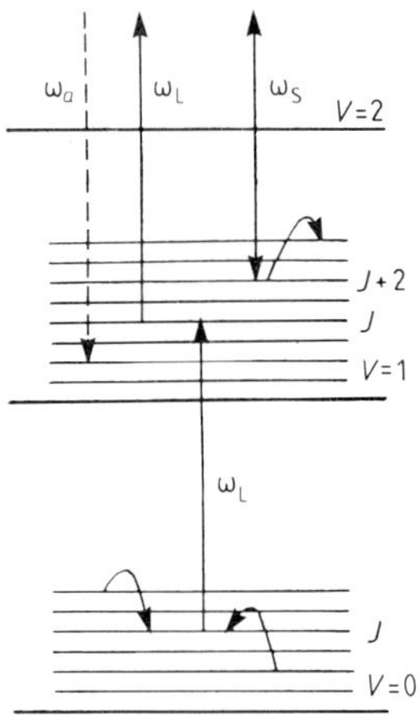

Fig.5.10. Scheme of vibration-rotation cascade transitions in SRRS.

the vibration-rotation levels [41].

Using the above obtained results , let us consider the mechanism of molecular excitation, involving a large number of rotational levels Let the pump frequency ω be in the resonance with the frequency of transition (v = 0, J) → (v = 1, J), where v, J are the vibration and rotation quantum numbers. Under the pump action the molecules are excited from (v = 0, J) level to (v = 1, J) level, then, due to the SRRS to the levels v = 1, J + 2, J + 4. The level (v = 0, J) is constantly refilled due to the transitions from other rotation sublevels that occur in Stokes and antistokes SRRS.

Let us estimate the pump intensity needed for the transition of molecule from the level (v = 1, J) to (v = 1, $J + \Delta J$) to occur during the pump pulse duration τ_{pump} . The necessary condition of this transition is the inequality

$$\sum_{k=1}^{\Delta J/2} t_0 (J_k) < \tau_{pump} , \tag{3.32}$$

where $t_0 (J_k)$ is the delay time of the SRRS pulse on the transition (v = 1, J_k) → (v = 1, $J_k + 2$). In numerical estimates we use the values of parameters, corresponding to SF_6 molecule. Take $\tau_{pump} = 150$ ns, $\Delta J = 20$. At the pressure $P = 0.03$ torr the concentration of molecules with $J \cong 63$ is $n_J \cong 10^{13}\ cm^{-3}$. The detuning $\omega_{ab} - \omega$ in the formula for Raman matrix $|r|$ (see (2.5a)) for the transition from the level $J + \delta J$ will be of order of $2B\delta J$, where $B \cong 0.027\ cm^{-1}$, $d_{b\alpha} = 0.3\ D$. Using (3.20), (3.32), we obtain the expression for the pump intensity needed for the cascade of transitions (v = 1, J) $\longrightarrow$ (v = 2, J + 20) to occur

$$|E_L|^2 = \frac{1}{\tau_{pump}} \frac{5\hbar c \ln (N v^{-1})}{2\pi \omega_s n_J\, l(d_{b\alpha} d_{\alpha a} / 4B\hbar)^2} \sum_{\delta J=1}^{10} (\delta J)^2 .$$

For $l = 10$ cm, A = 1 cm^2, $\omega_s \cong 2 \times 10^{14}\ s^{-1}$ we have $I_L \cong 10^{11}\ W/cm^2$. Note, that at higher levels of excitation of molecules, where the levels density strongly increases, the transitions due to SRRS will occur at still lower intensity.

5.3.5. *Influence of pump depletion on superradiance in Raman scattering of light.* In Sec. 5.2 we have briefly discussed the influence of the pump depletion on the SRRS process. Here we consider this question in more details [42]. We assume as before that initially all atoms are in the ground state with the mean polarization equal to zero. The pump field is given by the expression (2.2). The Stokes field we now take in the form of plane wave

$$\mathbf{E}_s(z, t) = \mathbf{e}_s\, [E_s^-(t) \exp(-i(\omega_s t - k_s z)) + c.c.] .$$

We neglect the dependency of the amplitudes $E_{s,L}(t)$ on z, assuming, that the radiation on the Stokes frequency occurs into single axial mode and $\tau_p c \gg l$.

We introduce the collective variables

$$\rho_-(t) = \sum_j \sigma_{j-}(t) \exp(i(\omega_{ab} t - (k_L - k_s) z_j)) ,$$

$$D = \sum_j \sigma_{j3} = \sum_j (\sigma_{jaa} - \sigma_{jbb}) , \tag{3.33}$$

where $\sigma_{j-} \exp \{[i\ [\omega_{ab}t - (k_L - k_s)\ z_j]\}$ is the slowly varying part of the non-diagonal element of density matrix of j-th atom, σ_{jaa} and σ_{jbb} are the populations of a and b -levels.

Using (2.6) and (2.7), we obtain the equations for the collective variables ρ_- and D :

$$\frac{\partial \rho_-}{\partial t} + \frac{1}{T_2} \rho_- = (ir^* / \hbar)\ E_L^- E_s^+ D , \tag{3.34}$$

$$\frac{\partial D}{\partial t} = - \frac{2\ i\ r^*}{\hbar} (E_L^- E_s^+ \rho_+ - E_L^+ E_s^- \rho_-) . \tag{3.35}$$

Here $r = (\mathbf{r}\ \mathbf{e}_s)$, $\mathbf{r}$ is given in (2.5a) .

The field amplitude E_L^- can be represented as a sum of two parts:

$$E_L^- (t) = E_{Lo}^- + \varepsilon_L^- (t) , \tag{3.36}$$

where E_{Lo}^-, ε_L^- are the solutions of the homogeneous and inhomogeneous Maxwell equations, E_{Lo}^- = const corresponds to the external pump field.

The reduced Maxwell equation for ε_L^- has the form

$$\frac{\partial \varepsilon_L^-}{\partial t} + \frac{1}{\tau} \varepsilon_L^- = - i\ 2\pi\ \omega\ V^{-1}\ r\ E_s^- \rho_- . \tag{3.37}$$

Here $\tau^{-1} = c\ /\ 2\ l$ is the phenomenological decay constant of the field. Since there is no the external field on the frequency ω_s , then $E_s^- = \varepsilon_s^-$ and equation for E_s^- has the form

$$\frac{\partial E_s^-}{\partial t} + \frac{1}{\tau} E_s^- = - i\, 2\pi\, \omega_s\, V^{-1}\, r^*\, E_L^-\, \rho_+ \,. \tag{3.38}$$

We assume, that τ^{-1} is the maximum of the inverse times of the problem and the following conditions hold

$$\tau^{-1} \gg \tau_p^{-1} \,, \quad t_0 \ll T_2 \,. \tag{3.39}$$

If the right inequality in (3.39) holds, the Eqs. (3.34) and (3.35) have the integral of motion

$$4\, \rho_- \rho_+ + D^2 = N^2 + 4N \equiv R^2 \,. \tag{3.40}$$

Due to the left inequality in (3.39) one can neglect the time derivatives $\partial \varepsilon_L^- / \partial t$ and $E_s^- / \partial t$ in the Eqs. (3.37) and (3.38). Then from (3.36) - (3.38) we obtain

$$E_L^-(t) = E_{L0} \,/\, [1 + 4\, \kappa \rho_-(t)\, \rho_+(t)] \,,$$

$$E_s^-(t) = - 2\pi\, i\, \omega_s \tau\, V^{-1}\, r^*\, \rho_-\, E_L^-(t) \,, \tag{3.41}$$

where $\kappa = \pi^2\, \omega_s\, \omega\, |r|\, \tau \,/\, V^2$. Substituting (3.41) into (3.34), (3.35) and using (3.40), we obtain the closed equation for $D\,(t)$:

$$\frac{\partial D}{\partial t} = \frac{\Omega\, (R^2 - D^2)}{[1 + \kappa\, (R^2 - D^2)]^2} \,, \qquad D\,(0) = - N \,. \tag{3.42}$$

Here

$$\Omega = 2\pi\, \omega_s |\mathrm{r}|^2\, |E_{L0}|^2 \tau \,/\, \hbar\, V \,.$$

The solution of (3.42) has the form

$$\ln \frac{N\,(R+D)}{R-D} = 2\,R\,\Omega\,t - 4\,\kappa\,N\,(D+N)$$

$$-\,2\,\kappa^2 N^2\,(D+N) + \frac{2}{3}\,\kappa^2\,N\,(D^3+N^3)\,. \tag{3.43}$$

The solution for $D\,(t)$ can be also represented in the standard form which is usually used for the description of the superradiant dynamics of the difference of the levels populations

$$D\,(t) = N \text{ th } [(t - t_0)\,/\,\tau\,]\,, \tag{3.44}$$

$$\tau_p = 1\,/\,\Omega\,N\,, \tag{3.45}$$

$$t_0 = \frac{1}{2}\,\tau_p\,[\ln N + 4\,\kappa\,N\,(D+N) + 2\,\kappa^2\,N^3\,(D+N)$$

$$-\,\frac{2}{3}\,\kappa^2\,N\,(D^3+N^3)\,]\,. \tag{3.46}$$

In (3.44) - (3.46) we put $R = N$ with the accuracy up to the terms of the order of N^{-1}.

The expression for the SRRS power $I_s = (2\,/\,\tau)\,V\,|E_s|^2\,/\,2\pi$, using (3.40) and (3.41), can be written in the form

$$I_s\,(t) = \frac{\hbar\omega_s}{2}\,\frac{\Omega\,(R^2 - D^2)}{[1 + \kappa\,(R^2 - D^2)]^2} = \frac{\hbar\omega_s}{2}\,\frac{\partial D}{\partial t}\,. \tag{3.47}$$

From (3.44) and (3.47) it follows, that the Stokes radiation is formed as a pulse with the duration τ_p, the maximum of which is reached at time moment t_0, when $D(t_0) \cong 0$ and the radiation intensity $I_s(t_0) \sim N^2$.

If one neglects in (3.46) the terms of order of κ, then it transforms into the expression for the delay time (3.20) , where $\nu = 1$. Thus, the condition of neglecting of the pump depletion is

$$\kappa N^2 \ll 1 , \qquad (3.48)$$

i.e. the number density of atoms has the upper bound, independent of the pump intensity. When the inequality (3.48) breaks down with the increase of N, the delay time starts to grow proportionally to N^3. This rapidly breaks down the condition of SRRS observation:$t_0 < T_2$. Thus, the taking into account of the pump depletion reveals the extra condition of the observation of SRRS. From (3.41) and (3.48) we have for the case $l < c T_2$

$$n < c / 2\pi \, l \, \omega \, |\mathbf{r}| = n_{max} . \qquad (3.49)$$

thus, the above analysis corroborates the estimate made in sec.5.2.4, where the qualitative investigation of the influence of the pump depletion on the SRRS has been carried out (compare (2.40), where $T_2 \longrightarrow l/c$.

5.4. Resonatorless bistability of optical transmission in two-photon and one-photon absorption

The problem of bistable behavior of optical transmission of the system of N two-level atoms, interacting with the resonance field is actively studied now [8,9]. Usually one considers the atomic system, placed inside resonator and interacting with a single field mode.The single mode approximation suitable for the resonator systems does not permit us to reveal sufficiently fully the collective character of the effect of bistability and to establish its connection with non-equilibrium phase transitions.

The approach, developed in this section, analogous to that, used in Sec.5.3 for the description of superradiance, enables one to study the resonatorless bistability. Here we show, that bistability occurs due to the interatomic interactions through the dipole radiation field, which qualitatively modifies the character of the saturation of atomic transitions under the action of the external field in comparison with the usual saturation of independent two-level atoms.

The consideration, carried out here and also in Chap.6 enables one to draw the conclusion, that the bistable behavior is inherent to the sufficiently dense ensembles of two-level atoms, interacting with each other through the "exchange" forces of different nature:the radiation field, dipole-dipole electrostatic interactions and exchange of phonons. The resonator is not needed for these effects to occur, also its absence strongly severe the necessary condition on the number density of atoms in the case of interatomic interactions through the radiation field. Thus, the cooperative effect of bistability is considered here to be the non-equilibrium phase transition in ensemble of N two-level atoms, induced by the light wave.

The consideration of this section is carried out for the case of bistability in two-photon absorption, when the sum of the frequencies of two external fields $\omega + \omega_s = \omega_0$, where ω_0 is the transition frequency of two-level atom . The equations, describing the effect of two-photon bistability, are obtained and an example of its possible experimental realization is considered. The results obtained are applicable to the case of one-photon case as well.

5.4.1. *Taking into account of collective radiation field in two-level atom equations.* Let us consider the system of N multilevel atoms , placed inside the volume V of arbitrary geometric shape. The system is under the action of the external light field, consisting of two waves

$$\mathbf{E}\,(\mathbf{r}_j,\, t) = \mathbf{e}_L\, E_L^-\,(\mathbf{r}_j,\, t) + \mathbf{E}_s^-(\mathbf{r}_j,\, t) + \text{c.c}$$

$$= \mathbf{e}_L\, E_L^-\, \exp\,(-\, i\omega t + i\,\mathbf{k}_L\mathbf{r}_j) + \mathbf{E}_{s0}^-\,(\mathbf{r}_j)\, \exp\,(-\, i\,\omega_0 t) + \text{c.c} \quad (4.1)$$

Here $|\mathbf{e}_L| = 1$, $\omega + \omega_{s0} = \omega_0 - \Delta$, where Δ is the detuning from the two-photon resonance, $\mathbf{r}_j$ is radius-vector of j-th atom.As a result of dipole two-photon reradiation the full field in the point $\mathbf{r}_j$ on each of the frequencies ω and ω_s consists of the field (4.1) and radiation field, created by all atoms. We assume the field with the frequency ω to be sufficiently strong, so that one can neglect the radiation field with this frequency and put E_L = const. The full field with frequency ω is given by the expression, analogous to (2.9) :

$$E_s^-\,(\mathbf{r}_j,\, t) = \mathbf{E}_{s0}^-\,(\mathbf{r}_j)\, \exp\,(-\, i\,\omega_s t) +$$

$$+ \frac{1}{c^2} \sum_{i(i \neq j)} \frac{1}{r_{jj}} \frac{\partial^2}{\partial t^2} [[\mathbf{P}_{si}(t') \, \mathbf{n}_{ij}] \, \mathbf{n}_{ij}] \, . \tag{4.2}$$

Here $t' = t - r_{ij} / c \equiv t - \tau_{ij}$ and the polarization is

$$\mathbf{P}_{si}(t) = - \mathbf{r} \, E_{\mathrm{L}}^{+} \sigma_{i-}(t) \exp(i\omega t - i \, \mathbf{k}_{\mathrm{L}} \mathbf{r}_j) \, , \tag{4.3}$$

where

$$\mathbf{r} = \hbar^{-1} \sum_{\alpha} [(\mathbf{d}_{b\alpha} \mathbf{e}_{\mathrm{L}}) \, \mathbf{d}_{\alpha a} / (\omega_{ba} + \omega) + \mathbf{d}_{b\alpha} (\mathbf{d}_{\alpha a} \mathbf{e}_{\mathrm{L}}) / (\omega_{b\alpha} + \omega_{\mathrm{s}})] \, .$$

The equations for the atomic variables, describing the processes of two-photon gain and absorption, are written down in the form, analogous to the case of RS (Sec.5.2) :

$$\frac{\partial \sigma_{j+}}{\partial t} - i \left(\omega_0 + \frac{i}{T_2} \right) \sigma_{j+} = - \frac{i}{\hbar} \mathbf{r} \, \mathbf{E}_{\mathrm{s}}^{+}(\mathbf{r}_j, t) \, E_{\mathrm{L}}^{+}(\mathbf{r}_j, t) \, \sigma_{j3}, \tag{4.4}$$

$$\frac{\partial \sigma_{j3}}{\partial t} + \frac{\sigma_{j3} - \sigma_{j3}^{0}}{T_1} = \frac{2i}{\hbar} (\mathbf{r} \, \mathbf{E}_{\mathrm{s}}^{+}(\mathbf{r}_j, t) \, E_{\mathrm{L}}^{+}(\mathbf{r}_j, t) \, \sigma_{j-} - \mathrm{c.c}) \, , \tag{4.5}$$

where $\sigma_{j3} = \sigma_{jaa} - \sigma_{jbb}$ is the population difference in j-th atom.

Let us introduce the slowly varying density matrix

$$\rho_{j-} = \sigma_{j-} \exp[i(\omega + \omega_{\mathrm{s}}) t - i \, \mathbf{k}_{\mathrm{L}} \mathbf{r}_j] \, . \tag{4.6}$$

Substituting (4.3) and (4.6) into (4.2) and neglecting there the derivatives of the slow variable, we obtain the expression for the full field on the frequency ω_{s}. Using this expression in Eqs. (4.) and (4.5) we obtain the following system of equations:

$$\frac{\partial \rho_{j+}}{\partial t} - i \left(\Delta - \frac{i}{T_2} \right) \rho_{j+} = \frac{i}{\hbar} \mathbf{d}^{*} \mathbf{E}_{\mathrm{s}}^{+}(\mathbf{r}_j) \, \sigma_{j3}$$

$$+ \frac{\Gamma}{2} \sum_{i(i \neq j)} b_{ij} \, \rho_{i+}(t - \tau_{ij}) \, \sigma_{j3}(t) \, , \tag{4.7}$$

$$\frac{\partial \sigma_{j3}}{\partial t} + \frac{\sigma_{j3} - \sigma^0_{j3}}{T_1} = \frac{2i}{\hbar}(\mathbf{d}\,\mathbf{E}^-_s(\mathbf{r}_j)\,\rho_{j+} - \mathrm{c.c})$$

$$- \Gamma G(\sum_{i(i\neq j)} b_{ij}\,\rho_{i+}(t-\tau_{ij})\,\rho_{j-}(t) + \mathrm{c.c.}) \,, \qquad (4.8)$$

where $\mathbf{d} = -\,\mathbf{r}E^*_L$ is the effective dipole moment, Γ is the two-photon decay constant $\Gamma = 2\,k^3_s\,\kappa_0 / \hbar$,

$$\kappa_0 = \langle -\,\mathbf{d}\,[[\,\mathbf{d}^*\,\mathbf{n}_{ij}\,]\,\mathbf{n}_{ij}\,]\rangle \,,$$

$$b_{ij} = \frac{\exp(-\,i\,k_s\,r_{ij})}{-\,i\,k_s\,r_{ij}} \qquad (4.9)$$

is the matrix of the effective interatomic interaction, arising due to the radiation field. The brackets in the expression for κ_0 means the averaging over the orientations of the effective dipole moment **d**.

In stationary regime one can put in the equations (4.7) and (4.8) $t - \tau_{ij} \cong t$. From the Eq. (4.7) it is seen, that the imaginary part of the interaction matrix determines the cooperative transition frequency shift and the real part - the renormalization of the decay constant T_2^{-1}. In the following we shall neglect the cooperative frequency shift, retaining only the real part of b_{ij} in (4.9).

Let us introduce the full population difference D :

$$D = \sum_j \sigma_{j3}(t) \;; \quad \sigma_{j3} \cong D\,(t)\,/\,N \,.(4.10)$$

We assume further, that T_1^{-1}, $T_2^{-1} \gg \Gamma$, then in Eqs. (4.7) and (4.8) one can extend the summation also over $i = j$. Taking this into account, we rewrite the system (4.7) and (4.8) in the form

$$\frac{\partial \rho_{j+}}{\partial t} + i\,(\Delta - i\,T_2^{-1})\,\rho_{j+}$$

$$= \frac{iD}{\hbar N} \mathbf{d}^* \mathbf{E}_s^+ (\mathbf{r}_j) + \frac{\Gamma D}{2N} \sum_j C_{ij} \rho_{i+}(t) , \tag{4.11}$$

$$\frac{\partial D}{\partial T} + \frac{D - D_0}{T_1} = \frac{2i}{\hbar} \sum_j (\mathbf{d} \mathbf{E}_s^- (\mathbf{r}_j) \rho_{j+} - \text{c.c.})$$

$$- \Gamma \sum_i \sum_j C_{ij} (\rho_{i-}(t) \rho_{j+}(t) + \rho_{i+}(t) \rho_{j-}(t)) , \tag{4.12}$$

where

$$C_{ij} = \sin k_s r_{ij} / (k_s r_{ij}) . \tag{4.13}$$

The equations, analogous to (4.11) and (4.12), describes also the ensemble of atoms, interacting through the electrostatic field (see Sec. 6.2) .

5.4.2. *Collective atomic variables and transmition bistability.* Let us pass over in Eqs. (4.11) and (4.12) to the collective variables. For this we introduce the eigenfunctions $\psi_q(\mathbf{r}_j)$ and the eigenvalues q of the matrix of the interatomic interaction C_{ij}, analogously to Sec.5.3. We define the collective variables in the following way :

$$\rho_+(q) = \sum_j \rho_{j+} \psi_q(\mathbf{r}_j) , \quad \varepsilon^+(q) = \sum_j \mathbf{d}^* \mathbf{E}_s^+ (\mathbf{r}_j) \psi_q(\mathbf{r}_j) . \tag{4.14}$$

With taking into account of (3.5) we have from (4.14) the inverse relations

$$\rho_{j+} = \sum_q \rho_+(q) \psi_q(\mathbf{r}_j) , \quad \mathbf{d}^* \mathbf{E}_s^+ (\mathbf{r}_j) = \sum_q \mathbf{d}_s^* \varepsilon^+(q) \psi_q(\mathbf{r}_j) . \tag{4.15}$$

Using (3.3), (3.5), (3.6) (4.14) and (4.15), from the Eqs. (4.11) and (4.12) we obtain the following system

$$\frac{\partial \rho_+(q)}{\partial t} + i (\Delta - i T_2^{-1}) \rho_+(q)$$

$$= \frac{iD}{\hbar N} \varepsilon^+(q) + \frac{\Gamma D}{2N} q \rho_+(q) , \tag{4.16}$$

$$\frac{\partial D}{\partial t} + \frac{D - D_0}{T_1} = \frac{2i}{\hbar} \sum_q (\varepsilon^-(q)\,\rho_+(q) - \varepsilon^+(q)\,\rho_-(q))$$

$$- 2\Gamma \sum_q q\rho_-(q)\rho_+(q). \qquad (4.17)$$

In quasistationary case from (4.16) we have

$$\rho_+(q) = \frac{DT_2}{\hbar N} \frac{\varepsilon^+(q)}{\Delta T_2 - i\,(1 - \Gamma q T_2 D / 2N)}. \qquad (4.18)$$

Substituting (4.18) into (4.17), we obtain the equation for the full population difference :

$$\frac{\partial D}{\partial t} + \frac{D - D_0}{T_1} =$$

$$= - \frac{4T_2 |d|^2}{\hbar^2 N} \sum_q \frac{|\varepsilon(q)|^2}{\Delta^2 T_2^2 + (1 - \Gamma q T_2 D / 2N^2)}. \qquad (4.19)$$

If one neglects in (4.19) the radiation field, putting $q = 0$, and takes into account, that $\sum_q |\varepsilon^+(q)|^2 = \sum_j |\mathbf{d}^* E_s^+(\mathbf{r}_j)|^2 = N|E_{s0}^+|^2|\mathbf{d}|^2$, where

$$\mathbf{E}_s^+(\mathbf{r}_j) = \mathbf{E}_{s0}^+ \exp(-i\,\mathbf{k}_s\mathbf{r}_j), \qquad (4.20)$$

then the Eq. (4.19) describes the effect of usual two-photon saturation of two-level atoms in the fixed field [32]. Taking into account of the term, proportional to q in (4.19), leads to the qualitative difference in the system behavior in the external field and causes the bistability effect.

Let us consider the particular case of the cylindrical volume. The characteristic geometrical parameters are $H = k_s l$, $h = k_s R$ ($H \gg h \gg 1$), where l is the cylinder length, R, A are the radius and the cross section area. All the largest eigenvalues of the interaction matrix C_{ij} are

$$q_0 = N\pi / H \gg 1, \quad F \ll 1; \quad q_0 = N / h^2, \quad F \gg 1,$$

where $F = A / \lambda_s L$ is the Fresnel number and the eigenfunctions, corresponding to the eigenvalues are approximately equal to

$$\psi_q(\mathbf{r}_j) = \left(\frac{2}{N} \right)^{1/2} \cos \mathbf{k}_s \mathbf{r}_j \, ,$$

$$\psi_q(\mathbf{r}_j) = \left(\frac{2}{N} \right)^{1/2} \sin \mathbf{k}_s \mathbf{r}_j \, , \quad (4.21)$$

where $\mathbf{k}_s = \mathbf{z}\, k_s$, $\mathbf{z}$ is the unit vector along the cylinder axis.

If the external field with frequency ω_s is given by (4.20), where $\mathbf{k}_s = \mathbf{k}_0$, then from (3.2) and (4.20) we obtain

$$\varepsilon^+(q) = \mathbf{d}^* \mathbf{E}^+_{s0} \left(\frac{2}{N} \right)^{1/2} (1 - i)\, \delta_{qq_0} \, . \qquad (4.22)$$

this means, that only the single mode (4.21) takes place in the process of two-photon absorption.

Using (4.22), from (4.19) we obtain the equation for the population difference in single mode case

$$\frac{\partial D}{\partial t} + \frac{D - D_0}{T_1} =$$

$$= - \frac{(4\, T_2\, |\mathbf{d}^*\, \mathbf{E}^+_{s0}|^2 / \hbar^2)\, D}{\Delta^2 T_2^2 + (1 - \Gamma\, q_0 T_2 D / 2N)^2} \, . \qquad (4.23)$$

Let us obtain now the equation for the output field with the frequency ω_s in dependence on the field amplitude of the input wave with the same frequency. From the right hand side of the Eq. (4.11) it is seen, that the field of dipole radiation ε^+_{sj} with the frequency ω_s in the point $\mathbf{r}_j$ can be written in the form

$$\mathbf{d}^* \overrightarrow{\varepsilon}_{sj}^{\,+} = \frac{i \hbar \Gamma}{2\, d^*} \sum_i C_{ij}\, \rho_{i+}(t) \; . \qquad (4.24)$$

Using (3.3), (3.5), (3.6), (4.14), (4.15) and (4.18),we express the field of the dipole radiation through the collective variables

$$\mathbf{d}^* \overrightarrow{\varepsilon}_{sj}^{\,+} = -\frac{i \hbar \Gamma}{2} \sum_q q\, \rho_+ (q)\, \psi_q (\mathbf{r}_j) =$$

$$= -\frac{i \Gamma D T_2}{2N} \sum_q \frac{q\, \varepsilon^+(q)\, \psi_q (\mathbf{r}_j)}{\Delta T_2 - i\,(1 - \Gamma q T_2 D / 2N)} \; . \qquad (4.25)$$

In a single mode case the expression (4.25) is considerably simplified and assumes the form

$$\mathbf{d}^* \overrightarrow{\varepsilon}_{sj}^{\,+} = -\, i\, \frac{\Gamma D T_2}{2N}\, \frac{q_0\, \mathbf{d}^* \mathbf{E}_s^+ (\mathbf{r}_j)}{\Delta T_2 - i\,(1 - \Gamma q_0 T_2 D / 2N)} \; , \qquad (4.26)$$

whence

$$D = \frac{(i \Delta T_2 + 1)\, \mathbf{d}^* \overrightarrow{\varepsilon}_{s0}^{\,+}}{(\Gamma q_0 T_2 / 2N)\, (\mathbf{d}^* \overrightarrow{\varepsilon}_{s0}^{\,+} + \mathbf{d}^* \mathbf{E}_{s0}^+)} \; , \qquad (4.27)$$

where ε_{s0}^+ is the amplitude of the dipole radiation field.In the following we shall consider for simplicity the resonance case $\Delta=0$.We introduce the abbreviations for the normalized amplitudes of the input and output fields :

$$y = \left[\frac{4 T_1 T_2}{\hbar^2} \right]^{1/2} \mathbf{d}^* \mathbf{E}_{s0}^+ \; ,$$

$$x = \left[\frac{4 T_1 T_2}{\hbar^2} \right]^{1/2} (\mathbf{d}^* \overrightarrow{\varepsilon}_{s0}^{\,+} + \mathbf{d}^* \mathbf{E}_{s0}^+) \; . \qquad (4.28)$$

With these designations the equation (4.23) in the stationary case is written in the form

$$D - D_0 = - y^2 \frac{D}{(1 - \Gamma q_0 T_2 D / 2 N)^2} . \qquad (4.29)$$

Substituting (4.27) into (4.29), we obtain the equation, determining the dependency of the amplitude of the output field on the amplitude of the input field

$$y = x + C x / (1 + x^2) , \quad (D_0 = - N) . \qquad (4.30)$$

The Eqs.(4.29) and (4.30) describe the bistable behavior of the two-photon transition in the medium with F » 1 or F « 1. In (4.30) we have designated

$$C = g_0 l = \frac{\Gamma q_0 T_2}{2} \equiv \frac{T_2}{\tau_c} . \qquad (4.31)$$

One can show, starting from (4.31), that for F » 1 the expression for g_0 coincides with the usual formula for the two-photon absorption (or gain) coefficient [32]. The condition $\tau_c > l / c$ determines the upper bound on the medium length

$$l < \left(\frac{c T_2}{g_0} \right)^{1/2} \qquad (4.32)$$

(One can neglect the propagation effects for such a length).

The dependency of the transmitted field amplitude on the amplitude of the input wave, corresponding to (4.30), for the case $C \gg 1$, is given in Fig.5.11.

The branches *OBC* and *DE* correspond to the stable stationary solutions and the branch *DC* - to the unstable ones. The unstable branch appears for $C > 8$. In this case when the input intensity of the field with the frequency ω_s is increased from $y = 0$ up to y_c, the amplitude of the transmitted field with frequency ω_s is determined by the branch *OBC*. With the further increase of y, the stationary value of x undergoes jumplike change from x_c to x_E and after that is determined by the branch *EF*. Under the subsequent decrease of the input field amplitude , the

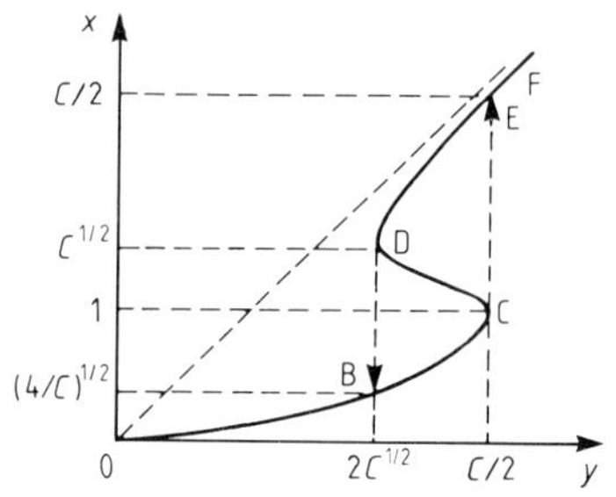

Fig.5.11. The hysteresis dependence of the input field amplitude x on the input field amplitude y, according to Eq.(4.30).

hysteresis dependence of the transmitted field amplitude on the amplitude of the input field takes place.

From the above analysis it follows, that in the pulsed external fields in quasistationary regime the effect of bistability takes place, if the following conditions hold

$$C > 8 \tag{4.33}$$

$$y > y_c \simeq \frac{C}{2}, \quad (C \gg 1). \tag{4.34}$$

Under the substitution $d \to d_{ab}$, $\Gamma \to 4\, d_{ab}^2\, \omega_0^3 / 3\, c^3$ the equations (4.30) and (4.29) describe the bistability effect in one-photon absorption.

Let us consider an example of the real system(cesium vapors), where the observation of two-photon transmission bistability is feasible. For F » 1 the conditions (4.33) and (4.34) are written in the form

$$C = 2\pi\, \omega_s\, |r|^2\, |E_L^+|^2\, n\, T_2\, l\, /\, \hbar\, c > 8\,, \tag{4.35}$$

$$|E_{so}^+| > |E_s^+|_c = \pi\, \omega_s\, |r|^2 \frac{n}{2\,c} \left(\frac{T_2}{T_1}\right)^{1/2} |E_L^+|^2\, l\,. \tag{4.36}$$

Since the above theory is valid in the given pump field (E_L^+) approximation, then besides the conditions (4.35) and (4.36) one extra condition $I_{s0} < I_L$ is needed. The scheme of cesium atom energy levels is shown in Fig.5.12, where the transition frequencies $\nu_{a\alpha} = 14340\ \text{cm}^{-1}$, $\nu_{\alpha b} = 11735\ \text{cm}^{-1}$

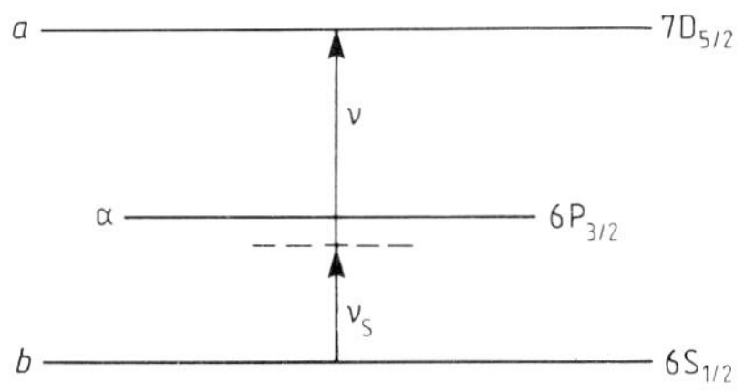

Fig.5.12. Caesium energy level diagram. ν is the frequency of intense field. The transmission bistability occurs on the frequency ν_s.

The matrix elements of the dipole moment were calculated in Ref.43: $d_{a\alpha} = 2.2$ D, $d_{ba} = 6.7$ D. The time $T_2 = 5 \times 10^{-10}$ s the pressure P = 0.1 torr ($n = 3.5 \times 10^{15}\ \text{cm}^{-3}$), $T_1 = 230$ ns (T_1 is the time of spontaneous transition $7D$ - $6P$). Let two fields act upon the ensemble of cesium atoms : the field of ruby laser with the frequency $\nu = 14401\ \text{cm}^{-1}$ with the intensity $I_L = 2 \times 10^7\ \text{W / cm}^2$ and the field of parametric generator of light with frequency $\nu_s = 11674\ \text{cm}^{-1}$, so that $\nu + \nu_s = \nu_{ab}$. Assuming $l = 0.5$ cm, we obtain then $r = 1.2 \times 10^{-21}$ ESU, $C = 22$, $|E_s^+|_c^2 =$

$4.5 \times 10^{-2} \, |E_L^+|^2$ Thus, the effect of bistability in two-photon absorption can be observed with the peak intensity of parametric generator slightly exceeding the value $I_c = 4 \times 10^4$ W /cm^2. The magnitude of the jump of the transmitted intensity is approximately equal to $C^2 \,/\, 4 = 10^2$ and the condition(4.32) holds. The characteristic duration of ruby laser and the parametric generator pulses is 30 - 100 ns. The time of setting in of the stationary regime $T_1 D_{st} \,/\, D_0$ [32] is less then T_1=230 ns at the intensities, corresponding to the branch *EF* in Fig. 5.11 .

5.5. Inversion filamentation in laser media due to interatomic interactions through radiation field

For the description of equilibrium phase transitions one can use the principle of free energy minimum ([5,6]) . This principle can be used also in the quasiequilibrium cases in the atomic systems, being under the action of laser light. If the energy of interatomic interaction through the radiation field depends on the order parameter, its spontaneous appearance at some critical value of control parameter minimizes the free energy.We use this law for the description of three different light induced phase transitions in Secs. 5.5 - 5.7. In this section we use this approach for the explaining from this point of view the phenomenon of the inversion filamentation in laser media [44].

This question is linked with the problem of grained structure of the intensity distribution in the radiation beam cross section in resonatorless laser systems (see [45,46] and ref.given therein and also Ref. 47). Its characteristic feature is the chaotic distribution of the grains in beam section, which is not reproduced from pulse to pulse. The microscopic cause of this phenomenon to our knowledge was not cleared.

It is natural to assume, that the observed picture appears due to the formation of the inversion threads inside the working medium. In this section we show, that with taking into account of the energy of interaction through radiation field the inverted medium of two-level atoms becomes unstable with respect to the transition into the state with the formation of thin inversion threads, extended in the direction of the cylindrical medium axis and chaotically distributed over the beam cross section. Such structure of spatial distribution of inversion minimizes the energy of effective interatomic interaction through the radiation field.

The appearance of the inversion threads is analogous to the phase transition with the stratification of liquid with the difference, that in the considered case the mean density of atoms remains constant and the stratification occurs in the subsystem of atomic excitations.

5.5.1. *Energy of effective interatomic interaction through radiation field.* Let us consider the system of N two-level atoms, confined in a macroscopically large volume V. The atoms are under the action of the laser field mode

$$\mathbf{E} = \mathbf{E}_0 \exp\left[-i\,(\omega t - kz)\right] + \text{c.c.} \tag{5.1}$$

The field (5.1) excites the coherent oscillations of the atomic transition dipole moments, which serve as the sources the reradiated spherical waves. Each atom interacts with the field of reradiation and the overall energy of dipole interaction with this field has the form

$$H = -\sum_i \left(P_i^* E_i + P_i E_i^* \right), \tag{5.2}$$

where P_i and E_i are the complex amplitudes. We use for the radiation field the formula (2.9), where make the substitutions $\mathbf{E}_S \longrightarrow \mathbf{E}_i$:

$$\mathbf{P}_{si} \longrightarrow \mathbf{P}_i \longrightarrow \mathbf{d}_i \sigma_{i-}, \; \sigma_{i-} = \rho_{i-} \exp(-i\omega t) \qquad (\mathbf{d}_i \equiv \mathbf{d}_{abi}), \tag{5.3}$$

where ρ_{i-} is the slowly varying part of the density matrix of j-th atom with the transition frequency ω_0. Then

$$\mathbf{E}_i = \sum_{j(j\neq i)} \frac{k^2}{r_{ij}} \rho_{j-} (t - \tau_{ij}) \left[\mathbf{n}_{ij} \left[\mathbf{d}_j \mathbf{n}_{ij}\right]\right],$$

and for the energy (5.2) we have

$$H = -\sum_{i\neq j}\sum \frac{k^2 \exp(ik\, r_{ij})}{r_{ij}} \rho_{i+}\rho_{j-}\, \mathbf{d}_i^* \left[\mathbf{n}_{ij}\left[\mathbf{d}_j \mathbf{n}_{ij}\right]\right] + \text{c.c.} \tag{5.4}$$

The non-diagonal elements of the density matrix ρ_{i-} are determined from the equations

$$\frac{\partial \sigma_{i-}}{\partial t} + i\left[\omega_0 - \frac{i}{T_2}\right]\sigma_{i-} = -\frac{i}{\hbar}\mathbf{d}_i^* D_i \mathbf{E}_i \quad (D_i \equiv \sigma_{i3}), \tag{5.5}$$

$$\frac{\partial D_i}{\partial t} + \frac{1}{T_1}(D_i - D_{i0}) = \frac{2i}{\hbar}[\mathbf{d}_i^* \sigma_{i+} - \mathbf{d}_i \sigma_{i-}]\mathbf{E}_i , \tag{5.6}$$

where $D_{i0} > 0$ obeys the normalization condition

$$\sum_i D_{i0} = D_0 , \tag{5.7}$$

and D_0 is determined by the pump intensity.

We consider the stationary case, for which from (5.3) and (5.5) we find

$$\rho_{i-} = -D_{i0}\frac{(\Delta^2 + T_2^{-2})\,E_0 d_i}{\hbar\,(\Delta^2 + \gamma_E^2)\,(\Delta - iT_2^{-1})}\exp(ikz_i) , \tag{5.8}$$

where $\Delta = \omega_0 - \omega$ is the detuning from the transition frequency, T_2^{-1} is the homogeneous level width, $\gamma_E = T_2^{-2}(1 + a E_0^2)$, $a = 4\,|d|^2\,T_2 T_1/\hbar^2$ is the saturation parameter.

The stationary regime is reached under the condition, that the radiated laser pulse duration τ_p exceeds the time of local equilibration, i.e. $\tau_p \gg \max\{T_2, T_{1eff}\}$, where $T_{1eff} = T_1(1 + a E_0^2)^{-1}$ is the effective constant of the population difference relaxation. This quasiequilibrium state can be described, using its free energy. Since the thermal energy $k_B T \sim 10^{-14}$ erg is much lesser then the energy of atomic excitations $\hbar\,\omega_0 \sim 10^{-12}$ erg, the free energy practically coincides with the internal energy.

Let us compare the uniform and the non-uniform distribution of the inverted atoms from the viewpoint of minimum of energy (5.4). For this we transform the Hamiltonian (5.4), taking into account of (5.8) Let us introduce the unit vectors along the pump field vector $\mathbf{E}_0 = \mathbf{e}\, E_0$ and along the vectors of the dipole moments $\mathbf{d}_i = d\, \mathbf{e}_i$ Using the abbreviation

$$J = \frac{E_0^2 k^3 d^4}{\hbar^2 (\Delta^2 + T_2^{-2})} \left[\frac{\Delta^2 + T_2^{-2}}{\Delta^2 + \gamma_E^2} \right]^2 \qquad (5.9)$$

for the effective constant of interatomic interaction through the radiation field, we obtain for the interaction energy

$$H = - \sum_i J_{ij}\, \theta_{ij}\, D_{i0}\, D_{j0}\,, \qquad (5.10)$$

where

$$J_{ij} = \frac{J}{k\, r_{ij}} \cos k\, (z_i - z_j - z_{ij}) \qquad (5.11)$$

and the angular factor is

$$\theta_{ij} = (\mathbf{e}\, \mathbf{e}_i)\, (\mathbf{e}\, \mathbf{e}_j)\, \mathbf{e}_i [\mathbf{n}_{ij}\, [\mathbf{e}_j \mathbf{n}_{ij}]] = \lambda\, (1 - (\mathbf{n}_{ij} \mathbf{e})^2)\,. \qquad (5.12)$$

In the case of crystals, when vectors $\mathbf{e}_i$, $\mathbf{e}_j$ and $\mathbf{e}$ are collinear the factor $\lambda = 1$; in the case of a gas, when the vectors $\mathbf{e}_i$ are distributed chaotically over the directions, factor $\lambda = 1 / 9$.

We note, that the sign minus in front of the expression for interaction energy (5.10) corresponds to the effective attraction between the atoms.

5.5.2. *Interaction energy in cylinder geometry.* We assume, that the medium is in the form of a cylinder with the radius $R \ll l$ - the cylinder length. Besides we assume that the following conditions hold

$$R / l, \quad \lambda / R, \quad r_m / \lambda \ll 1\,, \qquad (5.13)$$

where $\lambda = 2\pi\, c / \omega$ is the light wavelength, r_m is the mean interatomic distance. The last inequality in (5.13) enables to pass over in (5.10) from the summation to the integration with the help of the substitution

$$\sum_i \longrightarrow \frac{N}{V} \int d\mathbf{r}_i \,, \tag{5.14}$$

where $V = \pi R^2 l$ is the working volume. Let us direct z - axis along the cylinder axis. It is natural to assume, that the inversion density is uniform along this axis, with the coordinates x and y being equivalent, so that $D_0 = D(x_i, y_i) = D(y_i, x_i)$. The normalization condition (5.7) assumes the form

$$D_0 = \frac{NL}{V} \int D(x,y)\, dx\, dy \,. \tag{5.15}$$

Writing down the corresponding integral for the energy (5.10), note, that, if one directs the x - axis along the field vector $\mathbf{E}_0$, then $(\mathbf{n}_{ij}\mathbf{e})^2 = (x_i - x_j)^2 / r_{ij}^2$, if instead the y - axis is directed along $\mathbf{E}_0$, then $(\mathbf{n}_{ij}\,\mathbf{e})^2 = (y_i - y_j)^2 / r_{ij}^2$. The energy value must not depend on the orientations of x and y - axes, so it is convenient to take the mean value of these two expressions and obtain as integrand $(x_i - x_j)^2 / 2 r_{ij}^2 + (y_i + y_j)^2 / 2 r_{ij}^2$ instead of $(\mathbf{n}_{ij}\,\mathbf{e})^2$. Defining also the expressions

$$A(\rho) = \int_{-l/2}^{+l/2} \frac{\cos k\,[z - (\rho^2 + z^2)^{1/2}]}{(\rho^2 + z^2)^{1/2}} \left[1 - \frac{\rho^2}{2(\rho^2 + z^2)}\right] dz \tag{5.16}$$

we obtain

$$H = -\frac{N^2 l\, J}{V^2 k} \int A(\rho)\, D(x_1,y_1)\, D(x_2,y_2)\, dx_1 dy_1 dx_2 dy_2 \,. \tag{5.17}$$

where $\rho^2 = (x_1 - x_2)^2 + (y_1 - y_2)^2$, $\quad z = z_1 - z_2$

Let us rewrite (5.16), using the variable substitution

$$\xi = k\,[z - (\rho^2 + z^2)^{1/2}].$$

Then we obtain

$$A(\rho) = -\int_{\infty}^{k\rho^2/l} \frac{\cos\xi}{\xi}\left[1 - \frac{2\,k^2\rho^2\xi^2}{(\xi^2 + k^2\rho^2)^2}\right] d\xi\,. \tag{5.18}$$

Introducing the notations

$$\mathrm{Ci}\left(\frac{k\rho^2}{l}\right) = \int_{\infty}^{k\rho^2/l} \frac{\cos\xi}{\xi}\, d\xi\,, \quad f(\xi) = 2\left(\frac{k\rho\xi}{\xi^2 + k^2\rho^2}\right)^2,$$

we represent (5.18) in the form

$$A(\rho) = -\,\mathrm{Ci}\left(\frac{k\rho^2}{l}\right) + \int_{\infty}^{k\rho^2/l} f(\xi)\,\frac{\cos\xi}{\xi}\, d\xi\,.$$

The function $\cos\xi / \xi$ is integrable in interval (ξ, ∞) where $\xi > 0$; $f(\xi)$, as follows from its definition, is non-negative and continuous for all $\xi \in (0,\infty)$ and has the upper bound equal to $\max f(\xi) = 1/2$ reached at the point $\xi = k\rho \gg k\rho^2/ l$. Under these conditions the generalized theorem about the mean is valid, according to which there is such value of $\xi_1 = \sigma k \rho$, where $\sigma \in (0,\infty)$, that

$$\int_{\infty}^{\xi} f(\xi)\,\frac{\cos\xi}{\xi}\, d\xi = f(\xi_1)\int_{\infty}^{\xi} \frac{\cos\xi}{\xi}\, d\xi \qquad (\xi > 0)\,,$$

with

$$0 < f(\xi_1) = \frac{2\,\sigma^2}{(\sigma^2 + 1)^2} < \frac{1}{2}\,.$$

Then for the function (5.18) we have

$$A\ (\rho) = -\frac{(\sigma^4 + 1)}{(\sigma^2 + 1)^2}\ \mathrm{Ci}\left[\frac{k\,\rho^2}{l}\right].$$

Defining the renormalized constant of the effective interaction

$$\tilde{J} = J\ \frac{\sigma^4 + 1}{(\sigma^2 + 1)^2} \tag{5.19}$$

with $$\frac{1}{2} \le \frac{\tilde{J}}{J} \le 1\ ,$$

we find for the interaction energy

$$H = \frac{N^2 l}{V^2 k}\ \tilde{J} \int \mathrm{Ci}\left[\frac{k\,\rho^2}{l}\right]\ D\ (x,y)\ D\ (x',y')\ dx\ dy\ dx'dy' \tag{5.20}$$

Now we must compare the energies of spatially uniform and non-uniform distributions of the excited atoms. If one finds, that the interaction energy is lower in the case of threadlike structure of the inversion , it will mean, that it is advantageous for the system to make a transition into such state.

5.5.3. *Minimization of interatomic energy under inversion threads formation.* If the inversion density is uniform, i.e. $D\ (x,y)$ = const, then from the normalization condition (5.15) it follows, that $D(x,y) = D_0\ /\ N$. Then the energy (5.20) assumes the form

$$H\ = -\frac{D_0^2\tilde{J}}{(k\ R)^2}\ I\left[\frac{k\ R^2}{l}\right]. \tag{5.21}$$

Here

$$I(x) = -\int_0^x \mathrm{Ci}(\xi)\, d\xi = \sin x - x \,\mathrm{Ci}\, x\ . \tag{5.22}$$

Let us assume, that the radiation beam breaks down into N_f threads with radius R_f each. Since the extended in z-direction threads do not interact with each other , they are distributed randomly in x,y - plane. The inversion density $D(x,y) = D_f$ = const inside the thread and is equal to zero outside. In this situation the overall energy of the system $H_1 = N_f H_f$, where H_f is the energy of the single thread. The normalization condition (5.15) yields the number of the threads

$$N_f = \frac{D_0}{N\, D_f} \left(\frac{R}{R_f} \right)^2 , \tag{5.23}$$

The energy of a single thread one determines with the help of the formula (5.20), whence

$$H_f = -\left(\frac{N\, R_f D_f}{k\, R^2} \right)^2 \tilde{J}\, I\left(\frac{k\, R_f^2}{l} \right) . \tag{5.24}$$

Thus the full energy is

$$H_1 = -\frac{N\, D_0 D_f}{(k\ R)^2}\, \tilde{J}\, I\left(\frac{k\, R_f^2}{l} \right) . \tag{5.25}$$

Evidently, if one substitutes $R_f \longrightarrow R$, $N_f \longrightarrow 1$, then one has $D_f \longrightarrow D_0/ N$ and from (5.25) we come back to (5.21).

The expression (5.25) contains two parameters D_f and R_f, which are chosen so that to reach the minimum value of the system energy. We note, that the function (5.22) has the infinite number of extrema, from which one must choose the largest positive one, so that the inequality **would** hold

$$I\left(\frac{k R_f^2}{l}\right) = abs \max I(x) > 0 \ , \tag{5.26}$$

In this case the energy (5.25) has the minimum. From (5.25) it follows uniquely that the infimum of energy with respect to parameter D_f is reached, when the inversion density inside the thread tends to the maximum possible value $D_f = 1$, i.e. when all atoms inside the thread are excited. The minimization of the energy (5.25) under the condition $\partial^2 H_1 / \partial R_f^2 > 0$ leads to the equation, determining the thread radius Ci

$$\mathrm{Ci}\left(\frac{k R_f^2}{l}\right) = 0 \ . \tag{5.27}$$

Out of all zeros of the integral cosine one must choose the one, that satisfy the condition (5.26), i.e. the first zero: $k R_f^2 / l = 0.62$. Thus for the thread radius we obtain, using the tables of integral cosine

$$R_f = 0.31 \ (\lambda \ l)^{1/2} \ . \tag{5.28}$$

We note, that the solution of the equation $\partial H_1 / \partial R_f = 0$ one can obtain also analytically, using the approximate expressions

$$\mathrm{Ci}(x) \cong C + \ln x - \frac{x^2}{4} \ ,$$

$$I(x) \cong (1 - C)\, x - x \ln x + \frac{x^3}{12} \ ,$$

which are valid for $x < \pi$; here $C = 0.577$ is Euler constant. The above expansions yield

$$R_f = \left[\frac{\lambda \ l}{2\pi} \exp(-c)\right]^{1/2} \cong 0.3 \ (\lambda \ L)^{1/2} \ ,$$

This expression approximates well the formula (5.28).

Thus, taking into account, that $abs \max I(x) = I(0.62) \cong 1$,one obtains for the interaction energy in the case of inversion filamentation˙

$$H_1 = - \frac{N D_0 \tilde{J}}{(kR)^2} . \qquad (5.29)$$

Comparing (5.21) and (5.29), we see, that $H_1 < H_0$, i.e. threadlike structure is advantageous, if the following inequality holds

$$\frac{D_0}{N} I \left(\frac{k R^2}{L} \right) < 1 . \qquad (5.30)$$

The mean difference of the number of the excited and non-excited atoms D_0 is always lesser then N, in addition, as we have seen, $I (k R^2 / l) < 1$ for $R > R_f$. Moreover, since $R_f \ll R$, one has $I (k R^2 / l) \ll 1$. It means, that the inequality (5.30) is always fulfilled.

Thus, the energy of thread structure of the inversion distribution is lesser then the energy of the homogeneous distribution. Physical causes of this are clear enough. In fact, as one can see from (5.20), the energy of the effective interaction is proportional to Ci $(k\rho^2/l) \longrightarrow -\infty$, when ρ tends to zero, i.e. the deep potential well appears , in which the excitations are localized, forming the threads. These threads are distributed chaotic in the cross section, because the function Ci (x) in fact is practically zero for $x > k R_f^2 / l$, so that the interaction between the threads is absent.

Let us make the numerical estimates, using the values of parameters, characteristic for the experiments [45,46]. Take $\lambda = 5 - 6) 10^{-5}$ cm, $R = (0.1 - 0.3)$ cm, $l = 20 - 50$ cm. Then $R / l \sim 10^{-2}$, $\lambda / R \sim 10^{-4}$, $r_m / \lambda \sim 10^{-3}$ and the inequalities (5.14) are fulfilled. The criterion of the thread structure advantage (5.30) is fulfilled as a strong inequality, because $k R^2 / l \sim 10^2$ and $I (10^2) \ll 1$. The radius of the thread, determined by the formula (5.28), $R_f \sim (1 - 2) \times 10^{-2}$ cm. The number of the threads is estimated with the help of (5.23), whence for $D_0 \cong N / 2$ we have $N_f \sim 10^2 - 10^3$. The values obtained agree well with the experimental data of Ref. [45-46]. Some experimental evidence in favor of the ideas about the occurrence of the inversion filamentation and corresponding theoretical results, obtained in Ref. 44 and discussed in this section, were given recently in [47].

5.6. Ferromagnetic phase transition in paramagnetic medium, induced by coherent light wave

The effective interatomic interaction through the radiation field under certain conditions can lead to the non-equilibrium ferromagnetic phase transition into the ferromagnetic state, which is considered in this section[48]. The development of the theory of laser-induced ferromagnetic phase transition is stimulated by observation of the effect of inducing the macroscopic magnetic moment (the electromotive force - EMF) under the action of the coherent light wave on atomic paramagnetic gas rubidium [49,50]. This effect is observed in particular with linear polarized wave, that excludes the possibility of explaining it as an usual optical orientation. The dependence of EMF on the frequency of light has the characteristic shape with the minimum at the one photon resonance frequency and two maximums, lying close to resonance (Fig. 5.13).

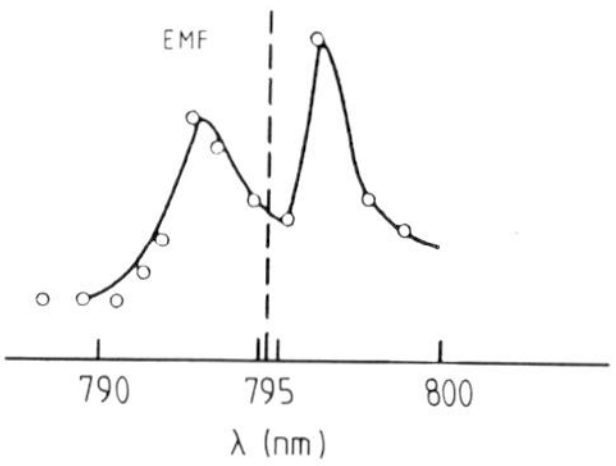

Fig.5.13. The EMF dependence on the excitation radiation wavelength. The dashed line shows the position of one photon resonance $5S_{1/2} \longrightarrow P_{1/2}$ [49].

EMF decreases with the temperature increase and disappears at some critical temperature T_c, depending on the pump intensity E^2 (Fig. 5.14).

All these features are explained by the theory of light-induced ferromagnetic phase transition.

To clarify the difference between this effect and the usual optical orientation we consider first the difference in action of the light waves

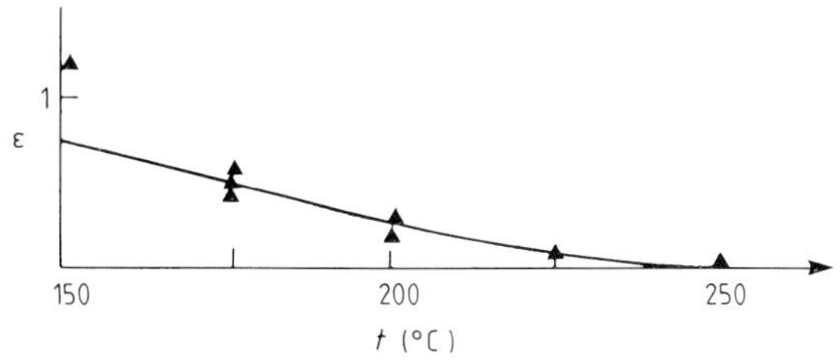

Fig. 5.14. The dependence of the normalized EMF ε on the temperature for $n = 3\ 10^{13}$ cm^{-3} [50]. The solid curve is the theoretical dependence according to Eq.6.27 ($\varepsilon = [(250^0 / T) - 1]$)

with linear and circular polarization. The action of the light wave with the circular polarization on paramagnetic media leads to the optical orienting of the magnetic moments with the appearance of the macroscopic constant magnetic moment . The type of light wave polarization plays the decisive role in this effect: the linear polarized wave doesn't yield the optical orienting (see Fig. 5.15). Due to the law of momentum conservation the absorption of photon with the circular polarization is accompanied by the atomic transition from the ground state ε_b,μ_{b_1} to the excited state ε_a,μ_{a_2}, i.e. the transition with the spin flip (μ_{b_1}, μ_{a_2} are the constant magnetic moments of the magnetic sublevels of atom). After the excitation atom relaxes from μ_{a_2},ε_a state to $\mu_{b_{1,2}},\varepsilon_b$ states with the rate γ_{21}, γ_{22}. Since $\gamma_{22} > \gamma_{21}$, atoms are gradually piling up in the ground state μ_{b_2},ε_b. Thus, the symmetry breaking of the magnetic sublevels filling in each atom is the consequence of the initial broken symmetry of the polarization of the pump light wave. One concludes, that the optical orientation effect is not cooperative in its nature. Since the wave with the linear polarization can be represented as the superposition of two waves with circular polarization (Fig. 5.15b), the above mechanism of optical orientation doesn't break down the symmetry of magnetic sublevels filling and macroscopic magnetic moment is not induced.

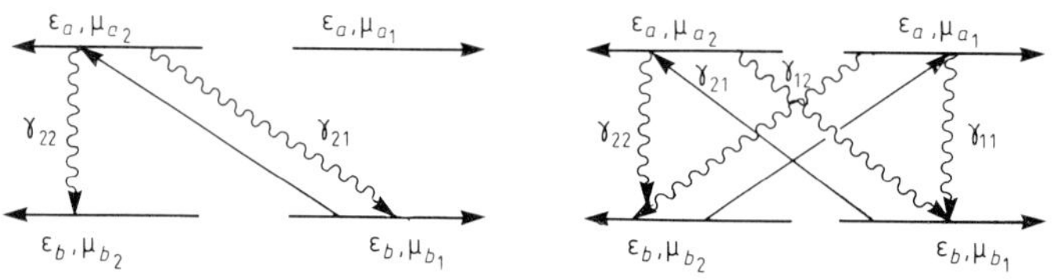

Fig.5.15. The scheme of electronic excitation and relaxation under the action of light wave: a) The circular polarization ($\gamma_{22} > \gamma_{21}$); b) The linear polarization ($\gamma_{22} = \gamma_{11} > \gamma_{21} = \gamma_{12}$).

Below we shall show, however, that in the cooperative system of paramagnetic two-level atoms, interacting through the radiation field, the symmetry of magnetic sublevels populations breaks down spontaneously at some critical value of the pump intensity E_c^2, which depends on temperature , i.e. the ferromagnetic phase transition occurs under the action of coherent wave with arbitrary(in particular linear) polarization.

As an order parameter serves macroscopic magnetization, which increases with the pump intensity increase (Fig. 5.16), analogously to the increase of the equilibrium magnetization with the temperature decrease. At constant pump intensity the order parameter appears at some critical value of temperature and increases with temperature decrease (Fig. 5.14).

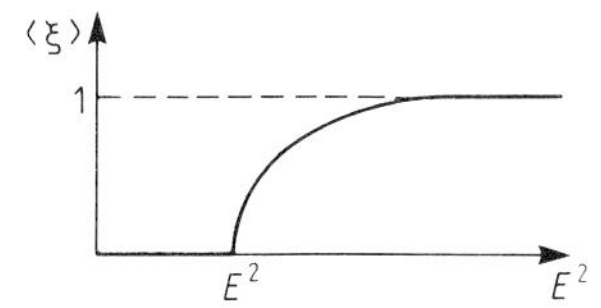

Fig.5.16. The dependence of the order parameter of ferromagnetic phase transition on the pump intensity.

As it has already been noted, the light-induced ferromagnetic phase transition occurs due to the effective interatomic interaction through radiation field in the system, which are acted upon by the coherent light wave. Due to the correlations of the transition dipole moments of atoms in different space points, induced by the spatially coherent light wave, the effective interatomic interaction occurs in macroscopically large volume. Thus, the macroscopic energy of this interaction turns out to be proportional to the medium length l, or to the absorption length l_a, if $l_a < l$. The phase transition occurs because the macroscopic energy of the interatomic interaction depends on the order parameter and is lowered with its appearance.

5.6.1. *Spontaneous appearance of magnetization under action of linear polarized light wave.* Let us consider the two-level atomic ensemble, enclosed in a macroscopically large volume in the form of cylinder, analogous to that considered in Sec.5.5. For definiteness we consider the transition $S_{1/2} \longrightarrow P_{1/2}$ in rubidium,on which the effect of inducing of the macroscopic magnetic moment was observed in particular in Ref.45,46. The upper and lower levels of this transition with the frequency ω_0, designated correspondingly by indexes a and b, are degenerate with respect to the magnetic number m. The states a_1, b_1 with $m = +1/2$ and a_2,b_2 with $m = -1/2$ correspond to the opposite directions of the projections of the magnetic moments on the quantization axis z (Fig. 5.15).

The macroscopic magnetic moment M in such a system is

$$M = n\,[\mu_a(\rho_{a1} - \rho_{a2}) + \mu_b(\rho_{b1} - \rho_{b2})]\ , \tag{6.1}$$

where n is a number density of atoms, $\rho_{a_{1,2}}$, $\rho_{b_{1,2}}$ are the populations of the corresponding states, μ_a, μ_b are the magnetic moments of the upper and lower states. Let us introduce the order parameters of the ferromagnetic phase transition as an averaged over the ensemble quantities :

$$<\xi> = <\rho^i_{a1} + \rho^i_{b1} - \rho^i_{a2} - \rho^i_{b2}>\ ,$$

$$<q> = <\rho^i_{a1} - \rho^i_{b1} - \rho^i_{a2} - \rho^i_{b2}>\ , \tag{6.2}$$

Then the mean macroscopic magnetic moment of gas is written in the the form

$$<M> = \frac{1}{2}\, n\, \mu_{ab} <\xi> + \frac{1}{2}\, n\,(\mu_a - \mu_b) <q>$$

$$\mu_{ab} = \frac{1}{2}\,(\mu_a + \mu_b)\ . \tag{6.3}$$

In equilibrium $<\xi>_0 = < q >_0 = 0$ and the magnetic moment is equal to zero.

Let the light wave with the linear x-polarization acts on this system. It can be represented as the sum of two circular polarized waves:

$$\mathbf{E}_j = \mathbf{e}_x E \exp\,[-i(\omega t - kz_j)] + \text{c.c.}$$

$$= \frac{1}{2}\,(\mathbf{e}_x + i\mathbf{e}_y)\, E \exp\,[-i(\omega t - kz_j)]$$

$$+ \frac{1}{2}\,(\mathbf{e}_x - i\mathbf{e}_y)\, E \exp\,[-i(\omega t - kz_j)] + \text{c.c.}. \tag{6.4}$$

The case of one circular polarized wave is included in (6.4).

Due to the selection rules for dipole transitions $\Delta\, m = +1, -1$ there is a pair of transitions a_1, b_2 and a_2, b_1, each of which serves as a source of radiated spherical waves. Expressing the field of dipole radiation in point $\mathbf{r}_i$ through the atomic dipole moments

$$\mathbf{E}_i(t) = \sum_j (k^2 / \mathbf{r}_{ij})\, \rho_{abj}(t - \tau_{ij})\, [\mathbf{n}_{ij}\, [\mathbf{d}\, \mathbf{n}_{ij}]],$$

where $\rho_{abj} \equiv \sigma_{j-}$, $\mathbf{d} \equiv \mathbf{d}_{ba}$, and using also the condition $l \gg R$, we write down the energy of interaction of atom with radiation field in point $\mathbf{r}_i$:

$$H_i = -\sum_p (\mathbf{d}_{a_1 b_2} \rho_{b_2 a_1 i} + d_{b_2 a_1} \rho_{a_1 b_2 i})\, \mathbf{E}_i$$

$$= -\, |d|^2 k^2 \sum_{j \neq i, p} \sum (\rho_{a_1 b_2 i} \rho_{b_2 a_1 i} \exp(-ik\, r_{ij}) / r_{ij} + \text{c.c.})\,, \tag{6.5}$$

where $\sum_p$ is the sum over the interchange of the indexes 1 and 2 (in the case of circular polarization only one term remains in $\sum_p$), ρ_{abi} are the non-diagonal elements of density matrix, $\mathbf{d}_{a_1 b_2} = \mathbf{d}_x + \mathbf{d}_y$ is the transition dipole moment $|d|^2 = d_x^2 + d_y^2$. To find $\rho_{a_1 b_2 i}$ and $\rho_{b_2 a_1 j}$ we use equations for the density matrix of two-level atom interacting with light field. In zero approximation this field coincides with the given pump field $\mathbf{E}_i$:

$$\frac{\partial \rho_{a_1 b_2 i}}{\partial t} + (i\omega_0 + 1 / T_2)\, \rho_{a_1 b_2 i} = -(i\mathbf{d}_{a_1 b_2} / \hbar)\, D_{12i}\, \mathbf{E}_i\,, \tag{6.6}$$

$$\frac{\partial D_{12i}}{\partial t} + \frac{1}{T_1} (D_{12i} - D^0_{12i})$$

$$= \frac{2\, i}{\hbar} (\mathbf{d}_{a_1 b_2} \rho_{b_2 a_1} - \mathbf{d}_{b_2 a_1 i} \rho_{a_1 b_2 i})\, \mathbf{E}_i\,. \tag{6.7}$$

Here $D_{12i} = \rho_{a_1 i} - \rho_{b_2 i}$, D^0_{12i} is the random (operator) source.

If one doesn't take into account of the radiation field, then

$$<D^0_{12i}> = <D^0_{21i}> = -\frac{1}{2}\,\text{th}\,(\hbar\omega_0 \,/\, 2k_B T) ,$$

i.e. the mean value is determined by the temperature alone. When the radiation field of other atoms is taken into account, the mean overall population difference in a given atom is still determined by formula

$$<D^0> = <D^0_{12i} + D^0_{21i}> = -\,\text{th}\,(\hbar\omega_0 \,/\, 2\,k_B\,T) ,$$

but the mean values $< D^0_{12i}>$ and $<D^0_{21i}>$must be now determined self-consistently, taking into account of the energy of interaction of atom with radiation field, since this energy is sensitive to the character of filling of states 1 and 2. For this , using the expression

$$\xi^0_i = D^0_{12i} - D^0_{21i} , \quad D^0 = D^0_{12i} + D^0_{21i} ,$$

we express operators D^0_{12i} and D^0_{21i} through D^0 and ξ. We have

$$D^0_{12i} = \frac{D^0}{2} + \frac{\xi^0_i}{2} , \quad D^0_{21i} = \frac{D^0}{2} - \frac{\xi^0_i}{2} . \tag{6.8}$$

From the definition (6.2) it is seen, that the operator ξ^0_i assumes two values

$$\xi^0_i = \pm 1 . \tag{6.9}$$

From (6.6) and (6.7), taking into account of the saturation effect, we obtain

$$D_{12i} = D^0_{12i}(\Delta^2 + T_2^{-2}) \,/\, (\Delta^2 + \gamma_E^2) ,$$

$$\xi = \xi^0(\Delta^2 + T_2^{-2}) \,/\, (\Delta^2 + \gamma_E^2) , \tag{6.10}$$

where $\Delta = \omega_0 - \omega$, $\gamma_E^2 = T_2^{-2}\,(1 + a\,E^2)$ and $a = 4\,|d|^2\,T_1\,T_2 \,/\, \hbar^2$ is the saturation parameter.

Let us introduce the complex amplitude $\tilde{\rho}_{a_1 b_2}$ according to the formula

$$\rho_{a_1 b_2 i} = \tilde{\rho}_{a_1 b_2} \exp(-i\omega t)$$

Then from (6.6) and (6.10) one has for the complex amplitude

$$\tilde{\rho}_{a_1 b_2 i} = - \frac{d_{a_1 b_2} E}{\Delta - i T_2^{-1}} D^0_{12i} \frac{\Delta^2 + T_2^{-2}}{\Delta^2 + \gamma_E^2} \exp(i k z_i) . \quad (6.11)$$

Substituting now $\rho_{a_1 b_2 i}$ and $\rho_{b_2 a_1 i}$ from (6.11) into (6.5) and using (6.10), (6,8), we obtain for the operator of the energy of interaction of i-th atom with the radiation field, created by all other atoms

$$H_i = -|d|^2 k^2 \left(\frac{2|d|^2 E^2}{\Delta^2 + T_2^{-2}} \right) \frac{\Delta^2 + T_2^{-2}}{\Delta^2 + \gamma_E^2} \times$$

$$\times \sum_j \frac{\cos [k [z_i - z_j) - r_{ij}]]}{r_{ij}} \left\{ \left(\frac{D^0}{2} + \frac{\xi_i^0}{2} \right) \left(D_2^0 + \frac{\xi_j^0}{2} \right) + \right.$$

$$\left. + \left(\frac{D^0}{2} - \frac{\xi_i^0}{2} \right) \left(\frac{D^0}{2} - \frac{\xi_j^0}{2} \right) \right\} . \quad (6.12)$$

Let us represent the operator $\xi^0_{i,j}$ as the sum of mean value and fluctuation

$$\xi^0_{i,j} = \langle \xi^0 \rangle + (\xi^0_{i,j} - \langle \xi^0 \rangle) .$$

Then in the mean field approximation, neglecting quadratic in fluctuations terms and evaluating the integral over the cylindrical volume, replacing $\sum_i \to n\int dV$, we obtain with the accuracy up to a constant, independent of $\langle\xi\rangle$, for the energy the following expression

$$H_i = -\alpha \left[-\frac{\langle\xi^0\rangle^2}{2} + \langle\xi^0\rangle\, \xi_i^0 \right], \tag{6.13}$$

$$\alpha = 4\pi\, n |d|^2 l\, k \frac{|d|^2 E^2}{\hbar^2 (\Delta^2 + T_2^{-2})} \left(\frac{\Delta^2 + T_2^{-2}}{\Delta^2 + \gamma_E^2} \right)^2 . \tag{6.14}$$

Under the condition, that the light pulse duration $\tau_{pump} > \tau_m$ - the relaxation time of population of magnetic sublevels, the order parameter distribution function has the Boltzmann form

$$f(\xi_i^0) = C \exp(-H_i / k_B T) .$$

Taking account of (6.13), we have

$$f(\xi_i^0) = \frac{\exp[\alpha \langle\xi^0\rangle\, \xi_i^0 / k_B T]}{2 \cosh(\alpha \langle\xi^0\rangle / k_B T)} . \tag{6.15}$$

calculating the mean value $\langle\xi_0\rangle = \mathrm{Sp}\,(f\, \xi_i^0)$, we find the following self-consistency condition

$$\langle\xi^0\rangle = \mathrm{th}\, [(\alpha / k_B T) \langle\xi^0\rangle] . \tag{6.16}$$

The equation for order parameter (6.16) is analogous to the Ising equation for magnetization in the equilibrium theory of ferromagnetism But in (6.16) the control parameter $\alpha / k_B T$ depends not only on temperature, but also on intensity and frequency of the exciting light wave. Its solution as a function of E^2 is schematically shown in fig.5.16. The values $\langle\xi^0\rangle \neq 0$ appear, when $E^2 > E_c^2$ and increase with the increase of E^2. For $E^2 \gg E_c^2$ the magnetization saturates: $\langle\xi^0\rangle \to \langle\xi^0\rangle_{max}$. At the same time another order parameter $\langle q\rangle = 0$, since the energy (6.13) doesn't depend on q .

Close to E_c^2 the order parameter $\langle\xi^0\rangle \ll 1$ and one can expand the right hand side of (6.16) in power series of parameter $\langle\xi^0\rangle$ and obtain the following equation for order parameter

$$\langle\xi^0\rangle\,(1 - \alpha / k_B T) = -(\alpha^3 / 3\,(k_B T)^3)\,\langle\xi^0\rangle^3 \,. \qquad (6.17)$$

The solution $\langle\xi\rangle \neq 0$ appears, when $\alpha / k_B T > 1$. From this condition one obtains at T = const the equation, which determines the critical intensity

$$1 - \frac{\alpha_c}{k_B T} \equiv 1 - \frac{4\pi\, d^2 n}{k_B T} \left[lk \frac{|d|^2 E_c^2}{\hbar^2 (\Delta^2 + T_2^{-2})} \right]$$

$$\times \left[\frac{\Delta^2 + T_2^{-2}}{\Delta^2 + \gamma_E^2} \right]^2 = 0 \,. \qquad (6.18)$$

On the other hand at E^2 = const the formula (6.18) determines the critical temperatureT_c.

The effective interatomic interaction arises also in two photon processes: two-photon absorption and Raman scattering. In these cases the expression for the interaction energy is still determined by formula (6.13), (6.14), where instead of d one must substitute the effective dipole moment of order of d $(dE/\hbar\,\Delta)$, where Δ is the detuning from resonance.

At $E^2 > E_c^2$ the macroscopic magnetic moment appears, which close to the threshold is described by the formula

$$\langle M\rangle = n\,\mu_{ab}\langle\xi\rangle$$

$$= n\,\mu_{ab} \left[\frac{\alpha / k_B T - 1}{\alpha^3 / 3\,(k_B T)^3} \right]^{1/2} \frac{\Delta^2 + T_2^{\,2}}{\Delta^2 + \gamma_E^2} \,. \qquad (6.19)$$

If the cell length $l < l_a$, where

$$l_a = \frac{\hbar\,(\Delta^2 + \gamma_E^2)}{2\pi\, n d^2 T_2^{-1} k} < l$$

then while determining the value of α from (6.14) one must substitute

$l \longrightarrow l_a$ and then obtains

$$\frac{\alpha}{k_B T} = \frac{2d^2 E^2}{\hbar k_B T\, T_2^{-1}} \, \frac{\Delta^2 + T_2^{-2}}{\Delta^2 + \gamma_E^2} \, .$$

As it follows from (6.19) the signal <*M*> decreases as the temperature increases.

In describing the response to the excitation with the resonance frequency one must take into account the pump line width $\delta\omega$, if it exceeds the field broadened transition line width γ_E . In the case of broad band Gaussian pump with the central frequency ω_1 one must substitute in (6.14)

$$E^2 \frac{(\Delta^2 + T_2^{-2})^2}{(\Delta^2 + \gamma_E^2)^2 (\Delta^2 + T_2^{-2})}$$

$$\equiv E^2 f(\omega) \longrightarrow \frac{E^2}{\sqrt{\pi}\,\delta\omega} \int_{-\infty}^{+\infty} \exp\left[- \left(\frac{\omega - \omega_1}{\delta\omega} \right)^2 \right] f(\omega)\, d\omega \, .$$

Then, for example for the case $T_2^{-1} \ll \delta\omega \leq \gamma_E$ and at resonance $\omega_0 = \omega_1$ we have in (6.14) the substitution

$$\frac{(\Delta^2 + \overline{T}^{-2})^2}{(\Delta^2 + \gamma_E^2)^2} \, \frac{1}{\Delta^2 + T_2^{-2}} \longrightarrow \delta\,\omega^2 / \gamma_E^4 \, . \qquad (6.20)$$

5.6.2. *Spontaneous appearance of EMF under action of linear polarized light wave.* In experiment one measures not magnetic moment <*M*> itself, but the EMF, induced in the coil, installed inside the cell with the paramagnetic gas.To compare the theory with the experiment one must know the rate of magnetic moment change, which determines the EMF:

$$\text{EMF} \equiv \varepsilon = - \frac{c^{-1}\, d\,\Phi}{dt} = - \frac{4\pi\, A\, m}{c} \, \frac{dM}{dt} \, , \qquad (6.21)$$

where A is the illuminated cross section area, m is the number of turns in the coil. The derivative dM / dt can be determined from the phenomenological Landau-Khalatnikov's equation for the order parameter ξ^0:

$$\frac{d\xi^0}{dt} = -\Gamma_0 \partial F / \partial \xi^0, \tag{6.22}$$

where Γ_0 is the phenomenological constant, $F = F(\xi^0)$ is the free energy of the system.

Using (6.13) we obtain

$$F(\xi^0) = -k_B T \ln \mathrm{Sp} \exp\left(-\frac{\alpha <\xi^0>^2}{2 k_B T} + \frac{<\xi^0> \xi^0_i}{k_B T}\right)$$

$$= \alpha \frac{<\xi^0>^2}{2} - k_B T \ln 2 \cosh \frac{\alpha <\xi^0>}{k_B T}$$

$$\cong \frac{\alpha}{2} <\xi^0>^2 \left(1 - \frac{\alpha}{k_B T}\right) + \frac{\alpha^4}{(k_B T)\ 12} <\xi^0>^4 . \tag{6.23}$$

From (6.22) and (6.23) we have the non-stationary equation for the order parameter:

$$\frac{d\xi^0}{dt} = \Gamma_0 \alpha \left(\frac{\alpha}{k_B T} - 1\right) \xi^0 - \frac{\Gamma_0 \alpha^4}{(k_B T)^3\ 3} \xi^{0^3} , \tag{6.24}$$

where $\alpha / k_B T > 1$ for $E^2 > E_c^2$. The equation for the stationary case, following from (6.24), naturally coincides with (6.17).

When ξ^0 is sufficiently small, one can retain only the linear term in (6.24).Then, taking into account of (6.19), we obtain

$$\frac{dM}{dt} = \Gamma_0 \alpha (\alpha / k_B T - 1) M \equiv \Gamma_0 \alpha (T / T_c - 1) M , \tag{6.25}$$

where we have introduced the critical temperature T_c, increasing with the temperature increase.

The EMF from (6.21) and (6.25) is written in the form

$$\varepsilon = -\,4\pi\Gamma_0 A\ m\ c^{-1}(T_c / T - 1)\ M\ \alpha\ . \qquad (6.26)$$

As one can see from (6.26), the EMF appears at $E^2 > E^2_c$ (or at constant pump intensity E^2 = const at $T < T_c$).

To evaluate ε we substitute into (6.26) $M = M_{max} = n\ \mu_{ab}$ and obtain

$$|\varepsilon| \le 4\pi\ n\ \mu_{ab}\Gamma_0\alpha\ (T_c / T - 1)\ A\ m = \text{const}\ (T_c / T - 1)\ . \qquad (6.27)$$

Note, that while the order parameter <M> obeys,in accordance with (6.19), to the typical square root dependency on the temperature, the rate of order parameter change: dM / dt, in accordance with (6.26), obeys qualitatively different dependency (compare Figs. 5.14 and 5.16). The dependency of ε on temperature was measured in Ref. 50. The experimental results of Ref. 50 are shown in Fig. 5.14 in comparison with the theoretical dependency (6.27). The critical temperature, determined from this comparison is $T_c \cong 250^0$ C. With the increase of E^2 the critical temperature, as it follows from (6.25) must increase.

As one can see from (6.27) the EMF ~ α ~ E^2 at n = const, besides the EMF ~ α ~ n at E^2= const. Both these theoretical conclusions are corroborated by experimental results of Ref. 50.

The dependence of the EMF on the light frequency is given by (6.27), where α is determined by (6.14). We have

$$\varepsilon = \text{const}\ [(\Delta^2 + T_2^{-2})/(\Delta^2 + \gamma_E^2)]^2 / (\Delta^2 + T_2^{-2})\ . \qquad (6.28)$$

whence it follows, that the maxima of the EMF lie in the points

$$\omega = \omega_0 \pm T_2^{-1}\ (\ aE^2 - 1\ )^{1/2}$$

The dependency (6.28) of ε on the light frequency with taking into account of the substitution (6.20) is shown in Fig.5.13. It corresponds to experimental results of Ref. 49.

Let us make the numerical estimates. Take $n = 10^{16}$ cm^{-3}, $k_B T = 2 \times 10^{-14}$ erg (T ~ 170^0 C), $|d|$ ~ 5×10^{-18} ESU, $T_2^{-1} = 10^9$ c^{-1}, Doppler width is $k\ (2\ k_B T / m)^{1/2}$ ~ 10^9 c^{-1} at k ~ 10^5 cm^{-1}, m ~ 10^{-22} g and γ ~ 3×10^8 c^{-1}. For $E^2 = 2 \times 10^3$ ESU (I ~ 10^6 W / cm^2), we obtain, that at

$|\Delta| = \gamma_E \sim 10^{12}\ c^{-1}$, $l = 10$ cm

$$\alpha \cong \frac{\pi n d^2}{k_B T} \gamma T_2 \frac{lk}{4} \cong 25 ,$$

and the critical condition (6.18) is fulfilled.

In resonance electronic Raman scattering the critical condition (6.18) can be fulfilled at still lower densities $n \sim 10^{13}$ cm^{-3}, since in this case

$$d \longrightarrow d \frac{d E}{\hbar T_2^{-1}} \cong 2 \times 10^2\ d , \quad \gamma \cong T_2^{-1} .$$

Let us now estimate the critical intensity at which the phase transition occurs.Assuming, that $aE_c^2 < 1$, we obtain from (6.7) for one-photon transition

$$E_c^2 = \frac{2 k_B T}{\pi n d^2} \frac{1}{lk} \frac{\hbar^2(\Delta^2 + T_2^{-2})}{d^2} .$$

At $\Delta = 0$ and for the values of other parameters used above, we obtain, that $E_c^2 \sim 10^{-4}$ ESU, i.e. $I_c \sim 10^{-1}$ W / cm^2.

5.7. Orientational phase transition in system of anisotropic molecules under action of light wave

The non-centrosymmetric molecules can interact with each other through the radiation field not only on the pump frequency ω (Secs. 5.5,5.6), but also on the frequency of second harmonic (SH), radiated by every molecule. The minimization of energy of intermolecular interaction through the SH radiation field with the appearance of order parameter can lead to phase transition. In this section we consider the light-induced orientational phase transition in the system of freely orienting molecules ,interacting through SH field. In Chapter 6 we shall consider the analogous phase transition in centrosymmetric solids, which occur due to the combined interactions through Lorentz electrostatic , lattice displacement and SH radiation fields.

In a number of substances (in liquid crystals, gases and liquids) the vectors of transition dipole moments of anisotropic molecules, directed along their long axes, are chaotically distributed over directions.In nematic liquid crystals they are oriented in two opposite directions along certain axis. The analogous situation arises under the illumination of the isotropic media by the powerfull light wave with linear polarization, when the transition dipole moments are oriented in two opposite directions along the polarization vector of the light wave. However this orientation doesn't violate the centrosymmetry of the medium, so that in this state the coherent SH generation (SHG) is impossible.

In the present section it is shown, that the orientational phase transition can be induced due to the action of powerfull coherent light wave, in which the vectors of transition dipole moments of different molecules are oriented in the same direction. It turns out to be, that the system of dipole moments can behave like the ferromagnetic spin system, with the effective intermolecular interaction through the field of dipole radiation on the frequency 2ω playing the role analogous to the "exchange" forces, orienting the spins [51]. If the molecules, in addition posses the constant dipole moment, the direction of which coincides with the direction of transition dipole moment then the orientation ordering is accompanied also by ferroelectric ordering [52].

Let us consider the system of N molecules in the cylindrical volume $V = Al$ $(l \gg A^{1/2})$, $n = N / V$. We assume, that the molecules are strongly anisotropic and the transition dipole moments are directed along their long axes. Besides we shall consider the simple case of preferential orientation along one of the axes (x-axis), for example, by the powerfull linear polarized light wave:

$$\mathbf{E} = \mathbf{e}_x (E \exp [- i\ (\omega t - kz)] + \text{c.c.}) \,, \qquad |\mathbf{e}_x| = 1. \tag{7.1}$$

The field (7.1) induces in molecule in point $\mathbf{r}_j$ the polarization $\mathbf{P}_j$ on the frequency 2ω, the complex amplitude of which is expressed through the quadratic susceptability of a molecule χ^2:

$$P_{xj}(2\omega) = \chi_j^2 E^2 \exp (2ikz) \equiv q_j \chi^2 E^2 \exp (2ikz_j) \,,$$

$$\chi^2 = \frac{1}{\hbar^2} \sum_{n,m,l} |d_{nm} d_{ml} d_{ln}| \, [(\rho_{nn} - \rho_{ll}) \, \zeta\, (\omega - \omega_{ln})$$

$$- (\rho_{ll} - \rho_{mm}) \, \zeta\, (\omega - \omega_{ml})] \, \zeta\, (2\omega - \omega_{mn}) \,, \tag{7.2}$$

where $\rho_{\alpha\alpha}$ are the populations of the corresponding levels, $\zeta\,(\omega - \omega_{kn}) = (\omega - \omega_{kn} + i\, T_2^{-1})$.

In writing of (7.2) it was taken into account, that the vector of the dipole moment of the transition between the levels m and n of a molecule in point $\mathbf{r}_j$ can be represented in the form $\mathbf{d}_{mnj} = |d_{mn}|\, q_j \mathbf{e}_x$, where the random number $q_j = +1, -1$ in dependence on the orientation of vector $\mathbf{d}_{mnj}$, in addition $q_j^3 = q_j$. In the following we assume, that the vectors $\mathbf{d}_{mnj}$ with different m and n are collinear.

The orientation number q_j, averaged over the ensemble:

$$<q_j> \equiv < q >$$

is the order parameter of the light-induced phase transition, considered here.

In a weak light field $<q> = 0$. We want to show, that in sufficiently strong coherent field the phase transition with the appearance of $<q> \neq 0$ occurs. For this we take into account that the polarization (7.2) serves as a source of radiation field of j-th molecule on the frequency 2ω. Expressing with the help of the solution of Maxwell equation the radiation field in the point $\mathbf{r}_j$ through $\mathbf{P}_j(2\omega)$ (analogously to Secs. 5.5, 5.6), we find the energy of interaction of i-th molecule with all other molecules :

$$H_i = - \mathbf{P}_i \mathbf{E}_i =$$

$$= -4k^2 \sum_{j(j\neq i)} [q_i q_j |\chi^2|^2 E^4 \exp [2ik (\mathbf{r}_{ij} - (z_i - z_j))]/r_{ij} + \text{c.c.}], \qquad (7.3)$$

where $r_{ij} = |\mathbf{r}_i - \mathbf{r}_j|$.

In the mean field approximation , substituting $q_i q_j \rightarrow q_i <q_j>$, evaluating in (7.3) the integral over the volume, we obtain

$$H_i = - 2\pi\, n\, |\chi^2|^2 E^4 l\, k <q>\, q_i \, . \qquad (7.4)$$

If the light pulse duration $\tau_{\text{pump}} > \tau_o$, where τ_o is the time of orientation relaxation, then the orientation distribution function has the form

$$f\,(q_i) = C \exp\,(- H_i / k_B T) \, ,$$

From the self-consistency condition

$$\langle q\rangle = \sum_{q_i=+1,-1} q_i f(q_i)$$

one obtains the order parameter equation

$$\langle q\rangle = \mathrm{th}\,(\alpha \langle q\rangle / k_B T) \qquad (7.5)$$

$$\alpha = 2\pi\, n\, |\chi^2|^2 E^4 l\, k / k_B T\ .(7.6)$$

The equation (7.5) is analogous to Ising equation (compare also Sec. 5.6). The solution of (7.5) as the function of E^2 is shown schematically in Fig.5.16. The non-zero values of order parameter $\langle q\rangle$ appears at $E^2 > E_c^2$ and $\langle q\rangle \longrightarrow 1$, when $E^2 \longrightarrow \infty$. Expanding the right hand side in (7.5) in power series of $\langle q\rangle$, we obtain from the condition of cancelling the coefficient at $\langle q\rangle$, the expression for critical intensity

$$E_c^2 = k_B T / (2\pi\, n\, |\chi^2|^2 l\, k)^{1/2}\ . \qquad (7.7)$$

For $E_2 > E_c^2$ the macroscopic quadratic susceptibility $n \langle\chi_j^2\rangle = n \langle q\rangle\, \chi^2$ appears, so that the coherent SHG becomes possible.

Let us make numeric estimates.Let us denote the minimum values of $\omega - \omega_{ln} = \Delta_1$ and of $2\omega - \omega_{mn} = \Delta_2$. Then from (7.2) we have

$$|\chi^2|^2 \cong d^6 / \,[\hbar^4\, (\Delta_1^2 + T_2^{-2})\, (\Delta_2^2 + T_2^{-2})].$$

In (7.7) $k\, l = k\, l_a = \hbar\, (\Delta_1^2 + T_2^{-2}) / 2\pi\, n\, d^2 T_2^{-1}$. Then from (7.7) we obtain

$$E_c^2 = [k_B \hbar^3\, T\, T_2^{-1}\, (\Delta_2^2 + T_2^{-2}) / d^4\,]^{1/2}.$$

For $k_B T \sim 10^{-14}$ erg ($T \sim 100\ K$), $d \sim 1$ D, $\Delta_2 \sim 10^{13}\ \mathrm{s}^{-1}$, $T_2^{-1} \sim 10^{11}\ \mathrm{s}^{-1}$, we have $I_c \cong 10^9\ \mathrm{W / cm}^2$.

The physical mechanism of this light-induced phase transition consists in the following. In the system of molecules, which is under the action of coherent light wave, the spontaneous (fluctuating) ordering of the orientations of transition dipole moments of subensemble of molecules occurs , so that $<q> \neq 0$. The overall SH radiation field of these molecules with the frequency 2ω acts on the non-linear SH polarization of i-th molecule, excited by the coherent pump with frequency ω and induces the rotation torque, which rotates the molecule in such a way, that the seeding value of $<q>$ is enhanced. Such positive feedback leads at $E^2 > E_c^2$ to the instability and the transition of the system in the state with ordering of $\mathbf{d}_{mnj}$ directions. We note, that the arising of the torque moment and the corresponding positive feedback is possible also in non-stationary case, when $\tau_{pump} < T_2, \tau_o$. Thus the orientational ordering of the above type can occur under the action of powerfull ultrashort ($\tau_{pump} \sim 10^{-11}$ - 10^{-13}s) laser pulses.

The above light-induced phase transition occurs due to the SH reradiation field. The molecules reradiate also on the fundamental frequency ω. However the interaction through this considerably more strong field doesn't lead to orientational phase transition, since $P_j(\omega) \sim q_j^2 = 1$, i.e. interaction energy doesn't depend on the orientation of transition dipole moments.

We note, that in Ref. [53] SHG was observed in nematic phase of liquid crystal. The above phase transition can be responsible for the observed threshold SHG in centrosymmetric molecular system.Recently a great interest was attracted to the effect of threshold SHG in centrosymmetric glass fibers. We consider this effect in Chapter 6.

5.8. Ferroelectric phase transition in equilibrium system of atoms, interacting through radiation field

Phase transitions considered in Section 5.2 – 5.7 are non-equilibrium effects, occurring due to the interaction through the field, radiated by atoms, excited by the external pump. If this external pump is switched off, atoms interact through the equilibrium electromagnetic field with the temperature playing the role of the control parameter. It is interesting to study the critical properties of such system. The situation here is quite analogous to the system of atoms, interacting through the equilibrium phonon field, in which, as it is known [54], the phase transitions of Jahn-Teller type occur.

In Ref.[55] it was discovered (see also [56]), that in equilibrium system of two-level atoms, interacting with single electromagnetic mode the phase transition is possible with the Bose condensation of photons, i.e. appearance of the mean number of photons in the mode $\langle a^+_\omega a_\omega \rangle \sim N$ where N is the number of atoms. For $T > T_c$ the quantity $\langle a^+_\omega a_\omega \rangle$ is determined by Plank's formula and doesn't depend on N. The necessary condition of this transition is the inequality $2\pi n d^2_{ab} / \hbar\omega_0 > 1$.

One must note, that the important question about the frequency of the photon condensat was not discussed in the initial work [55] and in a number of subsequent works it was claimed that in this phase transition the condensation of photons with the atomic transition frequency, or the frequency of photon mode occur. In connection with this the phase transition was called the transition into "superradiant state".

The question about the frequency of Bose-condensat was considered in Ref. [57,58]. In these works it was shown, that the soft polariton mode exists in the spectrum of collective excitations, the frequency of which cancels at $T = T_c$. This means that the frequency of Bose condensat is equal to zero and the macroscopic polarization on zero frequency spontaneously arises, i.e. the ferroelectric phase transition takes place.At present the ferroelectric character of equilibrium phase transition in the system of two-level atoms, interacting through the equilibrium electromagnetic field seems to be solidly established [59-62].

In the literature (see, for example [63]) one can encounter the assertion, that phase transition, discovered in [55], is the consequence of the use of the reduced Hamiltonian-the so called Dicke Hamiltonian. In this section we consider this phase transition on basis of full Hamiltonian of the system -atoms plus field. Such consideration enables one to reveal the influence of different approximations, adopted in choosing the Hamiltonian, on phase transition. The treatment of this section reveals the connection between the Dicke Hamiltonian and the concept of the acting field.

5.8.1. *Spectrum of equilibrium polaritons. Soft polariton mode.* To reveal the meaning of the approximations, adopted in description of "superradiant" phase transition, we write down the initial Hamiltonian of the system of atoms, interacting with radiation field, in the form

$$H = \int_V \frac{\omega_0}{2} D(\mathbf{r})\, dr - \int_V \mathbf{P}^{\perp} \mathbf{D}\, dr +$$

$$+ \frac{1}{8\pi} \int_V [\mathbf{D}^2 + \mathrm{rot}^2\mathbf{A}]\, d\mathbf{r} + 2\pi (1 - \beta) \int_V (\mathbf{P}^{\perp})^2\, dr \,. \quad (8.1)$$

Here $D(\mathbf{r}) = \Sigma_i D_i \delta(\mathbf{r} - \mathbf{r}_i)$ is the operator of density of atomic energy levels population difference,

$$\mathbf{P}^{\perp} = \mathbf{P} + \frac{1}{4\pi} \int \frac{\text{grad div } \mathbf{P}(\mathbf{r})}{|\mathbf{r}' - \mathbf{r}|} d\,\mathbf{r}'$$

is the operator of the transverse polarization, $\mathbf{P} = \sum_i \mathbf{d}(\sigma_{i-} + \sigma_{i+})\,\delta(\mathbf{r} - \mathbf{r}_i)$ is the operator of polarization of medium, $\mathbf{d} = \mathbf{e}_d\, d$ is the matrix element of dipole moment ($\mathbf{d} \equiv \mathbf{d}_{ab}$), $\mathbf{D} = \mathbf{E} + 4\pi\, \mathbf{P}^{\perp}$ is the induction operator, β is the parameter of the Hamiltonian, introduced by us: $0 \le \beta \le 1$. At $\beta = 1$ the Hamiltonian (8.1) is the full Hamiltonian of the considered system, taken in the dipole approximation [64.65]. At $\beta = 1$ the Eq. (8.1) yields the Dicke Hamiltonian (see below the Eq.(8.2)).

Introducing the operators of creation a_k^+ and annihilation a_k of field quanta in the volume V, so that

$$\mathbf{A}(\mathbf{r}) = \sum_{\mathbf{k}} \mathbf{e}_{\mathbf{k}} (2\pi \hbar c / k V)^{1/2} (a_{\mathbf{k}} + a^{+}_{-\mathbf{k}}) \exp(i\mathbf{k}\mathbf{r}) ,$$

$$\mathbf{D}(\mathbf{r}) = i \sum_{\mathbf{k}} \mathbf{e}_{\mathbf{k}} (2\pi \hbar c / V)^{1/2} (a_{\mathbf{k}} - a^{+}_{-\mathbf{k}}) \exp(i\mathbf{k}\mathbf{r}) , \quad |\mathbf{e}_{\mathbf{k}}| = 1,$$

we write down (8.1) in equivalent form, convenient for the comparison with the approximations, adopted in the literature:

$$H = \frac{1}{2} \hbar\, \omega_0 \sum_i D_i + \Sigma_{\mathbf{k}} \hbar\, k\, c\, a^{+}_{\mathbf{k}} a_{\mathbf{k}}$$

$$- i \sum_{\mathbf{k},j} \varepsilon_{\mathbf{k}} N^{-1/2} (\sigma_{i-} + \sigma_{j+}) (a_{\mathbf{k}} - a^{+}_{-\mathbf{k}}) \exp(i\mathbf{k}\mathbf{r}_j)$$

$$+ V^{-1}\, 2\pi (1-\beta) \sum_{\mathbf{k},j,l} [(\mathbf{d}\,\mathbf{e}_{\mathbf{k}})^2 (\sigma_{j+} + \sigma_{j-})(\sigma_{l+} + \sigma_{l-})] \exp[i\mathbf{k}(\mathbf{r}_j - \mathbf{r}_l)] ,$$

$$\varepsilon_{\mathbf{k}} = (2\pi\, n\, d^2\, \hbar\, k\, c)^{1/2} , \quad n = N / V . \qquad (8.2)$$

Note once more, that for $\beta = 1$ from (8.2) one obtains the approximate expression, which in a single mode approximation coincides with the Dicke Hamiltonian, used in Ref.55,56.

Using the commutation rules

$$[D\ (\mathbf{r}),\ \mathbf{P}^2(\mathbf{r}')] = -\ (2i\ /\ \omega_0)\ \dot{\mathbf{P}}^{\perp}\mathbf{P}^{\perp}\delta\ (\mathbf{r} - \mathbf{r}'),$$

$$[D\ (\mathbf{r}),\ \mathbf{P}^{\perp}\ (\mathbf{r}') = -\ (2\ i\ /\ \omega_0)\ \dot{\mathbf{P}}^{\perp}\delta\ (\mathbf{r} - \mathbf{r}')\ , \qquad (8.3)$$

we obtain with the help of (8.1) the equation of motion of the system operators :

$$\frac{\partial D}{\partial t} = \frac{2}{\hbar\ \omega_0}\ \frac{\partial\ \mathbf{P}^{\perp}}{\partial t}\ (\mathbf{E} + \beta\ 4\pi\ \mathbf{P}^{\perp}), \qquad (8.4)$$

$$\left[\frac{\partial^2}{\partial t^2} + \omega_0^2\right] \mathbf{P}^{\perp}$$

$$= -\ \frac{2\omega_0 D}{\hbar}\ \mathbf{d}\ [\mathbf{d}\ (\mathbf{E} + \beta\ 4\pi\ \mathbf{P}^{\perp})]\ , \qquad (8.5)$$

$$\frac{\partial^2 \mathbf{E}}{\partial t^2} - c^2 \Delta\ \mathbf{E} = -\ 4\pi\ \frac{\partial^2 \mathbf{P}^{\perp}}{\partial t^2}\ . \qquad (8.6)$$

It is seen from (8.4) and (8.5), that the variation of the parameter β in the Hamiltonian (8.1) means the variation of the acting field. At $\beta = 0$ the field, acting on atom is equal to Maxwell field **E**, determined from the solution of Maxwell equation (8.6). At $\beta = 1/3$ the acting field is equal to the Lorentz field: $\mathbf{E}_{act} = \mathbf{E} + (4\pi/3)\ \mathbf{P}^{\perp}$. At last, at $\beta = 1$ (Dicke model) the acting field coincides with the induction: $\mathbf{E}_{act} = \mathbf{E} + 4\pi\ \mathbf{P}^{\perp}$. Thus, we have cleared out, that the approximate Dicke Hamiltonian, used in different works, is in its essence, the particular model of acting field, analogous to Lorentz acting field model.

Let us consider on the basis of (8.4) – (8.6) the spectrum of collective polariton excitations. Taking into the account of the possibility of the phase transition with the appearance of non-zero mean values, we represent the operators of the system in the form

$$D\ (t) = n\ \langle D\rangle + D_1(t)\ ,$$

$$\mathbf{P}\ (t) = \langle\mathbf{P}\rangle + \mathbf{P}_1(t)\ , \qquad \mathbf{E}\ (t) = \langle\mathbf{E}\rangle + \mathbf{E}_1(t)\ .$$

From (8.5) far from the critical point ($T < T_c$) we obtain the connection between the mean values of **P** and D in the form

$$\langle\mathbf{P}\rangle = -\ (2\ n\ \mathbf{d}\ /\ \hbar\omega_0)\ \langle D\rangle\ [\mathbf{d}\ (\mathbf{E} + \beta\ 4\pi\ \mathbf{P})])\ . \qquad (8.7)$$

From (8.4) - (8.6) one obtains also the equations for the fluctuations. Performing the temporal and spatial Fourier transformation of these equations with taking into account of (8.7), we obtain the characteristic equation for the frequencies of the collective excitations. The solution of this equation has the form

$$(\omega_{\mathbf{k}}^{\pm})^2 = \frac{1}{2}\ [k^2c^2 + \tilde{\omega}_0^{\ 2} - \omega_0^2\ \lambda\ \ (\mathbf{e}_{\mathbf{k}}\mathbf{e}_{\mathbf{d}})\ \langle\ D\ \rangle] \quad \pm$$

$$\pm\{\ \frac{1}{4}\ [k^2c^2 + \tilde{\omega}_0^{\ 2} - \ \omega_0^2\ \lambda\ (\mathbf{e}_{\mathbf{k}}\mathbf{e}_{\mathbf{d}})^2\langle\ D\ \rangle]^2\ - k^2c^2\ \tilde{\omega}_0^{\ 2}\}^{1/2}\ .(8.8)$$

Here vector $\mathbf{e}_{\mathbf{k}}$ is defined by the relation $\mathbf{P}_{\mathrm{k}}^{\perp} = \mathbf{e}_{\mathrm{k}}\ P_{\mathrm{k}}^{\perp}$,

$$\lambda = 8\pi\ nd^2\ \ /\ \hbar\ \omega_0\ \ ,$$

$$\tilde{\omega}_0^2 = \omega_0^2\ [1 + \langle P\rangle^2\ /\ n^2d^2\langle D\rangle^2\ + \beta\ \lambda\ \langle D\rangle]\ .(8.9)$$

If the parameter $\beta = 0$, then the polariton spectrum shows no critical peculiarities. If $\beta > 0$, then at $\langle D\rangle < 0$ (equilibrium system) and $\langle P\rangle = 0$, i.e. for $T > T_c$ the frequency $\tilde{\omega}_0 \longrightarrow 0$, when $T \longrightarrow T_c$. The value of the critical temperature is determined, as it follows from (8.9), from the equation

$$|\langle D\rangle|_{c} = \mathrm{th}\ (\hbar\ \omega_0/\ 2k\ T_B) = 1\ /\ \beta\lambda\ . \qquad (8.10)$$

At $T = T_c$ the frequencies of the lower polariton branch ω_{k}^{-} ($\mathbf{k} \neq 0$) cancel (see Fig. 5.17).

In the point $k = 0$ for $T \longrightarrow T_c$ the derivative

$$\partial\,\omega_k^{-}\ /\ \partial\ (kc) = = 1\ \ /\ (4\pi\ \chi_0)^{1/2} \longrightarrow 0,$$

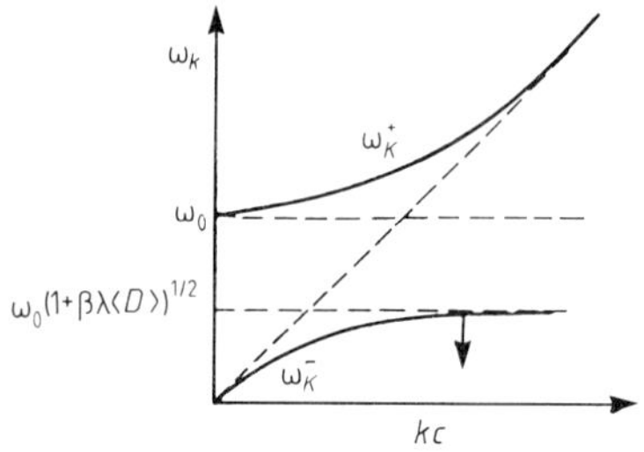

Fig.5.17. The spectrum of polariton excitations in two-level atoms system for $T > T_c$, ω_k^- is the soft mode frequency.

where

$$\chi_0 = \lambda \mid \langle D \rangle \mid \; / \; [4\pi \; (1 + \beta \; \lambda \; \langle D \rangle)]$$

is the static limit of the polarizability of atomic system. Thus, at $T = T_c$ the system becomes unstable in respect to the appearance of the macroscopic polarization on zero frequency.

5.8.2. *Temperature dependence of static polarization.* To obtain the equation for the mean polarization, which is the order parameter of the phase transition considered, we use the mean field approximation. In this approximation the Hamiltonian (8.1) assumes the form

$$H = \hbar \, \omega_0 (\; \frac{1}{2} \sum_i D_i + \frac{1}{4} \beta \, N \, \lambda \, \langle R \rangle^2 - \beta \, \lambda \, \langle R \rangle \; \frac{1}{2} \sum_i \, (\sigma_{i+} + \sigma_{i-}) \;) \quad (8.11)$$

where

$$\langle R \rangle \; = \; \langle \sigma_{i+} \; + \; \sigma_{i-} \rangle \; , \qquad \langle \mathbf{P} \rangle \; = \; n \; \mathbf{d} \; \langle R \rangle \; .$$

The last term in (8.11) plays the ordering role only under the condition $\beta > 0$. Thus, at $\beta \leq 0$ phase transition is impossible. This conclusion is in accordance with the fact, that at $\beta = 0$ there is no soft

mode in the spectrum of collective excitations.
We evaluate the statistical sum for the Hamiltonian (8.11). The free energy is

$$F = - k_B T \ln Z = \beta \lambda N \frac{<R>^2}{4}$$

$$- k_B T \ln \left[2 \cosh \frac{\hbar \omega_0 (1 + \beta^2 \lambda^2 <R>^2)^{1/2}}{2 k_B T} \right]^N$$

$$\cong F^0_{at.} + \frac{N \beta \lambda}{4} (1 - \beta \lambda |<D>|) <R>^2 + B <R>^4, \quad (8.12)$$

where $F^0_{at.}$ is the free energy of the system of N non-interacting atoms, B is positively defined function of T.

From the expansion of F in power series of order parameter <R> it is seen, that at $T = T_c$ the coefficient at $<R>^2$ cancels. Due to this the states with <R> = 0 are unstable or $T < T_c$. The equation for the order parameter $<R>$ follows from the condition of minimum of free energy $\partial F / \partial <R> = = 0$ and has the form

$$\frac{1}{\beta \lambda} = \frac{\text{th} \, [(\hbar\omega_0 / 2 k_B t) (1 + \beta^2 \lambda^2 <R>^2)]}{(1 + \beta^2 \lambda^2 <R>^2)^{1/2}} . \quad (8.13)$$

The population difference per one atom is obtained from the condition

$$< \sum_i D_i > = N <D> = \frac{\partial F}{\partial (\hbar \omega_0 / 2)} .$$

With taking into account of (8.12) we find

$$<D> = - \frac{\text{th}(\hbar \omega_0 / 2 k_B T) (1 + \beta^2 \lambda^2 <R>^2)^{1/2}]}{(1 + \beta^2 \lambda^2 <R^2>)^{1/2}}$$

$$
= \begin{cases} \text{th} \dfrac{\hbar \omega_0}{2 k_B T}, & T \geq T_{c.} \\ 1 / \beta \lambda , & T \leq T_{c.} \end{cases}
$$

The dependence of <P> as a function of T is shown in Fig. 5.18.

We note, that due to (8.9) the frequency of the upper polariton mode with $\mathbf{k} = 0$ starts to grow with the temperature decrease in the region $T < T_c$ i.e. with increase of the order parameter <P>.This can be recorded with the help of electronic Raman scattering in the same way as it is usually done in the studies of Jahn - Teller effect [54].

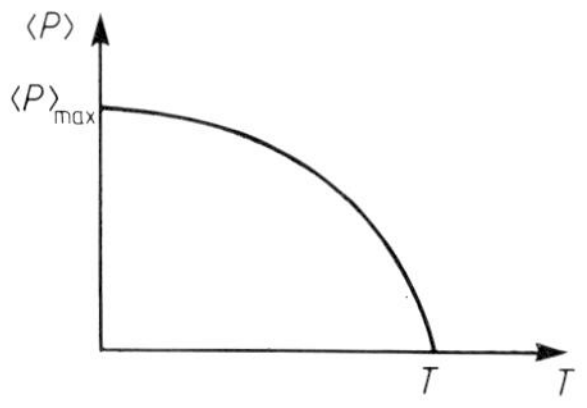

Fig.5.18. The dependence of static macroscopic polarization on temperature. $P_{max} = dn\,(1 - \beta^{-2} \lambda^{-2})^{1/2}$.

In conclusion, the consideration of this section shows, that in the model of the system of two-level atoms, interacting through the acting field the phase transition is possible with the appearance of the macroscopic polarization on zero frequency. As it was noted in [59] this phase transition can play the essential role in ferroelectric electronic ordering under the conditions of high number density of atoms.

CHAPTER 6
LIGHT-INDUCED PHASE TRANSITIONS IN SYSTEMS WITH INTERATOMIC INTERACTIONS THROUGH ELECTROSTATIC OR LATTICE DISPLACEMENT FIELDS

6.1. General characteristics of light-induced phase transitions in systems with electrostatic or phonon interactions

In this chapter we shall study the models of the light-induced phase transitions in the systems of two and three-level atoms, interacting in solids through steady dipole-dipole and electrostatic fields and through the field of lattice displacement. The case of semiconductor is also included into the consideration.

In Sec.6.2 we investigate the system of atoms with the constant dipole moments $\mathbf{d}_i$.We assume, that these atoms are embedded into two sublattices with the oppositely directed constant dipole moments $\mathbf{d}_i$,interacting with each other through dipole-dipole electrostatic forces. In its ground state such system is centrosymmetric, so that the steady macroscopic polarization $\langle \mathbf{P} \rangle = 0$. This model is realized, for example, in the case of doping ions Cr^{3+} in the host lattice of Al_2O_3-working medium of ruby laser or under orientation of dipoles in two opposite directions by action of powerful linear polarized laser wave. We shall show, that when this system is illuminated with the coherent laser light with a certain sign of detuning from the atomic transition frequency, depending on the inversion sign, at certain critical value of the pump intensity E_c^2 the light induced phase transition occurs with the appearance of steady polarization of the medium.This phase transition is connected with the soft relaxation mode ($T_1^{-1*} \equiv \gamma^* \longrightarrow 0$). Its necessary condition

$$2 \pi n d^2 / \hbar \gamma_{ab} > 1 \qquad (6.1)$$

can be compared with Eq.(5.1.4). Note, that in (6.1),instead of the transition dipole moment the constant dipole moment d occurs, that reflects the change of the nature of dominant interatomic interaction in comparison with Eq.(5.1.4). We study in Sec.6.2 also the case of interaction through the lattice displacement field in the same atomic ensemble, which can exceed in strength the dipole-dipole interaction. The order parameter of this light-induced transition is the static displacement of lattice, which appears, when the intensity of laser light, acting on the system, exceeds the critical value.

These phase transitions of both types occur due to the spontaneous breaking of the symmetry of saturation of atomic transitions in the ion sublattices. In Sec.6.3 we consider another possible light- induced phase transition with spontaneous breaking of atomic transition saturation symmetry in the system of $AL_2O_3{:}Cr^{3+}$ type and spontaneous appearance of the electric current, leading to charge separation and appearance of electrostatic field, directed in opposite to the electric current direction. This transition is accompanied by the working frequency shifts of opposite signs in two ions sublattices, i.e. by the splitting of the working line(for example *R*-line in ruby).The phase transition in ruby with the splitting of R-line under the action of external laser pump was experimentally observed in Ref.[1,2](see also Ref.3). We use the theory of light-induced phase transition, developed in Sec. 6.3 in the interpretation of the main experimental results of Ref.[1,2,3].

In Sec.6.4 the same system of non-centrosymmetric atoms with constant dipole moments, but now oriented in one and the same direction, i.e. the doped non-centrosymmetric crystal, is considered. It is known, that the illumination of such a system by the external light wave gives rise to the appearance of static polarization due to the redistribution of the populations of ground and excited energy levels with different values of constant dipole moments [4]. This zero frequency polarization increases with the light wave intensity increase. It is shown, that in a sufficiently dense ensemble of doping atoms the change of the polarization with the pump intensity increase bears the bistable character.The necessary condition of bistability, imposed on the atomic number density, is analogous to the condition (5.1.5). This phase transition occurs due to the soft population relaxation mode $\gamma^{*} \longrightarrow 0$. The jumplike changes of static polarization at $E^2 = E_c^2$ is accompanied by the jumps of the optical absorption, or gain coefficient.

The analogous bistability takes place, when the quasiresonance light wave acts upon the system of two-level atoms, or two-band semiconductor with the electron-phonon interaction. This effect is studied in the model of interaction with optical and acoustic lattice displacement fields in Sec.6.5.

In Sec.6.6 it is shown, that the interaction of doping atoms in solids through the strong electrostatic and lattice displacement fields in the presence of the powerful laser wave can give rise to the light-induced order-disorder phase transition with the ordering of constant dipole orientations.

We consider the mechanism of such phase transition, leading to the threshold generation of second harmonic in centrosymmetric solids. It is important to note, that the additive , relatively weak interaction through SH radiation field , considered in Sec.5.7, plays an important role in this process, arraying the dipoles into the grating with the alternation of their orientations in two opposite directions along the vector of light wave polarization.This automatically ensure the phase matching requirements in SHG, giving rise to the effective SHG. Beside order-disorder transition mechanism we consider also the mechanism of SHG due to spontaneous breaking of vibronic transitions saturation symmetry. The theory of threshold SHG, developed on basis of those mechanisms, is used for the interpretation of the experimental results, obtained in studies of recently discovered [5] phenomenon of threshold effective SHG in glass fibers [5-10].

In the above considerations we have implicitly assumed, that all instabilities and corresponding phase transitions occur in the bulk of the medium.In strongly absorbing semiconductors , when the frequency of the excitation light $\omega > E_g / \hbar$ (E_g is the forbidden gap), the optical absorption length may be as small as 10^{-5}- 10^{-6} cm.In this case laser-induced phase transitions occur in the subsurface layer and this introduces quite new interesting features in the picture of phenomenon.

In Sec.6.7 we study the novel laser-induced electron-deformation-thermal instability(EDTI) and resulting semiconductor-metal phase transition involving superstructure formation on the surface of semiconductor. The appearance of EDTI is due to the modulation of the band gap by deformation, by generation of non-equilibrium carriers, and by heating of a surface layer. Under certain conditions the generation of static ordered structures occur. Development of EDTI creates, depending on the geometry of the pump field and the symmetry of the surface, coupled surface fields of the strain, non-equilibrium carrier density, and temperature in the form of one-dimensional gratings, concentric rings, and radial-ring structures. The periods of these structures and the depth of their penetration into a medium are determined. It is shown, that the EDTI may induce a semiconductor- metal phase transition accompanied by the formation of ordered surface structures of different phases. This theory of the EDTI is used to interpret a whole range of recent experimental data on superfast structural phase transitions and the formation of ordered structures of different phases and of the surface relief created as a result of the interaction between laser pulses and the semiconductor.

In conclusion (Sec.6.8) we briefly review the two novel classes of non-resonant light-induced instabilities and phase transitions- the interferential instabilities of the surface relief and the diffusional-deformational instabilities, occurring on the surface of solids ,and compare them with the resonance light-induced phase transitions, considered in Chapters 5 and 6 .

6.2. Phase transition with appearance of static polarization, or static lattice displacement in centrosymmetric medium under action of light

The phase transitions, considered in the chapter 5, occur due to the interatomic interactions through radiation field. If atoms posses also the static dipole moments, another type of interaction can play the dominant role in some cases - static dipole-dipole interaction, which can be described with the help of static acting Lorentz field. This interaction is shown bellow to lead to the phase transitions of novel type under the variation of the intensity or the frequency of the external light wave, serving as a control parameter of non-equilibrium phase transition.

In this section we consider two types of such phase transitions. To the first belongs the threshold appearance of static polarization in centrosymmetric medium , composed of atoms with constant dipole moment in such a way, that the average value of dipole moment of the system in the initial state cancels .We study this phase transition in the model of the system of non-centrosymmetric doping atoms in host lattice, where the doping atoms occupy two different positions (denoted by indexes *A* and *B*), in which their vector dipoles are opposite to each other, so that the medium is macroscopically centrosymmetric. The important example of such a medium is the traditional laser material-ruby (Al_2O_3 lattice, doped with Cr^{3+} ions). In crystal, besides the interaction through static Lorentz field there is also interatomic interaction through the field of lattice displacement. In this case as an order parameter serves the static lattice displacement, appearing at the critical value of the light pump intensity. Both types of interactions are incorporated into the model considered here and are treated in a unified way. The comparison of different types of interactions through the radiation field, electrostatic Lorentz field, exchange of phonons shows the universal character of effects considered here.

Similar to Sec.5.4 this phase transition occurs due to the saturation of the atomic energy transitions , but its mechanism differs from the that, considered in Sec.5.4. In this case when the control parameter E^2 exceeds the critical value E_c^2 the spontaneous breaking of the symmetry of saturation of electronic resonance transitions in A and B ions occurs. It is connected with laser-induced soft modes: the population relaxation soft mode (the effective constant of population relaxation $\gamma_*^* \to 0$), or vibration soft mode with optical vibration frequency $\Omega_0^* \to 0$. For the first time this phase transition was considered in Ref.11, where the expression for critical intensity of light field E_c^2 was obtained, the analogous expressions were later derived in more elaborated models of this phase transition [12-14]. In Sec.6.3 it is shown, that the spontaneous breaking of the saturation symmetry can lead to the spontaneous appearance of photovoltaic current in the centrosymmetric media [15,16].

It is interesting to note, that the non-equilibrium phase transition with breaking of the saturation symmetry turns out to be formally analogous (in respect to the equations structure) to the equilibrium order-disorder phase transition, considered in Ref.17.

The laser-induced phase transition of the second type consists in appearance of static polarization in the system, composed of centrosymmetric atoms, possessing no constant dipole moments. (gas of the centrosymmetric molecules, or system of centrosymmetric doping atoms). The equilibrium ferroelectric phase transition in such a system was considered in Sec.5.8. In Sec. 6.2.2 it is shown, that the non-equilibrium light-induced phase transition with appearance of static polarization is possible in the medium of centrosymmetric three-level systems with two closely lying upper (or lower) levels, or with two degenerate levels. This phase transition is considered in the case of cooperative interaction through electrostatic Lorentz field.

6.2.1. *Laser-induced phase transition in macroscopically centrosymmetric ensemble of noncentrosymmetric two-level systems.* We consider the model of N_{ab} non-centrosymmetric two-level atoms with transition frequency ω_{ab}, interacting with the electric field E and with the optical vibration mode with wave vector $q = 0$. We denote the normal phonon mode coordinate by U. The equations of such a model has the following form

$$\left(\frac{\partial}{\partial t} + i\,\omega_{ab} + T_2^{-1} \right) \sigma_{j-} = - \frac{i}{\hbar} \mathbf{d}_{abj} D_j \mathbf{E} + \frac{i}{\hbar} \mathbf{d}_j \sigma_{j-} \mathbf{E}$$

$$+ \frac{i}{\hbar} C_{abj} D_j U - \frac{i}{\hbar} C_j \sigma_{j-} U \,, \tag{2.1}$$

$$\frac{\partial D_j}{\partial t} + \frac{1}{T_1} (D_j - D_j^0) = \frac{2i}{\hbar} (\mathbf{d}_{abj} \sigma_{j+} - \mathbf{d}_{baj} \sigma_{j-}) \mathbf{E}$$

$$- \frac{2i}{\hbar} (C_{abj} \sigma_{j+} - C_{baj} \sigma_{j-}) U \,, \tag{2.2}$$

$$\left(\frac{\partial^2}{\partial t^2} + \Omega_0^2 \right) U = - \frac{1}{MN} \sum_j (C_{abj} \sigma_{j+} + C_{baj} \sigma_{j-})$$

$$- \frac{1}{2\,MN} \sum_j C_j D_j \,. \tag{2.3}$$

Here $D_j = \sigma_{aaj} - \sigma_{bbj}$, σ_{j+}, σ_{j-} are the elements of the density matrix of atom in the point $\mathbf{r}_j$; $\mathbf{d}_{abj}$, $\mathbf{d}_{aaj}$, $\mathbf{d}_{bbj}$ ($\mathbf{d} = \mathbf{d}_{aaj} - \mathbf{d}_{bbj}$) are the matrix elements of the dipole moment. The transition frequency ω_{ab} and the equilibrium (stationary) value of the population difference D^0 don't depend on j; M is the reduced mass of the cell, N is the number of the cells in crystal; C_{abj}, C_{aaj}, C_{bbj} ($C = C_{aaj} - C_{bbj}$) are the matrix elements of the potential of interaction of the phonon mode with j–th two–level system. On the coefficients C_j and $\mathbf{d}_j$ we impose the following conditions

$$\sum_j C_j = 0 , \tag{2.4a}$$

$$\sum_j \mathbf{d}_j = 0 , \tag{2.4b}$$

The conditions (2.4) are fulfilled, for example, in the case of two–level systems, modeling the vibrations of protons in two sublattices with asymmetric potentials [14] and also in the ensemble of atoms with random distribution of dipole moments directions (molecular gas[12] or doping Cr^{3+} ions in Al_2O_3 [13]). In macroscopic sense these media are centrosymmetric.

The electric field vector is composed of two parts

$$\mathbf{E} = (\mathbf{e}_L E_L \exp(-i\omega t) + \text{c.c.}) + \vec{\varepsilon}, \tag{2.5}$$

where E_L is the given laser field. The low frequency part of the acting field

$$\vec{\varepsilon} = \beta\, 4\pi\, \mathbf{P}_D,$$

where $\mathbf{P}_D$ is the low frequency part of polarization

$$\mathbf{P} = \mathbf{P}_D + \mathbf{P}_{ab} = \frac{1}{2V} \sum_j \mathbf{d}_j D_j + \frac{1}{V} \sum_j (\mathbf{d}_{abj} \sigma_{j+} + \mathbf{d}_{ba} \sigma_{j-}), \tag{2.6}$$

where V is the crystal volume.

We represent the variables as the sums of the mean values and fluctuations

$$\sigma_{j-} = < \sigma_{j-} > + \sigma_{j-}(\omega) \exp(-i\omega t) + \delta \sigma_{j-}(t) ,$$

$$D_j = < D_j > + \delta D_j(t),$$

$$\overrightarrow{\varepsilon} = \overrightarrow{\varepsilon}_0 + \delta \overrightarrow{\varepsilon}(t) ,$$

$$U = U_0 + \delta U(t) . \tag{2.7}$$

We assume, that in the absence of laser light the mean values of static lattice displacement U_0, of acting field $\overrightarrow{\varepsilon}$, and $< \sigma_{j-} >$ are equal to zero. Let us show the possibility of occurrence of light–induced phase transition with the appearance of non-zero values of U_0, $\overrightarrow{\varepsilon}_0$ and $< \sigma_{j-} >$ (order parameters).

From (2.1), using the condition $\Omega_0 \ll \omega \sim \omega_{ab}$ we obtain

$$\sigma_{j-}(\omega) = - \frac{(\mathbf{d}_{abj}\, \mathbf{e}_L)\, E_L}{\hbar \Delta + C_j U - \mathbf{d}_j \overrightarrow{\varepsilon} - i \hbar T_2^{-1}} , \tag{2.8}$$

where $\Delta = \omega_{ab} - \omega$.

Substituting (2.8) into (2.2), we obtain the effective equation for D_j :

$$\frac{\partial D_j}{\partial t} + \frac{1}{T_1}(D_j - D^0) = - \frac{T_1^{-1} \hbar^2 T_2^{-2}\, a\, | E_L |^2 D_j}{(\hbar \Delta + C_j U - \mathbf{d}_j \overrightarrow{\varepsilon})^2 + \hbar^2 T_2^{-2}} , \tag{2.9}$$

where a is the saturation parameter. From (2.1), (2.9), (2.6) one obtains the following equations for the mean values :

$$< D_j > = D^0 \frac{(\hbar \Delta + C_j U_0 - \mathbf{d}_j \overrightarrow{\varepsilon}_0)^2 + T_2^{-2} \hbar^2}{(\hbar \Delta + C_j U_0 - \mathbf{d}_j \overrightarrow{\varepsilon}_0)^2 + \hbar^2 \gamma_E^2} , \tag{2.10}$$

$$U_0 = - \frac{U_0}{N} \sum_j \frac{2 \mid C_{abj} \mid^2 < D_j >}{\hbar \omega_{ab} M \Omega_0^2}$$

$$+ \frac{\varepsilon_0}{N} \sum_j \left[\frac{C_{abj} (\mathbf{d}_{baj} \mathbf{e}_0) + \text{c.c.}}{\hbar \omega_{ab} M \Omega_0^2} < D_j > \right]$$

$$- \frac{1}{2 M N \Omega_0^2} \sum_j C_j < D_j > , \tag{2.11}$$

$$\varepsilon_0 = - \frac{\varepsilon_0}{V} \sum_j \left[\frac{8\pi \beta \mid \mathbf{d}_{abj} \mathbf{e}_0 \mid^2}{\hbar \omega_{ab}} < D_j > \right]$$

$$+ \frac{U_0}{V} \sum_j \left[4\pi \beta \frac{(C_{abj} (\mathbf{d}_{baj} \mathbf{e}_0) + \text{c.c.}}{\hbar \omega_{ab}} < D_j > \right]$$

$$+ \frac{2\pi \beta}{V} \sum_j (\mathbf{d}_j \mathbf{e}_0) < D_j > . \tag{2.12}$$

Here $\vec{\varepsilon}_0 = \mathbf{e}_0 \varepsilon_0$, $|\mathbf{e}_0| = 1$.

In the case of centrosymmetric two–level systems $C_j = 0$, $\mathbf{d}_j = 0$ and from Eqs. (2.11) and (2.12) we obtain the condition of the instability of zero solution $U_0 = 0$, $\varepsilon_0 = 0$:

$$(1 + K_1 D^0)(1 + K_2 D^0) - (K_{12} D^0)^2 = 0 , \tag{2.13}$$

where

$$K_1 = \frac{1}{N} \sum_j \left[2 \frac{\mid C_{abj} \mid^2}{\hbar \omega_{ab} M \Omega_0^2} < D_j > / D^0 \right] ,$$

$$K_2 = \frac{4\pi \beta}{V} \sum_j \left[\frac{2 \mid \mathbf{d}_{abj} \mathbf{e}_0 \mid^2}{\hbar \omega_{ab}} < D_j > / D^0 \right] ,$$

$$K_{1,2} = \left[\frac{4\pi \ \beta}{V N M \Omega_0^2} \right]^{1/2} \sum_j \left[\frac{C_{abj} (\mathbf{d}_{baj} \ \mathbf{e}_o) + \text{c.c.}}{\hbar \omega_0} <D_j>/D^0 \right] .$$

In equilibrium

$$D_j = D^0 = - \text{th} (\hbar \omega_{ab} / 2 k_B T)$$

and from (2.13) one can obtain the expression for the critical temperature of the equilibrium phase transition. If the interatomic interaction through the lattice displacement is the strongest one, i.e.

$$K_1 \gg K_2 , K_{12},$$

then critical temperature T_c is determined from the equation

$$1 + K_1 D^0 = 0 ,$$

which yields expression, coinciding with the corresponding formula for T_c in vibronic theory of phase transitions [18,19].

If the electrostatic interactions prevails, i.e.

$$K_2 \gg K_{1,} K_{12},$$

then T_c is determined from the equation

$$1 + K_2 D^0 = 0,$$

which coincides with the condition of the equilibrium ferroelectric phase transition in the system of two-level atoms (compare the formula (5.8.10)).

The effect of action of laser field on the medium of two-level centrosymmetric systems, as it is seen from (2.13), where

$$\frac{|<D_j>|}{|D^0|} < 1,$$

consists in the decreasing of the T_c value.

A qualitatively different type of light–induced phase transition is possible in the medium of the non-centrosymmetric two–level systems. It is determined by the last terms in the right hand side of eqs. (2.11) and (2.12). As it will be seen from the following, at $C \sim C_{ab}$, $aE_L^2 \sim 1$ and $|\Delta| \sim T_2^{-1}$ these terms turn out to be of order of $\omega_{ab} T_2 \gg \mathbf{1}$, comparing with other terms in the right hand side, which we shall neglect.

Let us consider the simplest model, satisfying the condition (2.4). Let the two–level atoms occupy the position A and B in the two sublattices, in which their permanent dipole moment $\mathbf{d}_j$ have the opposite directions: $(\mathbf{d}_j \mathbf{e}_0) = \pm d$ and in addition the constants C_j are of opposite signs: $C_j = \pm C$. We introduce the dimensionless parameters

$$\xi = \frac{U_0 C - \varepsilon_0 d}{\hbar T_2^{-1}} , \qquad \delta = \Delta T_2 , \quad \eta = aE_L^2 > 0 ,$$

$$K = - D^0 \frac{n_{ab}}{4 \hbar T_2^{-1} n} \left[\frac{C^2}{M \Omega_0^2} + \beta 4\pi n d^2 \right] , \qquad (2.14)$$

where $n_{ab} = N_{ab} / V$, $n = N / V$.

Then from (2.10) – (2.12) we obtain the equation for the combined order parameter

$$\xi = K \left[\frac{(\delta + \xi)^2 + 1}{(\delta + \xi)^2 + 1 + \eta} - \frac{(\delta - \xi)^2 + 1}{(\delta - \xi)^2 + 1 + \eta} \right] . \qquad (2.15)$$

Nontrivial solution of Eq.(2.15)has the form

$$(\xi^2)_{1,2} = - (1 + \eta - \delta^2) \pm [4 K \delta \eta - 4 \delta^2 (1 + \eta)]^{1/2} . \qquad (2.16)$$

If the initial system is not inverted, i.e. $D^0 < 0$ and thus $K > 0$, then the real solution for ξ^2 is obtained under the condition $\delta > 0$. Thus, under the action of laser light with the frequency ω on the non-inverted system of doping atoms the phase transition occurs only in the case $\omega < \omega_{ab}$. If the system is inverted initially, i.e. $D^0 > 0$ and $K < 0$, then the phase transition takes place under the condition $\delta < 0$, i.e. $\omega > \omega_{ab}$.

The positive solutions for ξ^2 correspond to the ordered state. The condition of positiveness of ξ^2 is the inequality $K > K_c = \delta + \delta^{-1} \geq 2$ at $D^0 < 0$. This is the necessary condition imposed on the cooperative parameter K , determining the interaction strength.

In the region, defined by the conditions $K_c < K$ and $K < \delta^3 / (\delta - 1)$ (see Fig. 6.1) the phase transition of the second kind takes place.

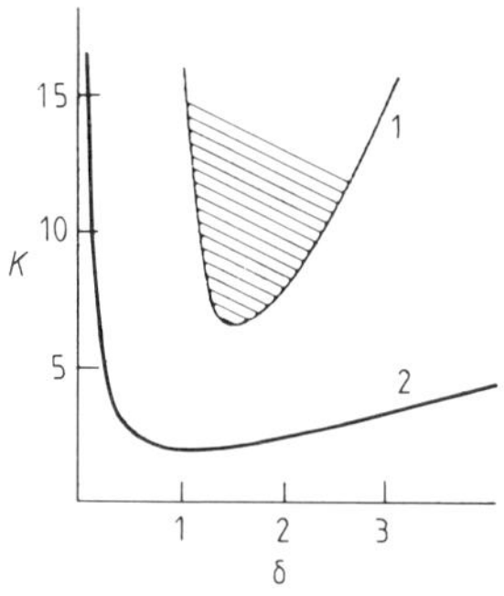

Fig.6.1. The regions of parameters K and δ, where the light–induced phase transition is possible. In the region between the curves 1 and 2 the phase transition of the second kind occurs, the dashed region corresponds to the phase transition of the first kind.

With the laser field intensity increase the ordered state appears at the value of the pump intensity $\eta = \eta_1$ and disappears at $\eta = \eta_2$ (see Fig. 6.2), where

$$\eta_{1,2} = \delta \{ 2K - K_c \mp 2 [K(K - K_c)]^{1/2} \} . \tag{2.17}$$

In the region $\delta > 1$, $K > \delta^3 / (\delta^2 - 1)$ when the pump intensity η is increased, the order parameter ξ appears in jump like manner at $\eta = \eta_1$; when the pump intensity is decreased it disappears at $\eta = \eta_0 = \delta / (K - \delta) < \eta_1$ (hysteresis, see Fig. 6.2b).

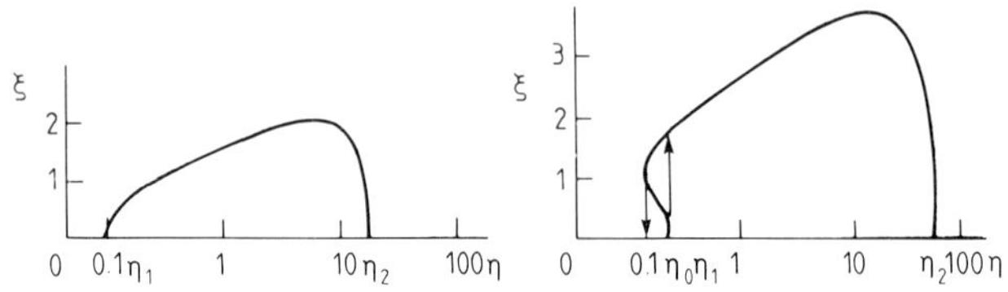

Fig.6.2. The dependence of the order parameter ξ on the pump intensity $\eta = aE_L^2$ (the solutions of Eq.(5.9.15): a – the second kind transition for $K > \delta^{-1} + \delta$ and $\delta > 0$; b – the first kind transition for $K > \delta^3 / (\delta - 1)$ and $\delta > 1$).

To obtain the order parameters U_0 and ε_0 we substitute (2.10) into (2.11) and (2.12), neglecting there again the terms with d_{ab} and C_{ab}. Taking into account of (2.14) and (2.15), we have

$$U_0 = - \xi \frac{C D^0 n_{ab}}{4 M K n \Omega_0^2} , \quad \varepsilon_0 = \xi \frac{\beta \pi n_{ab} d D^0}{K} \quad (2.18)$$

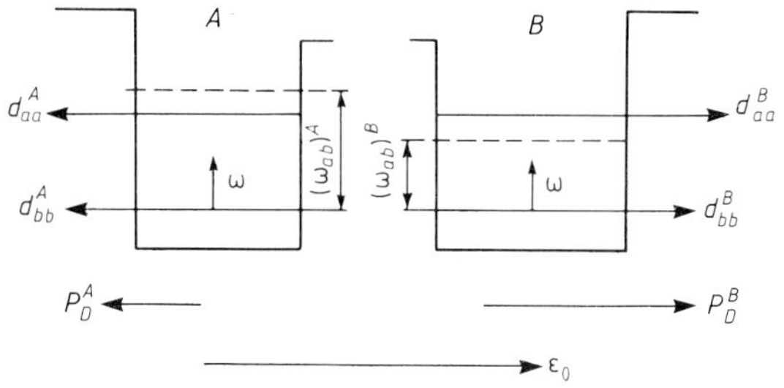

Fig.6.3. The physical mechanism of the polarizational laser–induced phase transition in the centrosymmetric medium.

The physical mechanism of the phase transition is explained in Fig. 6.3. Here the non-centrosymmetric potentials of the ions *A* and *B* and two resonance energy levels in each potential well are shown. The permanent dipole moment vectors $\mathbf{d}^A_{aa,bb}$ and $\mathbf{d}^B_{aa,bb}$ as well as the vectors of the dipole moment change under the excitation $\mathbf{d}^A = \mathbf{d}^A_{aa} - \mathbf{d}^A_{bb}$, $\mathbf{d}^B = \mathbf{d}^B_{aa} - \mathbf{d}^B_{bb}$ are directed oppositely to each other in ions of *A* and *B* type. Let due to a fluctuation the static field $\vec{\varepsilon}_0$ has arisen in the medium. It can be directed either parallel or antiparallel to $\mathbf{d}^B$. Assume, that it is parallel to $\mathbf{d}^B$. We write down the expressions for the polarization vectors of the ions subsystems $\mathbf{P}^A_D$ and $\mathbf{P}^B_D$, retaining only the non-equilibrium parts, arising due to the action of light field. From (2.6) and (2.10) we obtain, defining $\mathbf{d}^B \equiv \mathbf{d}$, for the case $D^0 < 0$:

$$\mathbf{P}_D = \mathbf{P}^B_D + \mathbf{P}^A_D = \mathbf{d}\, n\, |D^0|\, \frac{a E_L^2}{2}$$

$$\times \left\{ \left[\left(\Delta - \frac{d\,\varepsilon_0}{\hbar} \right)^2 + \gamma_E^2 \right]^{-1} - \left[\left(\Delta + \frac{d\,\varepsilon_0}{\hbar} \right)^2 + \gamma_E^2 \right]^{-1} \right\} \quad (2.19)$$

In the case $\Delta > 0$, the effective detuning, determined by the linear Stark shift of the levels , turns out to be larger in *A*–ions, then in *B*–ions. Due to this the resonance transition in *B*–ions are saturated more strongly, then in *A*–ions and from (2.19) one can see, that $|\mathbf{P}^B_D| > |\mathbf{P}^A_D|$.

Thus, the appearance of fluctuating field $\vec{\varepsilon}_0$ gives rise, due to the breaking of the saturation symmetry, to the appearance of additive polarization vector, directed along the seeding field $\vec{\varepsilon}_0$, which enhances the magnitude of $\vec{\varepsilon}_0$. Due to this positive feedback the instability sets in for $\Delta > 0$ and the system undergoes the transition into the state with non-zero value of static polarization. If the initial detuning has the opposite sign ($\Delta < 0$), then the fluctuation $\vec{\varepsilon}_0$ is , on the contrary, suppressed and the phase transition doesn't occur. We note, that the above consideration is valid at $D^0 < 0$. Thus, the necessary condition of the phase transition is the inequality $\Delta D^0 < 0$, i.e. one needs $\Delta > 0$ for $D^0 < 0$, or $\Delta < 0$ for $D^0 > 0$.

6.2.2. *Appearance of static polarization in medium of centrosymmetric three-level systems.* Let us now consider the ensemble of the systems with three energy levels $|a>$,$|b>$ and $|c>$ (Fig. 6.4). We assume that the transition frequency $\omega_{ab} \ll \omega_{bc}$, the case $\omega_{ab} = 0$ is included as well. For the sake of simplicity we limit ourselves with the consideration of only dipole interaction with electric field. We assume, that the matrix elements of dipole moment $\mathbf{d}_{ac}$, $\mathbf{d}_{bc} \neq 0$ and are the same for all systems and also $\mathbf{d}_{aa} = \mathbf{d}_{bb} = \mathbf{d}_{cc} = \mathbf{d}_{ac} = 0$. The light–induced phase transition, considered bellow, exists also in the case of arbitrary distribution of direction of the vectors $\mathbf{d}_{ab}$, in particular under the isotropic distribution of the directions of $\mathbf{d}_{ab}$ of different systems,

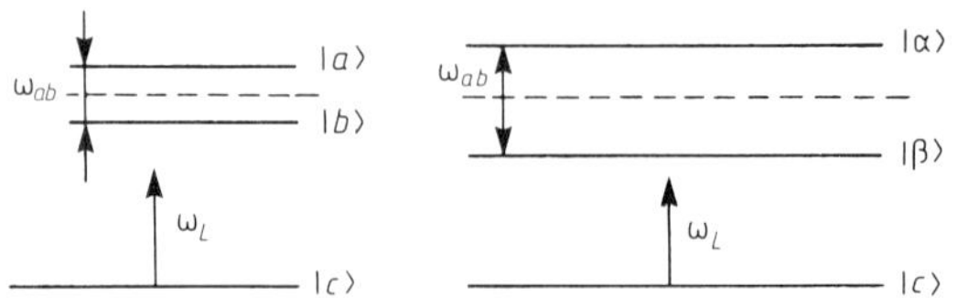

Fig.6.4. The three–level system with closely lying upper levels $|a>$ and $|b>$ in laser field with frequency ω_L : a – the static field $\varepsilon = 0$; b – the static field $\varepsilon \neq 0$ and $\omega_{\alpha\beta} = [\,\omega_{ab}^2 + (\,2\,|d_{ab}|\,\varepsilon / \hbar\,)^2\,]^{1/2}$.

because, as it will be shown, the appearing diagonal elements of the dipole moment are not changed under the substitution $\mathbf{d}_{ab} \to -\mathbf{d}_{ab}$ (see the formula (2.24)).

The Hamiltonian of the considered three–level system has the form

$$H = \omega_{ab}\frac{1}{2}(\rho_{aa} - \rho_{bb}) - (\omega_{bc} + \frac{\omega_{ab}}{2})\rho_{cc} - (d_{ab}\rho_{ba} + \text{c.c.})\vec{\varepsilon}$$

$$- (\mathbf{d}_{bc}\rho_{cb} + \text{c.c.})\vec{\varepsilon}, \qquad (2.20)$$

where the energy is measured from the middle of the energy separation between the levels $|a>$ and $|b>$, $\vec{\varepsilon} = \mathbf{e}_0\varepsilon$ is the acting static field, which, as it will be shown, can spontaneously arise in the medium under the action of laser field(ε is the order parameter), ρ_{aa}, ρ_{bb}, ρ_{ab}, ρ_{bc} are the matrix elements of the density matrix.

Let us assume, that the field ε arose as a fluctuation. It intermixes and shifts the levels $|a>$ and $|b>$, but, under the condition

$$\mathbf{d}_{bc}\vec{\varepsilon} \ll \hbar\,\omega_{bc}, \quad \omega_{ab} \ll \omega_{bc} \qquad (2.21)$$

almost doesn't influence the $|c>$–state.

We diagonalize Hamiltonian (2.20) with taking into account of the Eq.(2.21). Then we obtain

$$H = [(\frac{\hbar\omega_{ab}}{2})^2 + |d_{ab}|^2\varepsilon^2]^{1/2}(\rho_{\alpha\alpha} - \rho_{\beta\beta})$$

$$- \hbar\,(\omega_{bc} + \frac{\omega_{ab}}{2})\rho_{cc}\,. \qquad (2.22)$$

Here $\rho_{\alpha\alpha}$, $\rho_{\beta\beta}$ are the elements of the density matrix of the new states

$$|\alpha> = \lambda\,|\alpha> - \lambda\,\sigma\,d_{ba}\varepsilon\,|b>\,, \quad |\beta> = \lambda\,\sigma\,d_{ab}\varepsilon\,|a> + \lambda\,|b>\,, \qquad (2.23)$$

where

$$d_{ab} = (\mathbf{d}_{ab}\mathbf{e}_0) , \quad \sigma = [(\frac{\hbar\,\omega_{ab}}{2})^2 + |d_{ab}|^2\,\varepsilon^2)^{1/2}\,]^{-1},$$

$$\lambda = (1 + \sigma^2\,|d_{ab}|^2\,\varepsilon^2)^{-1/2} .$$

Using (2.23), we obtain the matrix elements of the dipole moment of the new states :

$$\mathbf{d}_{\beta\beta} = -\,\mathbf{d}_{\alpha\alpha} = (\mathbf{d}_{ab}d_{ba} + \mathbf{d}_{ba}d_{ab})\,\lambda^2\sigma\,\varepsilon ,$$

$$\mathbf{d}_{\alpha\beta} = \lambda^2\mathbf{d}_{ab} - \lambda^2\sigma^2 d_{ab}^2\varepsilon^2\mathbf{d}_{ba},$$

$$\mathbf{d}_{\alpha c} = -\,\lambda\,\sigma\,d_{ab}\varepsilon\,\mathbf{d}_{bc}, \quad \mathbf{d}_{\beta c} = \lambda\,\mathbf{d}_{bc} . \tag{2.24}$$

Thus, due to the presence of the field ε there appears the static polarization

$$\mathbf{P} = n_3\mathbf{d}_{\beta\beta}\,(\rho_{\beta\beta} - \rho_{\alpha\alpha}) \tag{2.25}$$

where n_3 is the concentration of the three–level systems and $\rho_{\alpha\alpha}$ and $\rho_{\beta\beta}$ depend on ε and the external laser field. From the definition of the acting field

$$\vec{\varepsilon} = \beta\,4\pi\,\mathbf{P}$$

and (2.25) we obtain the closed equation for the order parameter

$$\varepsilon = \beta\,8\pi\,n_3\,|d_{ab}|^2\,\lambda^2\sigma\,[\rho_{\beta\beta}\,(\varepsilon) - \rho_{\alpha\alpha}\,(\varepsilon)]\,\varepsilon . \tag{2.26}$$

Let the frequency of laser field

$$\mathbf{E}_L = \mathbf{e}_L E_L\,\exp\,(-\,i\omega_L t)\,+\text{c.c} , \tag{2.27}$$

is close to the frequency ω_{bc}. We write down the equations for the density matrix of the new states in the presence of the laser field (2.27) :

$$\left(\frac{\partial}{\partial t} + \gamma_{\alpha c} + i\,\omega_{\alpha c}\right)\rho_{\alpha c} = -\frac{i}{\hbar}\,\mathbf{d}_{\alpha c}\mathbf{E}_L\,(\rho_{\alpha\alpha} - \rho_{cc})\,,$$

$$\left(\frac{\partial}{\partial t} + \gamma_{\beta c} + i\,\omega_{\beta c}\right)\rho_{\beta c} = -\frac{i}{\hbar}\,\mathbf{d}_{\beta c}\mathbf{E}_L\,(\rho_{\beta\beta} - \rho_{cc})\,. \quad (2.28)$$

Here

$$\omega_{\alpha c} = \frac{\omega_{ac} - \omega_{ab}}{2} + \left[\left(\frac{\omega_{ab}}{2}\right)^2 + |d_{ab}|^2\frac{\varepsilon^2}{\hbar^2}\right]^{1/2}. \quad (2.29)$$

$$\omega_{\beta c} = \frac{\omega_{bc} + \omega_{ab}}{2} - \left[\left(\frac{\omega_{ab}}{2}\right)^2 + |d_{ab}|^2\frac{\varepsilon^2}{\hbar^2}\right]^{1/2}. \quad (2.30)$$

Further,

$$\frac{\partial\rho_{\alpha\alpha}}{\partial t} + \gamma_\alpha(\rho_{\alpha\alpha} - \rho^0_{\alpha\alpha}) + \gamma(\rho_{\alpha\alpha} - \rho_{\beta\beta})$$

$$= \frac{i}{\hbar}\,(\mathbf{d}_{\alpha c}\rho_{\alpha c} - \mathbf{d}_{c\alpha}\rho_{\alpha c})\,\mathbf{E}_L, \quad (2.31)$$

$$\frac{\partial\rho_{\beta\beta}}{\partial t} + \gamma_\beta(\rho_{\beta\beta} - \rho^0_{\beta\beta}) + \gamma(\rho_{\beta\beta} - \rho_{\alpha\alpha})$$

$$= \frac{i}{\hbar}\,(\mathbf{d}_{\beta c}\rho_{c\beta} - \mathbf{d}_{c\beta}\rho_{\beta c})\,\mathbf{E}_L\,,$$

$$\frac{\partial\rho_{cc}}{\partial t} + \gamma_c(\rho_{cc} - \rho^0_{cc}) = \frac{i}{\hbar}\,(\mathbf{d}_{c\alpha}\rho_{\alpha c} - \mathbf{d}_{\alpha c}\rho_{c\alpha})\,\mathbf{E}_L$$

$$+ \frac{i}{\hbar}\,(\mathbf{d}_{c\beta}\rho_{\beta c} - \mathbf{d}_{\beta c}\rho_{c\beta})\,\mathbf{E}_L\,. \quad (2.32)$$

In Eqs.2.28) – (2.32) we have phenomenologically introduced the relaxation
constants γ, γ_i, γ_{ij} $(i,j = \alpha, \beta, c)$ and retained only resonance terms.

Assuming for the simplicity sake, that $\rho^0_{\alpha\alpha} = \rho^0_{\beta\beta} = 0$, we obtain the stationary solutions :

$$\rho_{\beta\beta} - \rho_{\alpha\alpha} = \frac{\rho^0_{cc}\,(A_\beta - A_\alpha)}{f_0 + f_1 A_\alpha + f_2 A_\beta + f_3 A_\alpha A_\beta}\,, \tag{2.33}$$

where

$$A_\alpha = \frac{2\,|\mathbf{d}_{\alpha c}\mathbf{e}_L|^2\,E^2_L}{\hbar^2\,(\Delta^2_\alpha + \gamma^2_{\alpha c})}\,\frac{\gamma_{\alpha c}}{\gamma_\alpha}\,,$$

$$A_\beta = \frac{2\,|\mathrm{d}_{\beta c}\mathrm{e}_L|^2\,E^2_L}{\hbar^2\,(\Delta_\beta + \gamma^2_{\beta c})}\,\frac{\gamma_{\beta c}}{\gamma_\beta}\,.$$

Here

$$f_0 = 1 + \frac{\gamma}{\gamma_\alpha} + \frac{\gamma}{\gamma_\beta}\,,$$

$$f_1 = 1 + \frac{\gamma_\alpha}{\gamma_c} + \frac{\gamma}{\gamma_\beta} + 2\,\frac{\gamma\,\gamma_\alpha}{\gamma_\beta\gamma_c}\,,$$

$$f_2 = 1 + \frac{\gamma_\beta}{\gamma_c} + \frac{\gamma}{\gamma_\alpha} + 2\,\frac{\gamma\,\gamma_\beta}{\gamma_\alpha\gamma_c}\,,$$

$$f_3 = 1 + \frac{\gamma_\alpha}{\gamma_c} + \frac{\gamma_\beta}{\gamma_c}\,,$$

$$\Delta_{(\alpha/\beta)} = \Delta \pm \left[\left(\frac{\omega_{ab}}{2}\right)^2 + \frac{|d_{ab}|^2\,\varepsilon^2}{\hbar^2}\right]^{1/2},$$

$$\Delta = \frac{1}{2}\,(\omega_{bc} + \omega_{ab}) - \omega\,.$$

Let us consider in more details the case of degenerate state $\omega_{ab} =$

0. Then in the Eq. (2.26) we have

$$|d_{ab}|^2 \lambda^2 \sigma\varepsilon = \frac{1}{2} |d_{ab}|$$

and it can be transformed to the form

$$\varepsilon = \beta\, 2\pi\, n_3 |d_{ab}| \rho^0_{cc} \frac{A_\beta - A_\alpha}{f_0 + f_1 A_\alpha + f_2 A_\beta + f_3 A_\alpha A_\beta}, \qquad (2.34)$$

where now

$$A_\alpha = |\mathbf{d}_{bc}\mathbf{e}_L|^2 E_L^2 \frac{\gamma_{\alpha c}}{\hbar^2 [(\Delta + |d_{ab}|\ \varepsilon / \hbar)^2 + \gamma^2_{\alpha c}]\ \gamma_\alpha},$$

$$A_\beta = |\mathbf{d}_{bc}\mathbf{e}_L|^2 E_L^2 \frac{\gamma_{\beta c}}{\hbar^2 [(\Delta - |d_{ab}|\ \varepsilon / \hbar\ )^2 + \gamma^2_{\beta c}]\ \gamma_\beta}.$$

Equation (2.34) for ε is analogous to Eq. (2.15), studied before. In the case $\gamma \ll \gamma_\alpha = \gamma_\beta \ll \gamma_c$ and $\gamma_{\alpha c} = \gamma_{\beta c}$ Eq. (2.34) acquires exactly the form (2.15), where now

$$K = -\ \rho^0_{cc} \frac{\beta\ 2\pi n_3 |d_{ab}|^2}{\hbar\ \gamma_{\alpha c}}, \qquad \delta = \frac{\Delta}{\gamma_{\alpha c}},$$

$$\eta = \frac{|\mathbf{d}_{bc}\ \mathbf{e}_L|^2 E_L^2}{\hbar^2\ \gamma_{\alpha c}\ \gamma_\alpha}, \qquad \xi = -\ \frac{\varepsilon\ |d_{ab}|}{\hbar\ \gamma_{\alpha c}}.$$

Because of that the solution of Eq. (2.34) in this case is completely analogous to the formula (2.16).

Thus, in the case, when $\omega_{ab} = 0$, as it follows from the analysis of the solution of eq. (2.16), the laser–induced transition with appearance of static electric field ε is possible. The dependence of ε on the light wave intensity (i.e. on η) is analogous to that, obtained in Sec. 6.2.1.

Similar light–induced phase transition is possible also in case , when $\omega_{ab} \neq 0$. As it follows from (2.26), the solution $\varepsilon = 0$ becomes unstable under the condition

$$1 = \frac{\beta\, 4\pi\, n_3\, |d_{ab}|^2}{[(\hbar\, \omega_{ab}\ /\ 2)^2 + |d_{ab}|^2\, \varepsilon^2]^{1/2}}$$

$$\times\ [\rho_{\beta\beta}\,(\varepsilon,E_L) - \rho_{\alpha\alpha}\,(\varepsilon,E_L)]\ . \tag{2.35}$$

Eq. (2.35) is the equation for the order parameter ε. From it with taking into account of (2.23), we obtain the critical value of the population difference at which the phase transition occurs :

$$(\rho_{\beta\beta} - \rho_{\alpha\alpha})_c = \frac{\hbar\, \omega_{ab}}{\beta\, 8\pi\, n_3\, |d_{ab}|^2}\ . \tag{2.36}$$

We note, that the eq. (2.36) coincides formally with the condition of equilibrium phase transition in the system of two–level atoms, interacting through the electromagnetic field (Sec.5.8). However in the non-equilibrium case in contradistinction to the equilibrium one, the population difference $\rho_{\beta\beta} - \rho_{\alpha\alpha}$ is created by the external pump. At $\omega_{ab} \sim 10^{12}\ s^{-1}$, $\beta \sim 1$, $n_3 \sim 10^{21}\ sm^{-3}$, $d_{ab} \sim 1$ D we have for the critical population difference $(\rho_{\beta\beta} - \rho_{\alpha\alpha})_c \sim 10^{-2}$.

The nature of instability in the case of centrosymmetric three–level systems is analogous to that considered in Sec. 6.2.1. The fluctuating field ε brings about the appearance of the diagonal elements of the dipole moment (2.24). The polarization (2.25), arising due to this, under the condition $(\rho_{\beta\beta} - \rho_{\alpha\alpha}) > 0$ enhances the seeding field ε, leading to the occurrence of the instability. The population difference $(\rho_{\beta\beta} - \rho_{\alpha\alpha})$ is created by laser radiation(see Eq.(2.33)). In the degenerate case to ensure the fulfillment of the condition $(\rho_{\beta\beta} - \rho_{\alpha\alpha}) > 0$ it is necessary to have $\Delta = \omega_{bc} - \omega > 0$ (see (2.34)).

As we have seen, the above considered light–induced phase transition with breaking of saturation symmetry in macroscopically centrosymmetric medium, leads to the shift of transition frequency, when the light field intensity exceeds its critical value. Under certain conditions it can occur in a laser medium itself (the consideration of Sec. 6.2.1, for example, is based on the model, realized in the working medium of ruby laser). Self–induced phase transition in a laser must lead to the working frequency shift at some critical value of intensity of generation, that is to the automodulation of the gain coefficient, which can essentially influence generation dynamics. Another manifestation of saturation breaking phase transition can be the optical rectification in centrosymmetric media, for example, in gases (see Rev.20). The frequency shifts and appearance of new lines under the action of laser light may be interesting also for the spectroscopy.

Note in conclusion, that the model with the interatomic interaction through the lattice displacement, considered in Sec. 6.2.1, can be generalized to the case of semiconductor (see Sec. 6.7).

6.3. Laser-induced phase transition with appearance of photovoltaic current and electrostatic field in centrosymmetric crystal: spontaneous breaking of saturation symmetry

In this section we consider another important type of phase transition, occurring due to the spontaneous breaking of the symmetry of saturation of atomic transitions in laser field. The effect consists in the threshold appearance of photovoltaic current in centrosymmetric crystals of ruby type, leading to appearing of strong electrostatic field and quasi-Stark splitting of electronic states. The theory of this phase transition [15,16] is used to interpret the results of studies of laser-induced phase transition , discovered in ruby [1,2].

The scheme of energy transitions in Cr^{3+}-ions in ruby (Al_2O_3: Cr^{+3}), pertaining to the experimental conditions of Ref.1,2 is shown in Fig. 6.5. Quantum of argon laser radiation hω (λ = 514.5 nm) excites the electron-vibration level $4T_2$ (the vibration levels of $4T_2$ state at low temperatures (T < - 100^0 C) are well resolved [21]). Afterwards, in time $\gamma_1^{-1} \sim 10^{-10}$ s atom makes the transition to the metastable state ^2E (denoted by index c). The transition to the ground state $^4A_2$ (the level b) occurs in time $\gamma^{-1} = 3.6\ W\ 10^{-3}$ s. The upward transition with the absorption of the second quantum ħω excites the electron into the band with the energy ~ 4.2 eV, appearing at sufficiently high concentration of doping ions [1], that gives rise to the photovoltaic current pulse (Fig.6.6).

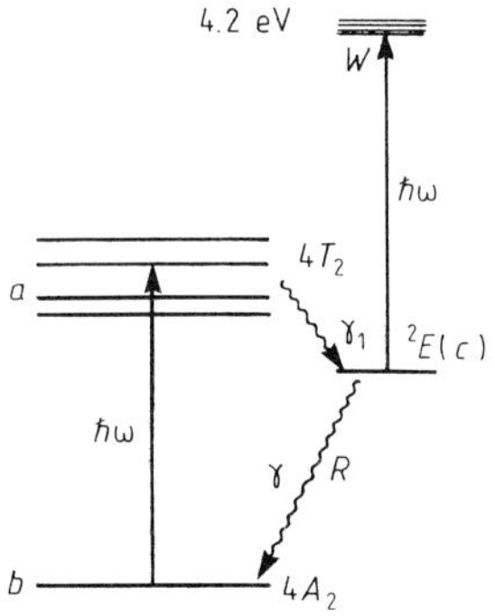

Fig.6.5. The scheme of transitions in Cr^{3+}-ions in AL_2O_3. The band with the energy ~ 4.2 eV appears at high doping concentration [1]

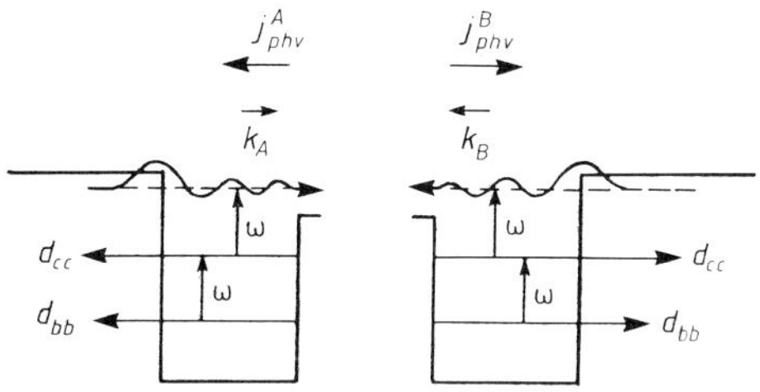

Fig.6.6. Appearance of photovoltaic current pulses under two-photon excitation into the states with the quasi momenta $\hbar k_A$ and $\hbar k_B$. In centrosymmetric medium (ruby) the currents j^A_{phv} and j^B_{phv} are directed oppositely to each other.

In Ref. [1] it was discovered, that under the steady illumination of ruby the splitting of the luminecscent *R*-line into two components occurs (i.e. the splitting of $^2E \longrightarrow 4A_2$ transition). This effect is observed only at sufficiently high concentrations of the doping ions, exceeding 0.25 %. The splitting $\Delta\nu$ disappears gradually with the further pump intensity ($I_L \sim E_L^2$) increase (see Fig.6.7).

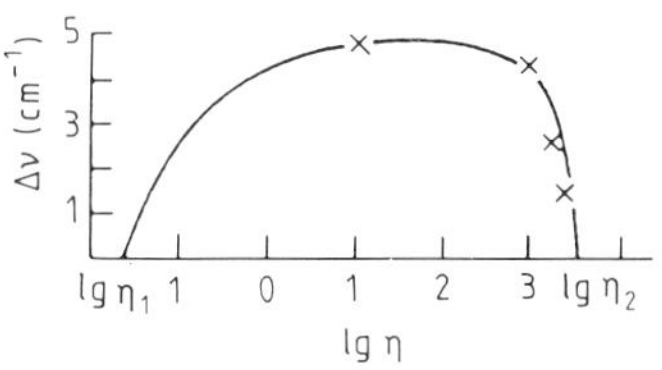

Fig.6.7. The dependence of the *R*-line splitting on the pump intensity.The solid line is calculated with the help of Eqs.(6.10.22) and (6.10.20) with K = 40, δ = 20, $q_0 = 10^{-3}$, T_2^{-1}=6 × 10^{11}s^{-1} (T=10 K), d_c / d = 0.02 and a = 5 10^2. The experimental points are from Ref.1.

The dependence of the rate of splitting increase on the pump intensity obeys the quadratic law $\partial\,\Delta\nu / \partial t \sim I_L^2$ in the region up to 10^2 W / cm^2, then the saturation follows.

Quasi-Stark splitting of *R*-line occurs due to the levels shift in electrostatic field, arising owing to the photovoltaic current j_{phv}. The Cr-ions in ruby occupies two positions *A* and *B* with oppositely directed dipole moments $\mathbf{d}^{A,B}$. In the ground state - $\mathbf{d}_{aa}^{A} = \mathbf{d}_{aa}^{B} \equiv \mathbf{d}_{aa}$, in the excited state - $\mathbf{d}_{bb}^{A} = \mathbf{d}_{bb}^{B} \equiv \mathbf{d}_{bb}$. In electrostatic field $\mathbf{E}_0$ the Stark shift of the

c-levels in ions of A type, equal to $\mathbf{d}_{cc}^{A}\mathbf{E}_0/\hbar$ has the sign opposite to the sign of shift in ions of B type: $\mathbf{d}_{cc}^{B}\mathbf{E}_0/\hbar = -\mathbf{d}_{cc}^{A}\mathbf{E}_0/\hbar$ which leads to the observed effect of R-line splitting.

The electrostatic field is created with the charge separation arising due to the photovoltaic current. This phenomenon is well known in non-centrosymmetric crystals like $LiNbO_3$ [22]. However ruby crystal is centrosymmetric, so that the photovoltaic current pulses generated in two-photon excitation of electrons into the band (Fig.6.5) from A and B ions must be directed oppositely to each other (see Fig.6.6) and the integral photovoltaic current cancels, if the number of A and B ions are equal to each other. In Ref.1 it was assumed, that there was initial asymmetry of the numbers of ions of A and B types, leading to occurring of photovoltaic current.

It was shown in Ref.2,3, however, that the arising of static field in ruby at constant pump intensity occurs only with temperature decrease bellow certain critical value (see Fig. 6.8).

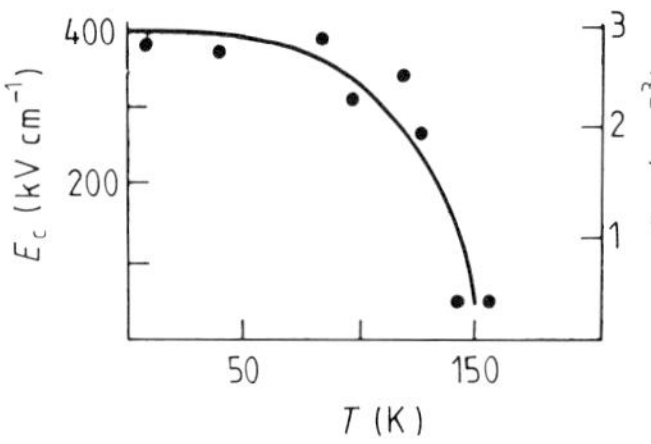

Fig.6.8. The dependence of the R-line splitting $\Delta\nu$ and the electrostatic field E_0 on the temperature for $I_L = 10^2$ W cm^{-2}. The experimental points are from Ref.[2]. The solid line is the theoretical dependence calculated with the help of (10.22), where $d_c/d = 0.012$;

For $T = 50$ K $\delta = 20$, $K = 40$, $\eta = 50$, $T_2^{-1} = 6 \times 10^{11}$ s^{-1};

for $T = 100$ K $\delta = 7{,}5$, $K = 20$, $\eta = 25$, $T_2^{-1} = 1.2 \times 10^{12}$ s^{-1}

for $T = 150$ K $\delta = 1$, $K = 10$, $\eta = 12.5$, $T_2^{-1} = 2.4 \times 10^{12}$ s^{-1}.

The temperature dependence of the magnitude of this field (or the magnitude of quasi-Stark splitting) has the shape typical to phase transitions of the second kind. The experimental data of Refs.2,3 thus show, that the photovoltaic current arises due to the spontaneous breaking of the symmetry of filling of metastable states ^2E in ions of A and B types.

Below we consider one of the possible mechanisms of spontaneous symmetry breaking of levels populations under the laser light action.

6.3.1. *Spontaneous breaking of saturation symmetry and arising of photovoltaic current in centrosymmetric crystal.* Let us consider the system of N doping atoms, comprising two subsystems with oppositely directed constant dipole moments, analogous to that, studied in Sec.6.2. The potentials of A and B ions are asymmetric along certain crystallographic direction (c-axis), correspondingly the constant dipole moments and photovoltaic current pulses are directed oppositely to each other (see fig. 6.6).

The photovoltaic currents along c-axis , created by the A and B ions can by written in the form

$$j^{A(B)}_{\mathrm{phv}} = n\,|e|\,W\,l\,\rho^{A(B)}_{cc}\,, \tag{3.1}$$

where $|e|$ is the carrier charge, l is the free path length, n is the concentration of the doping atoms (we assume $n^A = n^B = n$), W is the probability of transition per unit time from the metastable state into the band, $\rho^{A(B)}_{cc}$ are the populations of the metastable states. The integral photovoltaic current along c-axis

$$j_{\mathrm{phv}} = n\,|e|\,W\,l\,(\rho^B_{cc} - \rho^A_{cc})\,. \tag{3.2}$$

In the case of symmetric filling of metastable states one has $\rho^A_{cc} = \rho^B_{cc}$ and from (3.2) it follows, that $j_{\mathrm{phv}} = 0$. Let us consider the developing of the instability , leading to the spontaneous arising of the photovoltaic current and static field.

We assume, that due to the fluctuation the asymmetry of filling of metastable states arises, so that $\rho^A_{cc} \neq \rho^B_{cc}$. This leads to the arising of the photovoltaic current j_{phv}. The full current consists of the sum of the photovoltaic current j_{phv} and the conduction current σE_0 :

$$j = n\,|e|\,l\,W\,(\rho^B_{cc} - \rho^A_{cc}) + \sigma E_0\,, \tag{3.3}$$

where E_0 is the static field, arising in the medium due to the charge separation. It is connected with the current by the following equation

$$\frac{1}{4\pi}\frac{\partial E_0}{\partial t} = - j = n\,|e|\,l\,W\,(\rho^A_{cc} - \rho^B_{cc}) - \sigma E_0\,. \quad (3.4)$$

Let us take now into account of the influence of the static field $\mathbf{E}_0$ in the equations of the three-level atom, corresponding to the scheme of fig. 6.5, assuming that $W < \gamma_1$:

$$\frac{\partial \rho^\alpha_{ab}}{\partial t} + i\,(\omega_{ab} - \frac{\mathbf{d}_\alpha \mathbf{E}_0}{\hbar})\,\rho^\alpha_{ab} + T_2^{-1}\rho^\alpha_{ab}$$

$$= - \frac{i\mathbf{d}_{ab}\mathbf{E}_L}{\hbar}\,(\rho^\alpha_{aa} - \rho^\alpha_{bb})\,, \quad \mathbf{d}_\alpha = \mathbf{d}^\alpha_{aa} - \mathbf{d}^\alpha_{bb}, \quad (3.5)$$

$$\frac{\partial \rho^\alpha_{aa}}{\partial t} = - \gamma_1\rho^\alpha_{aa} + \frac{i}{\hbar}\,(\mathbf{d}_{ab}\,\rho^\alpha_{ba} - \mathbf{d}_{ba}\,\rho^\alpha_{ab}\,)\,\mathbf{E}_L\,, \quad (3.6)$$

$$\frac{\partial \rho^\alpha_{bb}}{\partial t} = \gamma\rho^\alpha_{cc} - \frac{i}{\hbar}\,(\mathbf{d}_{ab}\rho^\alpha_{ba} - \mathbf{d}_{ba}\rho^\alpha_{ab})\,\mathbf{E}_L\,, \quad (3.7)$$

$$\frac{\partial \rho^\alpha_{cc}}{\partial t} = \gamma_1\rho^\alpha_{aa} - \gamma\rho^\alpha_{cc}\,. \quad (3.8)$$

Here ρ^α_{ab} is the non-diagonal element of the density matrix ($\alpha = A,B$) corresponding to the resonance transition ω_{ab}; ρ_{aa}, ρ_{bb}, ρ_{cc} are the populations of the corresponding states (see Fig. 6.5), γ, γ_1 are the transition rates. It is seen, in full analogy with Sec.6.2 , that, the arising static field,owing to the linear Stark effect, modifies the transition frequencies differently in ions A and B. Let the seeding fluctuation field $\mathbf{E}_0$ be directed along $\mathbf{d}^B$ (Fig. 6.9).

Then we have

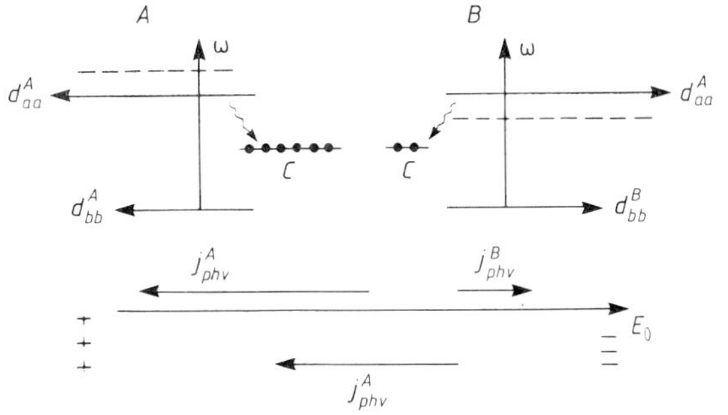

Fig.6.9. The physical mechanism of the light-induced phase transition with the photovoltaic current appearance. Due to the different signs of the linear Stark shifts, the asymmetry of the metastable c-levels in A and B ions is arises leading to the appearance of photovoltaic current.

$$\frac{\mathbf{d}^A \mathbf{E}_0}{\hbar_\hbar} = - \frac{d E_0}{\hbar_\hbar} , \quad \frac{\mathbf{d}^B \mathbf{E}_0}{\hbar_\hbar} = \frac{d E_0}{\hbar^\hbar} .$$

We find the stationary solution of the system of equations (3.5) - (3.8):

$$\rho_{cc}^{A (B)} = \frac{\overline{T}_2^{-2} \; a \, E_L^2}{(\omega_{ab} - \omega \pm d \, E_0 \, / \, \hbar)^2 + T_2^{-2} \; (1 + a \, E_L^2)} ,$$

$$a = 2 \, \frac{d_{ab}^2}{\hbar \, T_2^{-1} \gamma} . \tag{3.9}$$

Let the light frequency $\omega > \omega_{ab}$ (Fig. 6.9). Then, as one can see from (3.9) the effective detuning from resonance in A-ions equal to $(\omega_{ab} - \omega + dE_0 /\hbar\hbar)$, is less, then the effective detuning in the B-ions, equal to $(\omega_{ab} - \omega - dE_0 /\hbar\hbar)$. Owing to this $\rho_{cc}^A > \rho_{cc}^B$. Then due to Eqs. (3.1) and

(3.2) the photovoltaic current j^{A}_{phv} and the integral photovoltaic current j_{phv} are directed against the field vector $\mathbf{E}_0$, that gives rise to the additive charge separation and growth of the field $\mathbf{E}_0$. This, in its turn, leads to the further decreasing of the detuning in A-ions and increasing of detuning in *B*-ions, i.e. to the increase of the populations difference of *c*-states in *A* and *B* ions. Thus, the positive feedback and instability arise, leading to the transition of the system into the stationary state with the presence of the photovoltaic current and static field.

6.3.2. *Critical intensities of appearing and disappearing of electrostatic field.* Substituting (3.9) into (3.4), we obtain the closed equation for the static field, describing the instability and the stationary state. In writing it down it is convenient to introduce the following dimensionless variables:

The dimensionless order parameter

$$\xi = \frac{d\, E_0}{\hbar\, T_2^{-1}} , \tag{3.10}$$

the parameter of the pump

$$\eta = a\, E_L^2 \tag{3.11}$$

and the detuning

$$\delta = \frac{\omega_{ab} - \omega}{T_2^{-1}} . \tag{3.12}$$

Then the equation for the static field acquires the form

$$\frac{\varepsilon}{4\pi} \frac{\partial \xi}{\partial t} = - \sigma \xi - W \frac{n\, |e|\, d\, l\, \eta}{\hbar\, T_2^{-1}} \left[\frac{1}{(\delta - \xi)^2 + 1 + \eta} - \frac{1}{(\delta + \xi)^2 + 1 + \eta} \right] , \tag{3.13}$$

where ε is the static permeability.

To find the critical intensities of appearing and disappearing of static field, we expand the right hand side of Eq. (3.13) into the power series of ξ , retaining only the linear terms in this expansion. Then we obtain

$$\frac{\varepsilon}{4\pi}\frac{\partial \xi}{\partial t} = -\sigma \xi - W \frac{n\, d^2_{eff}}{\hbar\, T_2^{-1}} \frac{\eta\, \delta\, \xi}{(1 + \eta + \delta^2)^2}, \tag{3.14}$$

where

$$d^2_{eff} = |e|\, l\, d\,.$$

It is seen from here, that the instability arises only when the initial detuning $\delta < 0$ in accordance with above qualitative consideration of the physical mechanism of instability. Besides one can see, that at relatively high intensity of the pump, when $\eta \gg 1 + \delta^2$, the effective decay constant of the field becomes negative and the field fluctuations decay.

To define the critical intensities of instability thresholds let us write down the expression for σ in the form

$$\sigma = \sigma_0 + \sigma_\eta\,, \tag{3.15}$$

where σ_0 is the dark conductivity, $\sigma_\eta = |e|\, \mu\, n\, (\rho^A_c + \rho^B_c)\, \tau\, \Omega\, \eta$, $W = \Omega\, \eta$, Ω = const, μ is the mobility, τ is the electron lifetime in the band. Using (3.9), one has

$$\sigma = \sigma_0 + \frac{\sigma_m \Omega\, \tau\, \eta^2}{\delta^2 + 1 + \eta} \equiv \sigma_m \Omega\, \tau \left[q_0 + \frac{\eta^2}{\delta^2 + 1 + \eta} \right], \tag{3.16}$$

where

$$\sigma_m = |e|\, \mu\, 2\, n, \quad q_0 = \frac{\sigma_0}{\sigma_m \Omega\, \tau}\,.$$

From (3.16) and the condition $\partial\xi / \partial t = 0$ we obtain two values of the critical intensity :

$$\eta_1 = (\delta^2 + 1)\left(\frac{q_0}{4\ |\delta|\ K}\right)^{1/2}, \quad \eta_2 = 4\ |\delta|\ K - \delta^2 - 1\ , \quad \eta_1 \ll \eta_2, \quad (3.17)$$

where we have introduced the cooperative parameter

$$K = \frac{n\ d_{eff}^2}{\hbar\ T_2^{-1}\ \tau\ \sigma_m} \ . \quad (3.18)$$

The state with $\xi = 0$ is unstable in the region of intensities $\eta_1 < \eta < \eta_2$ where the static field grows. In the stationary state the conductivity current and photovoltaic current cancel each other and $\partial\xi\ /\ \partial t = 0$. Then from (3.13) one has the stationary equation for the order parameter :

$$\xi = \frac{K\ \eta^2}{q_0 + \eta^2\ (\delta^2 + 1 + \eta)^{-1}}$$

$$\times \left(\frac{1}{(\delta + \xi)^2 + 1 + \eta} - \frac{1}{(\delta - \xi)^2 + 1 + \eta} \right), \quad (3.19)$$

The solution of this equation has the form

$$\xi = [\delta^2 - (1 + \eta) + 2\ |\delta|\ (K\ |\delta|^{-1}\eta^2$$

$$\times \frac{1}{q_0 + \eta^2\ (\delta^2 + 1 + \eta)^{-1}} - 1 - \eta)^{1/2}]^{1/2} \ . \quad (3.20)$$

The plot of the dependency $\xi = \xi\ (\eta)$ according to Eq. (3.20) (see also (3.22)) is presented in Fig. 6.7. The static field has the non-zero values in the region $\eta_1 < \eta < \eta_2$ with the values η_1 and η_2, obtained from the condition $\xi = 0$, coinciding with (3.17). The appearance of non-zero values of ξ occurs due to the spontaneous breaking of the symmetry of saturation of transitions in A and B-ions: when $\xi \neq 0$, then $\rho_c^A \neq \rho_c^B$. .At high values of intensity η, when $\eta > \eta_2$, the symmetry of saturation is restored, since, as it follows from (3.9), $\rho_c^A \rightarrow \rho_c^B \longrightarrow 1$ for $\eta \longrightarrow \infty$, so that $\xi \longrightarrow 0$ (see Fig. 6.7).

Under the condition $\eta^2 / (\delta + 1 + \eta) \gg q_0$ from (3.14) we obtain for the time of establishing of the static field (τ_{est}):

$$\frac{\partial \xi}{\partial t} = \tau_{est}^{-1} \xi ,$$

$$\tau^{-1}_{est} = \frac{4\pi \sigma_m \Omega \tau \eta^2}{\varepsilon (\delta^2 + 1 + \eta)} \left[\frac{4 \, |\delta| \, K}{(1 + \delta^2 + \eta)} - 1 \right]. \qquad (3.21)$$

It is seen from here, that at $\eta < (1 + \delta^2)$ the growth rate of the static field is proportional the intensity square and for $\eta > 1 + \delta^2$ the saturation of the growth rate occurs. Both these conclusions are in agreement with the experimental results of Ref.1.

Let us now make some numeric estimates and comparison with experimental results. The saturation parameter $a = 4 \cdot 10^2$ ESU at $d_{ab} = 2\times10^{-1}$ D, $\gamma = 3\times10^2$ s^{-1}, $T_2^{-1} = 6 \times 10^{-11}$s, $T = 10$ K. Take further $\delta = |\Delta| / T_2^{-1} = 20$, which corresponds to the value of $|\Delta|$, equal to the half of the distance between the adjusting vibronic levels 4 T_2 [78]. At these values $E^2_{sat} \sim \delta^2 / a \sim 1$ ESU i.e. the saturation intensity $I_{sat} \sim 10^3$ W/cm^2, which corresponds to the experimental data of Ref.1. We take further $\sigma_0 \sim 10^{-2}$ s^{-1} in analogy with the data for $LiNbO_3$ [22]. The values μ and τ for the band 4.2 eV, appearing in strongly doped Al_2O_3-crystals [1] are unknown. We note, that even for the well studied case of $LiNbO_3$ values of μ, given in the literature differs from each other on 4 orders of magnitude($\mu \sim 3 \times 10^{-1}$ - 3×10 ESU [22]). Taking for example $\mu = 3 \times 10^{-1}$,for $n = 10^{21}$cm^{-3} we have $\sigma_m \cong 10^{11}$s^{-1}, so that at $\tau = 10^{-12}$s , we have $\sigma_m \tau \sim 10^{-1}$. Then at $d = 3$ D [1], $l \sim 1$ A , $T_2^{-1} \sim 6 \times 10^{11}$s^{-1} we obtain for the cooperative parameter $K \cong 50$. Using further the estimate of the absorption cross section from the metastable state ^2E

[1], we have $\Omega \sim 10$ and from (3.16) obtain $q_0 \sim 10^{-2}$. Substituting into (3.17) $K = 40$, we have with the above obtained values of parameters for the critical intensity of appearance of static field $I_{c1} \sim 1$ W / cm^2. The critical intensity of the disappearance of static field $I_{c2} \sim 3.5 \times 10$ W / cm^2, that corresponds to the experimental data of [1].

Fig. 6.7 shows the dependence of the splitting $\Delta\nu$ on pump intensity, calculated with the help of the formula

$$\Delta\nu = \frac{\xi}{\pi c T_2} \frac{d_c}{d}, \qquad (3.22)$$

where ξ is given by (3.20), d_c is the dipole moment of metastable state, in comparison with the experimental data of Ref.1. It is seen, that at chosen values of parameters the theory can satisfactory reproduce the experimental dependency.

To obtain the temperature dependence of $\Delta\nu$ with the help of (3.22) it is necessary to know the temperature dependence $T_2 = T_2(T)$ and $\omega_{ab} = \omega_{ab}(T)$, which are absent in the literature. To demonstrate, that this theory can be adequate to experiment we use the temperature dependencies $\gamma_{bc} = \gamma_{bc}(T)$ and $\omega_{bc} = \omega_{bc}(T)$ from [23] and use in (3.22) the relative dependencies $\delta = \delta(T)$, $K = K(T)$ and $\eta = \eta(T)$. The results are given in Fig. 6.8 in comparison with the experimental data of Ref. [2]. It is seen, that such a rough procedure can nevertheless satisfactory describe the temperature dependency $\Delta\nu = \Delta\nu(T)$, experimentally observed in Ref.2.

6.3.3. *Spontaneous appearance of electrostatic field in case of continuum of excited states.* If the vibronic levels 4 T_2 are overlapping and form the continuous band (band 4 T_2 in Fig. 6.5), then the electronic transition occurs into the continuum of states. We generalize the above mechanism of spontaneous appearance of static field on this continuum case.

The populations of the metastable levels, without taking into account of the saturation, can be written in the form

$$\rho_c^{A,B} = \frac{\alpha_{A,B}(\omega)\, E_L^2}{n\, 2\pi \hbar\, k\, \gamma} \equiv \tilde{a}\, E_L^2 \equiv \tilde{\eta}\,, \qquad (3.23)$$

where $k = \omega / c$, c is the light velocity, $\alpha_{A,B}(\omega)$ [cm^{-1}] are the optical absorption coefficients by the ions of A and B types :

$$\alpha_{A,B}(\omega) \sim \sum_i \frac{1}{\omega_{abi} - \omega \pm d E_0 / \hbar} = \frac{1}{2} \alpha \left(\omega \mp \frac{d E_0}{\hbar}\right) ,$$

where ω_{abi} is the frequency of transition between the ground state b and i-th excited state in quasicontinuum, $\alpha(\omega)$ is the absorption coefficient of the whole system of doping ions.

The equation for the static field is written analogously to (3.13) in the form

$$\frac{\varepsilon}{4\pi} \frac{\partial E_0}{\partial t} = - \sigma E_0 + \frac{W |e| l E_L^2}{4\pi \hbar k \gamma}$$

$$\times \; [\alpha(\omega - d_0 E / \hbar) - \alpha(\omega + d E_0 / \hbar)]$$

$$\simeq \left(- \sigma - \frac{\tilde{\Omega} d^2_{eff} n}{\hbar} \frac{1}{\alpha} \frac{\partial \alpha}{\partial \omega} \tilde{\eta}^2 \right) E_0 , \qquad (3.24)$$

where

$$W = \tilde{\Omega} \tilde{\eta} , \quad \tilde{\Omega} = 4 \times 10^4 \ \mathrm{s}^{-1} \ (\tilde{a} \sim 10^{-1} , \quad \tilde{\Omega} = \Omega \frac{a}{\tilde{a}} .$$

The instability occurs under the condition $\partial \alpha / \partial \omega < 0$, which is analogous to the condition $\delta < 0$ in the case of transitions between the discrete levels. Using in (3.24) $\sigma = \sigma_0$, we obtain the expression for the critical intensity of the instability

$$\tilde{\eta}_1 = \left[\frac{\sigma_0 \hbar \alpha}{\tilde{\Omega} \, |\partial\alpha / \partial\omega| \, d_{eff}^2 \, n} \right]^{1/2} . \qquad (3.25)$$

Assuming, that $|\partial\alpha / \partial\omega| / \alpha \sim \Gamma^{-1} \sim 10^{-14}$ s, where Γ is the band width, we have the estimate (compare with (3.17)):

$$\tilde{\eta}_1 \cong (\sigma_0 \hbar \Gamma \; / \; d_{eff}^2 n \, \tilde{\Omega})^{1/2} \sim 3 \times 10^{-3},$$

i.e. $I_1 \sim 10$ W / cm^2. For the full description of the continuum case one must take into account the saturation and obtain the Eq.for ξ, analogous to (3.20).

From the above consideration it follows, that the instability with the appearance of the photovoltaic current is quite analogous to the polarization instability, considered in sec.6.6.2. However, in the case of excitation of the ground state electronic subsystem with discrete resonance transition frequency ω_{ab} the polarization instability occurs under the condition $\omega < \omega_{ab}$(see the discussion after Eq.(2.6a)), whereas the photovoltaic current instability takes place at $\omega > \omega_{ab}$(see (3.14)). This difference in the signs of the detuning enables one to identify the particular instability.

6.4. Bistability of static polarization, appearing under laser illumination of noncentrosymmetric doped crystals

Let us now consider the polarization bistability, which can take place under laser illumination of centrosymmetric media. As it is known [4], under the laser illumination of crystals, doped with N ($N \gg 1$)non-centrosymmetric two-level atoms with the transition frequency ω_{ab} possessing the constant dipole moments, all oriented in one and the same direction, the static polarization appears. It occurs due to laser-induced change of populations of upper ρ_{aa} and lower ρ_{bb} atomic levels with different dipole moments $\mathbf{d}_{aa}$ and $\mathbf{d}_{bb}$. The polarization vector of such a medium is written in the form

$$\mathbf{P} = n\,(\mathbf{d}_{aa}\,\rho_{aa} + \mathbf{d}_{bb}\,\rho_{bb}) = \frac{n\,\mathbf{d}\,D}{2} + n\,\frac{1}{2}\,(\mathbf{d}_{aa} + \mathbf{d}_{bb}) \quad (4.1)$$

where $D = \rho_{aa} - \rho_{bb}$ is the population difference, n is the number density of doping atoms, $\mathbf{d} = \mathbf{d}_{aa} - \mathbf{d}_{bb}$. Note, that this effect is different from the optical rectification effect, taking place due to the non-linearity of two-level atoms [24].

In this section it is shown, that under illumination of the sufficiently strongly doped crystals by the quasiresonance light wave with the frequency ω and detuning $\omega_{ab}\ \omega = \Delta > 0$ the static polarization change in dependence on the light pump intensity I_L has the bistable character:at certain critical value of the pump intensity I_c the static polarization undergoes the jumplike change , with the subsequent decrease of the pump intensity at $I_L = I_{c1} < I_c$ the static field undergoes the jumplike decrease [25]. Such bistability occurs due to the saturation of the transitions in atoms and cooperative interaction between them through the Lorentz field.

This effect of bistability of static polarization is closely connected with the effect of light -induced phase transition, with the saturation symmetry breaking, considered in Sec.6.2. The principal difference between them consists in that the phase transition occurs in the system of chaotically oriented permanent dipoles. Because of this the order parameter (i.e. static polarization) appears only under the exceeding of certain critical value of pump intensity, and it is equal to zero for $I_L < I_c$. In the considered case of initially orientationally ordered system of permanent dipoles the static polarization appears at arbitrary small values of the pump intensities and undergoes the jumplike increase at $I_L = I_c$. The jump of the static field is accompanied by jumplike frequency shift, which partially compensates the initial detuning. Because of that the transmission (or the gain) of the excitation radiation must also undergoes the jumplike changes at $I_L = I_c$, I_{c1}.

We note, that such bistability of optical transmission is analogous in its manifestation to the bistability of transmission in the system of two-level atoms, interacting through the radiation field in resonator[26], or in the resonatorless medium (Sec.5.4) with the difference, that under the pump intensity increase the transmission is jumplike decreased (see bellow). However the physical mechanism of the bistability considered here is different from mechanisms of usual absorptive bistability considered in [26] and Sec.5.4. This can be seen from the fact, that the latter effect doesn't depend on the sign of the detuning Δ (in particular, Δ can be equal to zero, see Sec.5.4). In contradistinction to this, the bistability, considered in this section in the non-inverted system with $D^0 < 0$ (where D^0 is the population difference in the absence of laser field) occurs only under the condition $\Delta > 0$. In the inverted system ($D^0 > 0$) the analogous bistability can take place only under the condition $\Delta < 0$.

6.4.1. *Bistability and hysteresis of light-induced static polarization.* Consider the system of N non-centrosymmetric two-level atoms with the permanent dipole moments,all oriented in one and the same direction. The equations for the non-diagonal element of density matrix $\sigma_- \equiv \rho_{ab}$ and the population difference D of one of these atoms in laser field E_L have the form

$$\frac{\partial \rho_{ab}}{\partial t} + i\,(\Omega_{ab} - \frac{i}{T_2})\,\rho_{ab} = -\,(\frac{i\mathbf{d}_{ab}}{\hbar})\,\mathbf{E}_L D + (\frac{i\,\mathbf{d}}{\hbar})\,\vec{\varepsilon}\,\rho_{ab} \quad (4.2)$$

$$\frac{\partial D}{\partial t} + \frac{D - D^0}{T_1} = \frac{2\,i}{\hbar}\,(\mathbf{d}_{ab}\,\rho_{ba} - \mathbf{d}_{ba}\,\rho_{ab})\,\mathbf{E}_L\,, \quad (4.3)$$

$$\mathbf{E} = \mathbf{E}_L \exp\,(-i\,\omega\,t) + \text{c.c.} \quad (4.4)$$

In Eq. (4.2) one has taken into account the acting Lorentz field $\vec{\varepsilon}$:

$$\vec{\varepsilon} = \beta\,4\pi\,\mathbf{P}\,, \quad (4.5)$$

where P is the static polarization. From (4.2) and (4.3), taking into account of (4.4), we find the stationary equation for the population difference

$$D = D^0 \frac{\tilde{\Delta}^2 + T_2^{-2}}{\tilde{\Delta}^2 + \gamma_E^2} . \qquad (4.6)$$

Here

$$\tilde{\Delta} = \Omega_{ab} - \omega - \frac{\mathbf{d}\ \vec{\varepsilon}}{\hbar} \equiv \delta - \frac{\mathbf{d}\ \vec{\varepsilon}}{\hbar} . \qquad (4.7)$$

Substituting (4.6) into (4.1), then (4.1) into (4.5) we obtain the closed equation for the static field $\vec{\varepsilon}$:

$$\vec{\varepsilon} = \beta\ 2\pi\ \mathbf{d}\ n\ D^0 \frac{\tilde{\delta}^2 + T_2^{-2}}{\tilde{\delta}^2 + \gamma_E^2} + \beta\ 2\pi\ (\mathbf{d}_{aa} + \mathbf{d}_{bb})\ n . \qquad (4.8)$$

Using the condition $\rho^0_{aa} + \rho^0_{bb} = 1$, we transform the right hand side of this equation, representing it as the sum of equilibrium and non-equilibrium parts

$$\vec{\varepsilon} = - \frac{T_2^{-2} a\ E_L^2}{\tilde{\delta}^2 + \gamma_E^2} \beta\ 2\pi\ n\ \mathbf{d}\ D^0 + + \beta\ 4\pi\ n\ (\mathbf{d}_{aa}\rho^0_{aa} + \mathbf{d}_{bb}\rho^0_{bb}) . \qquad (4.9)$$

From (4.9) it is seen, that at $E_L = 0$ the static field is equal to its equilibrium value $\vec{\varepsilon} = \vec{\varepsilon}_0 = \beta\ 4\pi\ n\ (\mathbf{d}_{aa}\rho^0_{aa} + \mathbf{d}_{bb}\rho^0_{bb})$.

We pass over to the variable

$$\zeta = \vec{\varepsilon} - \vec{\varepsilon}_0 \qquad (4.10)$$

Then from (4.9) we have

$$\vec{\zeta} = - \beta\, 2\pi\, n\, \mathbf{d}\, D^0 \frac{a\, E_L^2}{(\tilde{\Delta}^2 + \gamma_E^2)\, T_2^2}\,, \tag{4.11}$$

$$\tilde{\Delta} = \Omega_{ab} - \frac{\mathbf{d}\, \vec{\varepsilon}_0}{\hbar} - \omega - \frac{\mathbf{d}\, \vec{\zeta}}{\hbar} \equiv \Delta - \frac{\mathbf{d}\, \vec{\zeta}}{\hbar}\,, \tag{4.12}$$

$$\Delta \equiv \Omega_{ab} - \frac{\mathrm{d}\, \vec{\varepsilon}_0}{\hbar} - \omega \equiv \omega_{ab} - \omega\,, \tag{4.13}$$

where Δ is the detuning of laser frequency from the true equilibrium value of the transition frequency $\omega_{ab} = \Omega_{ab} - \mathbf{d}\, \vec{\varepsilon}_0 / \hbar$. Let us introduce the dimensionless parameter of the static field

$$\xi = \frac{2\, \mathbf{d}\, \vec{\zeta}}{\hbar\, \Delta}\,, \tag{4.14}$$

and define

$$d = |\mathbf{d}|\,, \quad \zeta = |\, \vec{\zeta}\, |\,.$$

From (4.11) one obtains the following equation for ξ :

$$\frac{1}{4}\xi^3 - \xi^2 + \frac{\Delta^2 + \gamma_E^2}{\Delta^2}\,\xi = \frac{\sigma\, T_2^{-2}}{\Delta^2}\, a\, E^2 \equiv F\,, \tag{4.15}$$

where the dimensionless parameter

$$\sigma = - \frac{\beta\, 4\pi\, n\, d^2\, D^0}{\hbar\, \Delta}$$

characterizes the strength of interatomic interaction.

Let do the graphic analysis of the equation (4.15) for $\Delta > 0$, $D^0 < 0$. The function

$$f(\xi) = \frac{1}{4}\xi^3 - \xi^2 + \frac{\Delta^2 + \gamma_E^2}{\Delta^2}\xi$$

has the extrema under the condition, that the equation

$$\frac{\partial f}{\partial \xi} = 0 = \xi^2 - \frac{8}{3}\xi + \frac{4}{3}\frac{\Delta^2 + \gamma_E^2}{\Delta^2} \tag{4.16}$$

has the real roots. Its solutions have the form

$$\xi_m^{(1,2)} = \frac{4}{3} \pm \frac{2}{3}\left(\Delta^2 - \frac{3\gamma_E^2}{\Delta^2}\right)^{1/2}. \tag{4.17}$$

Thus, under the condition

$$\Delta^2 > 3\gamma_E^2 \tag{4.18}$$

the function $f(\xi)$ has the shape, shown in Fig. 6.10.

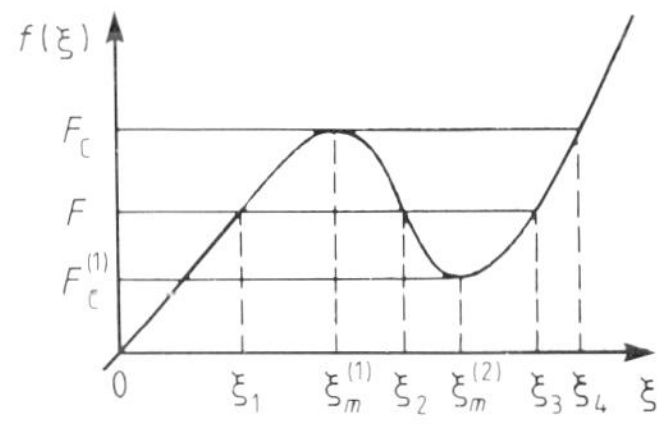

Fig.6.10. The graphical analysis of Eq.(6.11.15).

The solutions of the equation (4.15) correspond to the intersection points of the curve $f\ (\xi)$ and the horizontal line $F = \sigma\ T_2^{-2} a\ E_L^2\ /\ \Delta^2$. In the region $F_c^{(1)} < F < F_c^{(2)}$ there are three solutions :ξ_1, ξ_2, ξ_3. One can show, that ξ_1 and ξ_3 are the stable solutions and ξ_2 is unstable one. In each of the regions $F > F_c^{(2)}$ and $F < F_c^{(1)}$ there exists one stable solution. The corresponding dependence of the stable solution on the pump intensity I_L is shown in Fig. 6.11.

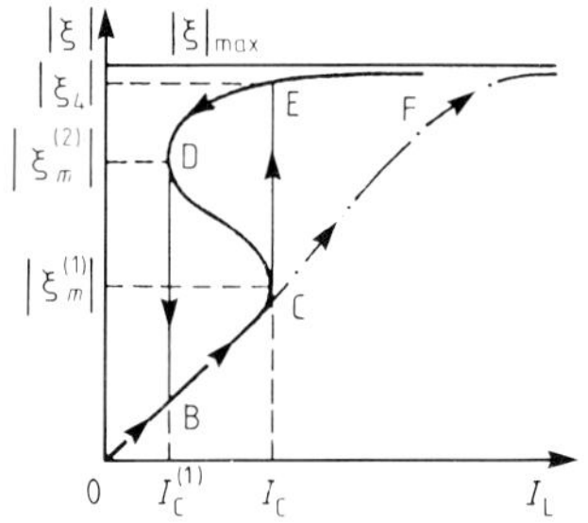

Fig.6.11. The dependence of the dimensionless parameter of static field $|\xi|$ from Eq.(6.11.14) on the laser field intensity I_L. The solid line is the solution of Eq.6.11.15) for $D^0\Delta < 0$; the dotted line is the solution of Eq.(6.11.15) for $D^0\Delta \geq 0$. Simultaneously the graph shows the dependence of the parameter of optical - (6.12.12), or acoustic- (6.12.28) deformations on I_L.

The magnitude of ξ increases continuously and monotonously when the pump intensity I_L is increased from zero value up to I_c, the latter being found from the condition

$$f\ (\xi_m^{(1)}) = F_c\ , \tag{4.19}$$

Then ξ, undergoing the jumplike change, assumes the value ξ_4 and with further increase of I_L reaches the saturated value

$$\xi_{max} = \beta\ 4\pi\ n\ d^2 |D^0|\ /\ \hbar\ \Delta\ ,$$

which is found from (4.11) and (4.14) in the limit $aE_L^2 \longrightarrow \infty$.

If now I_L is decreased, the backward down jump occurs at $I_c^{(1)} < I_c$ (hysteresis), the $I_c^{(1)}$ value being found from the condition

$$(\xi_m^{(2)}) = F_c^{(1)}\ . \qquad (4.20)$$

As an example we consider quantitatively the particular case $\Delta^2 = 4 \times \gamma_E^2$. From (4.17) then one has $\xi_m^{(1)} = 1$, $\xi_m^{(2)} = 5\ /\ 3$. From (4.19) and definition of the parameter F -(4.15) one finds for $a\ E_L^2 < 1$

$$I_c \cong \frac{c}{4\pi\ \sigma\ a}\ \Delta^2 T_2^2 = \frac{2\pi\ c\ \hbar\ T_2^{-1}}{\pi^2 \beta\ n\ d^2\ |D^0|\ a} \equiv \frac{4\ c}{\pi\ \bar{\sigma}\ a}\ . \qquad (4.21)$$

For the expression (4.21) to be consistent with the adopted condition $a\ E_L^2 < 1$, it is necessary, that the following inequality holds :

$$\bar{\sigma} = \frac{\beta\ 2\pi\ n\ d^2\ |D^0|}{\hbar\ T_2^{-1}} > 8\ . \qquad (4.22)$$

As one can see from fig. 6.11, the jump magnitude $\delta\xi = \xi_4 - \xi_m^{(1)} \geq \xi_m^{(2)} - \xi_m^{(1)} = 2\ /\ 3$.

Thus, taking into account of (4.14), one obtains, that at $I_l = I_c$ the static field undergoes the jumplike increase

$$\delta\zeta \geq \frac{\hbar\ \Delta}{3\ d} = \frac{2\ \hbar\ T_2^{-1}}{3\ d}\ .$$

We note, that with Δ increase and corresponding increase of I_c the jump magnitude $\delta\xi$ increases.

6.4.2. *Fluctuations, soft relaxation mode and nessesary condition of bistability.* We want now to show, that the above considered effect of bistability is connected with the soft relaxation mode. Substituting (4.1) into (4.5), taking the derivative with respect to time and using (4.3), we obtain the equation for the variable $\vec{\varepsilon}_D = \beta\, 2\pi\, n\, \mathbf{d}\, D$:

$$\frac{d\,\vec{\varepsilon}_D}{dt} = -\,T_1^{-1}\,(\vec{\varepsilon}_D - \vec{\varepsilon}_D^{\,0})$$

$$-\,a\,E_L^2\,T_2^{-2}\,T_1^{-1}\,\vec{\varepsilon}_D\,/\,[(\delta - \mathbf{d}\,\vec{\varepsilon}/\hbar)^2 + T_2^{-2}]\,. \qquad (4.23)$$

The stationary solution of (4.23) coincides with (4.6). Let us consider the fluctuation of the quasistatic field $\vec{\varepsilon}_D$ from the mean stationary value

$$\vec{\varepsilon}_D\,(t) = \langle\vec{\varepsilon}_D\rangle + \delta\,\vec{\varepsilon}_D\,(t)\,.$$

Linearizing (4.23) with respect to $\delta\,\vec{\varepsilon}_D$, we obtain the equation for the fluctuation

$$\frac{d\,\delta\,\vec{\varepsilon}_D}{dt} = -\,\gamma_{\mathrm{eff}}\,\delta\,\vec{\varepsilon}\,, \qquad (4.24)$$

where the effective relaxation constant

$$\gamma_{\mathrm{eff}} = T_1^{-1}\,\frac{\tilde{\Delta}^2 + \gamma_E^2}{\tilde{\Delta}^2 + T_2^{-2}}\left[\,1 + \frac{\beta\,4\pi\;n\;d^{\,2}\,T^{-2}\,a\,E_L^2}{\hbar\;(\tilde{\Delta}^{\,2} + \gamma_E^2)^2}\,\tilde{\Delta}\,D^{\,0}\right] .(4.25)$$

If $D^0 < 0$, then, under the condition $\tilde{\Delta} = \omega_{ab} - \omega - \mathbf{d}\,\vec{\zeta}/\hbar > 0$ (i.e. $\Delta >$

0), $\gamma_{eff} \longrightarrow 0$ for $E^2 \longrightarrow E_c^2$, i.e. the soft relaxation mode exists. The cancellation of γ_{eff} means, that the stationary state with the mean value equal to $<\varepsilon_D>$ becomes unstable, the fluctuation starts to grow exponentially in time and the system makes a jumplike transition into new stationary state. At $D^0 > 0$ (the inverted system) the instability occurs at $\Delta < 0$.

The critical value of intensity E_c^2 we find from the condition $\gamma_{eff} = 0$ with the help of eq. (4.11). Then one obtains the following condition

$$2\,\mathbf{d}\,\overrightarrow{\zeta_c}\;\frac{(\tilde{\Delta} - \mathbf{d}\,\overrightarrow{\zeta_c} / \hbar)}{\hbar\,/(\tilde{\Delta}_c^2 + \gamma_E^2)} = 1 \qquad (4.26)$$

where $\overrightarrow{\zeta_c}$ is the critical value of the order parameter (4.10). Using the dimensionless order parameter $\xi = 2\,\mathbf{d}\,\overrightarrow{\zeta} / \hbar\,\Delta$ - (4.14), we have from (4.26) the critical condition

$$\xi_c^2 - 8\,\xi_c / 3 + 4\,(\Delta^2 + \gamma_{E_c}^2) / 3\,\Delta^2 = 0\ ,$$

which exactly coincides with (4.17). Thus, we have shown, that in the points ξ_c, where the system undergoes the jumplike transitions into new states, the relaxation constant of population difference γ_{eff} - (4.25) cancels.

From (4.25) one can find also the necessary condition of bistability. Letting $a\,E_L^2 \longrightarrow \infty$ in the expression in the brackets, one obtains the following condition :

$$\beta\,4\pi\,n\,d^2\,|D^0|\,/\,\tilde{\Delta}_{max} > 1\ , \qquad (4.27)$$

where

$$\tilde{\Delta}_{max} = \Delta - d\,\zeta_{max} / \hbar = \Delta - 2\pi\,\beta\,n\,d^2\,|D^0|\,/\,\hbar\ .$$

From (4.27) we obtain the following necessary condition imposed on the cooperative bistable system

$$\beta\,6\pi\,n\,d^2\,|D^0|\,/\,(\hbar\,|\Delta|) >> 1\ . \qquad (4.28)$$

Let us find also more general, then the estimation (4.21) expression for the critical intensity. From (4,17) for $\Delta^2 \gg 3\,\gamma_E^2$ we have $\xi_c = = 2/3$, ,then from (4.11) one obtains

$$a\,E_c^2 = (\Delta\,T_2)^2\,\hbar\,\Delta\,/\,\beta\,6\pi\,n\,d^2\,|D^0|\,.$$

At $n = 5 \cdot 10^{20}$ cm^{-3}, $d \sim 1$ D, $|D^0| \sim 1$, $\beta \sim 1/3$, $\Delta \sim 10^{12}$ s^{-1} we have $a\,E_c^2 \cong 30$.

6.4.3. *Bistability of optical transmition and gain.* The dimensionless coefficient of absorption ($D^0 < 0$) or gain ($D^0 > 0$) of the system of doping atoms is determined by the expression

$$\mathrm{Im}\,\chi^1 = -\,\frac{d_{ab}^{\,2}\,n\,D^0}{\hbar}\;\frac{T_2^{-1}}{(\omega_{ab} - \omega - \mathbf{d}\,\overrightarrow{\zeta}\,/\,\hbar)^2 + \gamma_E^2}\,, \qquad (4.29)$$

where the parameter ζ, defined with taking into account of (4.14), by Eq.(4.15), reveals bistability as a function of the pump intensity (Fig. 6.11). At $I_L = I_c$ the jumplike increase of ζ leads to the jumplike decrease of the effective detuning (Fig. 6.12) and due to (4.29) to the jumplike increase of the absorption coefficient. With the subsequent decrease of I_L the corresponding jumplike decrease of the absorption coefficient at $I_L = I_c^{(1)}$ takes place.

The same behavior has the gain coefficient , when $D^0 > 0$, but in this case the fulfillment of the condition $\Delta < 0$ is needed. In fact, the Eq. (4.15) is not changed, so the ζ behavior is still described by Fig.6.11. However, due to (4.11) now one has $\zeta < 0$ and so the denominator in (4.23), where now the initial detuning $\Delta = \omega_{ab} - \omega < 0$, coincides with the denominator in the case on non-inverted system ($\Delta > 0$, $D^0 < 0$). Thus, under the condition $\Delta < 0$ the gain coefficient must undergo the jumplike changes at $I_L = I_c$ and $I_L = I_c^{(1)}$.

6.4.4. *Discussion of physical mechanism of instability and numerical estimates.* The above considered effect of the static polarization bistability takes place due to the positive feedback and the corresponding instability appearing in the following way. As it follows from (4.1) the magnitude of the static polarization and consequently of the acting field (4.5) depends on the distribution of atoms over the energy levels(i.e. on the population difference D^0). As it is seen from (4.9) the magnitude of the electrostatic acting field ε increases with the pump intensity I_L increase(i.e. with D decrease). Due to (4.12) this leads at $\Delta > 0$ to the decrease of the effective detuning from the resonance, i.e. to ever strong two-level systems saturation. The appearing positive feedback leads at $D^0\Delta < 0$ to the instability and the jumplike transition of the system to a new stationary state.

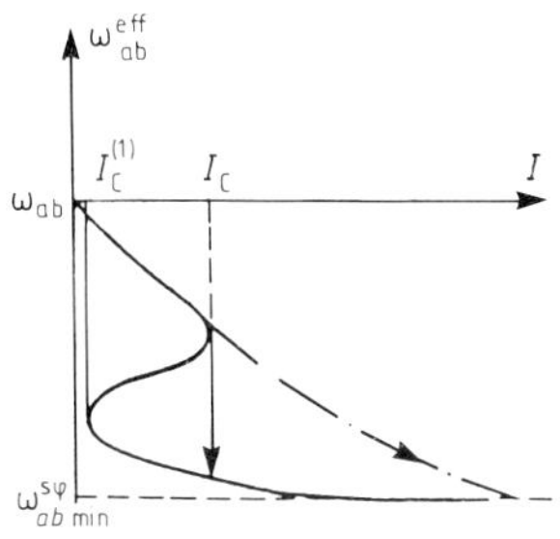

Fig.6.12. The dependence of the effective transition frequency $/\omega_{ab}^{eff} = \omega_{ab} - \vec{\mathbf{d}}\ \xi\ /\ \hbar$ on the pump intensity for $D^0\Delta < 0$ (solid line) and for $D^0\Delta \geq 0$(dotted line), $\omega_{ab}^{min} = \omega_{ab} - \beta\ 2\ \pi\ n\ d^2 |D^0|\ /\ \hbar$.

The above considered bistability effect is possible only in a sufficiently dense system of atoms(the condition (4.28)). Note, that this condition on the number density of atoms is considerably more strong then the condition of the optical bistability in resonator

$$\frac{2\pi\ n\ d_{ab}^2 \tau\ \omega_{ab}}{\hbar\ T_2^{-1}} > 1\ ,$$

where τ is the photon lifetime , because $\tau\omega_{ab} \gg 1$.

Let us do some numerical estimates using the values of parameters, characteristic to $LiNbO_3$ or $LiTaO_3$ crystals , doped with Cr^{+3} or Nd^{+3} ions:the concentration $n = 10^{20}\ cm^{-3}$ and the permanent dipole moment $d = 1$

D. With $d_{ab} \sim 10^{-2}$ D, $T_1^{-1} \sim 10^3\ s^{-1}$, $T_2^{-1} \sim 10^{10}\ s^{-1}$, $|D^0| \sim 1$ we obtain $\bar{\sigma} = 20$, $a = 40$, we have the critical pump intensity $I_c \sim 5$ W / cm^2 and the static field jump magnitude $\delta\ \varepsilon \sim 3 \cdot 10^3$ V / cm.

The effect of bistability considered here is connected with the shift of the energy levels in the external electrostatic field (well studied in doped [27] and molecular [28] crystals. The role of this external field is played here by the acting Lorentz field, which is changed under the electronic excitation. Note, that in a number of molecular crystals the change of the constant dipole moment under the electronic excitation is 10 - 20 D, that, together with the high densities of atoms in this media $n \sim 10^{22}$ cm^{-3}, makes it feasable observation of bistability effect in these media.

6.5. Laser-induced lattice bistability and dielectric permeability jumps in electron-phonon system

In this section we generalize the treatment of Sec.6.4 to the model of bistability in the system of two-level atoms, interacting through the optical or acoustic displacement fields.

6.5.1. *Bistability of optical lattice displacement.* As it has been already noted in Sec.6.2 , in vibronic system consisting of the two-level(or two-band) electron subsystem with the transition frequency ω_{ab}, interacting with the vibration optical mode, the equilibrium phase transition takes place with the appearance at some critical temperature of the static lattice displacement, accompanied by the electronic transition frequency change. The ferroelectric phase transition in crystals [18,19] and Jan-Teller transition in doping atoms or in molecules [29] are the examples of such transitions. The necessary condition of equilibrium ferroelectric phase transition, following from (2.13), has the form

$$C^2 / \hbar\, M\, \Omega_0^2 \omega_{ab} > 1\ , \tag{5.1}$$

Phase transition in vibronic system can be induced by the action of laser radiation. The particular case of vibronic system, defined by the condition (2.4), was considered in Sec.6.2. In the present section we shall not assume, that the condition (2.4) holds and show, that in general case laser radiation induces the lattice bistability, if the light frequency obeys the condition $D^0\ (\omega_{ab} - \omega) \equiv \equiv D^0\Delta < 0$. At some critical value of intensity of laser radiation $I_L = I_c$

the static lattice displacement U undergoes the jumplike increase, the magnitude of which depends on the magnitude of ΔD^0. With subsequent decrease of intensity down to the value $I_L = I_c^{(1)} < I_c$ the jumplike decrease of U occurs. Such sudden changes of U are accompanied by the changes of the electronic transition frequency ω_{ab} and the difference of the populations of electronic levels D . This picture is quite analogous to the bistability in the system with dipole-dipole interaction (Sec.6.4). The phase transition, considered here may be either of local character, with U being the local vibration coordinate (of molecule, for example), or it may be of collective character, then U is the normal phonon mode coordinate. In the local case the necessary condition of bistability has the form

$$C^2 / \hbar\, M\, \Omega_0^2 T_2^{-1} > 8 \, . \tag{5.2}$$

In collective case, when there are n_{ab} two-level atoms per unit volume, the condition of phase transition is more severe

$$n_{ab} C^2 / (n\, \hbar\, M\, \Omega_0^2 T_2^{-1}) > 8 \, , \tag{5.3}$$

where n is the number of the cells per unit volume. We note, that in the case of semiconductor $n_{ab} = n$ and the condition of bistability coincides with the local one - Eq.(5.2).

The optical manifestation of the effect considered here is the bistability of either optical transmission ($D^0 < 0$), or the gain ($D^0 >> 0$).

Let us consider the system of N_{ab} two-level atoms with the transition frequency Ω_{ab} , interacting with the optical lattice vibrations through the optical deformation potential. In the following we want to consider simultaneously two cases :

a) the local vibrations in the vicinity of two-level center with the frequency Ω_0;

b) isolated mode of crystal optical vibration with the frequency Ω_0 and wave vector $\mathbf{q} = 0$.

Let us write down the equations, describing the model, in the form, analogous to Eqs.(2.1) - (2.3), dropping the index j and retaining only the terms, relevant to the considered effect :

$$(\partial/\partial t + i\Omega_{ab} + T_2^{-1})\, \rho_{ab} = (-\, i d_{ab} E_L/\hbar)\, D - (iCU/\hbar)\, \rho_{ab} \, , \tag{5.4}$$

$$\partial D/\partial t + (D - D^0) / T_1 = (2i/\hbar)\, (d_{ab}\rho_{ba} - d_{ba}\rho_{ab})\, E_L \ , \tag{5.5}$$

$$(\partial^2/\partial t^2 + \Omega_0^2)\, U = - (C_{aa}/M)\, \rho_{aa} - (C_{bb}/M)\, \rho_{bb} \ . \tag{5.6}$$

Here E_L is the laser field, given by eq. (4.4). In the case of local vibrations $\rho_{aa} + \rho_{bb} = 1$, M is the reduced mass of the crystal sell. In the case of interaction with the optical mode $\rho_{aa} + \rho_{bb} = n_{ab}$, $M = M_0 N$, where M_0 is the mass of the cell, N is the number of the cells in crystal.

The Eqs. (5.4) - (5.6) in the case of local vibrations describe also the vibronic interaction in individual molecules under the action of laser light. In the case of collective vibrations these equations are easily generalized to the case of two-band semiconductor.

Taking into account that

$$\omega \sim \Omega_{ab} \gg \Omega_0, \quad T_1 \gg T_2, \tag{5.7}$$

from (5.6) we obtain

$$\partial D / \partial t + (D - D^0) / T_1 = a\, |E_L|^2 T_2^{-2} D \; \frac{1}{T_1}$$

$$\times \left\{ \left[\left[\left(\Omega_{ab} - \omega + \frac{CU}{\hbar} \right)^2 + T_2^{-2} \right] \right] \right\} , \tag{5.8}$$

$$(d^2 / \partial t^2 + \Omega_0^2)\, U = - CD / 2\, M - (C_{aa} + C_{bb}) / 2\, M \ ,(5.9)$$

$$a = 4\, |d_{ab}|^2 / \hbar^2 T_2^{-1} T_1^{-1} \ .$$

We introduce new parameters

$$\omega_{ab} = \Omega_{ab} - (C / \hbar\, M\, \Omega_0^2)\, (C_{aa}\rho_{aa} + C_{bb}\rho_{bb}) \ ,$$

where ω_{ab} is the equilibrium value of the transition frequency, renormalized by the electron-phonon interaction :

$$\Delta = \omega_{ab} - \omega\ , \quad \alpha \equiv T_2^{-1}\Delta^{-1}\ , \tag{5.10}$$

$$b = b_0 \equiv C^2 / 2M\hbar\, T_2^{-1}\Omega_0^2\ , \tag{5.11}$$

and pass over to the dimensionless lattice displacement coordinate

$$\xi = (C / \hbar\,\Delta)\ [U + (C_{aa}\rho^0_{aa} + C_{bb}\rho^0_{bb}) / M\,\Omega_0^2]\ . \tag{5.12}$$

The expression in the square brackets in (5.12), as one can verify with the help of (5.9), is the nonequilibrium lattice displacement, occurring due to the deviation of D from the equilibrium (or created by the external pump) value D^0. In the stationary case from (5.8) and (5.9) we obtain the equation for the stationary value of ξ :

$$\xi^3 + 2\,\xi^2 + [1 + \alpha^2\,(1 + a\,|E_L|^2)]\,\xi - b\,D^2\alpha\,a\,|E_L|^2 = 0\ . \tag{5.13}$$

The cubic equation (5.13) has two or three different real roots under the fixed value of $|E_L|^2$ when the following conditions hold

$$\alpha^2\,(1 + a\,|E_L|^2) < 1/3\ , \tag{5.14}$$

$$b\,D^0\,(\frac{1}{3} - \alpha^2)\,\alpha < -\,8/27\ . \tag{5.15}$$

Thus, eq.(5.13)has the multivalue solution, if $\alpha D^0 < 0$, i.e. if $D^0 < 0$, then $\Delta > 0$ and if $D^0 > 0$, then $\Delta < 0$. Besides, from (5.15) we obtain the necessary condition of bistability $|D^0|\,b > 4$, i.e.

$$|D^0|\ C^2 / \hbar\,M\,\Omega_0^2 T_2^{-1} > 8\ . \tag{5.16}$$

The dependence $|\xi|$ on the pump intensity I_L under the above conditions is shown in fig.6.11 ($\xi < 0$, if $D^0\alpha < 0$). When I_L is increased up to the critical value I_c the ξ value is monotonously increased along the *OBC*-branch. At $I_L = I_c$ $|\xi|$ jumps to the upper branch *DEF* and reaches the maximum value

$$|\xi|_{max} = |D^0|\ b\ |\alpha|$$

with the further increase of I_L. Under the subsequent decrease of I_L the jumplike transition to the lower branch occurs at $I_L = I_c^{(1)} < I_c$. Let estimate the critical intensity and the magnitude of the jump

$$\delta\,|\xi| = |\xi_4| - |\xi_m^{(1)}|$$

under the condition that

$$|\xi| \ll |D^0|\, b\, |\alpha|\ , \quad \alpha^2 \ll 1\ . \tag{5.17}$$

In this case one can neglect the term $\alpha^2 a\, |E|^2\, \xi$ in the left hand side of (5.13), so that the equation (5.13) acquires the form

$$f\,(\xi) = b\, D^0 \alpha^3 a\, |E_L|^2\ , \tag{5.18}$$

where the function

$$f\,(\xi) = \xi^3 + 2\,\xi^2 + (1 + \alpha^2)\,\xi$$

has the extrema in the points

$$\xi_m^{(1)} = -\,\frac{1}{3} - \frac{\alpha^2}{2}\ , \quad \xi_m^{(2)} = -\,1 + \frac{\alpha^2}{2}$$

(see Fig. 6.12), i.e. in (5.17) one has

$$|\xi| = 1, \quad |D^0|\, b\, |a| \gg 1.$$

The extremum value is $f\,(\xi_m^{(1)}) = -\,4\,/\,27$. From (5.18) we obtain the following condition, determining the critical intensity

$$|D^0|\, b\, |\alpha|^3\, a\, |E_L|^2 > 4\,/\,27\ , \tag{5.19}$$

whence one obtains

$$I_c = \frac{c\,|E_c|^2}{2\pi} = \frac{2\,c}{27\pi}\ |D^0|\, b\, |\alpha|^3\, a\ .$$

Using the values $\xi_m^{(1)}$ and $\xi_m^{(2)}$ one can estimate the magnitude of the lattice displacement jump:

$$\delta\,|\xi| = |\xi_4| - |\xi_m^{(1)}| \geq |\xi_D| - |\xi_m^{(1)}| = 2/3\,. \qquad (5.20)$$

Let us now study the physical mechanism of the instability. Substituting

$$D = \langle D^0\rangle + \delta D, \quad \xi = \xi + \delta\xi$$

into Eqs. (5.8) and (5.9) and linearizing them in respect to δD and $\delta\xi$, we obtain the system of equations for the fluctuations δD and $\delta\xi$. Seeking the solutions of these Eqs. in the form

$$\delta D,\ \delta\xi, \sim \exp\,(-\,i\omega t)\ ,$$

we obtain the following characteristic equation for ω:

$$\omega^3 + iT_1^{-1}\omega^2 - \Omega_0^2\omega - i\theta_0 = 0\,, \qquad (5.21)$$

$$\theta_0 = T_{1\mathrm{eff}}^{-1}\Omega_0^2\,(1 + 2\,b\alpha^3 a\,|E_L|^2 \times$$

$$\times\, D^0\,\frac{1+\xi}{[(1+\xi)^2 + \alpha^2\,(1 + a\,|E_L|)^2]}\,,$$

where

$$T_{1\mathrm{eff}}^{-1} = T_1^{-1}\,\frac{(1+\xi)^2 + \alpha^2\,(1 + a\,|E_L|^2)}{(1+\xi)^2 + \alpha^2}\,.$$

Its solution in the vicinity of $\theta_0 = 0$ (the soft mode) under the condition $\Omega_0 \gg T_{1\mathrm{eff}}^{-1}$ has the form

$$\Omega_1 \cong \Omega_0 - iT_{1\mathrm{eff}}^{-1}/2\,, \quad \Omega_2 = -\,i\theta_0/\Omega_0^2\,. \qquad (5.22)$$

The solution Ω_1 corresponds to the stable lattice vibrations. The frequency Ω_2 is the renormalized relaxation mode, corresponding to the relaxation of the populations difference D (if $E_L = 0$, then $\Omega_2 = -iT_1^{-1}$) It is cancelling of Ω_2 that leads to the instability. From, the condition $\Omega_2 = 0$ and Eq. (5.13) we obtain the following equation for the determination of those values of ξ, at which the instability sets in :

$$\xi^3 + (1 - \frac{3}{2} bD^0\alpha)\,\xi^2 - 2\,bD^0\alpha\,\xi - \frac{1}{2} bD^0\,(1 + \alpha^2) = 0 . \qquad (5.23)$$

It is easy to verify, that the equation for determination of the values $\xi_m^{(1)}$ and $\xi_m^{(2)}$ (see fig.6.10), obtained from (5.13), is identical to (5.23). Thus, one can conclude, that the laser-induced electronic relaxation mode is responsible for the bistability of the vibronic system.

Let us do the numerical estimates. In the case of rather strong electron-phonon interaction one has

$$b = C^2 n_{ab} / (\hbar M \Omega_0^2 T_2^{-1} n) \sim T_2 \omega_{ab} n_{ab} / n \sim 1 - 10^2 ,$$

for $\omega_{ab} \sim 10^{15} s^{-1}$, $T_2^{-1} \sim 10^{12} - 10^{10}\ s^{-1}$, $n_{ab} / n \sim 10^{-3}$. Let $b = 10^2$, $T_2^{-1} \sim 10^{10} s^{-1}$, $\alpha = T_2^{-1}\Delta^{-1} \sim 10^{-1}$, $|D^0| \cong 1$. Then the inequality (5.17) is fulfilled and from (5.19) we have $a\,|E_c|^2 \sim 1$. Taking $T_1^{-1} \sim 10^3 s^{-1}$, $d_{ab} \sim 10^{-19}$ ESU, we obtain $a \sim 40$ and $I_c \sim 10$ W / cm^2 . Note, that with the increase of Δ the critical value of intensity, as it is seen from (5.19). is increased proportionally to Δ^3.

6.5.2. *Bistability of acoustic lattice vibrations.* Quite analogous bistability effect occurs in the acoustic subsystem under the action of laser radiation on the electronic subsystem. In this case the electron-phonon interaction is described by the Hamiltonian

$$H = \theta \text{ div } \mathbf{U} , \qquad (5.24)$$

where θ is the acoustic deformation potential, $\mathbf{U}$ is the displacement vector of lattice acoustic vibrations. Now instead of eqs (5.4) and (5.6) we have the following equations for the system of two-level atoms

$$(\frac{\partial}{\partial t} + i\Omega_{ab} + iT_2^{-1})\,\rho_{ab} = - \frac{i}{\hbar} d_{ab} E_L D - \frac{i}{\hbar} \bar{\theta}\,\rho_{ab} \text{ div } \mathbf{U} , \qquad (5.25)$$

$$\frac{\partial^2 \mathbf{U}}{\partial t^2} = s^2 \Delta\,\mathbf{U} + n_{ab} \rho^{-1} \text{ grad } (\frac{\bar{\theta}}{2} D + \frac{\theta_{aa} + \theta_{bb}}{2}) , \qquad (5.26)$$

where s is the sound velocity, ρ is the medium density, $\rho_{aa} + \rho_{bb} = 1$, $\bar{\theta} = \theta_{aa} - \theta_{bb}$. The equation for $D = \rho_{aa} - \rho_{bb}$ coincides with (5.5). It is seen from the comparison of (5.25) and (5.4), that these equations are transformed into each other under the substitution $CU \rightleftarrows \theta \operatorname{div} \mathbf{U}$. From (5.26) in the stationary case, when $\partial^2\mathbf{U}/\partial t^2 = 0$ we obtain the equation for the medium deformation

$$s^2 \operatorname{div} \mathbf{U} = -\frac{\bar{\theta}}{2} n_{ab} D - \frac{\theta_{aa} + \theta_{bb}}{2\rho} n_{ab} . \tag{5.27}$$

This equation is completely analogous to the stationary equation (5.9). Let us introduce the dimensionless parameter of acoustic strain

$$\xi = \frac{\bar{\theta}}{\hbar\Delta} \frac{\operatorname{div} \mathbf{U} + (\theta_{aa}\overset{0}{\rho}_{aa} + \theta_{bb}\overset{0}{\rho}_{bb})\, n_{ab}}{\rho\, s^2} . \tag{5.28}$$

The expression in the square brackets, as is seen from (5.27), is the nonequilibrium strain, appearing due to the deviation of D from $\overset{0}{D}$. From (5.27), (5.28) , (5.25) and (5.26) we obtain the equation for ξ, coinciding with (5.13), where now

$$b = b_a \equiv \frac{\bar{\theta}^2 n_{ab}}{2\rho\hbar T_2^{-1} s^2} , \tag{5.29}$$

and all other parameters are the same as in (5.13). All the further treatment is completely analogous to that of the preceding section after the formula (5.13). Thus, to describe the acoustic bistability, one can use the formulas (5.14) - (5.22) and the corresponding conclusions of sec.6.5.1 are valid with the substitution of the parameter b, defined now by the formula (5.29), i.e. substitution

$$\frac{C^2 n_{ab}}{nM\Omega_0^2} \rightarrow \frac{\bar{\theta}^2 n_{ab}}{\rho\, s^2}$$

in each formula. Note, that at the same values of parameters the constant, determining the strength of electron-phonon interaction in the case of acoustic bistability b_a - (5.29) ~ 10 - 10^3 is approximately one order of magnitude larger then the analogous parameter b_0 - (5.11) in the case of optical vibrations. In the case of semiconductor in (5.29) one must substitute $n_{ab} = n$.

6.5.3. *Light-induced jumps of dielectric permeability.* As one can see from (5.12) and (5.14), the jumps of $|\xi|$ leads to the change of the transition frequency ω_{ab} and the corresponding jumplike changes of the dielectric permeability of the doping system. From (5.4), (5.8) and (5.12) we have the following expression for the polarizability of the system of doping atoms

$$\chi^{(1)}(\omega) = -d^0 n_{ab} |d_{ab}|^2 \frac{1 + \xi + i\alpha}{\hbar\Delta\,[(1 + \xi)^2 + \alpha^2(1 + aE_L^2)]}, \qquad (5.30)$$

where ξ can be as an optical - (5.12), as an acoustic - (5.28) lattice displacement, d_0 is the difference of the populations in an atom in the absence of laser field. In the case of local vibrations $d^0 \equiv D^0$ and in collective case $d^0 = D^0/N_{ab}$. In particular, from (5.23) it follows, that due to the jumps of $|\xi|$ the coefficient of the absorption Im $\chi^{(1)}$ ($d^0 <$ 0) or of the gain ($d^0 >$ 0) undergoes the sudden jumplike changes.We note, that in laser media this can lead to the appearance of spiking regime of laser generation.

6.6. Laser-induced phase transitions and threshold second harmonic generation in centrosymmetric media

In Sec.5.7 we have considered the orientational light-induced phase transition in the system of dipole molecules, occurring due to the interaction through the second harmonic field. In solids, as it was shown in preceding sections of chapter 6, besides the radiative interactions, there exists also strong interaction between the doping centers through the acting electrostatic field and (or) the lattice displacement. In this section we show, that these interactions in the presence of intense laser field can lead to the light-induced phase transitions with the threshold appearance of the second order macroscopic permeability, accompanied by threshold second harmonic generation (SHG) in the initially centrosymmetric media. The interaction through the second harmonic field, studied in Sec.5.7, ensuring the automatic fulfillment of the phase matching condition, is shown to play the important role in the process of threshold effective SHG in centrosymmetric medium.

We consider here two models of phase transitions leading to threshold

SHG. In Sec.6.6.1 the theory of light-induced order-disorder phase transition and SHG in the system of doping centers is developed and in Sec.6.6.2 we apply the model of spontaneous breaking of saturation symmetry with appearance of electrostatic field in the system of centers, developed in Sec.6.3, to the problem of SHG. It is shown that both models can account for experimental results obtained in studies of recently discovered [5,6] threshold effective SHG in centrosymmetric glass fibers [30].

Let us assume, that in the centrosymmetric medium there are (or are created by the light pump) noncentrosymmetric centers with the permanent dipole moments $\mathbf{d}_i$, which can be reoriented by the thermal excitation and in the equilibrium state are distributed chaotically over their directions, so that the mean value $\langle \mathbf{d}_i \rangle = 0$. Such a situation for example may be in centrosymmetric glass fibers [10].

Let the powerful laser field with linear polarization

$$\mathbf{E} = \mathbf{e}_x \, [E \exp(-i\omega t + ikz) + \text{c.c.}] \,, \quad |\mathbf{e}_x| = 1; \; k \equiv k(\omega) \,, \qquad (6.1)$$

act on this medium. Under the action of this light wave the dipoles are oriented preferably in two opposite directions along the x-axis, which we denote by (+) and (–). We consider bellow the idealized case of full orientation, i.e. there are centers of two types: centers of A-type with dipole moments oriented parallel to $\mathbf{e}_x$ and centers of B-type with dipole moments oriented antiparallel to $\mathbf{e}_x$.

6.6.1. *Order-disorder phase transition and SHG.* Analogously to Sec.5.7 we introduce the random orientational parameter q_i, equal to + 1, if $\mathbf{d}_i$ is parallel to the vector $\mathbf{e}_x$ and $q_i = -1$, if $\mathbf{d}_i$ is antiparallel to $\mathbf{e}_x$ and define $d_i = q_i d$, where $d = |\mathbf{d}_i|$.

Every center has the quadratic polarizability

$$\beta^2 = q_i \beta^2$$

but in the absence of dipole direction ordering the macroscopic quadratic permeability

$$\chi^2 = \langle q_i \rangle \, \beta^2 n = 0 \,, \qquad (6.2)$$

where n is the concentration of the centers.

We shall show below, that at the certain critical intensity of the pump $I_L = I_c$, due to the dipole-dipole interaction through electrostatic Lorentz field

$$\vec{\varepsilon} = \beta\, 4\pi\, n \,\langle \mathbf{d}_i \rangle$$

where β is the constant of acting Lorentz field,and through the lattice displacement U, laser-induced orientational phase transition occurs with the order parameter $\langle q \rangle \equiv \langle q_i \rangle$, in which the dipoles tend to array in an unique direction. Since, due to the extra interaction through the SH radiation field (Sec.5.7), the spatially oscillating, small additive orienting forces arise, the dipoles are arrayed into the grating with the periodic alternation of (+) and (–) directions, which leads to the appearance of the $\chi^{(2)}(z)$ - grating automatically fulfilling the phase matching conditions , that ensues the effective SHG.

The light pump (6.2) creates the centers in the host medium in accordance with the equation

$$\frac{dn}{dt} = w - \Gamma n \,, \tag{6.3}$$

where $w = \sigma E^2$ is the probability of the center creation per unit of time in a unit volume (σ = const), Γ^{-1} is the center excited state lifetime. The Eq. (6.3) describes the appearance of the centers of two types: the excited states of initially present doping centers, or the pump-induced new centers, for example, the color centers. For the following consideration it is unimportant what type of centers is created by the pump in accordance with Eq.(6.3).

Let us define the mean number of the centers with vectors $\mathbf{d}_i$, directed along the vector $\mathbf{e}_x$, through n_+ and with $\mathbf{d}_i$, directed oppositely to $\mathbf{e}_x$ - through n_- ; then $n_+ + n_- = n$. In a sufficiently weak pump field, when $I_L \ll I_c$, the order parameter of the phase transition

$$\langle q_i \rangle \equiv D = \frac{n_+ - n_-}{n} = 0,$$

which corresponds to disordered state.

The dipoles are reoriented due to the thermally activated passage over the potential barrier W, the magnitude of which, without taking into account of the cooperative dipole-dipole interaction,is determined by the interaction of the center with the medium and is decreased with the excitation of the center: $W_{ex} \ll W_g$, where W_g and W_{ex} are the activation energies in ground and excited states correspondingly.

The cooperative character of the dipole system stems from the two factors: the interaction with the electrostatic self-consistent field $\vec{\varepsilon} = \mathbf{e}_x \varepsilon$ and the interactions with the self-consistent lattice displacement field $\mathbf{U} = \mathbf{e}_x U$. The energy of the electrostatic interaction is

$$H_{i\varepsilon} = - q_i \, d \, \varepsilon$$

To describe the interaction through the lattice displacement we use the simple model of centers, interacting with optical phonon mode with wave vector $q = 0$ and frequency Ω_0, described in details in Sec.6.2.. In this case the energy of interaction of i-th dipole with the field of lattice displacement is

$$H_{iu} = q_i \, C \, U ,$$

where C is the interaction constant.
The equation for U has the form analogous to Eq. (2.3), where one must neglect the terms with σ_{+j} and σ_{-j}, substitute $C_j = C$ and $D_j = q_j$. Then from Eq.(2.3) in adiabatic approximation ($\partial^2 U / \partial t^2 = 0$), we obtain

$$U = - \sum_i q_i \frac{C}{\Omega_0^2 M \, N} ,$$

where N is the number of the cells in volume V, M is the cell mass. Using this formula in expression for H_{iu}, one obtains in the mean field approximation ($q_i q_j \longrightarrow q_i \langle q \rangle \equiv q_i \, D$) :

$$H_{iu} = - q_i \frac{C^2 \, n}{\Omega_0^2 M \, n_0} , \quad n_0 = \frac{N}{V} .$$

When one takes into account the cooperative interaction, the symmetry of transitions is broken : for the (+) $\longrightarrow$ (–) transition the barrier height is equal to U_+, and for (–) $\longrightarrow$ (+) transition the barrier is U_- :

$$U_{\pm} = U_{ex} \pm U_0 D \; . \tag{6.4}$$

Here

$$U_0 = \beta \, 4\pi \, d_{eff}^2 \, n \; ;$$

$$d_{eff}^2 = d^2 + \frac{C^2}{\Omega_0^2 M \, n_0} \; ,$$

where $0 < \beta \leq 1$.

The processes of thermally activated dipole reorientations are described by the following kinetic equations

$$\frac{\partial n_+}{\partial t} = - \frac{n_+}{\tau_{+-}} + \frac{n_-}{\tau_{-+}} \; ;$$

$$\frac{\partial n_-}{\partial t} = - \frac{n_-}{\tau_{-+}} + \frac{n_+}{\tau_{+-}} \; , \tag{6.5}$$

where

$$\tau_{+-} = \tau^{-1} \exp \left(- \frac{U_+}{k_B T}\right); \quad \tau_{-+} = \tau^{-1} \exp \left(- \frac{U_-}{k_B T}\right);$$

τ is the constant, T is the temperature.

From (6.5) one obtains the kinetic equation for the order parameter

$$\frac{\partial D}{\partial t} = \frac{2}{\tau_{ex}} \cosh \gamma D - \frac{2}{\tau_{ex}} D \cosh \gamma D \; , \tag{6.6}$$

where

$$\tau_{ex} = \tau^{-1} \exp \left(- \frac{U_{ex}}{k_B T}\right); \quad \gamma = \frac{U_0}{k_B T} \; . \tag{6.7}$$

The system of equations (6.3) and (6.6) describes the laser-induced phase transition in the cooperative system of dipole centers. In the following, assuming the fulfilling of the condition $\Gamma > \tau_{ex}^{-1}$, we obtain from (6.3)

$$n = \frac{w}{\Gamma} = \frac{\sigma E^2}{\Gamma} . \tag{6.8}$$

Now let us analyse the system of Eqs. (6.3), (6.8). At first we consider the stationary state, installing at times $t \gg \tau_{ex}$. From the Eq.(6.6) we obtain the following stationary equation for the order parameter

$$D = \text{th}\, \gamma\, D . \tag{6.9}$$

The Eq.(6.9) is again the Ising type equation, analogous to that, obtained in Sec.5.7 in the case of interaction through SH field (see Eq.(5.7.5). Now the coefficient γ depends on the pump intensity through the number of excited centers n (see (6.7), (6.4)) and (6.8)).

The value $D \neq 0$ appears at $I_L > I_c$ and $D \longrightarrow 1$ for $I_L \longrightarrow \infty$, that corresponds to the orientation of all dipoles in unique direction, i.e. ordered state.

Expanding the right hand side in (6.9) into power series of D, one obtains from the condition of cancelling of the coefficient at D, the critical condition, determining the value of I_c:

$$\gamma = \gamma_c = 1 ,$$

i.e.

$$\beta\, 4\pi\, d_{eff}^2 n_c = k_B T . \tag{6.10}$$

Now let us take into account, that i-th dipole interacts with the integral field of SH radiation , created by all others centers. The energy of this interaction one can find analogously to Sec.5.7 :

$$H_1^{(2)} = -\, k^2 (2\omega) \sum_{j \neq i} \{q_i q_j\, |\beta^{(2)}|^2 E^4 \frac{1}{r_{ij}} \exp\{i\, [r_{ij}\, k(2\omega)$$

$$- (z_i - z_j)\, 2k\, (\omega)]\} + \text{c.c.}\} , \tag{6.11}$$

where $k\,(2\omega)$ is the SH wave number ; $r_{ij} = |\mathbf{r}_i - \mathbf{r}_j|$ is the distance between i-th and j-th centers. From (6.11) it follows, that the integral Hamiltonian of the center interactions through SH in the cylinder of length l and radius a ($l \gg a$) in the mean field approximation ($q_i q_j \longrightarrow$ $<q(\mathbf{r})><q(\mathbf{r}')>$) has the form

$$H^{(2)} = - k^2 (2\omega) \mid \beta^2 \mid^2 E^4 n^2 \int\int_{VV} [<q(\mathbf{r})> <q(\mathbf{r}')> \times \frac{1}{R} \cos \Delta k (z - z')] \, d\mathbf{r} \, d\mathbf{r}' , \tag{6.12}$$

where $\Delta k = k(2\omega) - 2k(\omega)$ is the wave numbers mismatch, $R = |\mathbf{r} - \mathbf{r}'|$. The Hamiltonian (6.12) is minimized, if one takes the order parameter in the form

$$<q(\mathbf{r})> \equiv D = \cos \Delta k z , \tag{6.13}$$

This means, that the additive interaction through the SH field tends to arrays the dipole into the grating with the alternation of their directions. When the dipoles concentration is near the critical value (6.10) the dipoles are easy to reorient and the weak additive forces due to the SH field in fact can array the dipoles into the grating, which leads to the appearing of the grating of macroscopic quadratic permeability

$$\chi^2 = D_0 \beta^2 n \cos \Delta k z , \tag{6.14}$$

automatically ensuring the fulfillment of the wave matching on the full length of interaction l. Now the Eqs (6.6) and (6.9) are the equations for the order parameter grating amplitude D_0.

Let us study the kinetics of the ordered state installation. Expanding the right hand side in (6.6) into power series of D_0 (substituting before $D \rightarrow D_0$), one obtains the eq. for D_0, describing the stage of exponential growth. The solution of the latter yields

$$D_0(t) = D_0(0) \exp \left[\frac{2}{\tau_{ex}} (\gamma - 1) t\right]$$

$$\equiv D_0(0) \exp \left(\frac{1}{2} G_{theor} t\right). \tag{6.15}$$

If $\gamma < 1$ the initial fluctuation $D_0(0)$ decays, if $\gamma > 1$ it starts to grow in time. At time $t \gg G^{-1}_{theor}$ the D_0 growth saturates and the stationary regime sets in, described by the Eq. (6.9).

With the appearance of the order parameter grating ($D_0 \neq 0$) in the medium there appears the grating of the macroscopic second order permeability χ^2 with the amplitude $\chi^2_0 = n\ D_0\beta^2$. Since the radiated second harmonic intensity

$$w_{2\omega} \sim |\chi^2|^2 ,$$

then from (8.15) it follows that in the regime of exponential growth the SH radiated intensity

$$w_{2\omega} \sim \exp\ [\ \frac{4}{\tau_{ex}} (\gamma - 1)\ t]\ . \qquad (6.16)$$

Let us compare the above theory with the experimental results on the SHG in centrosymmetric media (glass fibers [5-10].). The theory of SHG must explain the following basic results :

1) The effective SHG occurs after the prolong (during the hours) irradiation of the centrosymmetric medium by the powerful laser pulses with the intensity W_ω [6].

2) The SH intensity $W_{2\omega}$ grows in the process of this irradiation according to the law

$$W \sim \exp\ (G_{exp}\ t\ ), \quad G_{exp} = 3.1\ (6.6\ W_w = 0.6)\ , \qquad (6.17)$$

where t is in hours, W_ω is in W [6]. This experimental result clearly shows the presence of the pump intensity threshold of SHG in centrosymmetric media.

3) The SHG efficiency is preserved after this irradiation at the same level during weeks [6].

4) In the case of linear polarized pump wave the SH wave is also linear polarized with the vector of polarization parallel to that of the pump [6].

5) The efficiency of SH conversion reaches high values of more then 1% [6], which means the autoinducing in the process of SHG of the χ^2-grating , since the intrinsic matching length in glass fibers is of order of 10^{-3} cm [7].

6) The installation of the fiber during the pump irradiation into the external electrostatic field ~ $10^3\ V\ /\ cm$ doesn't influence the SHG efficiency [8], on the other hand the electrostatic field ~ $10^6\ V\ /\ cm$

sharply diminishes the SHG efficiency down to 10^{-3} % [9].

The formula (6.16) predicts the following kinetics of the SHG growth

$$w_{2\omega} \sim \exp\,(G_{\rm theor}\, t)\ ;$$

$$G_{\rm theor} = \frac{4}{\tau_{\rm ex}} \left[\frac{\beta\, 4\, d^2_{\rm eff}}{k_{\rm B} T} \frac{\sigma\, E^2}{\Gamma} - 1 \right] . \qquad (6.18)$$

It is this functional dependency $w_{2\omega} = w_{2\omega}\,(E^2,\, t)$ that is experimentally observed [6] (compare (6.17) with (6.18)). Equating (6.17) and (6.18) one obtains $\tau_{\rm ex} \sim 2$ hr. With taking into account of (6.18) the formula (6.18) can be represented in the form

$$G_{\rm theor} = 4\, \tau^{-1}_{\rm ex}\, \left(\frac{n}{n_{\rm c}} - 1 \right),$$

where the critical concentration of the pump generated centers

$$n_{\rm c} = \frac{k_{\rm B}\, T}{\beta\, 4\pi\, d^2_{\rm eff}} \sim 5 \times 10^{17}\ {\rm cm}^{-3}\ ,$$

for $d \sim 3 \cdot 10^{-18}$ ESU, $\beta \sim 1$, $T \sim 3 \times 10^2$ K, $C \sim 10^{-4}$ erg / cm,

$\Omega_0 \sim 10^{12}\ {\rm s}^{-1}$, $M \sim 10^{-22}$ g, $n_0 \sim 10^{22}\ {\rm cm}^{-3}$.

Since $\chi^2_{\rm j} \sim d_{\rm i}\, (d_{\alpha\beta})_{\rm j}\, (d_{\alpha\beta})_{\rm k}$, where $d_{\alpha\beta}$ is the transition dipole moment and $P^{(2)}_{\rm i} = \chi^2_{\rm j}\, E_{\rm j} E_{\rm k}$, the vector $\mathbf{P}^{(2)} \parallel \mathbf{d}$, i.e. $\mathbf{P}^{(2)} \parallel \mathbf{E}$, that corresponds to the experiment.

In the external electrostatic field the dipole acquires the additive energy $H_{{\rm iE}_0} = d_{\rm i} E_0 \sim 3 \times 10^{-17}$ erg at $E_0 \sim 3 \times 10^{3}$ V / cm and the energy $H'_{{\rm iE}_0} \sim 3 \times 10^{-14}$ erg at $E'_0 \sim 3 \times 10^6$ V/cm. Besides, from (6.11) for the energy of interaction with all other dipoles through the SH radiation field one has the estimate

$$H^{(2)}_{\rm i} \sim 8\pi\, (k^2\, a^2)\ |\ \beta^2 |^2\ E^4\ \ln\, (ln^{1/3})\ ; \qquad (6.19)$$

To estimate β^2 we take into account, that the transitions in the centers occur between the vibronic levels with the distance between them equal to $\Delta \sim 10^{12}\ s^{-1}$, so that β^2 increases in comparison with the nonresonant value $\beta^2_{nonres} \sim 10^{-31}$ ESU by the factor $\omega / \Delta \sim 10^3$. At $a \sim 10^{-3}$ cm, $E^2 \sim 10^7$ ESU, $k \sim 6 \times 10^4\ cm^{-1}$, $n \sim 10^{21}\ cm^{-3}$, and $l \sim 10$ cm one has for the energy of interaction through the SH radiation field $H_i^{(2)} \sim 10^{-16}$ erg. Thus, the following hierarchy of the interaction energies exists in the system considered :

$$k_B T \sim \beta\, 4\pi\, n_c d^2_{eff} \sim 10^{-14}\ \text{erg} \sim H'_{iE_0} > H_i^{(2)} > H_{iE_0} .$$

These inequalities explain the above mentioned effects of the electrostatic field influence on SHG efficiency. In the weak field E_0 the interaction energy $H_{iE_0} < H_i^{(2)}$ so that this field can not break down the χ^2- grating, on the other hand the strong field with $H'_{iE_0} > H_i^{(2)}$ arrays the dipoles in a unique direction, breaking down the χ^2- grating, which leads the sharp fall of the SHG efficiency. At last, since the transition to the ground state occurs without the dipole flip, the state , arising during the fiber irradiation,is preserved also after its end. Since the orientation relaxation time in the ground state $\tau_g \gg \tau_{ex}$ the χ^2 - grating must be preserved for a long time.

Thus, the mechanism of laser-induced order-disorder phase transition may serve as a possible mechanism of threshold SHG.

6.6.2. *Spontaneous breaking of vibronic transitions saturation symmetry and SHG.* In this Section we consider another possible microscopic mechanism of threshold SHG in centrosymmetric media caused by the spontaneous breaking of the symmetry of saturation of transitions in centers irradiated by laser light [61].

As was shown in Sec.6.3 this symmetry breaking is accompanied by threshold appearance of photovoltaic current in centrosymmetric crystals leading to arising of strong electrostatic field. We show in this Section, that this field leads to inducing the effective quadratic permeability χ^2 and that the action of weak second harmonic field plays the important role in the vicinity of phase transition, leading to formation of spatially oscillating electrostatic field, i.e. spatially oscillating χ^2-grating, automatically satisfying phase matching condition.

Consider the same system of noncentrosymmetric centers in centrosymmetric medium irradiated by laser wave (6.1) as in Sec.6.1.1.We assume the same scheme of relevant vibronic transitions in the system of laser-field oriented A(+) and B(-) centers as was considered in Sec.6.3 (see Fig.6.5). This scheme describes now the excitation and rapid intraband relaxation with subsequent ionization of vibronic centers.

It was shown in Sec.6.3 that in the strong laser field with intensity exceeding threshold value η_1(Eq.3.17) the symmetry of occupation of *c*c-states (now bottom states of excited band) in A (+)and B(-)-centers is spontaneously broken and nonzero electrostatic field appears in centrosymmetric medium. This field is described by Eq.(3.20).

Let us make numerical estimates of characteristic parameters of phase transitions using formula of Sec.6.3 describing now the case of vibronic transitions in centers. The saturation parameter (Eq.3.9) is now $a = 10^{-4}$ ESU at $d_{ab} = 10^{-2}$ D, $\gamma = 10^7\ s^{-1}$, $T_2^{-1} = 10^{12} s^{-1}$. Take further for frequency detuning, given by the Eq.3.12 the value $\delta = |\Delta| / T_2^{-1} = 10$, which corresponds to the value of $|\Delta|$, equal to the the distance between the adjusting vibronic levels. At these values $E^2_{Lsat} \sim \delta^2 / a \sim 10^6$ ESU i.e. the saturation intensity $I_{Lsat} \sim 10^9$ W/cm^2. Evaluate now critical intensities of appearance and disappearance of electrostatic field according to formula (3.17). We take for glass $\sigma_0 \sim 10^{-4}\ s^{-1}$ that is two orders of magnitude smaller than the corresponding value for crystalline $LiNbO_3$ (Ref.22). $\mu = 3 \times 10^{-1}$, µor $n = 10^{19} cm^{-3}$ we have $\sigma_m \cong 10^9 s^{-1}$, so that at $\tau = 10^{-12}$s , we have $\sigma_m\tau \sim 10^{-3}$. Then at d = 3 D, $l \sim$ 1 A , $T_2^{-1} \sim 10^{12} s^{-1}$ we obtain from (3.18) for the cooperative parameter $K \cong 50$. At $\Omega \sim 10$ (compare Sec.6.3.2) from (3.16) we obtain $q_0 \sim 10^{-2}$. Substituting $K = 40$ into (3.17) and using above values of parameters, we obtain for the critical intensity of appearance of electrostatic field $I_{c1} \sim 10^8$W/cm^2. The critical intensity of the disappearance of electrostatic field $I_{c2} \sim 5\ 10^{11}$W/cm^2. For the characteristic time of establishing of electrostatic field we have from (3.21),using $\eta \sim 1$ and $\varepsilon \sim 10$, the value $\tau_{est} \sim 10^3$ s, that by order of magnitude corresponds to characteristic times of SHG rise under fiber irradiation [5,6]. At last for the maximum electrostatic field we obtain from (3.20) and (3.10) the following estimate E_{0max} ~ $\hbar|\Delta| / d$. At d =d~ 1D we have E_{0max} ~ 10^5 V/ cm .

The SH nonlinear polarization, induced by electrostatic field E_0,is expressed through cubic permeability χ^3 by formula

$$P \quad P((2\omega) = \chi^3 E_0 E_L E_L = \chi^2_{eff} E_L E_L, \tag{6.20}$$

where

$$\chi^2_{eff} = \chi^3 E_0 \tag{6.21}$$

is the effective quadratic permeability. Thus, the spontaneous appearance of electrostatic field in centrosymmetric medium due to saturation symmetry breaking is accompanied by threshold SHG. Since vectors $\mathbf{E}_L$ and $\mathbf{E}_0$ are directed along x-axis, the vector $\mathbf{P}(2\omega)$ and consequently vector $\mathbf{E}_2$ are also directed along x-axis.

Now we want to show that allowance for the initially weak SH field

$$E_2(zz)=E_2\exp(-i2\omega t+ ik(2\omega)z) + c.c. \tag{6.22}$$

in equations for density matrix (3.5)-(3.8) leads to appearance of spatially oscillating electrostatic field $E_0(z)$ and correspondingly to formation of χ^2_{eff}-grating, automatically satisfying phase matching condition.

In fact, using (3.5)-(3.8) and introducing $\rho_{ab} = \tilde{\rho}_{ab}\exp(-i\omega t)$, one can easily show, that taking into account of (6.22) leads to appearance of spatially inhomogeneous addition $D(z)$ to population difference on zero frequency

$$D = \rho_{aa} - \rho_{bb} = D_0 + D(z),$$

where E_0 is spatially uniform solution and

$$D \quad D(z) = \frac{d\, d_{ba} E_L E_2^* T_2^{-1} \tilde{\rho}_{ab}}{\gamma_1 \hbar^2 \, [\, (\omega_{ab} - 2\omega)^2 + T_2^{-2}]} \cos \{ [k(2\omega) - 2k(\omega)]zz \} \tag{6.23}$$

Using (6.23) in (3.5) and taking into account that $\mathbf{E}_2^* \sim \mathbf{P}(2\omega)^* \sim \mathbf{E}_0$, we obtain that in the expression for the Stark shift of the frequency ω_{ab} in

Eq.(3.5) and in subsequent formulas in Sec.6.3 one must substitute

$$\mathbf{d} \longrightarrow \mathbf{d}(z) = \mathbf{d} \; \{ \; 1 + \text{const} \cos[i\Delta kz] \; \}, \qquad (6.24)$$

where $\Delta k = k(2\omega) - 2\, k\mathrm{k}(\omega)$ is phase mismatch. Thus instead of (3.14) one has the following equation for electrostatic field near the threshold of instability

$$\frac{\varepsilon\partial\xi}{4\pi\partial\, tt} = -\,\sigma\xi \; - C_1\{1 + C_2\cos\,[\Delta k\; z]\}\;\xi, \qquad (6.31)$$

where C_1 and C_2 are independent of zz and can be expressed through parameters of the problem. Thus the growth rate is spatially modulated that leads to growth of spatially oscillating electrostatic field $E_0\;(z)$ and due to (21) to corresponding growth of effective χ^2–grating, automatically satisfying wave matching condition for effective SHG. The amplitude of $E_0(z)$-grating is determined by formula (3.20). To destroy this grating one must apply the external electrostatic field of order of this amplitude (10^5–10^6V/cm)

Thus we see that both microscopic mechanisms of threshold effective SHG in centrosymmetric medium with noncentrosymmetric centers, irradiated by powerful laser light (order-disorder phase transition and spontaneous breaking of saturations symmetry) can account for the above listed experimental data on the threshold exponential growth of SHG, threshold intensity and growth rate values, SH polarization and high efficiency due to automatic phase matching and also on external electrostatic field influence. Further experimental studies are needed to select the relevant mechanism of SHG in centrosymmetric glass fibers. We stress the fact that above considered mechanisms of threshold SHG may reveal themselves also in other centrosymmetric solids (ruby, for example) as well as in liquids and gases.

6.7. Laser-induced electron-deformation-thermal instability and semiconductor-metal phase transition involving superstructure formation.

6.7.1. *Introduction.* Up to now we studied the systems with discrete energy levels in the bulk of the working volume. New interesting features appear when one goes over to the study of the light - induced phase transitions on the surfaces of solids with the continuous electronic energy spectrum. Although the detailed studies of the possibilities opened in this case goes beyond the scopes of this book , we consider in this Section one important example of light-induced phase transition on the surface of semiconductor [31-33]. The short review of other problems of light-induced surface phase transitions, being actively investigated now, is given in Sec.6.10.

The laser-induced surface phase transitions are interesting to the applied problems of pulsed laser modification of surface layers of semiconductors which are currently attracting much attention [34]. Phase transitions induced by laser pulses are due to a change in the state of the lattice subsystem, whereas laser radiation of photon energy $\hbar\omega > E_g$ (where E_g is the width of the band gap) interacts with the electron subsystem. The electron-phonon interaction and the corresponding transformation of the absorbed energy thus play a key role in the effects mentioned above.

The usual mechanism by which the absorbed energy is transferred to the lattice consists of electron-electron, electron-phonon, and phonon-phonon relaxation stages with relaxation times τ_{rel} of the order of a picosecond. The resultant spatial distribution of the lattice changes should reflect a monotonic reduction of the radiation intensity away from the center of a laser-illuminated spot and an exponential reduction of the absorbed energy with depth in the investigated medium (with a characteristic scale at least equal to the linear absorption length γ_0^{-1}).

However, some recent experimental results obtained using high-power ultrashort laser pulses cannot be explained on the basis of these standard representations. Thus, instead of a monotonic pattern, frequently found that ordered surface structures are formed. For example, irradiation with picosecond laser pulses of a semiconducting phase of VO_2 creates a spatially periodic surface pattern representing an alternation of metallic and insulating phases, and if the laser field has axial symmetry, it is found that concentric rings appear at the center of the spot and radial rays are formed at the periphery (a "star" pattern is observed), whereas when the field has slab geometry it is found that one-dimensional gratings are obtained (Fig.6.13) [35].

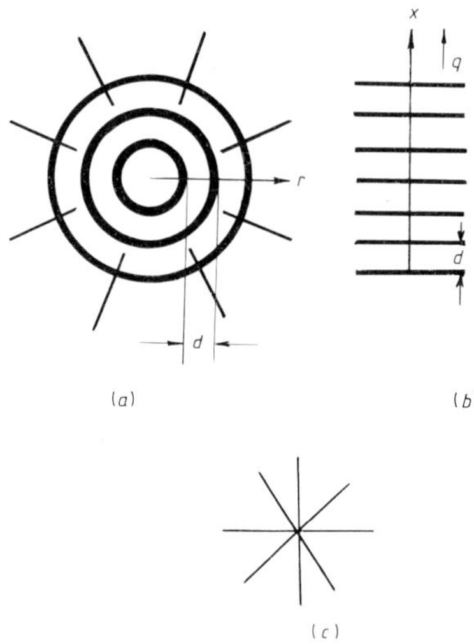

Fig.6.13. The geometry of the spatially periodic surface fields of strain, non-equilibrium carriers concentration and temperature arising due to EDTI: a)the "star-"structure and b - radial rays structure for the axial-field and isotropic surface; b)"grating" with the q-direction being selected by the one dimensional geometry of the pump field or by the crystallographic axis; c) radial rays structure.

Pulsed laser irradiation of the surface of a semiconductor also gives rise to concentric ring structures representing an alternation of crystalline and amorphous phases [36]. Some result support a nonlinear mechanism of transfer of the laser energy to the lattice at high pump intensities. For example, a reduction in the absorption length by an order of magnitude compared with γ_0^{-1} is reported in Ref. 37. It is shown in Ref. 38, that high-power femtosecond pulses induce disorder in the crystalline lattice in subsurface layer of a semiconductor during a pulse, i.e., in a time much shorter than τ_{rel}.

These experimental observations can be interpreted in terms of the formation of laser-induced instabilities on the surface of a semiconductor, leading to a periodic change in the resultant state of the lattice along the semiconductor surface and a strong increase in the effective optical absorption coefficient.

We shall consider in this section two new laser-induced instabilities on the surface of a semiconductor: an electron-deformation instability

(EDI) and a deformation-thermal instability (DTI), the theory of which can account for the formation of the ordered structures mentioned above. These two instabilities can occur simultaneously, giving rise to an electron-deformation-thermal instability (EDTI) [33].

The physical mechanism of the EDTI is as follows. The surface deformation ξ = div **U** (**U** is the displacement vector of the medium under consideration), as and temperature T, all of which modulate spatially the width of the band gap:

$$E_g = E_{go} + \theta\xi - \beta_n n - \beta_T T \,, \tag{7.1}$$

where E_{go} is the equilibrium value of the band gap: $dE_g / d\xi = \theta$; $\theta = \theta_{cc} - \theta_{vv}$; θ_{cc} and θ_{vv} are the deformation potentials of the conduction and valence bands, respectively. The phenomenologically introduced coefficient $\beta_n = |\partial E_g / \partial n| > 0$ describes the reduction in E_g because of the breaking of covalent bonds when electrons are transferred from bonding states in the valence band to antibonding states in the conduction band [39], while the coefficient $\beta_T = |\partial E_g / \partial T| > 0$ allows for the reduction in E_g as a result of heating [40]. Modulation of E_g results in modulation of the interband optical absorption coefficient, causing accordingly an additional modulation of n and T. The resultant forces $\mathbf{F}_n = \theta$ grad n, and $\mathbf{F}_T = - K\alpha$ grad T (K is the bulk modulus and α is the thermal expansion coefficient) maintain initial displacements of the lattice, giving rise to an instability of the amplitude of the surface strain ξ and also of the carrier density n and the temperature T when a certain critical intensity of the laser pump radiation is exceeded. In this section we consider the solution of the boundary value problem of the development of the EDTI on the surface of a semiconductor under strong optical absorption conditions. A general dispersion equation is obtained for the EDTI and its solution determines the rate of the exponential growth (with time) of the Fourier amplitudes of the coupled fields of ξ, n, and T on the wave vector $q[\lambda = \lambda\ (q) = \lambda' + i\ \lambda'']$. It follows from this equation that three qualitatively different instabilities can appear under the EDTI conditions. The first is the instability of surface acoustic waves ($\lambda'' \neq 0$, $\lambda' > 0$) which is initiated by thermal surface acoustic waves. The second is an instability (softening) of the acoustic surface waves frequencies. The third is the instability of surface static deformations ($\lambda'' = 0$, $\lambda' > 0$). In the last case the instability begins from the initial fluctuations of the the temperature or carrier density. The present section represents a detailed study of the last instability.

We shall show that the EDTI can create complex ordered configurations of the coupled surface fields of ξ, n, and T (in the form of gratings, rings, rays, "stars", and radial-ring cells formed by the intersection of rings and rays). The periods of these structures are determined as a function of the parameters of the investigated material and of the intensity and duration of the laser pulses, and of how far they penetrate from the surface into the material. A nonlinear steady-state EDTI regime is considered, which becomes stabilized because of nonlinear Auger recombination of carriers (and also because of an optoacoustic non-linearity). The steady-state values of the Fourier amplitudes of the strain, carrier density, and temperature are determined as a function of the wave vector q and of the pump radiation intensity. It is shown that the modulation amplitude of the band gap width along the surface, in accordance with the mechanism described by Eq. (7.1) may in principle reach values at which E_g vanishes in a spatially periodic manner, i.e., a laser induced semiconductor-metal transition accompanied by the formation of ordered surface structures is induced. In Sec.6.7.7 we apply the conception of EDTI for developing the model of ultrafast structural phase transition in GaAs induced by femtosecond laser pulse and discuss relevant experimental results.

Allowance for the deformations (strains) and for the influence of the surface in the present paper is a fundamentally novel feature of the theory of a laser-induced insulator-metal phase transition, which can explain the formation of copmlex periodic structures on the surface. Numerical estimates of the parameters of these structures showed that the EDTI mechanism may responsible for their formation.

6.7.2. *Closed system of equations describing modulation of nonequilibrium carrier density, temperature, and displacement vector in medium.* We shall consider a two-band semiconductor filling the half-space $z > 0$ and assume that a laser wave is incident normally on the $z = 0$ surface :

$$\mathbf{E}\ (\mathbf{r},t) = \mathbf{E}\ \exp\ (-i\omega t + ikz) + \text{c.c.} \tag{7.2}$$

Here $\mathbf{E} = \mathbf{E}\ (\mathbf{r})\ f\ (t)$; $f\ (t) = 1$ when $0 \le t \le \tau_p$; $f\ (t) = 0$ when $t < 0$, $t > \tau_p$; τ_p is the duration of a laser pulse; and $\mathbf{r} = \{x,\ y\}$.

In the case under discussion ($\varepsilon' \gg 1$, $\varepsilon' \gg \varepsilon''$) the equation for the temperature of the medium is (see Ref.37)

$$\frac{\partial T}{\partial t} - \chi\ \Delta\ T = \frac{2\ \omega\ \varepsilon''}{\pi\ c_V \varepsilon'}\ |\mathbf{E}|^2\ \exp\ (-\gamma z)\ , \tag{7.3a}$$

where χ is the thermal diffusivity; c_V is the specific heat per unit volume; ε' is the real part of the permittivity of the medium; $\gamma = \omega \varepsilon'' / c (\varepsilon')^{1/2}$ is the optical absorption coefficient; and c is the velocity of light in vacuum.

The equation for the density of non-equilibrium carriers can be written in the form

$$\frac{\partial n}{\partial t} - D \Delta n + \frac{n}{\tau} + \gamma_A n^3 = \frac{2 \varepsilon''}{\pi \hbar \varepsilon'} |\mathbf{E}|^2 \exp(-\gamma z), \qquad (7.3b)$$

where D is the carrier diffusion coefficient, τ is the linear recombination time, and γ_A is the nonlinear Auger recombination constant.

The equation for the displacement vector of the medium is the form

$$\frac{\partial^2 \mathbf{U}}{\partial t^2} = c_t^2 \Delta \mathbf{U} + (c_l^2 - c_t^2) \text{ grad div } \mathbf{U} + \sum_{j=n,T} f_j \text{ grad } Y_j \qquad (7.4)$$

where c_l and c_t are the longitudinal and transverse velocities of sound, $f_n = \theta / \rho$, $f_T = - K \alpha / \rho$, and ρ is the density of the medium. 1It follows from Eq. (7.1) that the optical constants ε'' and ε' depend on n, T, and ξ. In the range of values of n, T, and ξ of interest to us the dependence of ε'' can be represented in the following form if we use the expression for the interband permittivity of a semiconductor $\varepsilon(\omega) = \varepsilon' + i\varepsilon''$ and allow for Eq. (7.1) :

$$\varepsilon'' = \varepsilon_0'' + \frac{\partial \varepsilon''}{\partial \omega} \frac{1}{\hbar} (\beta_n n + \beta_T T - \theta \xi) = \varepsilon_0'' + \varepsilon_1'' . \qquad (7.5)$$

A similar dependence can be written down also for ε'. However, since in range of frequencies of interest to us we have $\varepsilon' \gg \varepsilon''$ ($\varepsilon' \sim 10$, $\varepsilon_0 \sim 10^{-1}$) and $\partial \varepsilon' / \partial \omega \sim \partial \varepsilon'' / \partial \omega$, we shall ignore the change in $\varepsilon' = \text{const}$. Therefore, we have $\gamma = \gamma_0 + \gamma_1$, where $\gamma_1 \propto \varepsilon_0''$.

We can see from Eq. (7.3) that, in the presence of a laser field, Eq. (7.5) should give rise to a relationship between the diffusing fields of n and T and the deformation (strain) of the medium.

We shall represent the required solutions in the form

$$n = n_0 + n_1, \quad T = T_0 + T_1, \quad \xi = \xi_0 + \xi_1, \quad \mathbf{U} = \mathbf{U}_0 + \mathbf{U}_1,$$

where n_0, T_0, ξ_0, and $\mathbf{U}_0$ are the solutions of Eqs.(7.3) and (7.4) subject to Eq.(7.5), obtained in the zeroth approximation with respect to n, T, and ξ under the appropriate boundary conditions at $z = 0$ (see below). The actual form of these solutions is unimportant for the purpose of the present treatment; the only significant factor is that these solutions are constant along the surface over distances $r \ll r_0$ of interest to us (r_0 is the radius of the laser beam) and they vary sufficiently slowly with time over intervals on the order of the instability development time (see below). We shall now consider the stability of the solutions, n_0, T_0, and ξ_0. Linearizing Eq. (7.3) and applying Eq. (7.5), we yield equations for $n_1 \equiv Y_{n1}$ and $T_1 \equiv Y_{T1}$, which can be written in a unified manner

$$\frac{\partial Y_{j1}}{\partial t} + \Delta_j(t)\, Y_{j1} = (\varepsilon_{j\xi}\, \xi_1 + \varepsilon_{jT}\, T_1 + \varepsilon_{jn}\, n_1) \times \exp(-\gamma_0 z), \tag{7.6}$$

where $j = n, T$; $\Delta_j(t) = -\chi_j\, \Delta + \tau_j^{-1}(t)$; $\chi_T = \chi$; $\chi_n = D$; $\tau_T^{-1} = 0$, $\tau_n^{-1} = \tau^{-1} + 3\,\gamma_A\, n_0^2(t)$. The coupling coefficients are described by the following expressions

$$\varepsilon_{ji} = 2\frac{\partial \varepsilon''_0}{\partial \omega} |\mathbf{E}|^2\, \sigma_{ji} / \varepsilon' \pi \hbar, \quad i = n, T, \xi\ , \tag{7.6a}$$

$$\sigma_{n\xi} = -\frac{\theta}{\hbar}, \quad \sigma_{nT} = \frac{\beta_T}{\hbar}, \quad \sigma_{nn} = \frac{\beta_n}{\hbar}, \quad \sigma_{T\xi} = -\frac{\omega\theta}{c_V},$$

$$\sigma_{TT} = \frac{\omega\beta_T}{c_V}, \quad \sigma_{Tn} = \frac{\omega\beta_n}{c_V}.$$

The boundary conditions for Y_{ji} can be written in the form

$$\left.\frac{\partial Y_{j1}}{\partial z}\right|_{z=0} = 0\ , \quad Y_{j1}\Big|_{z=\infty} = 0\ . \tag{7.7}$$

It should be noted that, in general, the boundary condition at $z = 0$ on a surface with relief is of the form

$$\frac{\partial Y_{ji}}{\partial z}\Big|_{z=0} + \frac{\partial^2 Y_{j0}}{\partial z^2}\Big|_{z=0} U_{1z}(0) = 0 ,$$

where $U_{1z}(0)$ is the displacement of the surface points $z = 0$ along z, which describes the surface relief. However, under the conditions of interest to us the additional term is small (representing 10^{-1} - 10^{-4} of the first term, see Sec.6.7.6), so that we shall ignore it and we the boundary conditions specified by Eq. (7.7). The nature of the boundary conditions for $\mathbf{U}_1$ depends on the symmetry of the laser field and on the required type of solution (see Sec.6.7.3).

The system of equations (7.4) (where $\mathbf{U} = \mathbf{U}_1$, $Y_j = Y_{ji}$) and (7.6) can be solved analytically in two limiting cases: a) weak optical absorption (bulk case); b) strong optical absorption when γ_0^{-1} is less than the depths of penetration of the surface material excitations, n_1, T_1, ξ_1 into the medium (surface case). We shall consider case b because it is of greater practical interest. We can then write Eq. (7.6) in a simpler form

$$\frac{\partial Y_{j1}}{\partial t} + \Delta_{j0}(t)\, Y_{j1} = (\varepsilon_{j\xi}\, \xi_1 + \varepsilon_{jT}\, T_1 + \varepsilon_{jn}\, n_1)\Big|_{z=0} \exp(-\gamma_0 z) . \qquad (7.8)$$

Here, $\Delta_{n0} = - D\,\Delta + \tau_{n0}^{-1}$, $\tau_{n0}^{-1} = \tau^{-1} + 3\,\gamma_A\, n_0^2(t)\big|_{z=0}$, $\Delta_{T0} = -\chi\,\Delta$.

6.7.3. *General dispersion equation for electron-deformation-thermal instability; formation of one-dimensional surface grating.* We assume that as result of breakdown of the symmetry of the field or medium there is some preferred direction x on the $\{x, y\}$ surface. Then, the boundary conditions on the $z = 0$ surface for the vector $\mathbf{U}_1$ can be written in the form (we write $\mathbf{U}_1 \equiv \mathbf{U}$)

$$\frac{\partial U_x}{\partial z} + \frac{\partial U_z}{\partial x} = 0$$

$$\sum_{j=n,T} \frac{f_j Y_{j1}}{c_1^2} + \frac{\partial U_z}{\partial z} + \left(1 - 2\beta\right)\frac{\partial U_x}{\partial x} = 0, \tag{7.9}$$

where $\beta = c_t^2 / c_1^2$. We look for the simultaneous solution of the system of equations (7.4) - (7.6) assuming **E** (**r**) = const in Eq. (7.2).

We specify the strain ξ_1 and the diffusion field y_{j1} at $z = 0$:

$$\xi_1 = A\,(t)\exp\left[iqx + \int_0^t \lambda\,dt\right],$$

$$Y_{j1} = A_j\,(t)\exp\left[iqx + \int_0^t \lambda dt\right], \tag{7.10}$$

where $A\,(t) = A\,(q,\,t)$ is a function proportional to the amplitude of the initial strain and $A_j\,(t)$ is a slow function of time. This solution describes the surface field of the strain, carrier density, and temperature in the form of one-dimensional gratings (Fig. 6.14b). The solution of Eq. (7.8) subject to Eq. (7.10) can be found in the form

$$Y_{j1} = [B_j\,(t)\exp(-\gamma_0 z) + C_j\,(t)\exp(-\delta_j z)]\exp\left[iqx + \int_0^t \lambda d\right]. \tag{7.11}$$

Substituting Eq. (7.11) into Eq. (7.8), allowing for the boundary conditions of Eq.(7.7) and Eqs.(7.11), (7.10), we find that if $\gamma_0 \gg \delta_j$ (i.e., if $C_j \gg B_j$), then

$$Y_{j1} = A\,\varepsilon_{j\xi}\exp\left[iqx + \int_0^t \lambda\,dt - \delta_j\,z\right] \times \left[\chi_j\,\gamma_0\,\delta_j\left(1 - \sum_{j=n,T}\frac{\varepsilon_{jj}}{\chi_j\gamma_0\delta_j}\right)\right]^{-1} \tag{7.12}$$

where

$$\delta_j = \delta_j\,(t) = \left[q^2 + \frac{\lambda + \tau_{j0}^{-1}\,(t)}{\chi_j}\right]^{1/2}. \tag{7.13}$$

The functions $f(t) = A_j(t), \delta_j(t), \lambda(t)$ obey the condition for slow change with time $f^{-1}(t)\,\partial f(t)/\partial t \ll \lambda$, which reduces to the condition that the functions n_0, T_0, and ξ_o vary little on the scale of λ^{-1} (adiabatic approximation). We now solve Eqs. (7.4) and (7.9). We represent the vector **U** in the form $\mathbf{U} = \mathbf{U}_l + \mathbf{U}_t$, where

$$\operatorname{div} \mathbf{U}_t = 0, \ \operatorname{rot} \mathbf{U}_l = 0 \,. \tag{7.14}$$

Equation(7.4) then yields equations for the vectors $\mathbf{U}_t$ and $\mathbf{U}_l$:

$$\frac{\partial^2 \mathbf{U}_\alpha}{\partial t^2} = c_\alpha^2 \, \Delta\, \mathbf{U}_\alpha + \delta_{\alpha,l} \operatorname{grad} \sum_{j=n,T} f_j Y_{jl} \tag{7.15}$$

where $\alpha = l, t$; $\delta_{\alpha,l} = 1$, if $\alpha = l$; $\delta_{\alpha,l} = 0$, if $\alpha = t$; Y_{jl} is given by Eq. (7.12). The solution of Eq. (7.15) is assumed to be in the form

$$\mathbf{U}_\alpha = (\mathbf{B}_\alpha \exp(-\kappa_\alpha z) + \delta_{\alpha,l} \sum_{j=n,T} \mathbf{D}_j \exp(-\delta_j z) \times \exp\left[iqx + \int_0^t \lambda dt\right] . \tag{7.16}$$

Substituting this solution in the form of Eq. (7.16) into Eq. (7.15) and allowing for the condition (14), we express the components B_{lx} and B_{lz} in terms of some constant M, and the components B_{tx} and B_{tz} in terms of a constant N. Then, using the conditions for self-consistency of the solutions described by Eq. (7.16) and of the expression for $\xi_1|_{z=0}$ given by Eq. (7.10), we can express A in terms of M :

$$A = \frac{\lambda^2 M}{c_l^2 \Phi} ,$$

$$\Phi = 1 - \sum_{n,T} \frac{R_j c_l^2 (q^2 - \delta_j^2)}{\chi_j \gamma_0 \delta_j \, [\lambda^2 + c_l^2 (q^2 - \delta_j^2)]} \left[1 - \sum_{j=n,T} \frac{\varepsilon_{jj}}{\chi_j \gamma_0 \delta_j} \right]^{-1} ,$$

$$R_n = \frac{2\,\theta^2}{\varepsilon'\pi\,\hbar^2\rho\,c_l^2}\,\frac{\partial\varepsilon''}{\partial\omega}\,|\mathbf{E}|^2 ,$$

$$R_T = -\frac{2\,K\,\alpha\,\theta\,\omega}{\varepsilon'\pi\,\hbar\,\rho c_V c_l^2}\,\frac{\partial\varepsilon''}{\partial\omega}\,|\mathbf{E}|^2 . \tag{7.17}$$

The solution with the vector **U** can be expressed in terms of M and N, and is of the form

$$U_x = \left[\kappa_t N \exp(-\kappa_t z) + iqM \exp(-\kappa_l z) - \lambda^2 Miq \sum_{j=n,T} \frac{\Phi_j}{\delta_j} \exp(-\delta_j z)\right] \exp\left[iqx + \int_0^t \lambda\, dt\right] , \tag{7.18}$$

$$U_z = \left[iq\,N \exp(-\kappa_t z) - \kappa_t M \exp(-\kappa_l z) + \lambda^2 M \sum_{j=n,T} \Phi_j \exp(-\delta_j z)\right] \exp\left[iqx + \int_0^t \lambda\, dt\right],$$

Here,

$$\kappa_{l,t} = \left(q^2 + \frac{\lambda^2}{c_{l,t}^2}\right)^{1/2} , \tag{7.19}$$

$$\Phi_j = R_j \left\{ \chi_j\,\gamma_0 \left[\lambda^2 + c_l^2(q^2 - \delta_j^2) \right]\right\}^{-1}$$

$$\times \left[1 - \sum_{j=n,T} \left\{ \frac{\varepsilon_{jj}}{\chi_j \gamma_0 \delta_j} - \frac{R_j c_1^2 \, (q^2 - \delta_j^2)}{\chi_j \gamma_0 \delta_j [\lambda^2 + c_1^2 \, (q^2 - \delta_j^2)]} \right\} \right]^{-1} . \quad (7.20)$$

Substituting Eq. (7.18) into Eq. (7.9), we obtain two homogeneous equations for M and N :

$$N\,(\kappa_t^2 + q^2) + 2\, iqM \left(\kappa_l - \lambda^2 \sum_{j=n,T} \Phi_j \right) = 0 \; ,$$

$$-\,2\, i\, \kappa_t q\, N + (\kappa_t^2 + q^2)\, M \left(1 - \lambda^2 \sum_{j=n,T} \frac{\Phi_j}{\delta_j} \right) = 0 \; . \quad (7.21)$$

Equating the determinant of the system (7.21) to zero, we obtain a general dispersion equation for the EDTI (which is valid also in the case of radial-ring structures, see Sec. 6.7.5) :

$$4\, q^2\, \kappa_t\, \kappa_l - (\kappa_t^2 + q^2)^2 = \frac{\lambda^2}{c_1^2} \sum_{j=n,T} R_j$$

$$\times \frac{4\, q^2\, \kappa_t\, \delta_j - (\kappa_t^2 + q^2)^2}{\chi_j \gamma_0 \delta_j\, (\lambda^2/c_1^2 + q^2 - \delta_j^2) - \sum_{j=n,T} [\lambda^2/c_1^2\, \varepsilon_{jj} + (R_j + \varepsilon_{jj})\, (q^2 - \delta_j^2)]} \; .$$

$$(7.22)$$

The dispersion equation (7.22) describes three qualitatively different types of the EDTI. We shall consider each of them separately and limit our treatment of Eq. (7.22) either to an allowance for the EDI for $R_n / D\delta_n \gg R_T / \chi\delta_T$ or to the DTI in the opposite limit.

a) Dynamic instability: $\lambda = \lambda' + i\lambda''$,, $\lambda' > 0$,, $\lambda'' \gg \lambda'$.

The dispersion equation (7.22) has been derived ignoring the viscosity of the medium. If we allow for the viscosity in Eq. (7.22) and also in the corresponding expressions for Φ_j, $\kappa_{l,t}$, and R_j, we have to replace $c^2_{l,t}$ with $c^2_{l,t}$ $(1 + \eta_{l,t}\ \lambda\ /\ \rho c^2_{l,t})$, where $\eta_t = \eta$ and $\eta_l = (4\eta/3) + \xi$ (η and ξ are the first and second viscosities). Then, separating in Eq. (7.22) the real and imaginary parts subject to the conditions $\lambda' + \tau^{-1}_{j0} / \chi_j \ll \lambda''^2 / c^2_{t,l}\, q^2$, $2\,\lambda'\chi_j / c^2_t \ll 1$, R_j, $\varepsilon_{jj} \ll \chi_j\, q\, \gamma_0$, and $\lambda''^2\ \eta\ \chi_j\ /\ \rho\ c^4_t \ll 1$, we obtain the usual expression for the frequency of the surface acoustic waves: $\lambda'' = \sigma\ c_t\ q$; $0.87 < \sigma < 0.95$ (Ref. 41). Using the imaginary part of Eq. (7.22) and applying the conditions $\lambda''\ /\ 2\ D\ q^2 \ll 1$, we obtain the following expression for the growth rate of surface acoustic waves in the EDI case :

$$\lambda' = -\ \frac{q^2 \sigma^3 \eta}{2\ \rho} - \frac{4\ (1 - \sigma^2) c^2_l}{D^2 q\ \sigma^2 \gamma_0}\ R_n\ . \tag{7.23}$$

The expression for the growth rate of a surface acoustic wave in the DTI case found subject to the condition $\lambda''\ /\ 2\ \chi\ q^2 \gg 1$:

$$\lambda' = -\ \frac{\sigma\ q\ \eta}{2\ \rho} + \frac{2\ q\ (1 - \sigma^2)^{1/2}}{\gamma_0}\ R_T\ . \tag{7.24}$$

It clear from Eqs. (7.18), (7.23), and (7.24) that the generation of surface acoustic waves ($\lambda' > 0$) due to the EDI is possible only if $\partial\varepsilon'' / \partial\omega < 0$, whereas in the case of the DTI surface acoustic waves can be excited if $\theta\ \partial\varepsilon''\ /\ \partial\omega < 0$. The critical value of the pump radiation is obtained from Eqs. (7.23) and (7.24) subject to the condition $\lambda' = 0$.

Equations (7.23) and (7.24) were derived ignoring the term in the sum in the denominator of the right-hand side of Eq. (7.22) on the assumption that ε'', $R_j \ll \chi_j\ q\ \gamma_0$. It therefore follows that changes in the band gap due to heating (ε_{TT}) and breaking of covalent bonds (ε_{nn}) play no significant role in the process of excitation of surface acoustic waves [but they may be significant in the generation of static structures (see bellow)] .

b) Softening of acoustic frequencies: $\lambda' < 0$, $\lambda'' \to 0$.

Separating Eq. (7.22) into real and imaginary parts subject to the conditions λ', $\lambda'' \ll qc_{t,l}$, $\lambda' / c_{t,l}$, $\lambda'' / c_{t,l} \ll q$,

$[(\lambda' + \tau_{j0}^{-1}) \ / \ \chi_j]^{1/2}$, λ', $\lambda'' \ll q^2 \chi_j$, we find that the pump renormalized acoustic frequencies are described by the expression

$$\lambda''^2 = \Omega_q^2 \left(1 - \frac{2\beta}{1 - \beta\left(q + (q^2 + \tau_{j0}^{-1} / \chi_j)^{1/2}\right)}\right.$$

$$\left. \times \frac{q R_j}{(q^2 + \tau_{j0}^{-1} \ / \chi_j)^{1/2} - R_j - \varepsilon_{jj}}\right),$$

$$\Omega_q = 2 \ (1 - \beta)^{1/2} \ qc_t \ / \ (3 - 2\beta)^{1/2}. \qquad (7.25)$$

We can see from Eq. (7.25) and (7.17) that softening of the acoustic frequencies occurs in the EDI case if $\partial\varepsilon''/\partial\omega > 0$, whereas in the DTI case this occurs if $\theta \ \partial \ \varepsilon'' \ / \ \partial\omega < 0$. The critical intensity is found from the condition $\lambda'' = 0$. In fact, the acoustic frequencies do not go to zero but to $\lambda'' \sim \sigma q^2 \eta^2 / 2\rho \ll \Omega_q c$. Laser generation of periodic static surface structures

Substituting $\lambda'' = 0$ in Eq. (7.22) and using the notation $\lambda' \equiv \lambda$, we find the following equation subject to the conditions

$$\lambda^2 \ / \ c_{t,l}^2 < q^2, \ (\lambda' + \tau_{jo}^{-1}) \ / \ \chi_j$$

$$(q + \delta_j)\ (\delta_j - q_{R\varepsilon}) = \frac{2\beta}{1-\beta} \ q_R \ q \ , \qquad (7.26)$$

where $q_R = R_j \ / \ \chi_j \gamma_0$, $q_{R\varepsilon} = (R_j + \varepsilon_{jj}) \ / \ \chi_j \ \gamma_0$. The solution of Eq. (7.26) is

$$\lambda = \chi_j \left\{ \left[\frac{(q + q_{R\varepsilon})^2}{4} + qq_R \ \frac{2\beta}{1-\beta} \right]^{1/2} + \frac{q_{R\varepsilon} - q}{2} \right\}^2$$

$$- \chi_j q^2 - \tau_{j0}^{-1} \ . \qquad (7.27)$$

The growth rate reaches its maximum value λ_{max} at point q_{max}, where

$$q_{max} = q_{R\varepsilon}\,[(5a^2 + 6a + 1)^{1/2} - (2a + 1)], \qquad (7.28a)$$

$$a = \frac{2\,\beta}{1-\beta}\,\frac{q_R}{q_{R\varepsilon}},$$

$$\lambda_{max} = \chi_j\, q_{R\varepsilon}^2\,\frac{1}{2}\,[(5a + 1)(5a^2 + 6a + 1)^{1/2}$$

$$- 11a^2 - 8a + 1]\,. \qquad (7.28b)$$

The dependence of λ on q according to the Eq.(7.27) is plotted for the EDI case in Fig.6.14a and for the DTI case in Fig.6.14b.

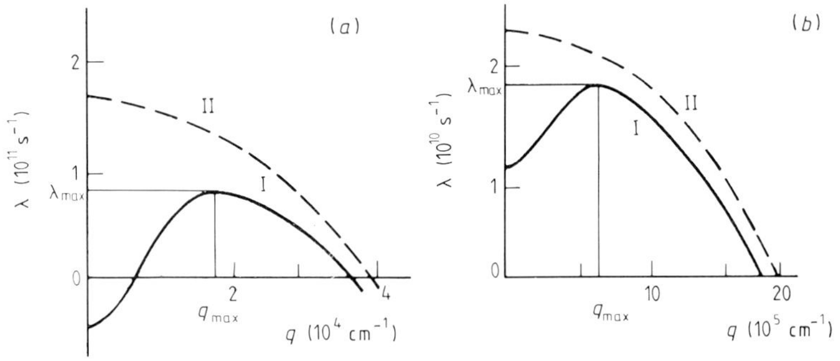

Fig.6.14. The dependence of the growth rates of the Fourier amplitudes of the static surface structures λ on the wave number of the Fourier harmonic q: a–EDTI, b-DTI. The curves I correspond to the condition $R_j \gg \varepsilon_{jj}$ and are calculated at the characteristic values of the semiconductor parameters from Sec.6.7.6.The curves II - the case $R_{jj} \ll \varepsilon_{jj}$(schematic).

If $R_j \gg \varepsilon_{jj}$ (deformation limit of the EDTI), the maximum of $\lambda\,(q)$ selects a dominant structure with $q = q_{max} \neq 0$, whereas for $R_j \ll \varepsilon_{jj}$ (diffusion limit of the EDTI) the maximum in the spectrum of $\lambda = \lambda\ (q)$ is reached at $q = 0$. However, even in the diffusion limit of the EDTI a dominant structure with $q \neq 0$ is selected. We can demonstrate this by considering first in greater detail the nature of the initial fluctuation in the EDTI case. At a moment $t = 0$ the boundary condition of Eq. (7.9) contains Y_n $(t = 0) = n_1\ (t = 0)$ and $Y_{T1}\ (t = 0) = T_1\ (t = 0)$, which are the Fourier amplitudes of the initial fluctuations of the population and temperature on the surface (when the simultaneous correlation function is $< Y_{j1}\ (t = 0)\ Y_{j1}\ (t = 0) >$ = const and is independent of q). Using Eq. (7.18) and of the relationship $N = -\ 2\ i\ q\ \kappa_1\ M\ /\ (\kappa_t^2 + q^2)$ that follows from Eq. (7.21), on the assumption that R_j, $\varepsilon_{jj} = 0$ at $t = 0$, and also taking from Eq. (7.27) the expression $\lambda = \lambda\ (t = 0) = -\ \chi_j\ q^2 - \tau_{j0}^{-1}$, we obtain from Eq. (7.9)

$$M = -f_j\ Y_{j1}\ (t = 0)\ /\ (1 - \beta)\ (\ \chi_j q^2 + \tau_{j0}^{-1})^2\ . \qquad (7.29)$$

Equations (7.18) and (7.29) together with the relationship between M and N, give the final solution for the vector **U** in the EDTI case when static structures are formed. Then, in the diffusion limit ($\varepsilon_{jj} \gg R_j$) the EDTI growth rate is given by the following expression which is deduced from Eq. (27):

$$\lambda = \chi_j\ (q_{R\varepsilon}^2 - q^2 - \tau_{jo}^{-1}\ /\ \chi_j)\ . \qquad (7.30)$$

Allowing for Eq. (7.29), we find that Eq. (7.17) yields an expression for the amplitude of the (strain) on the surface in Eq. (7.10) at time t :

$$A = A\ (q,t) = -\ \frac{\lambda^2 \quad f_j\ Y_{j1}(t = 0)}{(\chi_j q^2 + \tau_{j0}^{-1})\ \Phi\ c_1^2\ (1 - \beta)}\ ,$$

where λ is given by Eq. (7.30) [if $\varepsilon_{jj} \gg R_j$, the function Φ of the system (7.17) is independent of q] ; allowing for this question in Eq. (7.10), we find that the resultant strain on the surface at the end of a laser pulse $t = \tau_p$ is

$$\xi\ (x, \tau_p) = \sum_q A\ (q, \tau_p)\ \exp\left[\ iqx + \int_0^{\tau_p} \lambda\ dt\ \right]\ .$$

The strain correlation function is

$$\langle \xi^2 (x) \rangle = \sum_q \langle A^2 (q, \tau_p) \rangle \exp \left(2 \int_0^{\tau_p} \lambda \, dt\right)$$

$$\sim \int_0^\infty dq \, \langle A^2 (q, \tau_p) \rangle \exp (2 \lambda \tau_p) \sim \int_0^\infty \xi_q^2 dq \, ,$$

where the spectral strain function is

$$\xi_q^2 \sim [(q_{R\varepsilon}^2 - q^2 - \tau_{j0}^{-1} / \chi_j) / (q^2 + \tau_{j0}^{-1} / \chi_j)]^4 \exp (2 \lambda \tau_p) \, ,$$

and λ is given by Eq. (7.30). The value $q = q_{max}$ at which the maximum of the spectral function is attained at time $t = \tau_p$ determines, under linear conditions, the period of a grating in the case when $\varepsilon_{jj} \gg R_j$. For example, in the DTI case ($\tau_j^{-1} = 0$), we have

$$q_{max} = q_{R\varepsilon}, \quad \text{for } q_{R\varepsilon} > (4\tau_p \chi)^{-1/2} \, , \tag{7.31a}$$

$$q_{max} = (4\tau_p \chi)^{-1/2}, \quad \text{for } q_{R\varepsilon} < (4 \tau_p \chi)^{-1/2} \tag{7.31b}$$

If we calculate the critical intensity necessary for the EDTI (found from the condition $\lambda = 0$) by expanding Eq. (7.22) in terms of the parameters $\lambda / c_{t,l} q$, $(\lambda + \tau_{j0}^{-1}) / \chi_j q^2 \ll 1$. We then obtain a soft relaxation mode :

$$\lambda = - (2\chi_j q^2 + \tau_{j0}^{-1}) + \frac{2q}{\gamma_0} \left(\frac{R_j}{1 - \beta} + \varepsilon_{jj} \right) , \tag{7.32}$$

where R_j is given by Eq. (7.17) and ε_{jj} by Eq. (7.5).

In the case of the EDI we obtain from Eq. (7.32) a soft carrier-density mode

$$\lambda = -(2Dq^2 + \tau^{-1}) + \frac{4q\,|\mathbf{E}|^2}{\gamma_0 \pi \hbar^2 \varepsilon'} \frac{\partial \varepsilon''}{\partial \omega} \left[\left| \frac{\partial E_g}{\partial n} \right| + \frac{\theta^2}{(1-\beta)\rho c_l^2} \right]. \tag{7.33a}$$

We can therefore see that the EDI appears only when the conditions $\partial \varepsilon'' / \partial \omega > 0$ is satisfied. This condition is obeyed by various semiconductors in the range of frequencies of generally available lasers. In the case of DTI, we can deduce from Eq. (7.32) a soft temperature mode :

$$\lambda = -2\chi q^2 + \frac{4q\,|\mathbf{E}|^2\,\omega}{\gamma_0 \pi \hbar \varepsilon' c_V} \frac{\partial \varepsilon''}{\partial \omega} \left[\left| \frac{\partial E_g}{\partial T} \right| - \frac{\theta K \alpha}{(1-\beta) c_l^2} \right]. \tag{7.33b}$$

We can see that the DTI can be achieved for different sings of $\partial \varepsilon'' / \partial \omega$ depending on the sign of θ and on the relative magnitudes of the first and second terms in the brackets of Eq. (7.33). The condition $\lambda = 0$ in Eqs. (7.33a) and (7.33b) yields the critical intensity E_c^2 for the appearance of the ETI and DTI, respectively .

It therefore follows from this section that if $|\mathbf{E}|^2 > E_c^2$, that specific Fourier amplitudes of coupled fields of static strain [Eq. (7.18)], of the non-equilibrium carrier density, and the temperature [Eq. (7.12)] appear on the surface exponentially with time and the process is characterized by a growth rate λ_{max} [Eq. (7.28b)]. Consequently, in the linear EDTI case we can expect structures in the form of one-dimensional gratings with a period

$$d = 2\pi / q_{max}, \tag{7.34}$$

where in the case $R_j \gg \varepsilon_{jj}$ the value of q_{max} is given by Eq. (7.28a), while for $R_j \ll \varepsilon_{jj}$, it is given by Eq. (7.31) .

6.7.4. *Nonlinear electron-deformation-thermal instability and possibility of semiconductor-metal phase transition.* We consider the nonlinear case for the EDI stabilized by nonlinear Auger recombination. The functions of interest are described at $z = 0$ by $n_1(0) = n_q$ exp

(iqx), $\xi_1(0) = \xi_q \exp(iqx)$, and $T_1(0) = T_q \exp(iqx)$ where $E_g(0) = E_{g0} + E_{g1} \exp(iqx)$. Under steady-state condition ($\lambda = 0$) near the threshold ($|\mathbf{E}|^2 \geq E_c^2$) the substitution $\tau_n^{-1} = \tau^{-1} + \gamma_A |n_q|^2$ into the Eq.(7.32) gives

$$n_1(0) = |n_q| - \frac{1}{\gamma_A^{1/2}} \left[\frac{2q}{\gamma_o} \left(\frac{R_n}{1 - \beta} + \varepsilon_{nn} \right) - 2 D q^2 - \tau^{-1} \right]^{1/2} . \tag{7.35}$$

The threshold for the appearance of steady-state values of n_q corresponds to E_c^2 and it naturally coincides with the EDI threshold of Eq. (7.33a). Far from the threshold ($|\mathbf{E}|^2 \gg E_c^2$) we similarly obtain from the Eq. (7.27)

$$n_q = \frac{1}{\gamma_A^{1/2}} \left\{ D \left[\left[\left(\frac{(q + q_{R\varepsilon})^2}{4} + q\, q_R \frac{2\beta}{1 - \beta} \right)^{1/2} + \frac{q_{R\varepsilon} - q}{2} \right]^2 - D q^2 - \tau^{-1} \right\} . \tag{7.36}$$

Therefore, the dependence of n_q on q repeats the dependence $\lambda = \lambda(q)$. Equations (7.10) and (7.12) yield the amplitude of a steady-state deformation wave on the surface

$$\xi_q = n_q (D \gamma_0 \delta_n - \varepsilon_{nn}) / \varepsilon_{n\xi}, \tag{7.37}$$

where $\delta_n = [q^2 + (\tau^{-1} + \gamma_A n_q^2) / D]^{1/2}$, while the amplitude of a steady-state temperature wave is

$$T_q = \frac{\varepsilon_{T\xi} D \delta_n}{\varepsilon_{n\xi} \chi q} n_q .$$

Let us obtain from(7.37) more explicit formula for Fourier-harmonic amplitudes n_q and ξ_q at $q = q_m$ and $\lambda = \lambda_m$. Using values given in Sec.6.7.6 for the case of picosecond excitation ($I \sim 10^9$ W/ cm^2, $\tau_p \sim 10^{-11}$s)for the parameters entering the expressions for δ_n-(7.13) and n_q (7.36) we have $\delta_n \cong q_m$ and

$$n_{q_m} \cong \lambda_m^{1/2} \gamma_A^{1/2} \tag{7.37a}$$

Then using eq.(7.6a) we have $\varepsilon_{nn} >> D\,\gamma_0 \delta_n$ so that from (7.37) $\xi_{q_m} = n_{q_m} \varepsilon_{nn} / \varepsilon_{n\xi}$, whence

$$\xi_{q_m} \cong \frac{\lambda_m^{1/2} \beta_n}{\gamma_A^{1/2} \theta} . \tag{7.37b}$$

Here λ_m is given by (7.28b).

If the Auger recombination is unimportant (for example in the case on the time scale of 10^{-13}s ,i.e.during the femtosecond pulse duration (see Sec.6.7.7) then the EDI stabilization occurs due to the acousto-optic non-linearity. In this case the nonlinear regime of EDI in the bulk of the superconductor was considered in Ref. [32]. Here we outline the derivation of Ref. [32] and present the results. We represent the strain corresponding to Fourier harmonic with $q=q_m$ and $\lambda=\lambda_m$ in the form

$$\xi = \xi_{q_m} \exp(-iq_m x) + \text{c.c.}$$

Then we expand ε'' as in Eq.(7.5) but now we take into account the cubic term $\xi_{q_m}^3$, aside from the linear one with respect to ξ_{q_m}. Then from ((7.3b) we find the stationary solution for the harmonic amplitude

$$n = n_{q_m} \exp(-iq_m x) + \text{c.c.}$$

and from (7.4) we have the connection

$$n_{q_m} = - c_l^2 \rho \xi_{q_m} / \theta .$$

From above two connections we find the Landau type equation for the order parameter ξ_{q_m}

$$\xi_{q_m} \left(1 - \frac{R_n}{\Gamma_m}\right) = \frac{R_n}{6\Gamma_m} \frac{\theta^2}{\hbar^2} \frac{\partial^3 \varepsilon'' / \partial \omega^3}{\partial \varepsilon'' / \partial \omega} \xi^3_{q_m}$$

where $\Gamma_m = D q_m^2$ and the pump parameter $R_n \sim E^2$ is given in Ref. 32. From this equation we obtain

$$\xi_{q_m} = \frac{}{|\theta|} \left| \frac{6\, \partial \varepsilon'' / \partial \omega}{\partial^3 \varepsilon'' / \partial \omega^3} \right| \left(1 - \frac{\Gamma_m}{R_n} \right)^{1/2} .$$

At $\varepsilon'' = \text{const.} (\omega - E_g / \hbar)$ we find for order parameter

$$\xi_{q_m} = 5 (\hbar \omega - E_g)\left(1 - \frac{\Gamma_m}{R_n} \right)^{1/2} / |\theta| .$$

Thus upon high exceeding of the threshold ($R_n >> \Gamma_m$)the amplitude of the deformation wave is

$$\xi_{q_m} = 5 (\hbar \omega - E_g) / |\theta| . \tag{7.37c}$$

This formula is used for the evaluation of strain in EDI under femtosecond excitation in Sec.6.7.7.

We now substitute Eq. (7.37) and the value of T_q in Eq. (7.1). Then, a wave of the renormalized band gap at $z = 0$ is obtained using an expression for ε_{nn} given by Eq. (7.6a) on the assumption that $|\mathbf{E}|^2 > E_c^2$ is described by

$$E_g(0) = E_{g0} - E_{g1}(q) \cos qx ,$$

$$E_{q1}(q) = D\,\delta_n\, n_q \left[\varepsilon' \pi \hbar^2 \gamma_0 / 2 \frac{\partial \varepsilon''}{\partial \omega} |\mathbf{E}|^2 \right.$$

$$\left. + \frac{\hbar\omega}{\chi q c_V} \left| \frac{\partial E_g}{\partial T} \right| \right] . \qquad (7.38)$$

Note that the term β_n in Eq. (7.1) is balanced exactly by a part of the second term on the right-hand side of Eq. (7.37), i.e., that effective renormalization in Eq. (7.38) occurs only due to the deformation and temperature waves. When the modulation amplitude in Eq. (7.38) obeys $E_{g1} \geq E_{g0}$, we find that a grating appears on the surface, which represents alternation of the metallic and semiconductor phases. In the $R_j \gg \varepsilon_{jj}$ case the period of this grating is given by the value of q_{max} of Eq. (7.28). In the opposite limit of $R_j \ll \varepsilon_{jj}$ the maximum, n_q in Eq. (7.36) is attained at $q = 0$. Then, in Eq. (7.38) we can sum over the surface modes :

$$E_g(0) = E_{g0} - \sum_q E_{g1}(q) \cos(qx) \equiv E_{g0} - [E_{g1}(x)]_{eff}. \qquad (7.39)$$

It follows from Eq. (7.36) subject to the condition $\varepsilon_{jj} \gg R_j$ that $n_q \propto (q_{R\varepsilon}^2 - q^2)^{1/2}$ and with the aid of Eqs. (7.39) and (7.38), where $\gamma_A n_q^2 \gg q^2 D$ and the second term in parentheses is retained, that

$$[E_{g1}(x)]_{eff} = \int_0^{q_{R\varepsilon}} dq \cos(qx) (q_{R\varepsilon}^2 - q^2) \sim \sin(q_{R\varepsilon} x). \qquad (7.40)$$

Therefore, for $\varepsilon_{jj} \gg R_j$, a grating with a period $d = 2\pi / q_{R\varepsilon}$ forms in the nonlinear EDTI case.

In addition to the instability of the Fourier amplitudes with $q \neq 0$ in the EDTI, the amplitudes of the deformation (strain), carrier density, and temperature fields for $q = 0$ also increase exponentially in time, which gives rise to a spatially homogeneous reduction of E_g, i.e., it increases the optical absorption coefficient. This could account for the experimental observation of a strong reduction in the optical absorption length in semiconductors at a high rate of excitation with light characterized by a photon energy $\hbar\omega$ much higher than E_g (Ref.[37]).

6.7.5.*Formation of radial-ring structures.* Let us assume that the laser field has axial symmetry (relative to the z axis). We seek a simultaneous solution of the system of equations (7.4), (7.7), and (7.8) subject to the boundary condition (7.9) written down in cylindrical coordinates, ignoring at first the dependence of **E** on r in Eq. (7.2) (the effect of the influence of the dependence of the laser field on the coordinate r will be discussed later). The process of solution of the problem in cylindrical coordinates represents repetition of the procedure in Sec. 6.7.3, provided that in the expressions of Eq. (7.10) we make the substitution $\exp(iqx) \longrightarrow J_m(qr) \cos m\varphi$, where J_m is a Bessel function of the first kind of order m, and m is an integer. We then find that the components of the displacement vector ($\alpha = r, \varphi, z$) are given by

$$U_\alpha = \Big[a_\alpha N \exp(-\kappa_t z) + b_\alpha M \exp(-\kappa_l z)$$

$$+ M \sum_{j=n,T} c_{\alpha j} \Phi_j \exp(-\delta_j z)\Big] \Psi_\alpha \exp\left[\int_0^t \lambda\, dt\right], \qquad (7.41)$$

where

$$a_z = q, \quad a_\varphi = -a_r = \kappa_t, \quad b_z = \kappa_l, \quad b_\varphi = -b_r = q, \quad c_{zj} = -\lambda^2,$$

$$c_{rj} = -c_{\varphi j} = \lambda^2 q / \delta_j .$$

$$\Psi_z = J_m(qr) \cos m\varphi ,$$

$$\Psi_r = \frac{1}{2}[J_{m-1}(qr) - J_{m+1}(qr)] \cos m\varphi ,$$

$$\Psi_\varphi = \frac{1}{2}[J_{m+1}(qr) + J_{m-1}(qr)] \sin m\varphi .$$

The dispersion equation for the EDTI deduced from the boundary conditions for **U** and from Eq. (7.41) is exactly identical with Eq. (7.22). Therefore, the expressions for q_{max} and λ_{max} [Eqs. (7.28a) and (7.28b), which separate the dominant structure in the linear regime, are valid also in the case of the radial-ring structures under consideration.

It is clear from Eq. (7.22) that surface harmonics with any value of m are characterized by the same growth rate. This degeneracy in m is a consequence of neglect of the dependence of the intensity of the laser field $\mathbf{E}^2$ on r in Eq. (7.2) .

We can find the parameters m by solving the problem of the growth of the EDTI assuming a Gaussian distribution of the laser radiation intensity. We shall do this by considering an example of another possible class of surface structures in the form of radial rays.

Let us assume that the pump field has a Gaussian distribution:

$$|\mathbf{E}|^2 = E_0^2 \exp(-r^2 / r_0^2) . \tag{7.42}$$

Then, Eq. (7.8) becomes

$$\frac{\partial Y_{j1}}{\partial t} + \Delta_j Y_{j1} = (\varepsilon_{j\xi} \xi_1 + \varepsilon_{jn} n_1 + \varepsilon_{jT} T_1)|_{z=0}$$

$$\times \exp(-\gamma_0 z - r^2 / r_0^2) , \tag{7.43}$$

where $\varepsilon_{j\xi}$, ε_{jn}, and ε_{jT} given by expressions in Eq. (7.6a) provided we replace $|E|^2$ with E_0^2. The boundary conditions for Y_{j1} are still the same. We specify ξ_1 and Y_{j1} at $z = 0$ in the form

$$\xi_1 = A \left(\frac{r}{r_0}\right)^m \exp\left(-\frac{r^2}{r_0^2}\right) \cos(m\varphi) \exp\left[\int_0^t \lambda \, dt\right] , \tag{7.44}$$

$$Y_{j1} = A_j(t) \left(\frac{r}{r_0}\right)^m \exp\left(-\frac{r^2}{r_0^2}\right) \cos(m\varphi) \exp\left[\int_0^t \lambda dt\right] . \tag{7.45}$$

Here, as in Eqs. (7.9) and (7.10), we are using A for the initial amplitudes and A_j are the slow functions of time, where m is an integer; we shall assume that $m \gg 1$. Then a maximum of the function $(r / r_0)^m$ exp $(-r^2 / r_0^2)$ lies at a point $r_{max} = r_0 (m / 2)^{1/2} \gg r_0$. In the range $r < r_0$ we can assume that $(r / r_0)^m \exp(-r^2 / r_0^2) \cong (r / r_0)^m$ and Eq. (7.43) can be written as follows :

$$\frac{\partial Y_{j1}}{\partial t} + \Delta_j Y_{j1} = (\varepsilon_{j\xi} A + \varepsilon_{jn} A_n + \varepsilon_{jT} A_T) \left[\frac{r}{r_0}\right]^m \cos m\varphi$$

$$\times \exp\left[-\frac{r^2}{r_0^2} - \gamma_0 z + \int_0^t \lambda\, dt\right] . \tag{7.46}$$

The solution of equations (7.46) subject to the conditions $r / r_0 \ll m$, and $\gamma_0 \gg m^{1/2} r_0^{-1}$ can be obtained in the form

$$Y_{j1} = \frac{\varepsilon_{j\xi} A \exp\left[-\tilde{\delta}_j z - r^2 / r_0^2 + \int_0^t \lambda\, dt\right] (r / r_0)^m \cos m\varphi}{\chi_j \gamma_0 \tilde{\delta}_j \left[1 - \sum_{j=n,T} \varepsilon_{jj} / \chi_j \gamma_0 \tilde{\delta}_{jj}\right]} , \tag{7.47}$$

where $\tilde{\delta}_j^2$ is given by Eq. (7.13) when q^2 is replaced with $\tilde{q}^2$:

$$\tilde{q}^2 = 4 (m + 1) / r_0^2 . \tag{7.48}$$

Using the Eqs. (7.4) and the boundary conditions for the vector **U** in cylindrical coordinates and following the solution procedure similar to that described in Sec.6.7. 3, and also allowing for the solutions Y_{j1} described by Eq. (7.47) and the conditions subject to which the solution of Eq. (7.46) is obtained, we find that the components of the displacement vector are described by

$$U_i = \frac{r^{m-1}}{r_0^m} \exp\left[-\frac{r^2}{r_0^2} + \int_0^t \lambda\, dt\right] \Psi_i$$

$$\times \left[a_i M \exp(-\kappa_l z) + b_i N \exp(-\kappa_t z)\right.$$

$$\left. + M \sum_{j=n,T} c_{ij} \Phi_j \exp(-\delta_j z)\right] , \tag{7.49}$$

where

$i = z, \varphi, r$; $\Psi_r = m \cos m\varphi$, $\Psi_\varphi = -m \sin m\varphi$, $\Psi_z = r \cos m\varphi$; $a_r = a_\varphi = 1$, $a_z = -\kappa_l$; $b_r = b_\varphi = -\kappa_t$, $b_z = q$; $c_{rj} = c_{\varphi j} = -\lambda^2 / \delta_i$,

$c_{zj} = \lambda^2.$

Here, $\kappa_{1,t}$ and Φ_j are given by Eq. (7.19) and (7.20), where q^2 is replaced with $\tilde{q}^2$ given by Eq. (48). The dispersion equation, obtained from (7.9) and(7.49) coincides with the dispersion equation (7.22) if q^2 is replaced with $\tilde{q}^2$, i.e., it specify λ (m). If the expression for q_{max} [Eq. (7.28a)] and λ_{max} [Eq (7.28b)] are modified by replacing q^2 with $\tilde{q}^2$, they give the value of m_{max} at which λ reaches its maximum :

$$m_{max} = (q_{R\varepsilon}^2 r_0^2 / 4) [5a^2 + 6a + 1)^{1/2} - (2a + 1)^{1/2}]; \quad (7.50)$$

the value of m_{max} given by Eq. (50) determines the number of rays in the dominant structure.

6.7.6. *Comparison with experiment and numerical estimates* .

As already mentioned, the symmetry of the structures formed as a result of the EDTI should be governed by the symmetry of the laser beam for an isotropic surface or by the symmetry of the crystal surface. We now obtain numerical estimates of the parameters and compare the theoretical predictions with experimental results. For typical values $c_V = 2 \cdot 10^7$ erg/cm^3 K , $K \cong 10^{12}$ erg/cm^3, $\alpha \cong 2 \times 10^{-5}$ K^{-1}, $\partial\varepsilon'' / \partial\omega \cong 10^{-15}$ s^{-1}, $\theta \cong 10^{-11}$ erg, $\rho \cong 5$ g cm^{-3}, $c_1 = 5 \times 10^5$ cm/s, $\beta = 0.4$, $\omega \simeq 4 \times 10^{15}$ s^{-1}, $|\partial E_g / \partial T| \cong 4 \times 10^{-4}$ eV/K, $|\partial E_g / \partial n| = 10^{-32}$ erg cm^3, $\chi = 0.1$ cm^2/ s, $\gamma_0 = 10^5$ cm^{-1}, $\gamma_A = 4 \times 10^{-31}$ cm^6/s, $\hbar\omega \cong 1$ eV, $D \cong 10^2$ cm^2/s we find from Eq. (7.17) that $R_T \sim R_n \sim \varepsilon_{nn} \sim \varepsilon_{TT} \sim 10^4$ - 10^5 $|\mathbf{E}|^2$ esu. We find from Eq. (7.33b) that the critical intensity for the appearance of the DTI is $I_c \cong 2 \times 10^6$ W / cm^2 when $q \cong 10^4$ cm^{-1}. The critical intensity for the EDI is $(D / \chi)\, \hbar\omega\, |\partial E_g/\partial T| / (c_V\, |\partial E_g / \partial n|) \cong 10$ times larger. It follows from Eq. (7.28) that in the EDI case we have $q_{max} \sim q_{R\varepsilon} \sim 10^{-2} |\mathbf{E}|^2$ esu $\cong 10^4$ - 10^5 cm^{-1} for $I = 6 \times 10^8$ W / cm^2, i.e., the period d of the structures is of the order of 1-5 μm. Equation (7.28b) yields $\lambda_{max} \sim D\, q_{max} \sim 10^{10}$ - 10^{11} s^{-1} for the growth rate. The amplitude of the renormalization E_g for the above values of the parameters is $E_{gl} \cong 1$ eV, i.e., the EDTI can rise to an insulator - metal phase transition. This conclusion and estimates of the structure periods as well as of their growth times are in agreement with the experimental results reported in Ref. 35, where laser pulses of $\tau_p = 5 \times 10^{-11}$ s duration were used. It was found that gratings of alternate metallic and semiconducting phases or structures of the "star" type formed on an isotropic surface of VO_2, depending on the symmetry of the laser field It follows from Eq. (7.50) that for $r_0 \sim 5 \times 10^{-3}$cm (Ref. 35), the number of rays in the "star" structure can be, $m = 50$, which agrees with the experimental results; moreover, we have $\lambda_{max} \tau_p \sim 5 > 1$.

These estimates allow us to justify the boundary conditions given in Eq. (7.7). If we use Eq. (7.17), we find from Eq.(7.12) an expression for $(\partial Y_{j1} / \partial z)|_{z=0}$ and from Eq. (7.18) an expression for $U_z(0)$. We shall assume that in the case of picosecond laser pulses we have $T_0(z,t) \cong T(0, t) \exp(-\gamma_0 z)$ and $n_0(z, t) = n_0(0, t) \exp(-\gamma_0 z)$, so that in the DTI case we obtain the following estimate

$$\frac{\partial^2 T_0}{\partial z^2} U_z \Big|_{z=0} \Big/ \frac{\partial T_1}{\partial z} \Big|_{z=0} \sim (\gamma_0^2 \chi \lambda^{-1}) [K \alpha T_0(0)] / \rho c_1^2 \sim 10^{-4} ,$$

whereas in the EDI case, the corresponding estimate is

$$\frac{\partial^2 n_0}{\partial z^2} U_z \Big|_{z=0} \Big/ \left(\frac{\partial n_1}{\partial z}\right)_{z=0} \sim (\gamma_0^2 D \lambda^{-1}) [\theta\, n_0(0)] / \rho c_1^2 \sim 10^{-1}$$

for $n_0(0) \sim 10^{21}\text{cm}^{-3}$, which confirms the validity of the boundary conditions given by Eq. (7.7) for the case in question.

In the experiments , described in Ref.[42], (111) surface of crystalline silicon was irradiated with laser pulses of wavelength $\lambda = 0.53$ µm, creating one-dimensional gratings of the surface relief in which the lines were parallel to one of the [110] axes. The dependence of the grating orientation on the crystallographic axes indicated that a deformation mechanism was responsible for the formation of these structures. The structure period was independent of λ , angle of incidence, and the polarization of the incident radiation, but was governed by the pulse duration. It was found that $d_1 \cong 3 \times 10^{-5}$ cm for $\tau_{p1} = 10^{-11}$s, whereas $d_2 \cong 6 \times 10^{-6}$ cm for $\tau_{p2} = 5 \times 10^{-12}$ s, i.e., the relationship $(d_1 / d_2) \cong (\tau_{p1} / \tau_{p2})^{1/2}$ was obeyed. In the case of Si one have $\varepsilon_{nn} / R_n \sim 10$, so that the period of the EDTI structures should be given by Eq. (7.31b), i.e.$d \sim \tau_p^{1/2}$, which is in agreement with the experimental result of [42].

We note, that besides the periodic surface structures formed due to the EDTI development qualitatively different surface structures are also formed under pulsed laser irradiation of the semiconductors.These interference structures of modulation of the of surface relief with the periods $d \propto \lambda_L$,where λ_L is the excitation light wavelength, may appear because of the diffraction of the incident wave on the surface relief. We can describe them by including, on the right-hand side of Eq. (7.4), interference sources of the EE_1 type, where $E_1 \sim U_{1z} |_{z=0}$ is the amplitude of the diffracted wave. These structures appeared in the solid phase due to an interference instability of the sublimation or due to generation of coupled diffracted waves and surface acoustic waves (see the reviews given in Ref.[43] and the book [44]). The growth rates of the interference instability as a function of q have very narrow maxima (see Ref.[43,44]), the positions of which are normally different from the positions of maxima of the EDTI increments, so that we can regard these instabilities as independent in different ranges of q. Therefore, we ignored the contribution of interference sources in Eq. (7.4) when considering the EDTI. The excellent reviews of the experimental and theoretical investigations of the interferential surface instabilities are given in [45, 46].

In contrast to the interference instability case, the characteristic scale of a structure formed because of the EDTI is not governed directly by the wavelength λ_L and its geometry is not related to the polarization of thc incidcnt radiation, so that these two types of structures are easily distinguished experimentally.

6.7.7. *Superfast phase transition into centrosymmetric crystalline state in GaAs under femtosecond laser excitation.* It has been a great current interest in the past few years to the phase transitions on the surface of solids under pulsed femtosecond laser excitation [61-66], when the pulse duration is comparable or less then the characteristic times of electron-phonon relaxation in solids. One indication of ultrafast phase transition on the Si surface under 100-fs pulsed laser excitation was revealed in Ref. [64] with the use of time-resolved linear and second-harmonic reflection.

In recent experiments [65] the temporal behaviour of linear and nonlinear optical reflection from the GaAs surface under 100-fs pulsed laser excitation was observed , which gave an evidence of the ultrafast phase transition to the new structural phase with properties different both from initial crystal and melted material. Transition occurs on time scale of 100 fs which is approximately one order of magnitude shorter than the time required for the electron relaxation to the bottom of the conduction band, the latter time being discussed in numerous experimental and theoretical studies [62,64,67,68]. Therefore, such kind of phase transition can not be considered as conventional melting following energy transfer from excited electrons to the lattice, and principally different mechanisms should be analysed.

In this Section we present the phenomenological model [69] describing in self-consistent manner the ultrafast dynamics of linear and nonlinear optical reflection from GaAs surface observed in [65]. It infers initial ultrafast transition to the nonequilibrium crystalline semiconducting phase with centrosymmetric lattice structure, which exists only during first 1 ps after excitation and is followed by melting of subsurface layer due to energy transfer to the lattice and its heating up to the normal melting point.

It was found in Ref.[69] that structural phase transition is possible under high electron plasma density and laser-induced stress in the subsurface layer achievable during strong femtosecond laser excitation due to the EDTI development. Analysis of lattice dynamics under laser excitation based on the equation for the amplitude of the zone-edge TA-phonon mode, reveals the time required for phase transition to be consistent with the value observed experimentally.

In so doing laser irradiation is assumed to induce **coherent** shift of the atoms from their initial positions to the new quasiequilibrium positions in centrosymmetric lattice, rather than the increase of amplitude of random atomic vibrations corresponding to the increase of lattice temperature.

The experiments [65] utilized high structural sensitivity of the Second Harmonic Generation (SHG) in reflection from the surface to obtain the dynamics of phase transition in GaAs subsurface layer under 100-fs laser excitation were performed. GaAs crystal is non- centrosymmetric (symmetry class $\bar{4}3m$) and hence exhibits strong second-order optical nonlinearity χ^2 in the electric dipole approximation. Therefore, laser-induced phase transition to any phase with centrosymmetric structure (crystalline or disordered or melted) is expected to cause an abrupt decrease of the SH intensity. In centrosymmetric phase the second-order susceptibility is nonzero only in quadrupole approximation thus yielding a value of susceptibility of several order of magnitude less than that in dipole approximation.

Note that symmetry relations governing the nonlinear susceptibility, relay to the long-range order with the characteristic length comparable to laser wavelength. In contrast, linear reflection in cubic crystals is determined by the band structure and therefore, relays to the short range order within of few coordination spheres.

Pump-and-probe experiments [65] revealed the complicated dynamics of linear and nonlinear reflection from GaAs at laser fluence three times of the melting threshold. The linear reflections reveals more gradual rise with characteristic time 1 ps up to the value characteristic of molten (metallic) material.

In contrast, SH reflection demonstrates an abrupt decrease with characteristic time of 100 fs followed by additional fast changes in the first 1 ps and eventual onset of low constant value.

The SH dynamics implies the structural phase transition to the centrosymmetric state being still semiconducting during first picosecond after excitation as it is follows from linear reflection behavior. Therefore, the problem which should be resolved are following: i) what's the nature of the state of material existing in the first picosecond after excitation and ii) how can the phase transition to this state occur on the time scale of 100 fs? Obviously, the second problem can not be resolved on the way of assuming lattice heating through spontaneous emission of phonons by photoexcited electrons, because the latter process takes at least several hundreds femtosecond.

There are principally two possibility regarding the structure of the short-living state of GaAs in the first picosecond. Since the SH is sensitive to the long-range order, the fast drop of the reflected SH might be due to loss of crystalline order over the distance of at least laser wavelength while the short-range order on the scale of two fist coordination spheres remains approximately unchanged and thus, the material being still semiconducting. This situation meets the major conditions of the "cold melting" hypothesis, which implies lattice disordering [64].

Here we consider the principally different possibility. We will show that the observed drop of SH intensity can be attributed to the transition to crystallinc ccntrosymmctric state which exists only during first picosecond after excitation. It will be shown that the time necessary for such kind of transition is not limited by characteristic time of spontaneous electron-phonon relaxation, and it can be of the order of 100 fs.

It is known, that in equilibrium state covalent (Si,Ge) and ion-covalent (GaAs, ZnO, InP, InSb) semiconductors undergo phase transitions to new phases (characterized by β-tin, NaCl or orthorhombic structure) when the applied pressure exceeds certain critical values [70]. In GaAs the transition into orthorhombic semiconductor phase occurs at the lowest value of reduction of the relative crystal cell volume dV/V =-0.17 (in Si the transition to β-tin structure occurs at dV/V =-0.32).

In highly excited semiconductors phase transition can occur at lower

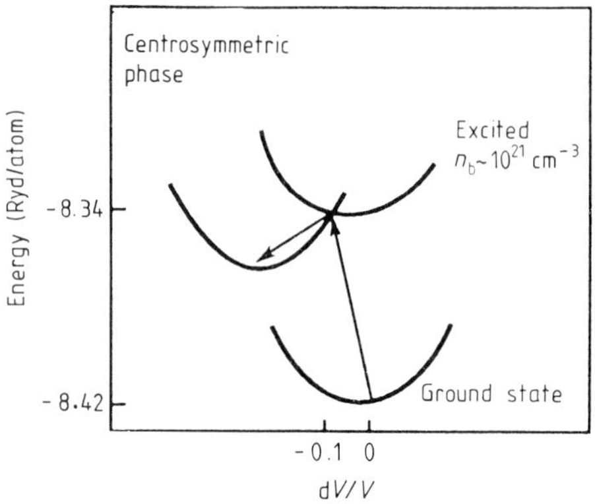

Fig.6.15. Energy diagram of GaAs. The arrows denote ultrafast transition to a new centrosymmetric phase, achieved via an excited state of GaAs.

pressures, and the nature of the final phase may change. Thus, it was shown theoretically in Ref.[71] that at the electron–hole plasma concentration of $n = 10^{22}$ cm^{-3} and electronic temperature $T_e = 10^4$K GaAs undergoes transition into a metallic centrosymmetric phase possessing the symmetry similar to that of NaCl at $dV/V = -0.1$ (see Fig.6.15).

It will be shown below that GaAs can undergo ultrafast phase transition to a new phase with the lattice symmetry different from the initial one under excitation by a strong femtosecond laser pulse realized in experiment of Ref.[65]. Using phenomenological description of the laser induced phase transition, the dependence of the semiconductor fundamental band gap E_g on the modulus of the relative change of crystalline lattice $|dV/V| = \xi$, and the electron-hole plasma concentration n can be written as (compare Eq.7.1)

$$E_g = E_{g0} + \theta\xi - \beta_n n , \qquad (7.51)$$

where E_{g0} is the equilibrium fundamental band gap, $\beta_n > 0$ and θ is the interband deformation potential. Note, that in (7.51) we have neglected the band gap renormalization due to laser heating, which is insignificant on the time scale of $t = 10^{-13}$ s under consideration.

At $E_g = E_{g1}$ the crystal undergoes phase transition to a new semiconductor phase, and at $E_g = 0$ - to a metal-like phase. Thus, the lines $E_g = E_{g1}$ and $E_g = 0$ divide the (n, ξ)- plane into three regions: the initial semiconductor-like phase (S_0), the new laser induced semiconductor-like phase (S_1) and solid metal-like phase (S_M) (Fig.6.16). In particular, it might happen that E_{g1}=0 and thus only semiconductor-metal phase transition takes place.

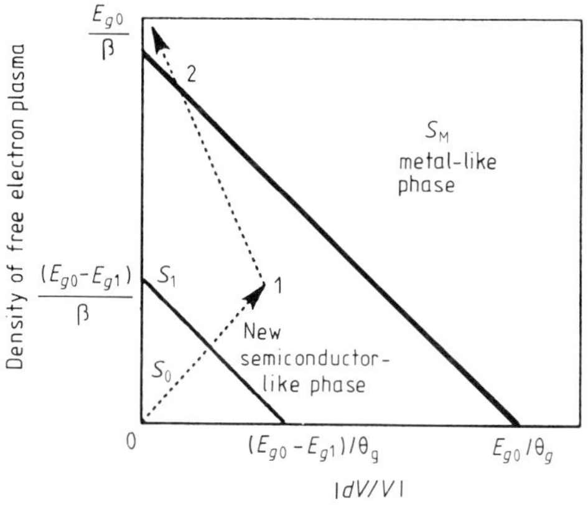

Fig.6.16. Schematic phase diagram of GaAs. The dotted line shows ultrafast transition from the initial state (0)to the centrosymmetric crystalline state (1) and the subsequent slower onset of the melt(2).

The initial equilibrium state corresponds to zero point of this diagram. The excitation pulse creates plasma with the density of about n =10^{22} cm^{-3} and lattice deformation of the order of ξ =0.1 (see below). The critical values n_c and ξ_c correspond to intersection of the initial semiconducting phase energy curve with that of the new semiconductor (E_{g1} $\neq$ 0), or metal (E_{g1} = 0) phase at the energy diagram of Fig.6.15. It is at this point that the transition to new symmetry takes place. This phase transition occurs as a displacive type II phase transition due to softening of the optical or acoustic phonon mode. It was shown in Ref.[72], for example, that under excitation of a dense (n_c > 10^{21} cm^{-3}) plasma the TA-acoustic mode at the edge of the Brillouin zone becomes unstable first. We assume that the amplitude of this unstable phonon mode obeys the following equation

$$\partial^2 U / \partial t^2 + \omega_0^2 (1 - \theta_0 \xi - \beta_{n1} n) U = - b |U|^2 U \quad (7.52)$$

and plays the role of the order parameter of this laser-induced displacive phase transition, where b is the phonon anharmonism constant,

$$\beta_{n1} = \beta_n (E_{g0} - E_{g1})^{-1}, \quad \theta_1 = \theta (E_{g0} - E_{g1})^{-1}.$$

From (7.52) we have, that the stationary state (new nonequilibrium phase) with the displacement

$$|U| = (\omega_0^2 / b)^{1/2} (\theta_1 \xi + \beta_{n1} n - 1) \quad (7.53)$$

is established in the time interval of

$$t = \omega_0^{-1} (\theta_1 \xi + \beta_{n1} n - 1)^{-1/2}. \quad (7.54)$$

Let us make the numerical estimates. The number density of plasma generated on the surface of the semiconductor by the end of the pulse (without taking into account diffusion and Auger-recombination: both are unimportant on a time scale $t = 10^{-13}$ s) is $n = \gamma(1-R)W/\hbar\omega w$, where the absorption coefficient of exciting radiation γ a=4 10^4 cm^{-1}, reflection coefficient $R = 0.3$, pulse energy density $W = 0.2$ J/cm^2, $\omega = 3 \times 10^{15}$ s^{-1}. With these parameters we have n = 2×10^{22} cm^{-3}.

Let us estimate now deformation of the subsurface layer.The light pressure yields the compression with ξx= $W / K\tau_p c \cong 10^{-3}$, $K = 5 \times 10^{11}$ erg/cm^3 is the bulk elasticity modulus of GaAs. The excitation of electron- hole plasma with the concentration $n = 10^{22}$ cm^{-3} leads to the crystal dilation $\xi = \sigma n = 2 \times 10^{-2}$, where σs= 2×10^{-24} cm^3 is the relative increase of the crystal volume accompanying generation of a single electron-hole pair.

The theory of electron-deformational instability (EDI) in the case of interband transitions in semiconductors , developed above, predicts ultrafast (t=10^{-13}s) formation of the surface grating of the plasma concentration n and coupled surface strain grating ξ with the period d, given by formula (7.34), of order of 10^{-5}cm ($W = 0.2$ J/cm^2)and amplitude ξ_{q_m}, given by Eq.(7.37c)

$$\xi = \xi_{q_m} \cos 2\pi x / d ,$$

$$\xi_{q_m} \cong (\hbar \omega - E_g) / |\theta| ,$$

where x is the coordinate along one of the crystallographic axes of the surface layer.

For($\hbar\omega - E_g$) = 1 eV, $|\theta|$q=10 eV the amplitude of the strain grating is ξx_{q_m} = 0.1. It follows from the EDI-theory that this grating penetrates from the surface into the bulk at the distance of the order of d = 10^{-5}cm, which is equal approximately to the SH absorption length . After the end of the pulse the grating disappears with a characteristic time t = d^2/D = 10^{-13}s, where D = 10^3cm^2/s is the diffusion coefficient of the nonequilibrium carriers under strong excitation.

Thus, it is seen that the femtosecond laser pulses, used in experiments can create dense plasma (n = 10^{22}cm^{-3}) in highly compressed regions (ξx= 0.1) of the subsurface layer . These values of n and ξ meet the condition of the laser- induced structural phase transition, discussed above. It follows from (7.54) that this phase transition is completed with a characteristic time $t < \omega_0^{-1} = \omega_{TA}^{-1} = 10^{-13}$ s^{-1}, where ω_T- ω_A is the transverse acoustic phonon frequency at the edge of the Brillouin zone .

Thus, we see that both the threshold condition of the phase transition and the requirement of its being sufficiently fast might be satisfied in experiments [65]. The above analysis leads to the conclusion that during the action of powerful femtosecond laser pulse the grating of alternating different phases S_o, S_1(or S_0, S_M) is formed with the period d = 10^{-5}cm, which disappears when the pulse is over. If we assume, that the symmetry of S_1 (S_M) phases is such, that the value of $\chi^2$of these phases is drastically reduced , then a strong decrease of the SHG intensity on a time scale of 10^{-13}s should be observed during the pulse, replaced by the rise of SHG immediately after the end of the pulse. If, for example, one takes $\chi^2_{S_1(S_M)}$ = 0 , then the SHG intensity drops by 75%, which indeed was observed in the experiment of [65].

Thus, the concept of ultrafast structural phase transition of GaAs into a new centrosymmetric phase can explain the ultrafast decrease of the SHG intensity observed experimentally in [65.

We note that recent results of [66, 73 are in agreement with the present model. It was found in [66], that the characteristic time of fast laser-induced phase transition in GaAs is strongly decreased with laser fluence increase. Indeed the expression (7.54) predicts the decrease of the characteristic transition time to new crystalline phase with increasing the level of excitation(i.e. with increasing of n and ξ).

6.8. Conclusions

The consideration, carried out in Chapters 5 and 6 , gives the review of the models of a number of novel light-induced phase transitions, which can be observed in the typical systems of quantum electronics - the media of two or three - level atoms, interacting through the electromagnetic, electrostatic or lattice displacement fields. It is possible that some of these effects, as it was seen from the comparison with experimental results, have been already observed.

To clear out the principal questions we restricted ourselves in the main with the consideration of the simplest (but quite realistic) systems with discrete levels in the bulk of the material. The single exception is Sec. 6.7, where the particular type of the light-induced phase transition on the surface of the two-band semiconductor was considered. In some sense it is the generalization of the two-level model of Sec. 6.2, where phase transition with the appearance of static bulk lattice displacement was considered.

Some other effect, considered for the case of discrete levels, can be easily generalized to the case of semiconductors, for example, the effect of lattice bistability (Sec.6.5). In this case the initial detuning $\Delta = \omega_{ab} - \omega > 0$, i.e. the energy of the light quantum must be smaller then the forbidden gap width E_g. In this connection we note, that the strong optical nonlinearities and bistability are observed under the action of light wave with the frequency, lying just bellow the interband absorption edge of semiconductor [45].

As it has been already noted in Sec.5.1, all of the light-induced phase transitions in the systems with discrete levels, considered in Chaps.5 and 6, are resonant.The temporal coherence of the exciting radiation is the necessary condition of their observation and its frequency must lie near resonance transition frequency.

Recently the two broad classes of nonresonant light-induced phase transition have been investigated. The exciting light frequency can be varied in a wide region without changing of the main features of observed phenomena. Such situation is realized under the action of the powerful light wave on the strongly absorbing surface of metals and semiconductors (in the latter case the energy of quantum of exciting wave must exceed the forbidden gap width).

To the first class belong the interferential instabilities(II) of the surface relief occurring due to the interference of the excited surface electromagnetic waves and exciting spatially coherent laser wave (see [46-49] and also [43,44]). The physical mechanism of the II consists in the following [48,49]. The impinging laser wave is diffracted on the initial fluctuations of the surface relief. As a result two diffracted surface wave are generated - the Stokes and the anti-Stokes ones. Due to the surface electromagnetic resonance , the diffracted surface waves with the wave vectors, the modulae of which are approximately equal to the modulus of the exciting wave, have the biggest amplitudes and are selected due to the resultant instability. The interference of these resonance diffracted waves with the exciting laser wave creates in the subsurface layer the spatially periodic (along the surface) optical field. Due to the optical absorption this leads to the appearance of the spatially periodic temperature field, giving rise to the appearance of spatially periodic forces of different nature, enhancing the initial modulations of the surface relief and leading to the II development. Due to the II the transient or permanent periodic structures of the surface relief are formed on the surfaces of the laser irradiated semiconductor or metal surfaces. The characteristic features of these surface gratings are the rigid linear dependence of their period on the excitation wavelength and the dependence of their orientations on the orientation of vector of polarization. The theory of the nonequilibrium light-induced phase transitions, developed in [46-49] explains the basic experimental results, obtained in studies of the periodic surface structures formation.

To the second class belong the laser-induced diffusional-deformational instabilities (DDI) on the surfaces of solids, leading to the nonequilibrium phase transitions and formation of the periodic surface structures [31-33,50-52].The coherence of the exciting radiation(either temporal or spatial) is unimportant in the case of DDI. These instabilities take place in the systems, in which some diffusional variable(the concentration of the nonequilibrium carriers or the temperature in semiconductors, concentrations of point or extended defects in solids)is modulated by the medium strain. On the other hand, the appearance of the spatially nonuniform field of the concentration of diffusional variable deforms the elastic continuum. As a result the positive feedback gives rise to the DDI , in which the certain Fourier-amplitudes of the coupled diffusion variable and strain fields grow exponentially in time. In contradistinction to the case of II-instability, the period of DD-structures(the periodic piles up of surface defect concentration and of surface coupled strain field) is not proportional to the wavelength of exciting wave, but is determined, as it was shown in Sec.6.7, by the interplay of the distribution of the pump energy over coupled diffusion-deformational modes and the rates of modes decay The prototype of the DDI - the laser- induced electron-deformation-thermal instability in the interband electronic transitions on the surface of semiconductors was considered in details in this book (Sec.6.7). In this example we have seen, that the coherence of laser radiation in fact is unimportant for DDI and the corresponding nonequilibrium phase transition is nonresonant-the frequency of the exciting light wave ω is only bound from bellow by the condition $h\,\omega >$

E_g, where E_g is the forbidden gap width. (The comprehensive review of recent experimental and theoretical studies of the DDI is given in [51].

The material, exposed in Chaps.5 and 6, is based on the original results and, of course doesn't cover the enormous field of light- induced phase transitions. We shall only mention some references, in which the studies were made of light-induced critical phenomena akin to that considered in this book. To them belong, for example, the light induced orientational phase transition in liquid crystals [53], the resonatorless bistability due to the local field [54], metal- dielectric [55] and metal-superconductor [56] phase transitions under the action of light. A number of the questions of the theory of equilibrium phase transitions in atomic systems with the electromagnetic interaction are considered in [57,58].The SR effect as a consequence of the dissipative instability was considered in [59]. Noise-induced (including light noise induced) phase transitions are studied in the book [60]).

The light-induced phase transitions and instabilities in the bulk and on the surface of solids is now very hot field of theoretical and experimental research. We hope, that contribution of this book will serve the purpose of further developing of investigations in this field.

APPENDICES

1. Classification of Dicke states

The Schrödinger equation for the two-level atom is

$$i\hbar \frac{da_s}{dt} = \sum_{s=1}^{2} H_{ss'}(t)\, a_{s'} , \qquad \text{(A.1)}$$

where $H_{ss'}$ is the reduced Hamiltonian of the atom interacting with the external (c-number) electromagnetic field.

In the Schrödinger representation the solution of the Eq. (A.1) can be expressed in terms of the unitary evolution operator $U(t)$

$$a_s(t) = \sum_{s'} U_{ss'}(t) a_{s'}(0) ,$$

where U(t) satisfies the initial condition $U_{ss'}(0) = \delta_{ss'}$ and obeys the equation

$$i\hbar \frac{dU_{ss'}}{dt} = \sum_{s''} H_{ss''}\, U_{s''s'} . \qquad \text{(A.2)}$$

The operator $U(t)$ is an unitary operator, i.e. for any vectors $a_s(t) = \sum_{s'} U_{ss'}(t)\; a_{s'}(0)$ and $b_s(t) = \sum_{s'} U_{ss'}(t)\; b_{s'}(0)$ the equation $\langle b|a\rangle = \sum_s b_s^*(t)\; a_s(t) = \sum_s b_s^*(0)\; a_s(0)$ holds. The later follows immediately from the equation $\frac{d}{dt}\langle b|a\rangle = 0$, which results from the Eq. (A.2) if we take into account the Hermitian property of the Hamiltonian $H_{ss'} = H^*_{s's}$.

Thus,the evolution of the two-level atom state is described by the unitary matrix $U_{ss'}$. A complete set of such unitary two-dimensional matrices forms the group $U(2)$.

In fact, we can always restrict ourselves by considering the unimodal matrices $U_{ss'}$ only. Really, in the quantum mechanics the state of a system is given by the vector, which is determined to the accuracy of the factor $e^{i\phi}$, where ϕ is a real phase. Therefore we can always multiply $U_{ss'}$ by $e^{i\phi(t)}$, where $\phi(t)$ is an arbitrary real function of time. In this case the matrix determinant multiplies by $e^{2i\phi(t)}$. From the condition of the matrix unitarity written in the form $\sum_{s''} U_{ss''} U^*_{s's''} = \delta_{ss'}$ or in the matrix form $U\,U^+ = 1$, it follows that $\det U \det U^* = |\det U|^2 = 1$, i.e. $\det U = e^{i\psi(t)}$. Assuming $\phi(t) = -\psi(t)/2$, we get $\det \tilde{U} = 1$, where $\tilde{U} = e^{i\phi(t)} U$. It is seen that the introduction of the new evolution operator $\tilde{U}(t)$ corresponds to the incorporating of the new Hamiltonian $\tilde{H} = H - h\, d\phi/dt$. Really, we get

$$i\hbar \frac{d\tilde{U}}{dt} = i\hbar\left[e^{i\phi} \frac{dU}{dt} + i \frac{d\phi}{dt} e^{i\phi} U \right] = H\tilde{U} - \hbar \frac{d\phi}{dt} \tilde{U} = \tilde{H}\,\tilde{U}\,.$$

It is always allowable to add the reduced Hamiltonian by an operator proportional to the unity operator. In Heisenberg representation the multiplication of U by factor $e^{i\phi}$ does not change the operator $A_H = U^{-1} A\, U$, and, hence, its mean value.

Thus, it can be assumed that $U(t)$ is a unitary matrix, the determinant of which is equal to unity, i.e. U belongs to the group of the unimodal unitary two-dimensional matrices $SU(2)$.

As $H(0) = i\hbar \dfrac{U(\Delta t) - U(0)}{\Delta t}$, then $U(\Delta t) \quad 1 - \dfrac{i}{\hbar} H(0)\, \Delta t$, and it follows, according to condition $\det U(\Delta t) = 1$, that $\mathrm{Sp}\{H\} = \sum_s H_{ss} = 0.$ Thus we can assume that the Hamiltonian H has the zero trace.

Let us define the general form of the two-dimensional unitary matrix, the determinant of which is equal to 1. Let $U = \begin{pmatrix} a & b \\ c & d \end{pmatrix}$. From the condition of unitarity

$$U\,U^+ = \begin{pmatrix} a & b \\ c & d \end{pmatrix} \begin{pmatrix} a^* & c^* \\ b^* & d^* \end{pmatrix} = \begin{pmatrix} 1 & 0 \\ 0 & 1 \end{pmatrix}$$

it follows, that

$$aa^* + bb^* = 1\,, \quad a^*c + b^*d = 0\,,$$

i.e. $c = -b^*d/a$.

In conjunction with the condition $\det U = ad - bc = 1$, it gives $(aa^* + bb^*)d/a = 1$, i.e. $d = a^*$, $c = -b^*$.

Finally, the matrix U becomes

$$U = \begin{pmatrix} a & b \\ -b^* & a^* \end{pmatrix} , \quad |a|^2 + |b|^2 = 1 .$$

There is a close relation between the matrices forming the $SU(2)$ group, and the three dimensional real orthogonal matrices, forming the orthogonal $SO(3)$ group, which is isomorfic to the three-dimensional rotation group [1].

The matrices of the orthogonal group (the rotation group) are completely determined by the matrices A_i $(i = 1,2,3)$ of the infinitesimal rotations about the axes x_i . If

$$T_3(\alpha) = \begin{pmatrix} \cos\alpha & \sin\alpha & 0 \\ -\sin\alpha & \cos\alpha & 0 \\ 0 & 0 & 1 \end{pmatrix}$$

is a matrix of rotation through angle α about axis x_3 , then

$$A_3 = \frac{\partial T_3(\alpha)}{\partial \alpha}\bigg|_{\alpha=0} = \begin{pmatrix} 0 & 1 & 0 \\ -1 & 0 & 0 \\ 0 & 0 & 0 \end{pmatrix} ,$$

and vice versa

$$T_3(\alpha) = \lim_{n \to \infty} \left(1 + \frac{\alpha}{n} A_3\right)^n = \exp(\alpha A_3) .$$

In the same manner, we can determine the matrices A_1 and A_2 , which are the matrices of the infinitesimal rotations about axes x_1 and x_2 . The three matrices A_i obey the commutation rules

$$[A_i, A_j] = -\varepsilon_{ijk} A_k ,$$

where ε_{ijk} is a completely antisymmetric tensor, the components of which equal $+1$ if the indexes ijk are the even rearrangement of the numbers 1,2,3 and equal -1 , if this rearrangement is odd.

Any rotation can be represented in the form

$$T = \exp(\alpha_1 A_1 + \alpha_2 A_2 + \alpha_3 A_3) , \qquad (A.4)$$

and α_i can be accepted as the parameters of the rotation group.

However, the Euler angles α , β , γ are often used as these parameters.

The correspondence between the two-dimensional unitary unimodal matrices and the orthogonal ones can be established by the following way [2]. Let us consider the known Pauli matrices

$$\sigma_1 = \begin{pmatrix} 0 & 1 \\ 1 & 0 \end{pmatrix}, \quad \sigma_2 = \begin{pmatrix} 0 & -i \\ i & 0 \end{pmatrix}, \quad \sigma_3 = \begin{pmatrix} 1 & 0 \\ 0 & -1 \end{pmatrix}.$$

Any two-dimensional Hermitian matrix with zero trace can be expressed in terms of these matrices with the real expansion coefficients:

$$H = x_1\sigma_1 + x_2\sigma_2 + x_3\sigma_3 = (\mathbf{rS}) = \begin{pmatrix} x_3 & x_1 - ix_2 \\ x_1 + ix_2 & -x_3 \end{pmatrix}.$$

If we transform the matrix H with the help of the unitary matrix U , the determinant of which is equal to 1 , than we get the matrix $H' = U\,H\,U^+$ with zero trace, because

$$\mathrm{Sp}\{H'\} = \mathrm{Sp}\{UHU^+\} = \mathrm{Sp}\{HUU^+\} = \mathrm{Sp}\{H\} = 0 .$$

Therefore, the matrix H' can be also written in the form of the linear combination of the Pauli matrices:

$$H' = \begin{pmatrix} a & b \\ -b^* & a^* \end{pmatrix} \begin{pmatrix} x_3 & x_1 - ix_2 \\ x_1 + ix_2 & -x_3 \end{pmatrix} \begin{pmatrix} a^* & -b \\ b^* & a \end{pmatrix} = \begin{pmatrix} x'_3 & x'_1 - ix'_2 \\ x'_1 + ix'_2 & -x'_3 \end{pmatrix}.$$

This gives the linear relationships between x'_i and x_i , and transformation matrix becomes orthogonal automatically. The complex numbers a and b will be the parameters of the orthogonal transformation. They obey the condition $|a|^2 + |b|^2 = 1$ and are called by the Cayley-Klein parameters of the orthogonal transformation. Thus, some orthogonal matrix T_U is put into correspondence to matrix U .

This correspondence obeys the rule that the product of the matrices U_2U_1 corresponds to the product of the appropriate rotations $T_{U_1}T_{U_2}$. This follows from the following transformations

$$U_2U_1(\mathbf{rS})(U_2U_1)^+ = U_2U_1(\mathbf{rS})U_1^+U_2^+ = U_2(T_{U_1}\mathbf{r}\ \mathbf{S})U_2^+$$

$$= (T_{U_2}T_{U_1}\mathbf{r}\ \mathbf{S}) = (T_{U_2U_1}\mathbf{r}\ \mathbf{S}) .$$

We can easily check that the matrices

$$T_1(\alpha_1) = e^{\alpha_1 A_1} , \quad T_2(\alpha_2) = e^{\alpha_2 A_2} , \quad T_3(\alpha_3) = e^{\alpha_3 A_3} ,$$

which are the matrices of the rotations through angles α_1 , α_2 , α_3 about the axes x_1 , x_2 , x_3 , correspond to the matrices

$$U_1(\alpha_1) = \begin{bmatrix} \cos(\alpha_1/2) & \mathrm{i}\,\sin(\alpha_1/2) \\ \mathrm{i}\,\sin(\alpha_1/2) & \cos(\alpha_1/2) \end{bmatrix} ,$$

$$U_2(\alpha_2) = \begin{bmatrix} \cos(\alpha_2/2) & \sin(\alpha_2/2) \\ -\sin(\alpha_2/2) & \cos(\alpha_2/2) \end{bmatrix} , \tag{A.5}$$

$$U_3(\alpha_3) = \begin{bmatrix} \exp(\mathrm{i}\alpha_3/2) & 0 \\ 0 & \exp(-\mathrm{i}\alpha_3/2) \end{bmatrix} .$$

Then, the matrices,

$$\frac{\partial U_k(\alpha_k)}{\partial \alpha_k}\bigg|_{\alpha_k=0} = \frac{\mathrm{i}}{2}\sigma_k ,$$

correspond to the matrices A_i of the infinitesimal rotations, and vice versa $U_k(\alpha_k) = \exp(\mathrm{i}\alpha_k\sigma_k/2)$.

Hence, the matrices $\mathrm{i}\sigma_k/2$ must obey the same commutation rules as the matrices A_k . We can directly check this fact using the explicit form of the Pauli matrices. So we get that the matrices $J_i = \sigma_i/2$ obey the usual commutation rules for the angular momentum operators:

$$[J_j, J_k] = \mathrm{i}\varepsilon_{jkl} J_l . \tag{A.6}$$

The rotation, determined by the Eulerian angles α , β , γ , can be represented in the following form

$$T(\alpha,\beta,\gamma) = T_3(\gamma)\; T_2(\beta)\; T_3(\alpha) .$$

This rotation is described by the matrix

$$U(\alpha,\beta,\gamma) = U_3(\gamma)\; U_2(\beta)\; U_3(\alpha) .$$

In the case of the two-level atom, the three-dimensional Euclidean space, in which the rotations T are considered, has not any physical sense. In contrast to this, for the spin systems this space is a real physical space, and the rotations correspond to the real spatial rotations of the spins or can be considered as the transformations of the coordinate axes. The matrices U, corresponding to these rotations, determine the transformation rules for the spin wave functions in rotations or transformations of the system of axes. These matrices form the spinor representation of the rotation group or the irreducible representation $j = 1/2$. In this case, the operators J_i are the momentum operators and they have the eigenvalues $\pm 1/2$, which correspond to the spin 1/2. In this representation, the operator $\mathbf{J}^2 = J_1^2 + J_2^2 + J_3^2$ is proportional to the unity two-dimensional matrix:

$$\mathbf{J}^2 = \frac{3}{4}\begin{bmatrix} 1 & 0 \\ 0 & 1 \end{bmatrix},$$

and has the eigenvalues $(1/2 + 1)/2 = 3/4$.

As a result of the formal analogy between the two-level atom and the spin system, it is convenient to describe the two-level system with the help of the spin operators σ_i or $J_i = \sigma_i/2$, which, as we have noticed, obey the momentum operator commutation rules. But we should not forget that the vectors $\begin{pmatrix} 1 \\ 0 \end{pmatrix}$ and $\begin{pmatrix} 0 \\ 1 \end{pmatrix}$, which are the eigenvectors of the operator J_3 and correspond to the eigenvalues $\pm 1/2$, are not to be the vectors of the states with the momentum projections $\pm 1/2$. In the case of the two-level atom they are the vectors of the upper and lower states, for which the quantum numbers of the projection momentum may not exist at all.

The states of the two-level system can be enumerated by the eigenvalues of the operator J_3 and designated by $|+1/2\rangle$ and $|-1/2\rangle$ or in more compact form $|+\rangle$ and $|-\rangle$ for the upper and lower states respectively.

For the purposes of the future analysis it will be useful to get the wave function of the state, which can be produced under the action of operator $U(\vartheta,\phi)$ on the lower state of a system. This operator corresponds to the rotation $T(\vartheta,\phi)$ through angle ϑ about the axis lying in the plane (x_1, x_2) and forming an angle ϕ with the axis x_2. The rotation $T(\vartheta,\phi)$ can be written in the form

$$T(\vartheta,\phi) = T(\phi,\vartheta,-\phi) = T_3(-\phi)\, T_2(\vartheta)\, T_3(\phi) .$$

Using the Eqs. (A.5) and (A.7) we get

$$U(\vartheta,\phi) = U(\phi,\vartheta,-\phi) = U_3(-\phi)\, U_2(\vartheta)\, U_3(\phi)$$

$$= \begin{bmatrix} \cos(\vartheta/2) & e^{-i\phi}\sin(\vartheta/2) \\ -e^{i\phi}\sin(\vartheta/2) & \cos(\vartheta/2) \end{bmatrix} . \qquad \text{(A.8)}$$

Hence, $<1/2|\, U(\vartheta,\phi)\,|-1/2> = e^{-i\phi}\sin(\vartheta/2)$ and $<-1/2|\, U(\vartheta,\phi)\,|-1/2> = \cos(\vartheta/2)$.

An arbitrary vector of the two-level system state can be written in the form

$$|\psi> = \begin{bmatrix} e^{-i\phi}\sin(\vartheta/2) \\ \cos(\vartheta/2) \end{bmatrix} ,$$

if we can neglect the phase factor. Therefore, an arbitrary state can be considered as the state $|\vartheta,\phi> = U(\vartheta,\phi)\,|-1/2>$, which is described by the angles ϑ and ϕ and can be produced from the lower state $|-1/2>$ with the help of the unitary operator corresponding to the rotation $T(\vartheta,\phi)$.

Let us now discuss the states of a collection of the two-level atoms. We shall start with the state (1.2.3), defined in the Chapter 1. For a given n these states form the basis of the representation of the rearrangements group $S(N)$. This representation is not irreducible [2]. It contains the irreducible representations (one by one) corresponding to the Joung schemes [1] with the two rows, where the length of the first one equals to n' , $n' + 1$,..., N , and n' is the greatest among the numbers n and $N - n$.

As a result of the application of the Joung symmetrizators to the functions of the type (1.2.3), it is produced the wave function antysymmetric in respect to $Q = N - n'$, $N - n' -1$,..., 1 pairs of variables. The dimensionality of the representation given by the Joung scheme with the length of the first row n' is

$$C_N^{n'} - C_N^{n'+1} ,$$

so the complete number of the functions is

$$(C_N^{n'} - C_N^{n'+1}) + (C_N^{n'+1} - C_N^{n'+2}) + ... + (C_N^{N-1} - C_N^{N}) + C_N^{N} = C_N^{n'} = C_N^{n} ,$$

as we discussed in Chapter 1.

Thus, the set of the states of the type (1.2.3) is replaced by their linear combinations, which form the irreducible representation of the rearrangement group $S(N)$. The application of this new set of the state vectors is oftenly useful, because any completely symmetric operator acts in the subspace of these vectors. So, the collective operators $R_i = \sum_l \sigma_i^{(l)}$, where the operators $\sigma_i^{(l)}$ acts as the Pauli matrix σ_i in the space of the lth atom states, transform the vectors belonging to some irreducible representation of the group $S(N)$ into the vectors belonging to the same representation and even to the same row of the representation of the rearrangement group.

The operator R_3 multiplies the vector (1.2.3) by $n - (N - n)$ and corresponds to the population difference of the upper and lower levels. At the same time, the operators $R^{\pm} = (R_1 \pm iR_2)/2$ convert the vectors (1.2.3) to the linear combinations of the vectors of the same type states, but with n greater by unity (operator R^+) and smaller by unity (operator R^-). If we will consider the action of these operators on the functions belonging to the rows of the irreducible representation of the group $S(N)$, then we get the functions belonging to the same rows of the same irreducible representation but with n increased or decreased by unity. The operators R_i , $R^{\pm}$, as any other completely symmetric operators, do not change the rearrangement symmetry of the wave functions.

For the wave function of N variables, which take the two values only, there is a remarkable property, that the rearrangement symmetry of the functions defines simultaneously the transformation rule of these functions under unitary transformations:

$$U = \prod_{l=1}^{N} U^{(l)} , \tag{A.9}$$

which are accomplished by the identical two-dimensional matrices $U^{(l)}$, the determinant of which is equal to unity, and every of which acts in the space of the states of the individual atom [2].

The matrices of the infinitesimal rotations under transformations of the (A.9) type are expressed in terms of the operators R_i:

$$\left.\frac{\partial U_i}{\partial \alpha_i}\right|_{\alpha_i=0} = \frac{i}{2} R_i \quad (i = 1,2,3) \ . \tag{A.10}$$

The connection between the rearrangement symmetry of functions and their transformation rules in the unitary transformations of the (A.9) type can be easily understood, if we take into account that the functions, belonging to the irreducible representation of the rearrangement group with the Joung scheme with the length of the shortest row Q (see Fig. A.1), are antisymmetric in respect to the Q pairs of indexes and symmetric in respect to the other $(N - 2Q)$ indexes. In unitary transformations, U, they are transformed as the spinors of the N-th order, which were traced over the Q pairs of indexes and are symmetric in respect to the rest indexes. They are transformed [1] in accordance with the $j = N/2 - Q$ representation of the $SU(2)$ group.

1	2	...	$N-Q-1$	$N-Q$
$N-Q+1$	...	N		

Fig.A.1. The Joung scheme for the functions which are transformed in accordance with the $j = N/2 - Q$ representation of the SU(2) group.

Thus, the length of the shortest row on the Joung scheme (Q in Fig. A.1) defines the $SU(2)$ group representation which prescribes the transformation rule for the symmetrized wave functions.

In fact, in the process of symmetrization of the functions with the binary arguments we obtain the irreducible representations of the direct product of the $S(N)$ and $SU(2)$ groups, i.e. the product of the irreducible representations of these groups. The quantum number j is integer or semi-integer in depend on the parity of N.

Thus, instead of the 2^N states of the (1.2.3) type we can introduce the complete set of the Dicke states $|jm\alpha\rangle$ for a collection of the two-level atoms. Here, the quantum numbers m and j enumerate the eigenvectors of the commuting operators J_3 and $\mathbf{J}^2 = J_1^2 + J_2^2 + J_3^2$, where the collective operators $J_i = R_i/2 = \sum_l \sigma_i^{(l)}/2$ obey the commutation rules for the momentum operators:

$$[J_j,J_k] = i\varepsilon_{jkl}J_l \; . \tag{A.11}$$

These commutation rules follow directly from the fact that the operators J_i, according to the Eq. (A.10), correspond to the infinitesimal rotations:

$$\frac{\partial U_k}{\partial \alpha_k}\bigg|_{\alpha_k=0} = iJ_k \; .$$

The validity of the commutation rules (A.11) can be checked directly if we use the definition of the operators $J_i = \sum_l \frac{\sigma_i^{(l)}}{2}$ and the commutation rules of Pauli matrices σ_i .

Thus

$$\mathbf{J}^2|jm\alpha\rangle = j(j+1)|jm\alpha\rangle \; , \; J_3|jm\alpha\rangle = m|jm\alpha\rangle \; , \; -j \le m \le j \; , \tag{A.12}$$

where j and m are simultaneously the integer or semi-integer numbers in depend on the parity of the number of atoms N. The maximal value of j is equal to $j_{\max} = N/2$. The operators J_i, and $J^{\pm} = J_1 \pm iJ_2$ will be widely used in this book along with the operators $R_i = 2J_i$ and $R^{\pm} = J^{\pm}$. The concurrent use of the operators J_i and R_i is explained by the fact, that the first ones obey the commutation rules (A.11) and have the well known matrix elements, while the operators R_i have the integer eigenvalues and the operator R_3 is the population difference operator.

The quantum number α enumerates the vectors relating to the different rows of the $S(N)$ group representation. As we have mentioned, this specific representation of the rearrangement group is determined by the quantum number j . Namely, the length of the shortest row of the Joung scheme, $Q = N/2 - j$, determines the representation. The first row of the Joung scheme has the length $N - Q = N/2 + j$. For a given j the quantum number α can accept n_j different values, where

$$n_j = C_N^{N/2+j} - C_N^{N/2+j+1} .$$

The complete number of vectors is

$$\sum_j (2j + 1)n_j = 2^N ,$$

as it should be.

It should be emphasized again that the quantum number j determines the $SU(2)$ group representation, according which the vectors $|jm\alpha>$ are transformed in unitary transformation (A.9), but it does not correspond to the real momentum of the system while it determines unambiguously the rearrangement symmetry of the state. Particularly, the case $j = j_{max} = N/2$ corresponds to the Joung scheme with the one row (the length of the second row is $Q = 0$), and, hence the completely symmetric states, associated with this case, transform in accordance with the unitary representation of the rearrangement group.

The operators $R^{\pm} = J^{\pm} = J_1 \pm iJ_2$ have the following non-zero matrix elements [1] for the states $|jm\alpha>$:

$$<jm\alpha|J^{+}|j(m-1)\alpha> = <j(m-1)\alpha|J^{-}|jm\alpha> = \sqrt{(j + m)(j - m + 1)} .\text{(A.13)}$$

Because the operators $J^{\pm} = R^{\pm}$ are the completely symmetric operators and commute with any rearrangement operator, they do not change the quantum number α_{+}. The quantum number j is not also changed, because the operators $J^{\pm}$ commute with $\mathbf{J}^2$. The same is true for the operator J_3 or $R_3 = 2J_3$.

In conclusion, notice that for a collection of the N three-level atoms the classification of the states is determined by the groups $SU(3)$ and $S(N)$, i.e. by the unitary three-by-three matrices and the rearrangement group. There is a close connection between the representations of these groups, and besides we need not consider for the $S(N)$ group the Joung schemes containing more than three rows.

2. Coherent atomic states

The set of the $|N/2,m>$ Dicke states forms a basis in the space of the symmetrized states of the $\sum P_N|\psi>$ type, where $|\psi>$ is an arbitrary vector of a state for a collection of the N atoms, and P_N is the rearrangement operators. Among these states the first place is occupied by the following state

$$|\psi\rangle = \prod_l |\psi\rangle^{(l)} , \quad l = 1,\dots,N ,$$

when all of the states $|\psi\rangle^{(l)}$ are identical.

As we have mentioned before, an arbitrary state of the isolated atom can be defined by the two parameters ϑ and ϕ and can be written in the form

$$|\vartheta,\phi\rangle^{(l)} = U^{(l)}(\vartheta,\phi)|-1/2\rangle^{(l)} . \tag{A.14}$$

Hence, for the case of N atoms the symmetric state defined by the two parameters ϑ and ϕ is

$$|\vartheta,\phi\rangle = \prod_l |\vartheta,\phi\rangle^{(l)} = U(\vartheta,\phi)|N/2,-N/2\rangle , \tag{A.15}$$

where

$$U(\vartheta,\phi) = \prod_l U^{(l)}(\vartheta,\phi) = U(\phi,\vartheta,-\phi) = U_3(-\phi)U_2(\vartheta)U_3(\phi) .$$

This state is called by a coherent atomic state. As a result of a completeness of the Dicke states in the space of the symmetrized states, the coherent atomic states can be expressed in terms of the symmetric Dicke states $|N/2,m\rangle$:

$$|\vartheta,\phi\rangle = \sum_m |N/2,m\rangle\langle N/2,m|\vartheta,\phi\rangle = \sum_m |N/2,m\rangle \sqrt{\frac{N!}{(N/2+m)!(N/2-m)!}}$$

$$\times \prod_l \langle 1/2|U^{(l)}(\vartheta,\phi|-1/2\rangle \prod_{l'} \langle -1/2|U^{(l')}(\vartheta,\phi)|-1/2\rangle , \tag{A.16}$$

$$1 \le l \le N/2 + m , \quad N/2 + m + 1 \le l' \le N .$$

In deriving (A.16) we have used the representation (1.2.5) for the vector $|N/2,m\rangle$. If the Eq. (A.8) is applied to the matrix elements $U^{(l)}(\vartheta,\phi)$, we get

$$|\vartheta,\phi\rangle = \sum_m \sqrt{\frac{N!}{(N/2+m)!(N/2-m)!}} \sin^{N/2+m}(\vartheta/2)\cos^{N/2-m}(\vartheta/2)$$

$$\times \exp[-i(N/2+m)\phi]\,|N/2,m\rangle\ . \qquad \text{(A.17)}$$

We can generalize the Eq. (A.17) if we introduce the coherent atomic states for the case $j \quad N/2$, when there is no the complete rearrangement symmetry [3]:

$$|\vartheta,\phi,j,\alpha\rangle = U(\vartheta,\phi)\,|j,-j,\alpha\rangle\ . \qquad \text{(A.18)}$$

This states have no such simple physical interpretation as the states (A.17). Nevertheless they can be effectively used to descibe the evolution of a system which starts from the state of the partial excitation $|\,|m|\ m\ \alpha\rangle$. The Eq. (A.17) is generalized in this case by the following implicit way [3]:

$$|\vartheta,\phi,j,\alpha\rangle=\sum_m \sqrt{C_{2j}^{j+m}}\ \sin^{j+m}(\vartheta/2)\cos^{j-m}(\vartheta/2)\exp[-i(j+m)\phi]\,|jm\alpha\rangle. \qquad \text{(A.19)}$$

The coherent atomic states are analogous to the Glauber coherent states for the harmonic oscillator or for the mode of the electromagnetic field. This analogy and its nature are discussed in detail by Arecchi et al [3]. As in the case of the Glauber states, the uncertainty relation for J_i takes the form of an equality in the case of the coherent atomic states. A set of these states possesses the property of the completeness, which is determined for the case of the vectors with the definite j and α by the following equation

$$\frac{2j+1}{4\pi}\int \sin\vartheta\, d\vartheta\, d\phi\ |\vartheta,\phi\rangle\langle\vartheta,\phi| = \sum_m |jm\rangle\langle jm| = 1\ . \qquad \text{(A.20)}$$

Certainly, this set of vectors depending on the two continuous parameters ϑ and ϕ are essentially overcompleted because the vectors $|\vartheta,\phi\rangle$ are the vectors in the $(2j+1)$ - dimensional space and, hence, the number of the linear independent vectors is finite and equal to $2j+1$. Overcompleteness is essentially higher here than in the case of the Glauber states, where the set of the vectors depending on the two parameters substitutes by the enumerate sequence of the vectors of the states with the given occupation numbers.

As in the case of the Glauber states, we can expand any vector with the given j and α into the coherent atomic states series:

$$|\psi\rangle = \frac{2j + 1}{4\pi} \int |\vartheta,\phi\rangle\langle\vartheta,\phi|\psi\rangle \sin\vartheta \, d\vartheta \, d\phi \,, \qquad \text{(A.21)}$$

where we have used the Eq. (A.20). Any arbitrary operator G can be represented in the form

$$G = \left(\frac{2j + 1}{4\pi}\right)^2 \iint |\vartheta,\phi\rangle\langle\vartheta,\phi|G|\vartheta',\phi'\rangle\langle\vartheta',\phi'| \sin\vartheta \sin\vartheta' \, d\vartheta \, d\vartheta' \, d\phi \, d\phi' \,.$$

There is always the diagonal representation of any operator:

$$G = \int |\vartheta,\phi\rangle\langle\vartheta,\phi| \; g(\vartheta,\phi) \sin\vartheta \, d\vartheta \, d\phi \,, \qquad \text{(A.21)}$$

and the function $g(\vartheta,\phi)$ is determined by the finite number of parameters, they are $(2j + 1)^2$ matrix elements of the operator G. Therefore, it is not surprised that the function $g(\vartheta,\phi)$ can be considered as a superposition of the spherical functions $Y_{lm}(\vartheta,\phi)$ with l varying from 0 to $2j$ and m varying from $-l$ to $+l$. The complete number of these functions equals exactly to $(2j + 1)^2$ [3].

REFERENCES

Introduction

1. Dicke R H 1954 Phys.Rev. **93** 99-110
2. Agarwal G S 1977 *Quantum Statistical Theories of Spontaneous Emission and their Relation to the Approaches* (Berlin: Springer)
3. Allen L and Eberly J H 1975 *Optical Resonance and Two-Level Atoms* (New York: Wiley)
4. Golenishchev-Kutuzov V A and Samartsev V V (eds) 1975 *Electromagnetic Superradiance* (Kazan: Tatgosizdat) [in Russian]
5. Skribanowitz N, Hermann I P, MacGillivray I C and Feld M S 1973 Phys.Rev.Lett. **30** 309-312
6. Trammel G T and Hannon I P 1975 Opt.Commun. **15** 325-329
7. Andreev A V, Ilinskii Yu A and Khokhlov R V 1977 Sov.Phys.JETP **46** 682-685
8. MacGillivray J C and Feld M S 1977 Appl.Phys.Lett. **31** 74-76
9. Szoke A, Daneu V, Goldhar E H and Kurnit N A 1969 Appl.Phys.Lett. **30** 280
10. Gibbs H M, McCall S L and Venkatesan T N C 1976 Phys.Rev.Lett. **36** 1135-1138
11. Gibbs H M, McCall S L and Venkatesan T N C 1980 Opt.Engin. **19** 463-468
12. Hopf F A and Bowden C M 1984 Phys.Rev. A **29** 2591-2596
13. Haken H 1978 *Synergetics*(Berlin: Springer)
14. Bonifacio R 1982 *Dissipative Systems in Quantum Optics. Resonance Fluorescence, Optical Bistability, Superfluorescence* (Berlin: Springer) p 53
15. Stenholm S 1973 Phys.Reports **6c** 1-121
16. Andreev A V, Emelyanov V I and Iliinskii Yu A 1980 Sov.Phys.Usp. **23** 493-514
17. Gross M and Haroche S 1982 Phys.Reports **93** 301-396
18. Bogolyubov N N, Plechko V N and Shumovskii A S 1983 Elementary Particles and Atomic Nuclear **14** 1443-1499 [in Russian]
19. Feld M S and MacGillivray J C 1981 *Coherent Nonlinear Optics* ed M S Feld and V S Letokhov (Berlin: Springer) p 7
20. Vrehen Q H F and Gibbs H M 1982 *Dissipative Systems in Quantum Optics. Resonance Fluorescence, Optical Bistability, Superfluorescence* (Berlin: Springer) p 11
21. Bowden C M, Howgate D W and Robl H R (eds) 1977 *Cooperative Effects in Matter and Radiation* (New York: Plenum)
22. Mandel L and Wolf E (eds) 1978 *Coherence and Quantum Optics, IV* (New York: Plenum)
23. Walther H and Rothe K W (eds) 1981 *Laser Spectroscopy, IV*

(Berlin: Springer)
24. Feld M S and Letokhov V S (eds) 1981 *Coherent Nonlinear Optics* (Berlin: Springer)
25. 1982 *Dissipative Systems in Quantum Optics. Resonance Fluorescence, Optical Bistability, Superfluorescence* (Berlin: Springer)
26. Trifonov E D 1980 *Theory Cooperative Coherent Effects in Emission* (Leningrad: A I Herzen LGPI Publ.House)

Chapter 1

1. Dicke R H 1954 Phys.Rev. **93** 99-110
2. Bonifacio R, Schwendimann P and Haake F 1971 Phys.Rev. A **4** (a) 302-313, (b) 854-864
3. Landau L D and Lifshits E M 1973 *Field Theory* (Moscow: Nauka) [in Russian]
4. Landau L D and Lifshits E M 1974 *Quantum Mechanics* (Moscow: Nauka) [in Russian]
5. Berestetskii V B, Lifshits E M and Pitaevskii L P 1968 *Relativistic Quantum Theory* (Moscow: Nauka) part 1
6. Scully M and Lamb W E 1967 Phys.Rev. A **159** 208-286
7. Lax M 1968 *Fluctuations and Coherence Phenomena in Classical and Quantum Optics* (New York: Gordon)
8. Narducci L M, Coulter C A and Bowden C M 1974 Phys.Rev. A **9** 829-845
9. Glauber R T and Haake F 1976 Phys.Rev. A **13** 357-366
10. Gronchi H and Lugiato L A (a) 1976 Phys.Rev. A **13** 830-852 (b) ibid **14** 502-506

Chapter 2.

1. Akhiezer A I and Berestetskii V B 1969 *Quantum Electrodynamics* (Moscow: Nauka) [in Russian]
2. Louisell W H 1964 *Radiation and Noise in Quantum Electronics* (New York: McGrow Hill)
3. Berestetskii V B, Lifshits E M and Pitaevskii L P 1980 *Quantum Electrodynamics* (Moscow: Nauka) [in Russian]
4. Allen L and Eberly J H 1975 *Optical Resonance and Two-Level Atoms* (New York: Wiley)
5. Tikhonov A N and Samarskii A A 1972 *Equations of Mathematical Physics* (Moscow: Nauka) [in Russian]
6. Agarwal G S 1971 Phys.Rev.A **3** 1783-1793
7. Andreev A V 1977 Sov.Phys.JETP **45** 734-739
8. Bonifacio R, Schwendimann P and Haake F 1971 Phys.Rev. A **4** (a) 302-313, (b) 854-864

9. Bonifacio R and Lugiato L A (a) 1975 Phys.Rev. A **11** 1507-1521, (b) ibid **12** 587-598
10. Andreev A V and Ilinskii Yu A 1981 Sov.J.Quantum.Electr.**11** 163-165, 1982 Bull.Acad.Sci.USSR, Phys.Ser.**46** 150-154
11. Andreev A V, Enaki N A and Ilinskii Yu A 1984 Sov.Phys.JETP **60** 229-233
12. Rehler N E and Eberly J H 1971 Phys.Rev. A **3** 1735-1751

Chapter 3.

1. Dicke R H 1954 Phys.Rev. **93** 99-110
2. Fain V M 1958 Usp.Fiz.Nauk **64** 273-313 [in Russian]
3. Banfi G and Bonifacio R 1978 Phys.Rev. A **12** 2068-2082
4. Ressayre E and Tallet A 1978 Phys.Rev. A **18** 2196-2203
5. Andreev A V 1977 Sov.Phys.JETP **45** 734-739
6. Andreev A V 1978 Sov.J.Quantum.Electr.**8** 476-481
7. Ressayre E and Tallet A 1977 Phys.Rev. A **15** 2410-2423
8. Nagibarov V R and Kopvillem U Kh 1968 Sov.Phys.JETP **27** 167-169
9. Ernst V and Stehle P 1968 Phys.Rev. **176** 1456-1479
10. Eberly J H and Rehler N E (a) 1969 Phys.Lett. A **29** 142-143, (b) 1970 Phys.Rev. A **2** 1607-, (c) Rehler N E and Eberly J H 1971 Phys.Rev. A **3** 1735-1751
11. Bonifacio R, Schwendimann P and Haake F (a) 1970 Phys.Rev. A **3** 509-511, (b) ibid 512-
12. Bonifacio R, Schwendimann P and Haake F (a) 1971 Phys.Rev. A **4** 302-313, (b) ibid 854-864
13. Agarwal G S (a) 1970 Phys.Rev. A **2** 2038-2046, (b) 1971 ibid **3** 1783-1793, (c) ibid 4 1791-1801
14. Arecchi F T and Courtens E 1970 Phys. Rev. A **2** 1730-1737
15. Andreev A V 1977 Sov.Tech.Phys.Lett. **3** 317-319
16. Andreev A V Galkin V Ya and Tikhomirov O Yu 1977 *Analysis and Interpretation of Physical Experiments* (Moscow: Izd.Mosc.Univ.) p 6 [in Russian]
17. Skribanowitz N, Hermann I P, MacGillivray I C and Feld M S (a) 1973 Phys. Rev. Lett. **30** 309-312, (b) 1975 *Laser Spectroscopy* (New York: Plenum) p 379
18. Emelyanov V I and Klimontovich Yu A 1976 Opt.Spectrosc. **41** 541-545
19. Andreev A V, Ilinskii Yu A and Khokhlov R V 1977 Sov.Phys.JETP **46** 682-685
20. Malikov R F, Malyshev V A and Trifonov E D 1980 *Theory of Cooperative Coherent Effects in Emission* (Leningrad: Nauka) p 3 [in Russian]
21. Bonifacio R and Lugiato L A (a) 1975 Phys.Rev. A **11** 1507-1521,

(b) ibid **12** 587-598

22. Kittel C 1971 *Introduction to Solid State Physics* (New York: Wiley)
23. (a) Andreev A V and Akhmanov S A 1981 Preprint no.26/1981 (Moscow: Phys.Dept.of Moscow State Univ.) [in Russian],
(b) Andreev A V, Akhmanov S A and Kov'ev E K 1983 Bull. Acad. Sci. USSR, Phys.Ser. **47** 21-24
24. Andreev A V and Ilinskii Yu A (a) 1975 Sov.Phys.JETP **41** 403-405, (b) 1976 ibid **43** 893-896, (c) 1975 JETP Lett. **22** 223-224
25. Andreev A V, Arutyunyan R V and Ilinskii Yu A 1979 Moscow Univ.Phys.Bull. **34** 45-51
26. Andreev A V, Tikhomirov O Yu and Shaiymkulov M O 1985 Sov.J.Quantum.Electr. **15** 1640-1643
27. Baldwin G S, Solem J C and Goldanskii V I 1981 Rev.Mod.Phys. **53** part 2 687-744
28. Mantsyzov B I and Kuzmin R N 1986 Sov.Phys.JETP **64** 37-44
29. Andreev A V and Ilinskii Yu A (a) 1981 Sov.J.Quantum.Electr. **11** 163-165, (b) 1982 Bull.Acad.Sci.USSR, Phys.Ser. **46** 150-154
30. Andreev A V, Galkin V Ya, Ilinskii Yu A and Tikhomirov O Yu 1978 *Analysis and Interpretation of Physical Experiments* (Moscow: Izd.Mosc.Univ.) p 60 [in Russian]
31. Tikhonov A N, Andreev A V, Galkin V Ya, Ilinskii Yu A and Tikhomirov O Yu 1979 *Algorithms and Programs for Analysis of Some Physical Problems* (Budapest: Hungarian Acad.Sci.) p 131 [in Russian]
32. Mattar F P, Gibbs H M, McCall S L and Feld M S 1981 Phys.Rev.Lett. **46** 1123-1126
33. Vrehen Q H F and Weduwe J J 1981 Phys.Rev. A **24** 2857-2860
34. Watson E A, Gibbs H M, Mattar F P, Cornier M, Claude Y, McCall S L and Feld M S 1983 Phys.Rev. A **27** 1427-1434
35. Drummond P D and Eberly J H (a) 1981 J.Opt.Soc. Amer. **71** 1588-1589, (b) 1982 Phys.Rev. A **25** 3446-3448
36. Laptev V D, Reutova N M and Sokolov I V 1983 Sov.J.Quantum.Electr. **13** 1372-1375
37. Andreev A V, Enaki N A and Ilinskii Yu A 1985 Theor.& Math.Phys. **64** 960-965
38. Andreev A V, Arutyunyan R V and Ilinskii Yu A 1981 Opt.Spectr. **50** 578-581
39. Florian R, Schwan L O and Schmid D 1982 Sol.State Comm. **42** 55-57
40. Rosenberger A T, DeTemple T A, Bowden C M and Sung C C 1978 J.Opt.Soc.Amer. **68** 700-
41. Rosenberger A T and DeTemple T A 1981 Phys.Rev. A **24**

868-882
42. Moi L, Goy P, Gross P, Raimond J M, Fabre C and Haroche S 1983 Phys.Rev. A **27** 2043-2064
43. Goy P, Moi L, Gross P, Raimond J M, Fabre C and Haroche S 1983 Phys.Rev. A **27** 2065-2081
44. MacGillivray I C and Feld M S (a) 1976 Phys.Rev. A **14** 1169-1189, (b) 1981 ibid **23** 1334-1349
45. Gross P, Fabre C, Pillet P and Haroche S 1976 Phys.Rev.Lett. **36** 1035-1038
46. Vrehen Q H F, Hikspoors H M J and Gibbs H M 1977 Phys.Rev.Lett. **38** 764-767
47. Gibbs H M, Vrehen Q H F and Hikspoors H M J 1977 Phys.Rev.Lett. **39** 547-550
48. Grubellier A, Libermann S and Pillet P 1978 Phys.Rev.Lett. **41** 1237-1240
49. Brechignac C and Cauzac P 1979 J.Phys.Lett.(Fr.) **40** 123-125
50. Cahuzac P, Sontag H and Toschek P E 1979 Opt.Commun. **31** 37-41
51. Grubellier A, Libermann S, Mayon D, Pillet P and Schweighofer M G 1982 Opt.Lett. **7** 16-18
52. Flusberg A, Mossberg T and Hartmann S R 1976 Phys.Lett. A **58** 373-374
53. Cremer C and Gerber G 1984 Appl.Phys. B **35** 7-10
54. Okada J, Ikeda K and Matsuoka M 1978 Opt.Commun. **26** 189-192
55. Grubellier A, Brechignac C, Cahuzac P and Pillet P 1979 *Laser Spectroscopy IV* (Berlin: Springer) p 480
56. Gross M, Raimond J M and Haroche S 1978 Phys.Rev.Lett. **40** 1711-1714
57. Florian R, Schwan L O and Schmid D 1984 Phys.Rev. A **29** 2709-15
58. Naboikin Yu V, Samartsev V V and Silaeva N B 1983 Bull.Acad.Sci.USSR, Phys.Ser. **47** 74-78
59. Zinovev P V, Lopina S V, Naboikin Yu V, Silaeva N B, Samartsev V V and Sheibut Yu E 1983 Sov.Phys.JETP **58** 1129-1133
60. Grubillier A, Libermann S, Mayot D and Pillet P 1983 Opt.Lett. **8** 105-107
61. Jaroszynski D A and King T A 1983 *Quantum Electronics & Electrooptics* (Hull: IOP) p 125
62. Kiselev Yu F, Prudkoglyad A F, Shumovskii A S and Yukalov V I 1988 Sov.Phys.JETP **67** 413-415
63. Bazhanov N A, Bulyanitsa D S, Kovalev A I, Polyakov V V, Trautman V Yu, Trifonov E D and Shvedchikov A V 1989 Sov. Phys. Solid State **31** 291-292

Chapter 4.

1. Andreev A V 1989 Sov.Phys.JETP **68** 903-907
2. McCall S L and Hahn E L 1967 Phys.Rev.Lett. **18** 908-
3. McCall S L and Hahn E L 1969 Phys.Rev. **183** 457-
4. Slusher R E and Gibbs H M 1972 Phys Rev A **5** 1634-
5. Lamb G L (Jr) 1971 Rev.Mod.Phys. **43** 99-
6. Poluektov I A, Popov Yu M and Roitberg V S 1974 Sov.J.Quantum Electron. 4 (a) 423-438, (b) 719-739
7. Golenishchev-Kutuzov V A, Samartsev V V, Solovarov N K and Khabibullin B M 1977 *Magnetic Quantum Acoustics* (Moscow: Nauka) [in Russian]
8. Manykin E A and Samartsev V V 1984 *Optical Echo Spectroscopy* (Moscow: Nauka) [in Russian]
9. Zakharov V E, Manakov S V, Novikov S P and Pitaevskii L P 1980 *Soliton Theory: Inverse Scattering Method* (Moscow: Nauka) [in Russian]
10. Nagibarov I A, Bogdanov E I and Deryugin I A 1986 *Dynamics of Quantum Systems* (Minsk: Nauka i Teknika) [in Russian]
11. Kryukov P G and Letokhov V S 1970 Sov.Phys.Usp. **12** 641-672
12. Shen Y R 1984 *Principles of Nonlinear Optics* (New York: Wiley)
13. Allen L and Eberly J H 1975 *Optical Resonance and Two-Level Atoms* (New York: Wiley)
14. Andreev A V, Ilinskii Yu A and Mkoyan A S 1989 Sov.Phys.JETP **69** 904-908
15. Andreev A V 1977 Sov.Tech.Phys.Lett. **3** 317-319
16. Andreev A V1990 Sov.Phys.Usp. **33** 997-1020

Chapter 5.

1. De Giorgio and V Scully M O 1970 Phys.Rev. A **2** 1170-1195
2. Graham R and Haken H 1970 Zs.Phys. **237** 31.
3. Haken H 1978 Synergetics (Springer-Verlag: Berlin Heidelberg New York)
4. Haken H 1985 Light vol 2 Laser Light Dynamics (North Holland Physics Publishing: Amsterdam NY Oxford Tokyo)
5. Brout R 1965 Phase transitions (Benjamin: New York)
6. Vaks V G 1973 Introduction to the microscopic theory of ferroelectrics (Moscow:Nauka)
7. Bogoliubov N N 1962 Problems of a Dynamical theory in statistical Physics In Stydies in Statistical Mechanics vol 1 ed J der Boer and G E Uhlenbeck (Interscience: NY)
8. Lugiato L A 1984 In Progress in Optics 21 **71** ed E.Wolf (North

Holland: Amsterdam)
9. Gibbs H M 1985 Optical Bistability: Controlling Light with Light (Academic Press: NY)
10. Bloembergen N 1965 Nonlinear Optics (NY Amsterdam: Benjamin)
11. Akhmanov S A, Drabovich K N, Suchorukov A P, and Chirkin A S 1970 Sov.Phys.JETP **59** 485-505
12. Ultrashort Light Pulses 1977 In Topics in Applied Physics vol 18 ed S L Shapiro (Springer Verlag: Berlin Heidelberg NY)
13. Akhmanov S A and Koroteev N I 1981 Methods of nonlinear optics in spectroscopy of light scattering (Moscow: Nauka in Russian)
14. Shen Y R 1984 The principles of nonlinear optics (John Wiley & Sons :NY Chichester Brisbane Toronto Singapore)
15. Yariv A 1975 Quantum Electronics (John Wiley and Sons Inc: NY London Sydney Toronto)
16. Sparks M and Wilson J H 1975 Phys.Rev. B **12** 4493- 4500
17. Sparks M 1975 Phys.Rev.Lett. **32** 450- 453
18. Sparks M 1975 Phys.Rev. A **11** 595- 606
19. Hagenlocker E.E. Minck R.W. and Rado W.G. 1967 Phys.Rev. **154** 226-246
20. Grun J B , McGuillan A K and Stoichev B P 1969 Phys.Rev. **180** 61-
21. Morozova E A 1977 Lebedev Physical Institute Proc.vol 99 100-144
22. Helly P L 1979 Phys.Rev. **A20** 372-374
23. Eimerl D 1979 Phys.Rev. **A 20** 369-371
24. Emelyanov V I and Klimontovich Yu L 1975 Sov.Phys.JETP **68** 929-940
25. Kasakova E K, Kraiskii A V and Zubov V A 1970 Lebedev Physical institute commun. vol 7 42 -46
26. Angelov I P 1985 Investigation of nonstationary RS in hydrogen Ph.D. Thesis Physics faculty of Moscow state University Moscow USSR
27. Rautian S G and Chernobrod B M 1977 Sov.Phys.JETP **72** 1342-54
28. Emelyanov V I and Seminogov V N 1979 Sov.Phys.JETP **78** 34-47
29. Emelyanov V I and Klimontovich Yu L 1976 Opt.Spectrosc. **41** 541-545
30. Andreev A V, Emelyanov V I and Ilinskii Yu A 1980 Sov.Phys.Usp. **23** 493-514
31. Emelyanov V I 1985 Proc.3 All Union Simp.on Light Echo and Coherent Spectroscopy,Kharkov,USSR,pp 78-79
32. Butilkin V S, Kaplan A E, Khronopulo Yu G and Yakubovich E I 1977 Resonance light-matter interactions (Moscow:Nauka in Russian)
33. Khanin L I 1975 Dynamics of quantum generators (Moscow:Sov.Radio

in Russian)
34. Zhdanov B V Investigation of dispersion of threshold nonlinear optical effects 1977 Ph.D.thesis Physics faculty of Moscow State University,Moscow, USSR
35. Grun J B, McGuillan A K and Stoicheff B P 1969 Phys.Rev. **180** 61-84
36. Venkin G V, Kuliuk L L and Maleev D I 1975 Sov.J.Quantum.Electr. **2** 2475-2486
37. Pivtsov V S, Rautian S G, Safonov V P, Folin K G, and Chernobrod B 1981 Sov.Phys.JETP **81** 468-479
38. Emelyanov V I 1981 The report on The 1-st All Union Simp.on Light Echo and Coherent spectroscopy, Kasan, USSR
39. Emelyanov V I, Kravtsov N V and Naumkin N I 1990 Sov.J.Quantum. Electr. **17** 1534-1535
40. Ressayre E and Tallet A 1977 Phys.Rev. A **15** 2410-2423
41. Emelyanov V I and Seminogov V N 1980 Moscow Univ.Phys.Bull. **21** 37-43
42. Emelyanov V I and Seminogov V N 1979 Sov.J.Quantum.Electr. **6** 635-638
43. Eicher H 1975 IEEE J.Quantum.Electr.**11** 121-130
44. Emelyanov V I and Yukalov V I 1986 Opt.Spectrosc.**60** 385- 394
45. Korolev F A, Odintsov A M, Turkin E G and Yakunin V P 1975 Sov.J.Quantum.Electr. **2** 413-421
46. Ischenko V I, Lisitsin V N, Razhev A M, Rautian S G and Shalagin A M 1974 JETP Lett. **19** 669-672
47. Pastor I and Guerra J M 1990 Appl.Phys. B **51** 342-349
48. Emelyanov V I 1980 Proc.10th All Union Conf.Coherent and Nonlinear Optics Part 1 Kiev pp 358-359
49. Badalyan A N, Dabagyan A A, Movsesyan M E and Movsesyan R E 1973 Bull.Acad.Sci.USSR Phys.Ser. **34** 304-309
50. Movsesyan M E, Movsesyan R E, and Khanbekyan A M 1985 Sov.J. Quantum.Electr. **12** 2486-2488
51. Emelyanov V I 1982 Sov.J.Tech.Phys. **52** 998-1000
52. Emelyanov V I and Zochdi Z 1982 Sov.J.Tech.Phys. **52** 1883-1894
53. Arakeljan S M, Grigoryan G L Narsisyan S G and Chilingaryan Yu S 1978 Sov.Phys.JETP Lett. **28** 282-285
54. Elliot R G, Harley R T,Hayes W and Smith S R P 1972 Proc.Roy.Soc.(London) A**328** 217-233
55. Hepp K, Lieb E H 1973 Ann.d.Phys. **76** 360-375
56. Wang Y K, Hioe F T 1973 Phys.Rev. A **7** 831-836
57. Emelyanov V I and Klimontovich Yu L 1976 Sov.J.Quantum.Electr. **3** 848-852
58. Emelyanov V I and Klimontovich Yu L 1976 Phys.Lett. A **59** 366-8

59. Elesin V F and Kopaev Yu V 1976 Sov.Phys.JETP Lett. **24** 78-81
60. Machviladze T M and Sarichev M E 1982 Lebedev Phys.Inst.Proc. **132** 188-238
61. Kimura M 1980 Sci.Reports of Kansave Univer.**25** 15-31
62. Kimura M 1981 Progr.Theor.Phys. **65** 435-450
63. Rzazewski K, Wodkiewicz K and Zakowez W 1975 Phys.Rev.Lett. **35** 432-435
64. Power E A and Zienau S 1959 Phil.Trans.Roy.Soc. **2-1** 427-447
65. Stenholm S 1973 Phys Rep. **6c** 1-121
66. Emelyanov V I and Yu L Klimontovich 1972 Phys.Lett. A **42** 303-4
67. Emelyanov V I 1973 Sov.Phys.Sol.State **15** 75-81
68. Emelyanov V I 1973 Sov.Phys.Sol.State **15** 2509-2510

Chapter 6.

1. Liao P, Glass A and Hamphrey L 1980 Phys.Rev. B **22** 2276-2281
2. Basun S A, Kaplyanskii A A, Feofilov S P and Furman A S 1984 Sov.Phys.JETP Lett. **39** 161-164
3. Basun S A, Kaplyanskii A A and Feofilov S P 1984 Sov.Phys.JETP **87** 2047-2063
4. Lines M E and Glass A M 1977 Principle and applications of ferroelectrics and related materials (Oxford: Clarendon press)
5. Osterberg U and Margulis W 1986 Opt.Lett. **11** 516-519
6. Osterberg U and Margulis W 1987 Opt.Lett. **12** 57-60
7. Farris M C, Russel P S, Fermann M E and Payne D N 1987 Electron.Lett. **23** 322-325
8. Misrahi V, Osterberg U and Krautschic C 1988 Appl.Phys.Lett. **53** 557-560
9. Bergot M V, Farries M C and Fermann M E1988 Opt.Lett. **13** 592-5
10. Stolen R H and Tom H W K 1987 Opt.Lett. **12** 585-588
11. Emelyanov V I 1977 Sov.Phys.Solid State **19** 3312-3317
12. Emelyanov V I and Klimontovich Yu L 1978 Sov.Phys.JETP Lett. **27** 4-7
13. Emelyanov V I and Indenbom M V 1979 Sov.Phys.Solid State **21** 668-695
14. Osipov M A 1979 Sov.Phys.Solid State **21** 1477-1487
15. Emelyanov V I 1985 Sov.J.Quantum.Electr. **12** 1729-1741
16. Emelyanov V I 1987 Bull.Acad.Sci.USSR Phys.Ser. **51** 52-57
17. Bostel G, Andrade P and Merten L 1975 Phys.Stat.Sol. B **70** 641-652
18. Bersuker I B, Vekhter B G and Ogurtsov I Ya 1975 Sov.Phys.Usp. **116** 605-610
19. Kristoffel N N and Konsin P I 1976 Sov.Phys.Usp. **120** 507-513

20. Morozov B H and Ayvasyan Yu M 1980 Sov.J.Quantum.Electr. 7 5-30
21. Grechushnikov B N and Feofilov P P 1955 Sov.Phys.JETP **29** 384-96
22. Fridkin V M and Popov B N 1978 **126** 657-663
23. McCamber D E and Sturge M D 1963 J.Appl.Phys. **34** 1682-1697
24. Shen Y R 1984 The Principles of Nonlinear Optics (John Wiley & Sons: NY Chichester Brisbane Toronto Singapore)
25. Emelyanov V I and Zochdi Z 1980 Sov.J.Quantum.Electron. **7** 1510-1521
26. Lugiato L A 1984 In Progress in Optics **21** 71 ed.E Wolf (North Holland:Amsterdam)
27. Kaiser W, Sugano S and Wood D L 1961 Phys.Rev.Lett. **6** 605-608
28. Koreneva L G, Zolin V F and Davidov B L 1975 Molecular crystals in nonlinear optics(Moscow:Nauka in Russian)
29. Bates R 1978 Phys.Reports **35**
30. Emelyanov V I and Taeshnikov A B 1991 Sov.J.Quantum Electr. **18** 266-268
31. Emelyanov V I 1985 Preprint 5 Phys.Faculty Moscow univer.Moscow USSR
32. Emelyanov V I and Uvarova I F 1986 Bull.Acad.Sci.USSR Phys.Ser. **501** 214-1219
33. Emelyanov V I and Uvarova I F 1988 Sov.Phys.JETP **67** 1662-1670
34. Materials Research Soc.Symp.Proc.1984 **23** (NY: North Holland)
35. Bugaev A A and Klochkov A V 1984 Sov.Phys.Sol.State **26** 2100-2
36. Baskin B L, Polyakov A A, Trukhin V N and Yarishetskii I D 1985 Sov.Tekh.Phys. Lett.**11** 517-520
37. Merkle K L,Uebbing R H, Baumgart H and Phillip F 1982 in Laser and electron beam interactions with solids ed Appleton B R and Celler G K (NY:North Holland) p 337
38. Tom H W, Aumiller G D and Brito-Cruz C H 1988 Phys.Rev.**60** 1438-1441
39. Kopaev V V, Kopaev Yu V and Molotkov C N 1983 Microelectronics **12** 499-501
40. Kopaev Yu V 1985 Zh.Tekh.Fiz.1985 **30** 1328-1343
41. Landau L D and Lifshits E M 1987 Theory of elastisity (Moscow:Nauka) Chap 3
42. Jost D, Luthy W, Weber H and Salathe R 1986 Appl.Phys.Lett **49** 625-628
43. Akhmanov S A, Emelyanov V I ,Koroteev N I and Seminogov V N 1985 Sov.Phys.Usp. **28** 1989 -2059

44. Akhmanov S A, Emelyanov V I and Koroteev N I 1990 Interaction of strong laser radiation with solids and nonlinear optical diagnostics of surfaces Teubner-Texte zur Physik Band 24 (Teubner Verlagsgesellschaft: Leipzig)
45. Smith S D 1984 Acta Phys.Austriaca **56** 75-91
46. Young J F, Preston J S, Van Driel H M and Sipe J E 1983 Phys. Rev.B **27** 1141-1172
47. Young J F, Sipe J E and Van Driel H M 1984 Phys.Rev.B**30** 2001-2015
48. Siegman A E and auchet P M 1986 IEEE J.Quantum.Electron.QE-**22** 1384-1403
49. Emelyanov V I, Seminogov V N Laser-induced instabilities of surface relief and modiffication of optical reflection and absorption of condensed matter In Physical foundations of Laser Beam Technology vol **1** (1988) pp 118-179; vol 3 (1989)pp 57-91 ed.Koroteev N I and Panchenko V Ya (VINITY: Moscow)
50. Emelyanov V I Instabilities and phase transitions with formation of ordered surface structures under the action of external energy beams 1988 Proc.Int.Symp.on Statist.Mechanics.Joint Institute for Nuclear Research D 17-88-95 Dubna, USSR pp 119-128
51. Emelyanov V I Nonlinear waves
52. Bagratashvilii V N, Banishev A F,Gnedoy S A and Emelyanov V I 1991 Appl.Phys. A **52** 438-444
53. Zeldovich B Ya,Tabiryan N V 1985 Sov.Phys.Usp. **147** 633-658
54. Hopf F A, Bowden C M, Louisell W H 1984 Phys.Rev. A **29** 2591-6
55. Keldish L V 1978 Moscow Univ.Phys.Bull. **19** 86-94
56. Elesin V F,Kopaev V V, Kopaev Yu V 1976 **71** 714-730
57. Bogolyubov N N Jr, Plechko V N and Shumovskii A S 1983 Element.Particles and Atomic Nuclear **14** 1443-1499
58. Machviladze T M and Sarichev M E 1982 Lebedev Phys.Inst.Proc. **132** 188-238
59. Zhelyaznikov V V, Kocharovskii V V, Kocharovskii Vl V 1984 Sov.Phys.JETP **87** 1565-1585
60. Horsthemke W and Lefever R 1984 Noise-Induced Transitions (Springer Verlag:Berlin Heidelberg NY Tokyo)

Appendices.

1. Landau L D and Lifshits E M 1974 *Quantum Mechanics* (Moscow: Nauka) [in Russian]
2. Wigner E P 1959 *Group Theory* (New York: Academic Press)
3. Arecchi F T, Courtens E, Gilmore R and Thomas H 1972 Phys.Rev. A **6** 2211-2237

Index